Textbook of
Work Physiology

Textbook of Work Physiology

Per-Olof Åstrand, M.D.

Associate Professor
Department of Physiology
Gymnastik-och Idrottshögskolan
(Swedish College of Physical Education)
Stockholm, Sweden

Kaare Rodahl, M.D.

Director, Institute of Work Physiology
Professor, Norwegian College of Physical Education
Oslo, Norway

McGraw-Hill Book Company

New York, St. Louis, San Francisco, London
Sydney, Toronto, Mexico, Panama

Textbook of Work Physiology

Copyright © 1970 by McGraw-Hill, Inc. All rights reserved.
Printed in the United States of America. No part of this
publication may be reproduced, stored in a retrieval system,
or transmitted, in any form or by any means, electronic,
mechanical, photocopying, recording, or otherwise, without
the prior written permission of the publisher.

Library of Congress Catalog Card Number 67-13508

ISBN 07-002405-7

890 MAMM 76543

To
Professor Erik Hohwü Christensen,

*who first introduced us to the field of work physiology.
It is to a large measure due to his encouragement and
continuous and active interest that the writing of this
book was undertaken.*

Preface

The purpose of this text is to try to bring together into one volume the various factors affecting human physical performance in a manner comprehensible to the physiologist, the physical educator, and the clinician. Contrary to most of the conventional textbooks of physiology, in which the emphasis is on the regulation of the various functions of the body at rest, the regulatory mechanisms studied during physical activity have been especially emphasized in this book. It is assumed that the reader has some knowledge of elementary physics and chemistry, as well as human anatomy and physiology. However, to facilitate the understanding of some of the physiological and biochemical events encountered during work stress and physical exercise, a certain amount of basic physiology and biochemistry has been included.

In the selection of the material, an attempt has been made to meet the modern needs of the student of physical education, at both the undergraduate and the postgraduate level. For the convenience of the reader certain sections of the text of a more advanced or specialized nature have been set in small print. More references have also been included than is customary in most textbooks. Inevitably, since the submission of our manuscript, new developments have taken place which we are not able to include in this edition.

We are aware of the fact that the curriculum in many physical education colleges does not permit such a comprehensive study of physiology as this book may entail. For this reason each chapter has been written as a fairly complete entity, relatively independent of the rest of the book. With this arrangement the book may also be useful for those students who wish to penetrate more deeply into a particular field or a limited area of study.

It is our hope that this text may be useful not only in the teaching of physical education but also in the teaching of clinical and applied physiology and that it may serve to stimulate the appreciation of the role of physical education for young and old, in health and disease.

Much of the unpublished material included in this book has been gathered in collaboration with our colleagues at Gymnastik-och Idrottshögskolan in Stockholm; the Division of Research at Lankenau Hospital, Philadelphia, Pennsylvania; the Institute of Work Physiology and the College of Physical Education, Oslo. Their kind cooperation is gratefully acknowledged. We have also benefited greatly from personal association and frequent discussions with our many colleagues in all these institutions. We are especially indebted to Dr. Irma Åstrand for her contributions to Chap. 11, to Dr. E. L. Bortz for his encouragement, and to Miss Mary Ethel Pew for her valuable support of many of the studies included in this volume. We are also very grateful for the technical assistance given us by Mrs. Carolyn Hyatt, Mrs. Karin Marina, and Mrs. Joan Rodahl in the preparation of the manuscript, and to Mrs. Dorothy Robinson for graphic arts.

The unpublished results of studies included in this book have been supported in part by the following grants: the Tri-centennial Fund of the Bank of

Sweden, the Swedish Sports Federation, the Swedish Association for the Prevention of Heart and Chest Diseases, the Swedish Medical Research Council, the Norwegian Medical Research Council, and the Norwegian Borregaard Research Fund.

Per-Olof Åstrand
Kaare Rodahl

Contents

Textbook of
Work Physiology

1
The Body at Work

chapter one

The Body at Work

In the simplest forms of animal life, such as the amoeba, all essential functions (metabolism, response to stimuli, movement, and reproduction) are developed in one single cell. Because of the cell's minute size, foodstuffs, waste products, electrolytes, and dissolved gases can be distributed within the cell and between it and its surroundings mainly by diffusion and osmosis.

In the evolution of higher organisms, their size increased. Different cells specialize in the performance of particular functions and build up tissues and organs. The individual cell (Fig. 1-1) is now removed from intimate contact with the animal's environment, which, furthermore, might have changed from water to air. This evolution creates the problem of transportation within the body and communication with the environment.

The human body consists of 50 to 70 percent water. Here the individual cell, like the amoeba, can bathe in fluid. The composition of this extracellular or interstitial fluid is of the utmost importance for the function of the cell. Its content of organic compounds, such as fatty acids, glucose, hormones, and enzymes, and of inorganic substances exerts a profound influence upon the cell in one way or another.

The main objective of most organ functions is to maintain the internal equilibrium of the single cell in spite of primary changes or disturbances in the animal's internal or external environment, in accordance with the "one for all and all for one" concept. A continuous exchange of materials between interstitial fluid and blood plasma is necessary for the normal function of the cell. With the exception of respiratory gases, the exchange of material between the body and the external environment can be intermittent because of the storage capacity of the tissues.

Higher animals are basically designed for mobility. Consequently, their locomotive apparatus and service organs constitute the main part of the total body mass. The shape and dimensions of the human skeleton and musculature are

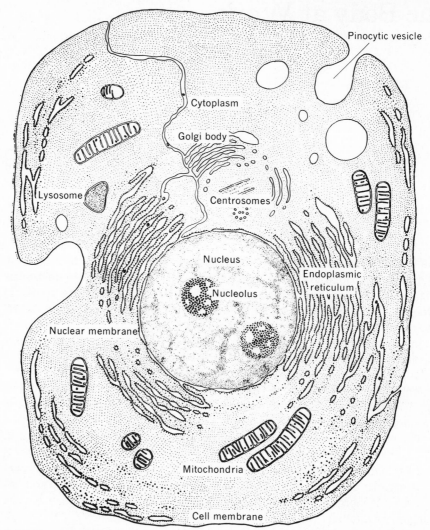

*Fig. 1-1 A typical cell. (From Jean Bracket, Scientific American, **205:3**, 1961.)
The cell membrane has the same general structure in all cells. It consists of lipoproteins and has a porous structure. (The pores are not shown in the diagram.)
Encapsulated within the boundaries of the cell membrane is the cytoplasm, which
contains a number of formed and dissolved elements, including enzymes which
support the anaerobic metabolic processes of the cell. The mitochondria are rod-
shaped bodies, surrounded by a double-walled membrane. The mitochondria may be
isolated from minced tissue by centrifugation, which makes it possible to analyze
the mitochondria chemically. The mitochondria, which take up oxygen, represent the
"powerhouse" of the cell. Here fuel and oxygen enter into energy-yielding processes
resulting in the formation of ATP. In the mitochondria the enzymes are fixed to
the organic structures as in an assembly-line operation. In the endoplasmic*

such that the human body cannot compete with a gazelle in speed or an elephant in sturdiness, but in diversity man is indeed outstanding.

The basic instrument of mobility is the muscle. It is unique in that it can vary its metabolic rate to a greater degree than any other tissue. In fact, the working skeletal muscles may increase their oxidative processes to more than 50 times the resting level (Asmussen et al., 1939). Such an enormous variation in metabolic rate must necessarily create serious problems for the working muscle cell, for while the consumption of fuel and oxygen increases fiftyfold, the rate of removal of heat, carbon dioxide, water, and waste products must be increased similarly. To maintain the chemical and physical equilibrium of the cells, there must be a tremendous increase in the exchange of molecules between intra- and extracellular fluid; "fresh" fluid must continuously flush the exercising cell. When muscles are thrown into vigorous activity, the ability to maintain the internal equilibria necessary to continue the work is entirely dependent on those organs which service the muscles. This dependence is especially true of the circulatory and respiratory organ functions which strive to keep, so to speak, the muscle cell in an indirect instantaneous contact with the surrounding air at all times.

Since the heat production may increase from about 1 kcal/min at rest to perhaps 50 kcal/min during maximal work, or from about 80 to several thousand watts, the temperature-controlling mechanisms must come into play to arrange for the excess heat to be transported from the muscles to the skin. Profuse sweating may cause a water and salt loss which secondarily may affect the circulation and the renal function. To restore the energy content of the body working at maximal capacity, up to four times more food must be digested daily than when the individual is at rest. During exercise many of the hormone-producing glands are involved in the regulation of metabolic and circulatory functions. Parts of the central nervous system are specialized in receiving sensory information from muscles and joints and sending impulses to the muscles. In the final analysis, all the external evidence of the activity of the brain is eventually manifested by muscular movement.

One fascinating aspect of the physiology of man at work is that it provides basic information about the nature and range of the functional capacity of different organ systems. Physiological and clinical studies on human beings

---

reticulum a network of canaliculi formed by a system of membranes may extend all the way from the outer surface of the cell to the membrane surrounding the nucleus. Through these canaliculi, substances may move from the outer membrane of the cell to the membrane of the nucleus. The dots that line the endoplasmic reticulum are ribosomes. These are the sites of protein synthesis. The nucleus contains the chromosomes, which contain genes and deoxyribonucleic acid and are the carriers of the hereditary factors. In cell division the pair of chromosomes, shown in longitudinal section (rods) and in cross section (circles), parts to form two poles of an apparatus that separates two duplicate sets of chromosomes.

cannot be restricted to a resting or basal condition, because the functional capacity of an organ can only be evaluated when the organ is subject to functional loads. A theory on the regulation of a function must consider and explain the adaptation to various physiological conditions, including muscular activity.

Manual labor, sometimes under adverse environmental conditions, still exists in all countries, and will probably always remain an essential part of society. Furthermore, individuals continue to find satisfaction and enjoyment in their leisure time through sports or other types of muscular activity. Important objectives of physiological research are to study the effect of various activities and environmental factors on different organ functions; to investigate the capacity of the individual to meet the demands imposed upon him; and finally, to determine how this capacity can be influenced by factors such as training and acclimatization (Fig. 1-2).

In a very broad sense, physical performance or fitness is determined by the individual's capacity for energy output (aerobic and anaerobic processes), neuromuscular function (muscle strength and technique), and psychological factors (e.g., his motivation and tactics). These factors play a more or less dominating role, depending upon the nature of the performance. In golf there is usually no need for a high energy output, but a good technique is essential. In a high jump, the muscular strength and technique are of primary importance. In long-distance running and cross-country skiing, the capacity of the aerobic energy-

Fig. 1-2   Factors constituting physical performance.

Physical Performance

Energy output

    Aerobic processes

    Anaerobic processes

Neuromuscular function

    Strength

    Technique

Psychological factors

    Motivation

    Tactics

yielding processes determines the speed to a high degree. The technique factor is more important in skiing than in running; psychological stamina and motivation are essential to fight the feeling of fatigue. Various activities, sports, and exercises can thus be analyzed by a variety of methods in order to determine the specific requirements for optimal performance.

REFERENCE

Asmussen, E., E. H. Christensen, and M. Nielsen: Die O_2-Aufnahme der Ruhenden und der Arbeitenden Skelettmuskein, *Skand. Arch. Physiol.*, **82**:212, 1939.

2

Energy Liberation and Transfer

contents

chapter two
Energy Liberation and Transfer

The physiology of muscular work and exercise is basically the chemistry and physics of the transformation of chemically bound energy into mechanical energy.

There are many similarities between the "human engine" and the combustion engine constructed by man. In the combustion engine, gasoline and air are introduced into the cylinder. The spark from the spark plug initiates the explosive combustion of the gas mixture. Chemical energy is transformed into kinetic energy and heat. The expansion of the gas forces the piston to move, and a system of mechanical devices can transfer this motion to the wheels. The motor is cooled by fluid or air to prevent overheating. The waste products are expelled with the exhaust. As this motor can work only in the presence of oxygen, its function is *aerobic*. When the gasoline tank is empty, the engine can no longer continue to run, since the operation of the combustion engine is dependent upon a continuous supply of fuel. In an automobile, the self-starter provides the energy for the first movements of the pistons. This energy comes from the electrical accumulator (battery); the starter can thus work in the absence of oxygen, or *anaerobically*. The stored energy of the battery is quite limited, however, so that the battery must be frequently recharged, or reloaded with electrical energy.

"Living organisms, like machines, conform to the law of the conservation of energy, and must pay for all their activities in the currency of metabolism" (Baldwin, 1967). In the human machine the muscle fibers are the "pistons" — when fuel and a spark that starts the breaking down of the fuel are available, part of the yielded energy can cause movement of the pistons. Heat and various waste products are produced.

In the following paragraphs we shall briefly summarize the chemical processes involved in the human machine, leaving out, for sake of simplicity, the more complicated steps of the reactions. For a more profound and complete treatment, it is recommended that the reader consult more detailed textbooks and reviews (Soskin and Levine, 1952; Needham, 1960; Keele and Neil, 1965; Mahler and Cordes, 1966; Baldwin, 1967).

OXIDATION

A classical equation for an energy-providing oxidative metabolic process is the oxidation of glucose:

$$C_6H_{12}O_6 + 6O_2 \xrightleftharpoons[\text{photosynthesis}]{\text{oxidation}} 6CO_2 + 6H_2O + \text{energy}$$

The glucose has a much higher energy level than CO_2 and H_2O, and energy is liberated by the oxidation and made available for other energy-requiring processes. (One mole of glucose can yield a maximum of 686 kcal of chemical energy.) The reaction is reversible. With chlorophyll as a catalyst, CO_2 and H_2O can combine again, the necessary energy for this reaction being provided by light. Thus the circle is completed: animals dissimilate or break down carbohydrates (catabolism), while plants can also assimilate or synthesize carbohydrates (anabolism).

Chemical compounds can be oxidized by the removal of hydrogen, i.e., by dehydrogenation. This can be expressed in general terms as follows:

$$AH_2 + B \rightarrow A + BH_2$$

The oxidized substance is the hydrogen donor (A), and the reduced compound is the hydrogen acceptor (B). In biological oxidation-reduction reactions, an intermediary carrier of hydrogen usually acts together with a catalyst (enzymes and coenzymes). One group of such enzymes can utilize molecular oxygen directly as the hydrogen acceptor (aerobic oxidation), and another group uses other hydrogen acceptors (anaerobic oxidation).

In the mitochondria of the muscle cell there is an organized system of enzymes, coenzymes, and activators bound into the structures. Each one of them specializes in carrying hydrogen atoms or electrons in just one range of the molecular chain in the substrate. Together they can catalyze the complete aerobic oxidation of the foodstuffs. In the cytoplasm of the muscle cell there are the catalysts required for an anaerobic breakdown of glycogen and glucose to pyruvate and lactic acid (glycogenolysis and glycolysis).

THE FUEL

The main source of energy for the muscular contraction is carbohydrate and fat from the food consumed. Special chemical compounds act as carriers of energy within the cell, from the fat and carbohydrate to the point where the biologically meaningful reaction takes place. Adenosine triphosphate (ATP) is one such substance of fundamental importance in biological exchanges of energy; creatine phosphate is another. The uptake or release of the phosphate from the compound is inherently associated with the gain or loss, respectively, of energy. Various phosphate compounds represent different energy levels. ATP is a high-energy phosphate compound; with the loss of one of the three phosphate radicals, it descends in a stepwise fashion on the energy scale to ADP (adenosine diphosphate) and eventually to AMP (adenosine monophosphate). Inorganic phosphate is on the lowest energy level. The phosphates also hold key positions in the intermediary metabolism of protein, carbohydrate, and fat. In other words, the cell's primary source of energy for metabolic processes is in the terminal phosphate group of ATP. This energy is used to drive the various reactions which require energy in order to proceed.

Figure 2-1 illustrates schematically the main steps involved in the processes providing the energy for muscular contraction and other biological processes. The high-energy phosphates, ATP and creatine phosphate, are stored in the tissues and are found in a relatively high concentration in the muscles. Of vital importance as the firsthand energy provider for the muscle is ATP as it breaks down to ADP; the trigger is the nerve impulse. Part of the energy released can be transformed to mechanical work, i.e., the contraction of the muscle fiber. Muscles are specialized in the conversion of chemical energy into mechanical work. Energy can also be transformed to electrical work, osmotic work, heat, and chemical synthesis, depending upon the characteristics of the activated cells.

ATP is the universal intracellular carrier of chemical energy. While there is a relatively high content of ATP in the muscles, the store is yet very limited. For maintained activity of the tissue, a very rapid resynthesis of the ATP from ADP and phosphate is therefore essential; the "battery must be recharged." For this purpose the high-energy creatine phosphate is mobilized, and the activity can continue, theoretically, until the supply of creatine phosphate is exhausted (creatine phosphate $+$ ADP $\rightleftharpoons$ creatine $+$ ATP). Normally, a third and much richer source of energy supply is available in glycogen. The low-energy phosphates set free are "dropped down" (notice the glucose-6-phosphate in Fig. 2-1), and energy is pumped in with the oxidation of glycogen. Somewhere along the line the regenerated high-energy phosphate is made available for ADP, thereby resynthesizing ATP and indirectly the creatine phosphate (Fig. 2-1). The creatine phosphate contributes to a great extent during the early stage of activity because it can react faster than glycolysis. In prolonged work, when the carbohydrate and fat metabolism is in full swing, the creatine phosphate compound might again retire into a reserve position, although this has not been definitely established (Cain et al., 1962). The creatine phosphate pool represents a labile reservoir of high-energy phosphate groups. It is regenerated from free creatine at the expense of ATP (see Fig. 2-1).

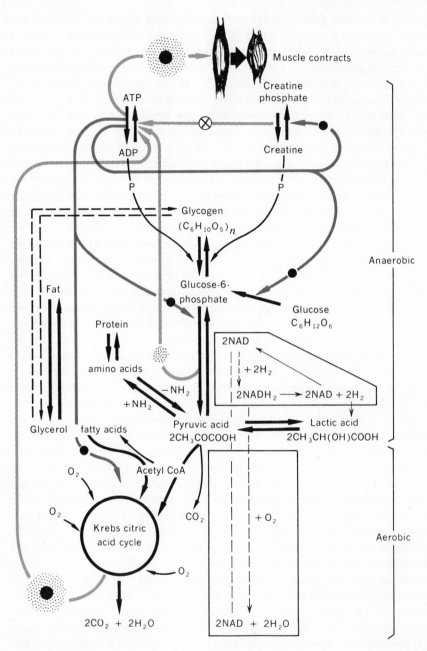

Fig. 2-1 Simplified schematic diagram showing energy liberation and transfer in the living cell.

Let us now consider in more detail the *anaerobic condition,* i.e., when oxygen is *not* at the disposal of the cell. The breakdown of glycogen or glucose to pyruvic acid can occur in the absence of oxygen, and this process provides high-energy phosphates. This anaerobic oxidation is made possible by the simultaneous reduction of a coenzyme, diphosphopyridine nucleotide (DPN), or to be more precise, nicotinamide adenine dinucleotide (NAD), which acts as a hydrogen acceptor (electron carrier). This reaction would soon stop, however, if NAD (now actually $NADH_2$) were not reoxidized. Here the formation of lactic acid comes into the picture. Pyruvic acid is reduced by taking over the hydrogen atoms from $NADH_2$, forming lactic acid. The oxidized NAD can now capture and carry more hydrogen "down"; more glycogen is oxidized, and more energy in the form of high-energy phosphates is provided for the transformation of ADP into ATP. Thus, the formation of lactic acid as such is not necessary for the delivery of energy, but it serves as a storehouse for hydrogen and thereby keeps the reaction going. Under anaerobic conditions, the accumulation of lactic acid or the exhaustion of the supply of glycogen or glucose may limit the cell activity.

Under aerobic conditions, the problem of oxidizing the $NADH_2$ is solved in a different way. For simplicity, Fig. 2-1 presents the complicated Krebs citric acid cycle without any formulas. Oxygen is the final hydrogen acceptor in this complete oxidation of the compounds.

The steps by which hydrogen atoms (or the electrons derived from them) pass down a series of hydrogen carriers (or electron carriers), the so-called respiratory chain, are hidden in the box. In this aerobic mechanism the flavoprotein-cytochrome system serves as enzymes. These intricate reactions have one important purpose: to generate new energy-rich compounds from energy-poor phosphate radicals.

At this point, a comment should be made on the *interplay between anaerobic and aerobic oxidation.* In mild or moderate exercise, the oxygen supply to the muscle cells is sufficient to reoxidize the reduced $NADH_2$ at the same pace as it is formed, oxygen being the final hydrogen acceptor; the pyruvic acid is completely oxidized. With increasing severity of exercise, the breakdown of glycogen speeds up, and the reduction of NAD is correspondingly speeded up. If the oxygen-transporting system now cannot provide enough oxygen to the cells, the pyruvic acid must play the role as hydrogen acceptor. Two-way traffic has a definitely higher capacity for transportation than the one-way system. Some of the coenzyme $NADH_2$ is reoxidized in the reaction through which lactic acid is formed. After the cessation of exercise, the metabolic "generators" continue to run for awhile to replenish the energy store in an immediately accessible form.

While it is the availability of oxygen in the cell that determines to what extent the metabolic processes can proceed aerobically or anaerobically, the exact regulatory mechanism is unknown. However, the intracellular enzymes engaged in the aerobic metabolism can still operate maximally at oxygen tensions in the order of about 1 mm Hg. If this is a critical value, the capillary oxygen pressure must be higher than 1 mm Hg to secure an effective diffusion gradient in order to supply the cell with the oxygen needed for the aerobic processes to proceed. An increase in concentration of the phosphate, ADP, and AMP, which may also be formed, will enhance the glycolysis.

Glucose has a lower energy content than glycogen; ATP must provide energy to "lift" it into the glycogen–pyruvic acid system (Fig. 2-1). The fatty acids also contribute significantly in the aerobic oxidation by entering the Krebs cycle. Even the amino acids, after deamination, can enter the Krebs cycle via pyruvic acid and be completely oxidized or synthesized to glycogen or fat. As a fuel for muscular contractions, however, the oxidation of proteins normally plays a very limited role.

The main steps in the energy exchange in the muscle cell can be summarized in the following way:

$$
\begin{aligned}
\text{Anaerobic} &\begin{cases}
(1) & \text{ATP} \rightleftharpoons \text{ADP} + \text{P} + \text{free energy} \\
(2) & \text{Creatine phosphate} + \text{ADP} \rightleftharpoons \text{creatine} + \text{ATP} \\
(3) & \text{Glycogen or glucose} + \text{P} + \text{ADP} \rightarrow \text{lactate} + \text{ATP}
\end{cases} \\
\text{Aerobic} \quad &\;\;(4) \quad \text{Glycogen and free fatty acids} + \text{P} + \text{ADP} + \text{O}_2 \rightarrow \\
&\hspace{10em} \text{CO}_2 + \text{H}_2\text{O} + \text{ATP}
\end{aligned}
$$

In this schematic presentation all quantitative aspects and the efficiency of the processes are disregarded. Note that only the reactions (1) and (2) are reversible in the muscle.

ENERGY YIELD

The *anaerobic* breakdown of a glucose molecule to lactate in living cells takes place according to the following simplified overall equation:

Glucose $+ 2$ phosphate $+ 2\,\text{ADP} \rightarrow 2\,\text{lactate} + 2\,\text{ATP}$

The change of free energy, ΔG^1, is -38 kcal/mole. In the simple breakdown of glucose to lactate (glucose $\rightarrow 2$ lactate) the ΔG^1 is -52 kcal/mole. Therefore, the breakdown of glucose by the fermentation mechanism accompanied by phosphorylation of two moles of ADP proceeds with a smaller decline in free energy or about 38 kcal/mole of glucose. Actually, generation of ATP from ADP and phosphate conserves energy, or at least 7 kcal/mole of ATP. Since two moles of ATP were formed, 14 of the available 52 kcal were captured by the ATP, or 27 percent of the potential energy. (In fact, four ATP are formed from the breakdown of one mole of glucose, but to drive this breakdown two ATP molecules enter the scheme, Fig. 2-1, as priming agents and the net gain will be the formation of two ATP from ADP and phosphate.)

The anaerobic breakdown of glycogen into lactic acid releases approximately 55 kcal for each six-carbon unit (glucose molecule) glycolized. One molecule of ATP is required for this reaction, and of the total of four ATP molecules formed, the net gain will be three molecules of ATP, which conserves about $3 \times 7 = 21$ kcal. The approximate efficiency of energy conservation is thus $(21/55) \times 100 = 38$ percent.

The complete *aerobic* oxidation of one mole of glucose provides 686 kcal of free energy. The overall equation for this oxidation is as follows:

$$\text{Glucose} + 38 \text{ phosphate} + 38 \text{ ADP} + 6 \text{ O}_2 \rightarrow 6 \text{ CO}_2 + 44 \text{ H}_2\text{O*} + 38 \text{ ATP}$$

Since the free energy is 686 kcal and 38 molecules of ATP are formed, each requiring a minimal input of 7 kcal/mole, the efficiency of energy conservation is approximately $(38 \times 7/686) \times 100 = 39$ percent.

Starting with glycogen, aerobic oxidation provides energy for the formation of one more ATP/mole of six-carbon unit, or altogether 39 ATP.

For palmitic acid a similar analysis gives schematically the following result:

$$\text{Palmitic acid} + 130 \text{ phosphate} + 130 \text{ ADP} + 23 \text{ O}_2 \rightarrow$$
$$16 \text{ CO}_2 + 146 \text{ H}_2\text{O} + 130 \text{ ATP}$$

ΔG^1 is 2340 kcal of which the 130 moles of ATP capture a minimum of $7 \times 130 = 910$ kcal. The efficiency will be $(910/2,340) \times 100 = 39$ percent.

These formulas indicate that one molecule of extra oxygen is consumed in the formation of about 6.5 ATP when glycogen is completely oxidized, but only 5.7 ATP when a fatty acid is oxidized. When the oxygen supply to a muscle becomes limited during heavy exercise, glycogen contributes relatively more to the energy yield than does fat (see Chap. 14). This apparently gives a better utilization of the oxygen transported to the muscle.

The efficiency of these processes is thus very high or some 30 to 40 percent. However, these figures are very likely only minimal ones. In fact, under intracellular conditions the reversible ATP $\rightleftharpoons$ ADP + phosphate reaction probably involves a change in the energy level that is higher than 7 kcal/mole ("the standard free energy" of hydrolysis of ATP to ADP). It may be as high as 10 to 12 kcal or even 16 kcal which gives an efficiency of the conservation of ATP energy as high as 55 to 87 percent. The mechanical efficiency of physical work, as estimated on the basis of the ratio of mechanical work to oxygen utilization, is at the most 20 to 25 percent.

It should be noted that when the glucose and glycogen are completely oxidized aerobically, the energy release available for the ATP formation is much greater than in the partial anaerobic breakdown. Thus, glucose can generate $38/2 = 19$ times more ATP per gram mole aerobically than anaerobically.

THE FATE OF LACTIC ACID

The anaerobic energy output is relatively modest compared with the aerobic energy output. Glucose can, for instance, provide almost 20 times more energy per gram mole aerobically than anaerobically. It is thus clear that oxygen is the key for unlocking the doors to the great energy stores of the living cells. It is there-fore the availability of oxygen to the working muscle cells which, in the final analy-

* The 38 extra moles of H_2O have to do with the ATP formation.

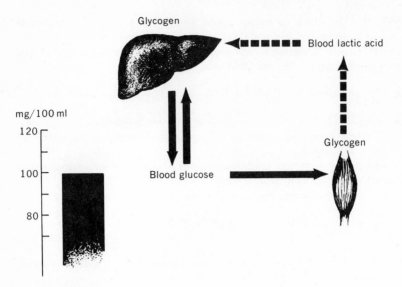

Fig. 2-2 Blood glucose can form glycogen in the liver and the muscles (enzyme hexokinase) but can only be released from the liver (enzyme phosphatase). The formation of glycogen is induced by insulin and by a rise in blood sugar. The formation of glucose is induced by epinephrine, a diabetogenic hormone, and a fall in blood sugar.

sis, determines endurance in prolonged physical work. On the other hand, the residual energy of lactic acid is by no means lost. The lactic acid diffuses into the interstitial fluid, enters the circulating blood, and, primarily in the liver, is eventually synthesized to glycogen or completely oxidized, e.g., in the heart muscle (Fig. 2-2).*

ENERGY CONTENT IN THE BODY AVAILABLE FOR MUSCULAR WORK

A given weight of an organic compound contains a fixed amount of potential energy locked up in the bonds between the atoms of its molecules. Knowing the amount of such available compounds within the body and their energy contents gives us the energy stores of the human machine. Table 2-1 summarizes data which, however, are approximate and subjected to large individual fluctuations, particularly the content of fat. In a well-trained man with a body weight of 75 kg, some 20 kg of muscles may be active during running.

It is quite evident that the available energy from ATP and creatine phosphate is very limited. During one day some 2000 to 5000 kcal may be spent, and during maximal exercise the energy demand may exceed 50 kcal/min. Therefore the supply from a breakdown of the total ATP would only cover a 1-sec effort and a generation of ATP from a breakdown of creatine phosphate would cover another few seconds of maximal

* For details see Chap. 9.

TABLE 2-1

	Concentration		Energy/mole, kcal	Total kcal in man body weight 75 kg, muscle weight 20 kg
	mmoles/100 g dry muscle	Per kg wet muscle		
ATP	2.5*	6 mmoles	10	1.2
Creatine phosphate	7.0*	17 mmoles	10.5	3.6
Glycogen	....	15 g†	700	1,200
Fat	....	....	.....	50,000

* E. Hultman, J. Bergström, and N. McLennan Anderson: Breakdown and Resynthesis of Phosphorylcreatine and Adenosine Triphosphate in Connection with Muscular Work in Man, *Scand. J. Clin. Lab. Invest.*, **19**:56, 1967.
† E. Hultman: Studies on Muscle Metabolism of Glycogen and Active Phosphate in Man with Special Reference to Exercise and Diet, *Scand. J. Clin. Lab. Invest.*, **19** (Suppl. 94), 1967.

effort. It is evident that the rapid and continuous energy production from the oxidation of glycogen and fatty acids is of extreme importance.

Direct studies of the changes in concentration of ATP and creatine phosphate during maximal exercise analyzed from muscle biopsies show that the concentration of creatine phosphate decreases toward 0 after a few minutes' work, but there was still about 60 percent of ATP left when the subject became exhausted (Hultman et al., 1967). (The few seconds it took to take the sample and stop all enzymatic activities may have been sufficient to replace some of the ATP.)

It is well established that a maximal speed can only be maintained for some seconds, e.g., in a 100-m dash, and the explanation could be that "rapid energy" is no longer available due to an exhaustion of creatine phosphate, eventually also of ATP. The maximal time for support from the somewhat slower glycolysis would be less than 1 min, mainly due to an accumulation of lactic acid in the muscle.

In the complete aerobic breakdown of glycogen, 1 liter of used oxygen provides 5.05 kcal. The circulatory capacity to transport oxygen during heavy exercise averages about 3.0 liters/min in a young man. The energy supply from ATP and creatine phosphate would theoretically substitute approximately 80 ml of oxygen/kg wet muscle (with the efficiency in the aerobic processes considered). With 20 kg of muscles working, the energy supplies would be equivalent to an oxygen uptake of 1.6 liters or 50 percent of the maximal oxygen uptake per minute.

SUMMARY

High-energy phosphate compounds represent the common currency for the transfer of energy within the living organism. The ATP-ADP system is the primary carrier of chemical energy in each and every cellular reaction. As quickly as ADP is formed it is rephosphorylated, and the metabolic generation of ATP provides a general mechanism for the coupling of energy-yielding and energy-requiring processes. If the process is anaerobic, ADP is rephosphorylated at the expense of glycolysis, lactate being formed, or by creatine phosphate. In aerobic conditions, ADP is rephosphorylated during oxidative phosphorylation by the mitochondria.

In the absence of oxygen the skeleton muscles can work only for short periods of time, and the total energy available is very limited compared with the aerobic work situation. Theoretically, ATP and creatine phosphate could alone cover the energy demand during a few seconds of heavy exercise.

REFERENCES

Baldwin, E.: "Dynamic Aspects of Biochemistry," 5th ed., Cambridge University Press, New York, 1967.

Cain, D. F., A. A. Infante, and R. E. Davies: Chemistry of Muscle Contraction: Adenosine Triphosphate and Phosphorylcreatine as Energy Supplies for Single Contractions of Working Muscle, *Nature (London)*, **196:**214, 1962.

Hultman, E.: Studies on Muscle Metabolism of Glycogen and Active Phosphate in Man with Special Reference to Exercise and Diet, *Scand. J. Clin. Lab. Invest.*, **19** (Suppl. 94), 1967.

Hultman, E., J. Bergström, and N. McLennan Anderson: Breakdown and Resynthesis of Phosphorylcreatine and Adenosine Triphosphate in Connection with Muscular Work in Man, *Scand. J. Clin. Lab. Invest.*, **19:**56, 1967.

Keele, C. A., and E. Neil: "Samson Wright's Applied Physiology," 11th ed., Oxford University Press, Fair Lawn, N.J., 1965.

Lehninger, A. L.: "Bioenergetics," W. A. Benjamin, New York, 1965.

Mahler, H. R., and E. H. Cordes: "Biological Chemistry," Harper & Row, Publishers, Incorporated, New York, 1966.

Needham, D. M.: Biochemistry of Muscular Action, in G. H. Bourne (ed.), "The Structure and Function of Muscle," vol. II, chap. 2, p. 55, Academic Press Inc., New York, 1960.

Soskin, S., and R. Levine: "Carbohydrate Metabolism," 2d ed., The University of Chicago Press, Chicago, 1952.

3

Muscle Contraction

contents

chapter three

Muscle Contraction

ARCHITECTURE

The previous chapter dealt with the principles of the energy-yielding reactions. Now we shall turn our attention to the question of how the chemical energy is transformed into mechanical work and describe the mechanical design of the muscle.

Standard textbooks of anatomy and histology should be consulted for detailed descriptions of the muscle. In this review we shall simply present a summary of the architecture of the muscle tissue, with some details of the working unit within the muscle. Of the three types of mammalian muscles, smooth muscles, heart muscle (striated involuntary muscle), and skeletal muscles (striated voluntary muscles), it is mainly the skeletal muscular tissue that will be discussed here.

The *muscle groups* as we know them by their Latin names, such as the brachialis, are mostly groups of muscle bundles that join into a tendon at each end. On the outside the bundles are covered by a fascia of fibrous connective tissue known as the *epimysium*. Each bundle is separately wrapped in a sheath of connective tissue called *perimysium*. The bundle is made up of thousands of muscle fibers, each embedded in a fine layer of connective tissue (*endomysium*) (Fig. 3-1). The various sheaths of connective tissue blend with the tendon in a way that is determined by function and space.

The functional unit within a muscle is the group of muscle fibers innervated by a single motor nerve fiber. The individual muscle fiber is, however, anatomically separated from the neighboring fibers by the endomysium. Therefore it can be considered as the working unit.

The amount of connective tissue (collagen fibers, elastic fibers, and other cells) varies in different muscles and in different species of animals. In man, the number of muscle fibers in a muscle group probably is finally established after the embryo has reached the age of 4 to 5 months (MacCallum, 1898).

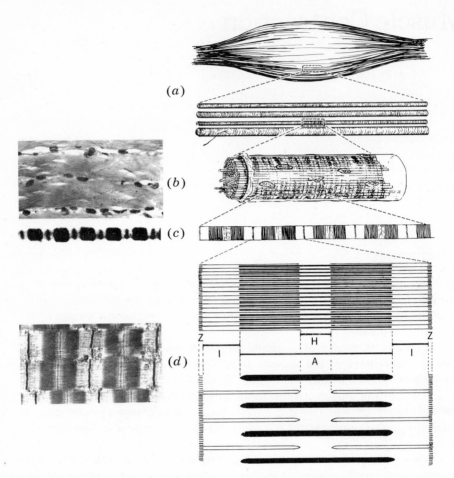

Fig. 3-1   Schematic drawing of striated muscle (right) with corresponding photomicrographs (left). The striated muscle (a) is made up of muscle fibers (b), which appear striated in the light microscope. The small branching structures at the surface of the fibers are the "endplates" of motor nerves, which signal the fibers to contract. Each single muscle fiber (c) is made up of myofibrils, beside which lie cell nuclei and mitochondria. In a single myofibril (d), the striations are resolved into a repeating pattern of light and dark bands. A single unit of this pattern consists of a Z line, then an I band, then an A band which is interrupted by an H zone, then the next I band, and finally, the next Z line. This repeating band pattern is due to the overlapping of thick and thin filaments (bottom part of diagram). (Redrawn from H. E. Huxley, Sci. Am., November, 1958.)

However, the thickness of a fiber can vary. At birth the fiber is about twice as thick as in the fourth fetal month, but has only one-fifth of the adult thickness (Lockhart, 1960).

The *skeletal muscle fiber* is a cylindrical, more or less elongated cell. Its thickness varies in different muscles or even in the same muscle, and may be from 10 to 100 μ. The length of the individual cell extends, in many muscles, all the way from the tendon of origin to the tendon of insertion. Whether this is the case in the longest muscles, such as the sartorius, is not clear. However, cells more than 30 cm long have been traced in this muscle. The muscle cell is multinucleated, with sometimes as many as several hundred nuclei in a single fiber (Walls, 1960).

Other elements in the cell are the sarcolemma, myofibrils, and sarcoplasm. The *sarcolemma* is a thin elastic noncellular membrane, less than 100 Å thick, enveloping the striated muscle fiber. Its structure is very similar to the internal membranes of other cells (nerve cells, Schwann cell, etc.). In some places the sarcolemma has tunnels, cavelike invaginations, or open vesicles. These structures might be morphological manifestations of active transport mechanisms (Walls, 1960). The sarcolemma also has remarkable electrical properties.

The contractile element of the cell consists of the *myofibrils*. The structure and function of these fibrils have been described by A. F. Huxley (1962) and H. E. Huxley and coworkers (1962, 1965). Methods which have been used in the study of these structures have ranged from electron microscopy and small-angle ray diffraction to light microscopy. Furthermore, biochemical studies of the protein components of skeletal muscles have contributed to the present understanding of muscular function (Needham, 1960; Szent-Györgyi, 1962; Lorand and Molnar, 1962).

In each muscle fiber (or cell) there are many myofibrils, each 1 to 3 μ thick, arranged parallel to one another. The individual myofibrils are aligned within the sarcolemma so that points with the same density lie at the same level, giving the appearance of disks crossing the whole thickness of the muscle fiber. This is illustrated in Figs. 3-1 and 3-2. Each repeat is called a *sarcomere* and is bordered by a narrow membrane called the *Z* line, which, like a disk, crosscuts the myofibril in units. In the middle region of the sarcomere there is a dark band, the *A* band (detected by using the deep-focusing position on the microscope). *A* stands for *anisotropic*, which refers to the optical property of the tissue. The alternate light bands are *isotropic*, or *I* bands. The *Z* line mentioned is located in the midst of the *I* band. The region in the center of the *A* band is called the *H* zone, being less refractile (lighter) than the rest of the *A* band.

This terminology may be summarized as follows (Fig. 3-2): Two bands of protein rods or filaments are distributed in the myofibrils in parallel order.

Fig. 3-2   Schematic drawing of the protein rods or filaments of the myofibril. The thick filaments consist of myosin; the thin filaments consist of actin.

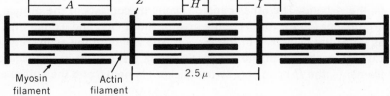

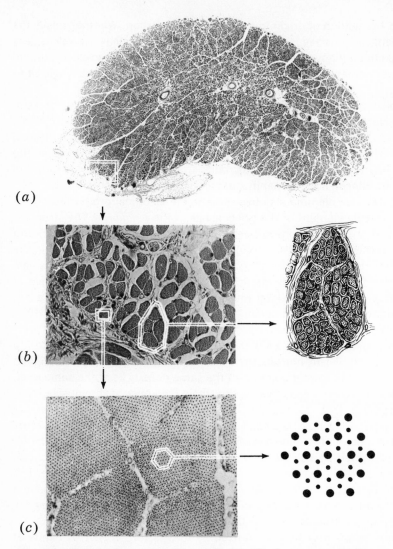

Fig. 3-3   Cross section of striated muscle (a). The larger magnification shows the individual fibers (b) and the fibrils (c), with the very regular pattern of thick and thin filaments.

Filaments of *myosin* (about 1.5 μ in length and 150 Å thick), arranged lengthwise, fill the *A* band. The other band of filaments consists of the protein *actin*. They are thinner (50 Å), run from the area of the *Z* line, and extend into the *A* band as far as the beginning of the *H* zone. The length of the actin filaments is about 1.0 μ on each side of the *Z* line. Evidently the *I* band is occupied only by thin filaments, the *H* band only by the thicker filaments, and the outer parts of the *A* band by both. A cross section of a muscle fiber in the overlapping zone reveals a regular pattern. The thick filaments lie about 450 Å apart with thin filaments in between, so that each of the thin ones is "shared" by three thick filaments (Figs. 3-2, 3-3).

CHANGES IN LENGTH

While the length of a muscle fiber is changing, by shortening or lengthening, the A band remains constant, but the I band changes its length. This is brought about by sliding movements of the arrays of thin filaments inward into the arrays of thick filaments. The length of the actin and myosin filaments remains constant. In a maximal contraction, the Z line may touch the ends of the two adjacent A bands, and eventually the filaments fold or crumple slightly. In such an isotonic contraction, the I band almost disappears (Fig. 3-4). In an isometric muscular contraction (no change in length), the length of the A and I bands remains constant, as do the lengths of the filaments.

The individual filament is built as a chain of several hundred molecules of protein. From the thick filaments, short lateral projections extend toward, and appear to touch, the thin filaments within the A band (Huxley and Hanson, 1960). It is suggested that in the moment of activation, cross-linkages are formed between actin molecules and the "protruding" myosin groups. The active actin

Fig. 3-4 Schematic drawing showing the arrangement of the filament with change in the length of the muscle. (a) shows the muscle stretched, (b) at resting length, and (c) shows the contracted muscle, when the thin filaments meet.

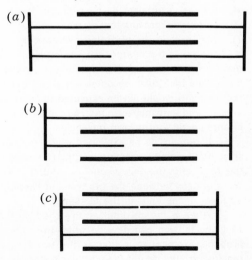

group then successively reacts with several linearly arranged myosin groups, and the actin filament travels alongside the myosin filament. If the muscle is working isometrically, the same molecular groups may react with one another several times, but eventually the actin filament will slip back.

By a special technique of electron microscopy, more detailed information about the structures have been obtained (Huxley, 1965). The myosin molecule seems to be asymmetric, with a "head" (of about one-sixth of the total length) and a "tail." The sites responsible for its enzymatic activity (see below) and its affinity for actin are located in its globular head (of heavy-meromyosin), and the sites responsible for its affinity to other adjacent myosin molecules are in its tail (of light-meromyosin). The molecules are oriented with their heads pointed in one direction along half of the filament, and in the opposite direction along the other half, leaving a projection-free region midway along their length (Fig. 3-5). The heads serve as the cross-bridges which are the only structural and mechanical linkage between the thick and thin filaments.

The thin filament has the form of a double helix, consisting of two chains of roughly globular subunits twisted around each other (like two twisted strings of beads). All actin molecules in a given filament are oriented in the same manner. They can all interact in identical fashion with a given myosin cross-bridge. Furthermore, the thin filaments on one side of a Z line are all similarly oriented, but on the opposite side of a Z line the orientation is reversed. With this arrangement of actin and myosin molecules in the two halves of an A band, we can expect the actin filaments on either side of the sarcomere to move in opposite directions, that is, toward each other in the middle of the sarcomere.

The alternating high and low points of the thin filaments (Fig. 3-5) suggest a general arrangement for the successive active sites on the filament to which the cross-bridges may attach themselves.

When the muscle shortens, the ends of the thin filaments will slide toward each other in the center of the A band, and they may even overlap (see bottom illustration on Fig. 3-5). (A dense zone appears in the center of the A band, and in a transverse section of this zone in an electron micrograph, there are twice as many thin filaments as in a relaxed muscle, proving that an overlap of actin filaments may occur during contraction.) This may explain the observed decrease in maximal tension generated by a muscle as it shortens. At the center of the thick filaments the projections are absent, and as the thin filaments from one Z line continue into the "wrong" part of the A band, the orientation of the molecules becomes abnormal from a functional point of view. In this region the filaments would not be expected to contribute to the development of tension by the muscle, and they may even interfere in a negative way with the interaction of the correctly oriented actin and myosin molecules.

It remains to be explained why a muscle can develop its maximal tension as it is stretched beyond its resting length. This does diminish the distance of overlap between the thick and the thin filaments, and therefore the more they slide apart the fewer cross-bridges between the two types of filaments are possible. The cross-bridges may be stronger at the ends of the filaments.

When a muscle fiber is stretched so that the length of the sarcomere exceeds 3.5 μ, no tension can develop when the muscle is stimulated (A. F. Huxley, 1962). The probable explanation is that as the two sets of filaments have ceased to overlap, the reacting groups do not "reach" each other.

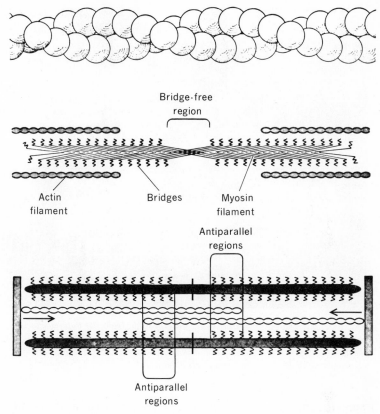

Fig. 3-5   *Structure of actin is represented by two chains of beads twisted into a double helix (top). The contact of actin and myosin might be made in the manner schematically illustrated in the middle part of the figure. The thin actin filaments at top and bottom are so shaped that certain sites are closest to the thick myosin filament in the middle. The heads of individual myosin molecules (zigzag lines) extend as cross-bridges to the actin filament at these close sites. At the bottom of the figure is shown double overlap of thin filaments from each side of the sarcomere, which would result if the sliding-filament hypothesis is correct. The tension would fall when thin filaments cross the center and interact with improperly oriented cross-bridges. (From H. E. Huxley, Sci. Am., 213(6):18, 1965.)*

Besides the contractile fibrils themselves, the mitochondria play an essential role in the energy-providing aerobic oxidation of the fuel (Fig. 3-6).

PLUMBING AND FUELING SYSTEM

The filaments just described are embedded in a fluid in which there are soluble proteins (such as myoglobin), glycogen granules, fat droplets, phosphate com-

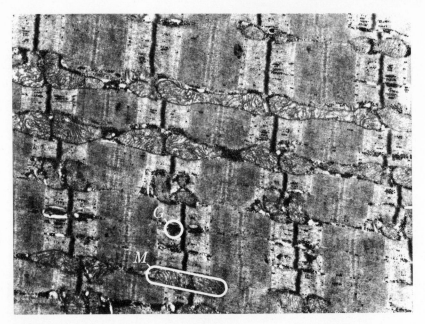

Fig. 3-6 Electron micrograph of striated muscle from the quadriceps (longitudinal section) in man (maximal oxygen uptake 5.03 liters/min). A small piece of muscle tissue was obtained by biopsy. Note the mitochondria (M) and glycogen granules (G). (By courtesy of L. Hermansen and R. Gustavsson.)

pounds, other smaller molecules, and ions. This aqueous phase is called the *sarcoplasmic matrix.*

Another fraction of the sarcoplasm is the sarcoplasmic reticulum (sarcotubules). It consists of elaborately anastomosing tiny tubular sacs, and channels of varying caliber extending parallel in the spaces between the myofibrils and transversely across the fiber. One or two such systems (triods) are present for each sarcomere. The membranes bounding the channels are similar in structure to the sarcolemma. It is presumed that the content thus enclosed consists of a proteinaceous fluid containing dissolved ions of a different concentration than in the fluid outside the reticulum. Therefore, an electrical potential difference might exist between the two compartments (Bennet, 1960). In mammals, part of the sarcoplasmic reticulum probably forms a network across the muscle fiber on either side of the Z line.

This system could explain how the excitation of a muscle, moving along the sarcolemma of the fibril, may penetrate into the fibrils. In fact, some channels start as inward extensions of the sarcolemma at each Z line. They could, therefore, conduct electrical messages and also serve as inlets through which fluid outside the cell could flow into the cell. The sarcoplasmic reticulum could also transport lactic acid to the surface of the cell for further transport by the bloodstream to the liver and other tissues. It has been shown (A. F. Huxley, 1962) that a reduction of the surface membrane potential at certain spots on the projection, close to the Z line, could lead to a shortening of the corresponding I band. Thus, the shortening spreads inward rather than to the adjacent sarcomeres. With the filaments lying parallel in the muscle fiber, a coordination in time within each sarcomere is certainly a wise design. Margreth et al. (1963) point out that the structural origin of the sarcotubular system together with the biochemical data strongly indicate that this system may play a primary role in the regulation of carbohydrate metabolism in relation to the contractile activity of skeletal muscle. The sarcoplasmic reticulum might be the plumbing and fueling system of the muscle, with structures which support, control, regulate, and excite the contractile material itself (Bennet, 1960).

BIOCHEMISTRY OF MUSCLE FIBER CONTRACTION

We have described how the changes in the muscle length can be induced by a sliding of the thin actin filaments past the myosin filaments (Fig. 3-3) and that this may occur by a cyclic forming and breaking of linkages between the actin molecule and the myosin molecule. What process causes the two sets of filaments to connect and disconnect such cross-linkages? Of the muscle protein more than 50 percent is myosin (M) and at least 25 percent actin (A). Neither protein, by itself, is contractile. *In vitro* these two proteins can, under certain conditions, form a complex protein, actomyosin, which can contract. Such experiments suggest that the cross-linkages are due to chemical reactions and that the forming of $A + M \rightarrow AM$ is essential for the contraction.

In the substrate ATP, Mg^{++} ions and Ca^{++} ions must be present if a contraction is to occur. For a break of the linkages and relaxation of the fiber, the substrate must be free from Ca^{++} and contain a chemically unspecified "relaxing factor" (Marsh factor), ATP, and Mg^{++}. There are several hypotheses for the series of events (Needham, 1960; Szent-Györgyi, 1962; Lorand and Molnar, 1962). Some theories are based on the assumption that the site of reaction is probably located where the lateral projections of the myosin filaments touch the actin (Huxley and Hanson, 1960), and that —SH groups are likely to be involved.

CONTRACTION

The energy for this interaction is transformed to the myosin by transphosphorylation of the ATP into an energy-rich phosphate bond of myosin.

$$M \cdot ATP \rightarrow M \sim P + ADP$$
$$M \sim P + A \rightarrow AM + P + \text{free energy transformed to tension or work}$$

(Needham, 1960). $\sim P$ = the high-energy phosphate bonds.

Thus, the ATP-ADP system, driven by the enzyme ATPase, provides the energy for the actomyosin machine. If the ATPase activity is allowed to continue, the reaction between actin and myosin may go on until the supply of ATP is depleted. The linkages then stay locked, and the muscle goes into sustained contraction. If ATP is then added, the muscle can relax.

The enzyme (ATPase) that splits ATP is actually myosin with Mg^{++} as an activator (ATPase is potentiated in its effect as actomyosin forms). An internal rearrangement of molecules within the filaments may occur by a successive reaction between the actin and myosin molecules. With an increase in ATP concentration by the arrival of fresh ATP, the links are broken.

$$\mathrm{ATP + AM \rightarrow M \cdot ATP + A}$$

If the ATPase is still active, the series of events can be repeated. The actin sites are now within the range of further myosin projections, and the actin filament slides along the myosin chain. If the muscle is made to work isometrically, the actin filament may slip back the moment the "locks" open, and the same links will act many times. With a low speed of contraction, the repeat of cycles occurs at longer intervals.

RELAXATION

ATP has obviously a twofold function: it is required for the contraction process as well as for the relaxation process. The probable explanation for its essential role in muscle relaxation is that this process is energy-consuming, and the ATP-ADP mechanism provides the energy needed.

$$\mathrm{S + ATP \rightarrow S \sim P + ADP}$$
$$\mathrm{S \sim P + AM \rightarrow S + A + M + P}$$

where S = a relaxing factor.

The relaxing factor is functionally attached to the sarcoplasmic reticulum. In the relaxed stage, no Ca^{++} ions are free; if Ca^{++} is added, the muscle contracts. Ca^{++} ions are essential for the formation of the actomyosin molecule (by activating the ATPase or inhibiting the relaxing factor). One is tempted to think that those ions are bound within the fiber at rest and then set free when the arrival of a nerve impulse depolarizes the membrane of the sarcoplasmic reticulum (or a separate system of small tubules, the T system). Then the calcium ions are recaptured by the sarcoplasmic reticulum. In some way or other, the reticulum may assume the role of an ion exchange which functions in a cyclic manner (Csapo, 1959; Lorand and Molnar, 1962). Thereby the activating substance will be released, trapped, released again, etc., in rhythm with the variation in electric potential.

SUMMARY

Chemically stored energy is converted into mechanical work through mediation of an enzyme system. The contraction of the muscle fiber is caused by a reaction between myosin and actin, the main protein compounds of the individual filaments within the muscle fiber. A fixation of the actin filaments to the myosin or a sliding of the filaments past each other can thus occur. The initial effect of the nervous stimulation of the muscle is still obscure. It presumably results in an inactivation of a relaxing factor and the activation of ATPase. As long as the stimulus is maintained, there is an enzymatic depletion of ATP, which provides the energy necessary to link the proteins to each other. When fresh ATP arrives or is formed, it works with the relaxing factor to break the links, and the events may start all over again.

REFERENCES

Bennet, H. S.: The Structure of Striated Muscle as Seen by the Electron Microscope, in G. H. Bourne (ed.), "Structure and Function of Muscle," vol. 1, p. 137, Academic Press Inc., New York, 1960.

Csapo, A.: Studies on Excitation-Contraction Coupling, *Ann. N.Y. Acad. Sci.*, **8**:453, 1959.

Huxley, A. F.: Skeletal Muscle, in K. Rodahl and S. M. Horvath (eds.), "Muscle as a Tissue," p. 3, McGraw-Hill Book Company, New York, 1962.

Huxley, H. E.: Muscular Contraction, in K. Rodahl and S. M. Horvath (eds.), "Muscle as a Tissue," p. 63, McGraw-Hill Book Company, New York, 1962.

Huxley, H. E.: The Mechanism of Muscular Contraction, *Sci. Am.*, **213**(6):18, 1965.

Huxley, H. E., and J. Hanson: The Molecular Basis of Contraction in Cross-striated Muscles, in G. H. Bourne (ed.), "Structure and Function of Muscle," vol. 1, p. 183, Academic Press Inc., New York, 1960.

Lockhart, R. D.: The Anatomy of Muscles and Their Relation to Movement and Posture, in G. H. Bourne (ed.), "Structure and Function of Muscle," vol. 1, p. 1, Academic Press Inc., New York, 1960.

Lorand, L., and J. Molnar: Biochemical Control of Relaxation in Muscle Systems, in K. Rodahl and S. M. Horvath (eds.), "Muscle as a Tissue," p. 97, McGraw-Hill Book Company, New York, 1962.

MacCallum, J. B.: On the Histogenesis of the Striated Muscle Fiber, and the Growth of the Human Sartorius Muscle, *Bull. Johns Hopkins Hosp.*, **9**:208, 1898.

Margreth, A., U. Muscatello, and E. Andersson-Cedergren: A Morphological and Biochemical Study on the Regulation of Carbohydrate Metabolism in the Muscle Cell, *Exp. Cell Research*, **32**:484, 1963.

Needham, D. M.: Biochemistry of Muscular Action, in G. H. Bourne (ed.), "Structure and Function of Muscle," vol. II, p. 55, Academic Press Inc., New York, 1960.

Szent-Györgyi, A. G.: Aspects of the Chemistry of Muscle Contraction, in K. Rodahl and S. M. Horvath (eds.), "Muscle as a Tissue," p. 68, McGraw-Hill Book Company, New York, 1962.

Walls, E. W.: The Micro-anatomy of Muscle, in G. H. Bourne (ed.), "Structure and Function of Muscle," vol. 1, p. 21, Academic Press Inc., New York, 1960.

4

Neuromuscular Function

contents

chapter four

Neuromuscular Function

The central nervous system (CNS) receives information about the outside world via exteroceptors reacting to light, sound, touch, temperature, or chemical agents, and interoceptors stimulated by changes within the body, e.g., *proprio-ceptors* (such as muscle spindles, tendon-end organs of Golgi, joint receptors, and vestibular receptors), *chemoreceptors*, and *visceroceptors* (Fig. 4-1). The CNS is equipped to receive, interpret, and handle information, and then to transform the result into muscular movements. Even the process of feeding information to the CNS may involve some muscular activity; for instance, the eye muscles are almost continuously active, especially while the individual is awake. Reaction to various stimuli includes gestures, speech, writing, and sometimes very vigorous muscular contractions, as in the case of an approaching danger. There is hardly any stimulus that does not, through reflexes, affect smooth muscles, heart muscle, or skeletal muscles.

A presentation of the physiology of work and exercise would justify a very detailed discussion of the function of the CNS. In fact, it is difficult to decide which aspects are more essential and which are less essential for the under-standing of muscular movements, athletic techniques, and training. However, we will limit our discussion mainly to the units more directly involved in the activation of the skeletal muscle. For a more comprehensive description of the function of the CNS, the reader should consult basic physiological textbooks.

THE MOTONEURON AND TRANSMISSION OF NERVE IMPULSES*

Anatomy

The task of the single nerve cell is to receive a message and then to pass it on to other cells; the message may be to a nerve cell, muscle cells, or other types

* The following is based mainly on the reviews by J. C. Eccles (1957, 1964, 1965).

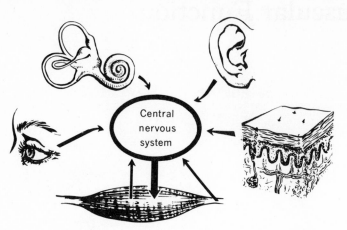

Fig. 4-1   The central nervous system receives information via afferent nerves from the various receptors. Impulses are transmitted through the "final common path," which is the only available route to the skeletal muscle.

of cells. Some nerve cells are specialized in "holding the message back" and preventing other cells from reacting to impulses (Fig. 4-2). It is a happy coincidence that the nerve cells which finally control the skeletal muscles (the motoneurons) are the most studied nerve cells in the animal. Consequently, these events are fairly well understood. (The physiology behind the capacity of nerve cells to store information for future use is, on the other hand, a well-kept secret.)

Principally, the motoneuron is similar to other nerve cells in its anatomy and physiology. The cell body, or soma, is approximately 70 μ across. A number of branching processes (dendrites) radiate from the cell, and these dendrites soon split up into terminal branches (Fig. 4-3). One long branch, the axon, or motor nerve fiber, leaves the spinal cord in the ventral root and runs to the muscle, a distance that may be 90 cm in man. A membrane (50 Å thick) covers the soma and the processes. Mitochondria in the cell indicate a high level of metabolic activity. Actually, the oxygen uptake of nerve cells is probably of the same order as that of maximally contracting skeletal muscles. In the processes, fine threads known as neurofibrils run parallel to the axis and are bathed in the axoplasm. The membrane, or axolemma, constitutes a barrier between the intra- and extracellular fluid. In the peripheral part, the axon is covered by a myelin sheath. For the nutrition of the dendrite and axon an uninterrupted connection with the cell body is essential, and there is a flow of axoplasm peripherally from the cell body. If the nerve fiber is cut, the peripheral part will degenerate.

Axonal terminals (synaptic knobs) of other nerves end at the surface of the soma and the basal regions of the dendrites (Fig. 4-3). Actually, more than

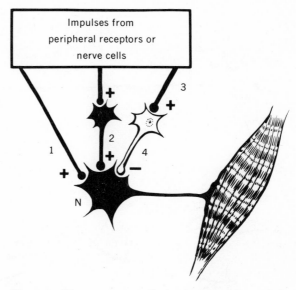

Fig. 4-2   The motoneuron, N, in the ventral horn of the gray matter of the spinal cord can be excited (+) directly (1) or via an interneuron (2). Thereby, an impulse is propagated in the nerve fiber, and the muscle is stimulated, causing muscular activity. However, other nerve terminals can prevent the motoneuron from responding to the exciting impulse. Schematically this is illustrated as follows: Nerve end (3) stimulates an interneuron nerve cell (4). But this cell inhibits (−) the motoneuron to react to the stimulation by nerve 1. The mechanisms involved are discussed in the text.

40 percent of the surface of a motoneuron may be covered by synaptic knobs from several hundred other nerves. The space between the membranes of the synaptic knobs and the contacted nerve cell (postsynaptic membrane) is about 200 Å. This space is named synaptic cleft.

Resting Membrane Potential

In the nerve cells, as in other cells, the chemical composition of the fluid on each side of the semipermeable cell membranes is different. In its composition the external fluid medium is an ultrafiltrate of blood, but the internal fluid contains a lower concentration of sodium and chloride ions and a higher concentration of potassium (ratio for Na^+ is about 150:15; for K^+, 5.5:150; for Cl^-, 125:9; the figures give the actual concentration in milliequivalents per liter). These differences in concentration of ions should cause a diffusion across the membrane if perfusibility and electrochemical gradients would permit this. The membrane does not, however, permit a free passage of ions. The cell contains a high con-

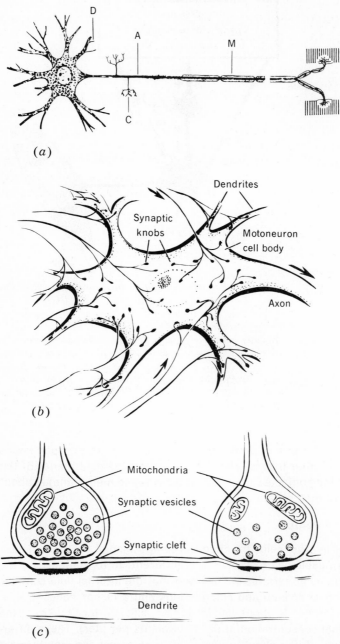

(a)

(b)

(c)

Fig. 4-3   (a) Motoneuron: A, axon; C, collateral; D, dendrite; M, myelin. (K. E. Schreiner and A. Schreiner.)
(b) A motoneuron cell body and its dendrites covered with synaptic

centration of protein anions which cannot escape through the membrane to the fluid outside the cell where the concentration of protein is low. Electric charges of opposite sign attract each other, and the excess protein anions inside the cell therefore attract cations. The membrane is much less permeable to Na^+ than to K^+; the tendency for K^+ to diffuse out of the cell according to a difference in concentration gradient is counteracted by the attraction from the protein anions inside the cell. Everything considered, with Cl^- in a diffusional equilibrium the K^+ outward flux is slightly greater than the inward flux, and the Na^+ inward, diffusional flux would be many times the outward flux. To maintain the difference in concentration of ions across the resting cell membrane, there must be a forced movement of Na^+ outward and K^+ inward, virtually equaling the diffusion in the opposite directions (Fig. 4-4). Actually, aerobic metabolic processes supply the energy to keep a sodium and a potassium pump going, whereby ions that diffuse or "leak" in or out of the cell are forced back. The net effect is a *higher concentration of anions in the interior of the soma and axon.* With a microelectrode inserted in the cell, the potential difference has been measured at about 70 mv over the membrane, the interior being negative with respect to the exterior. The cell contains nondiffusible protein; a disturbance of the resting potential difference would essentially occur if the sodium and potassium ions were allowed to move according to the diffusion gradients. Such a flux of ions across the surface membrane should disturb the resting potential of 70 mv. Hence, an inward flow of positive ions, e.g., sodium, would decrease the potential and cause a *hypopolarization;* an outward flow of positive ions, e.g., potassium, would give rise to a *hyperpolarization.* Actually such fluxes of ions occur normally when the nerves terminating at the motoneuron are stimulated.

Facilitation—Excitation

Let us first consider the events taking place when the axon (Fig. 4-2, 1) is stimulated. The electron microscope has revealed the construction of the synaptic junction: the synaptic knob of about 2 μ diameter, the synaptic cleft, and the subsynaptic membrane covering the motoneuron. Within the synaptic knob there are numerous vesicles approximately 300 Å in diameter (Fig. 4-5). They probably contain the chemical substance that is responsible for the transmission of the impulse across the synaptic junction. The arrival of an impulse at the synaptic knob causes a synchronous ejection of many vesicles. A corresponding

---

knobs. These knobs are the terminals of the impulse-carrying nerve fibers (axons) from other nerve cells.
(c) When stimulated, the synaptic knobs deliver a chemical transmitter substance into the synaptic cleft, where it can act on the surface of the opposite nerve cell membrane. The chemical transmitter substance is stored in numerous vesicles. (b and c from "The Synapse" by Sir John Eccles. Copyright © January, 1965 by Scientific American, Inc. All rights reserved.)

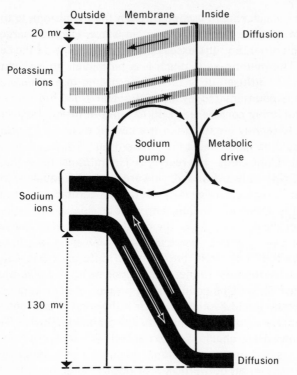

Fig. 4-4   Schematic diagram showing the K^+ and Na^+ fluxes through the surface membrane of the nerve cell in the resting state. The slopes in the flux channels across the membrane represent the respective electrochemical gradients. The voltage drop across the surface membrane is about 70 mv, with the inside being negative. Under these conditions, chloride ions diffuse inward and outward at equal rates. But the concentration of sodium and potassium has to be maintained by some sort of a metabolic pump. The negative potential inside the nerve cell membrane is 20 mv short of the equilibrium potential for potassium ions. Therefore, the pump has to actively prevent an outward diffusion of potassium ions. For sodium ions, the potential across the membrane is 130 mv in the other direction. To keep the sodium out, a very active pump is needed. (Modified from Eccles, 1965.)

number of quanta of transmitter substance is liberated. The transmitter substance diffuses readily to receptors on the subsynaptic membrane which are specifically affected by the substance. With a microelectrode inserted in the cell, the effect of this event can be recorded: there is a *hypopolarization* which can be assigned to an inflow of sodium ions. Eventually the active sodium pump will restore the resting membrane potential of −70 mv by extracting the Na⁺

ions. However, if the membrane potential is reduced to a critical value, about −60 mv, there is suddenly a free passage through the membrane of sodium ions, and they move in the direction of their concentration gradient, causing a *depolarization* of the membrane. The equilibrium potential for Na⁺ ions will actually be about +60 mv, with the interior of the motoneuron 60 mv positive to the extracellular fluid. However, the potential charge or height of the recorded "spike" (*action potential*) will not be 70 + 60 = 130 mv, but only about 100 mv, for immediately after the beginning of the influx of sodium there is an outward flow of potassium ions along their electrochemical gradient. Consequently, the number of positive ions within the cell will again decrease. The efflux of potassium ions actually begins early during the "rising" limb of the spike and reduces the spike potential. Within 1 msec the resting membrane potential of −70 mv is almost restored by trading potassium ions for sodium ions across the cell membrane. The movement of ions back to where they belong is now effected by a "slow" process requiring metabolic energy. Actually following the action potential there is a prolonged phase (15 to 100 msec) when the permeability of the neuronal membrane for potassium is still increased, and during this phase there is an after-hyperpolarization adding up to 5 mv to the resting membrane potential of −70 mv (Fig. 4-6).

The events just described have far-reaching effects: The electrochemical charges initiated by the transmitter mechanism will, if exceeding the given threshold of about −60 mv, cause similar sequential changes in the ionic permeability of the membrane of the whole nerve axon. This traffic is associated with the passage of a change in membrane potential (an action potential) during activity along the axon cell interior, changing from −70 mv to about +30 mv with respect to the exterior (Fig. 4-6). A propagated nerve impulse, traveling at a speed of up to 100 m/sec, will reach the muscle and stimulate it to contract.

Fig. 4-5 Synaptic vesicles. Unfilled dots denote transmitter substance (e.g., acetylcholine), filled dots symbolize an inactivator (e.g., cholinesterase). (Redrawn from Eccles, 1957.)

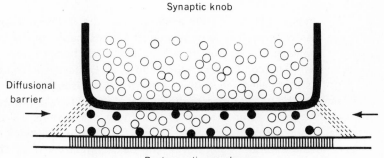

Synaptic knob

Diffusional barrier →

←

Postsynaptic membrane

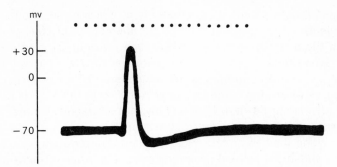

Fig. 4-6   Changes in membrane potential during activity along an axon. The interior voltage changes from −70 mv to about +30 mv with respect to the exterior. Note the hyperpolarization after the "spike." Time base, 2,000 cps. (Modified from Kuffler et al., 1951.)

The resting cell membrane is highly impermeable to negative ions and to sodium ions, but apparently freely permeable to potassium. Due to the steep concentration gradient for potassium (Fig. 4-4), it tends to leak out of the nerve at a high rate. The consequence is that the inside of the cell becomes electrically negative with respect to the outside. An equilibrium is actually reached when the tendency for potassium to diffuse outward is balanced by the electric field that is created. At this point the inside of the cell and axon is about −70 mv in relation to the external fluid. In this way the resting potential is created (Baker, 1966). When the membrane is affected by the transmitter substance or the potential becomes reduced, its permeability to sodium increases suddenly, and these ions can diffuse inwards due to both the concentration and voltage gradient. Evidently no source of energy other than the difference in ion concentrations on the two sides of the membrane is needed to amplify an electric impulse and propagate it along the axon (Baker, 1966). However, with repeated impulse traffic there would be a gain of sodium and loss of potassium, and the concentration gradient on which the nerve function depends would soon become destroyed. The sodium pump, using metabolic energy derived from ATP, extrudes sodium ions from soma and axon in exchange for potassium ions, and restores the membrane potential.

The hypopolarization after a stimulus of the different excitatory nerves may not be strong enough to depolarize the postsynaptic membrane. It is then *subliminal* and gives rise only to a local potential change under the activated synaptic knob. This condition is referred to as an *excitatory postsynaptic potential* (EPSP); it is also called *facilitation*. After a few milliseconds the sodium pump has restored the normal membrane potential of −70 mv and a new impulse of the same strength at an interval of 5 or more milliseconds will merely repeat the local hypopolarization say to −63 mv, the interior still being negative with respect to the exterior. If, however, the next stimulus arrives at the synaptic knob close to the preceding volley and initiates an additional ejection of transmitter substance while the subsynaptic membrane is still hypopolarized, say to −66 mv, the new sodium influx will be superimposed. This will decrease the potential to about −59 mv. This voltage charge can evoke a depolarization and generate a propagating spike potential in the axon. If the single stimulus is below the minimal strength, successive stimuli can thus elicit a depolarization by a summation of the receptor potential. In this case it is called a *temporal summation*. With several synaptic knobs terminating close together on a nerve cell, a simultaneous excitation may deliver the quanta of transmitter substance sufficient to produce the sieve for sodium and potassium on the subsynaptic membrane. This is an example of *spatial summation*.

In some synapses, acetylcholine (ACh) is identified as the *transmitter substance*. Some sort of barrier keeps it fairly effectively within the synaptic cleft, but some of the molecules will disappear by diffusion. More important for its gradual inactivation is a breakdown to much less active choline and acetic acid by an enzyme, cholinesterase, present in the postsynaptic structures (Fig. 4-5). Apparently the combination between ACh and the specific receptor reduces the electrical resistance of the subsynaptic membrane, causing a hypo- or depolarization. After 1 to 10 msec (for some cells up to 50 msec) the ejected ACh is inactivated, however, and the resting membrane potential is gradually restored by the ion pumps. It is to be noted that both the strength of the stimulus of the nerve and the rate of application of the stimulus are of critical importance for the membrane potential. This is readily explained, as the delivered quanta of ACh are continuously destroyed, and its concentration on the subsynaptic membrane may not become high enough for an effective hypopolarization. (Actually at rest there is a release of a few quanta of ACh in a random manner, causing a minute local membrane potential change. The passage of a nerve impulse produces, however, a synchronous ejection of a great number of such quanta, and finally the subsynaptic nerve cell does not have the capacity to "defend" itself against its depolarizing effect.)

If the stimulus is adequate to elicit a propagated spike, the magnitude of the spike potential of a single nerve fiber is independent of the strength of the stimulus; the nerve responds to the utmost of its ability. The answer to a stimulus is an all-or-none response.

During the spike potential, neither excitability nor conductivity is present in the nerve; it is in an *absolute refractory period*. For the next few milliseconds there exists a *partial refractoriness* during which time a stronger stimulus is necessary to evoke an action potential. The normal excitability is gradually restored in the course of about 100 msec. Nerves with large diameters recover 90 percent of the normal state within 1 msec, with the consequence that they can generate impulses at a frequency of 1,000/sec. The intermittent impulse traffic in nerves is due to the refractoriness, and the recovery of the nerve cell is certainly enhanced by this mechanism. During the stimulus, only a tiny fraction of the available sodium and potassium ions is exchanged across the cell membrane. The increased potassium permeability actually accelerates the recovery of the membrane potential, so that it is ready to propagate another impulse. The sodium and potassium pump are then very effective to restore the "unbalanced balance" characterizing the resting condition.

Summary   When an action-potential spike reaches the terminal of a synaptic knob, the postsynaptic membrane is converted by a chemical transmitter substance, which is released from the knob, into a sieve, first permitting an inward flux of sodium ions, then an outward flux of potassium. This "short circuit" depolarizes the postsynaptic membrane and generates an impulse, carried by similar ion movements, that propagates along the axon. The transmitter is inactivated, and the aerobic metabolism provides energy to extract the sodium ions from the intracellular fluid and to bring back the potassium ions. Thus, the normal voltage difference of 70 mv across the cell membrane is gradually restored. If the stimulus of a single impulse is not strong enough to evoke an action potential, it can provide a local hypopolarization (EPSP). This can add to the hypopolarization produced by preceding (temporal) or concurrent (spatial) impulses, and hence the threshold level for initiating an impulse in the effected nerve may be reached, this threshold being about 10 to 18 mv above the resting potential in the spinal motoneurons (Fig. 4-7).

Inhibition

From a functional viewpoint, there is a different type of neuron with short axons ending in synaptic connection with motoneurons or other nerve cells, forming interneurons. When stimulated, their synaptic knobs liberate a chemical transmitter which initiates an increased flux of potassium and chloride ions, but the membrane retains its high degree of impermeability to sodium ions. The result is an outward current and an increase in the voltage difference between the interior of the cell membrane and its exterior, i.e., a *hyperpolarization*. It rises to a summit in 1.5 to 2 msec after its onset and fades exponentially. During this phase a stronger stimulus is needed to discharge an impulse in the motoneuron; i.e., an *inhibitory postsynaptic potential* (IPSP) has developed. The IPSP can bring the voltage to −80 mv, which is the equilibrium potential of the inhibitory ionic mechanism. The further the membrane is removed from the equilibrium potential of −70 mv, the larger is the IPSP. Recovery would occur as the intracellular potassium ion concentration is restored by diffusion and by the operation of a potassium pump.

A movement of chloride ions also contributes to the IPSP. The nature of the chemical transmitter substance that causes the subsynaptic membrane to become a sieve whose pores permit the passage of potassium and chloride ions is presently unknown. In Fig. 4-2, nerve 3 represents such an inhibitory interneuron. Within a motoneuron there is a spatial as well as temporal summation of inhibition, as was the case for excitatory stimuli. The IPSP is actually a mirror image of the EPSP, differing in its longer latency and shorter time constant of decay. The total duration for the inhibition by the IPSP is 2 to 3 msec at the most.

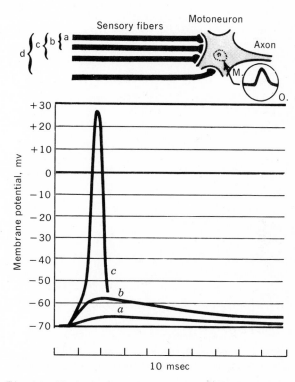

Fig. 4-7 Excitation of a motoneuron can be studied by stimulating the sensory fibers that send impulses to it. A microelectrode (M.) implanted in the motoneuron measures the changes in the cell's internal electric potential. These changes, the excitatory postsynaptic potentials (EPSPs), appear on oscilloscope (O.). It is assumed here that one to four sensory fibers can be activated. When only one fiber is activated (a), the potential inside the motoneuron shifts only slightly. When two fibers are activated (b), the shift is somewhat greater. When three fibers are activated (c), the potential reaches the threshold at which depolarization proceeds swiftly, and a spike appears on the oscilloscope. The spike signifies that the motoneuron has generated a nerve impulse of its own. When four or more fibers are activated (d), the motoneuron reaches the threshold more quickly (spike not shown). (Modified from Eccles, 1965.)

Summary A specific transmitter substance causes an inhibitory synaptic response on areas of the motoneuron by changing the ionic flux of potassium and chloride ions through the subsynaptic membrane. During the hyperpolarization caused by this ion flux, a stronger excitatory action is necessary for initiation of a depolarization, evoking the discharge of an impulse. The terminology is summarized in Fig. 4-8.

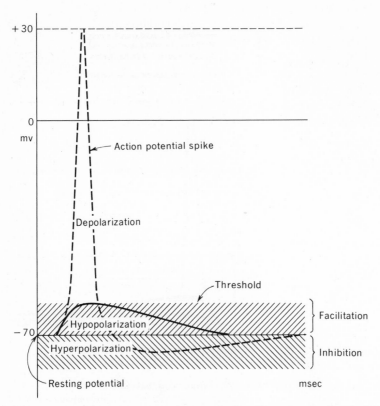

Fig. 4-8 Summary of the terminology used when discussing the electrical activity in a nerve cell under various conditions.

Inhibition and Excitation

Figure 4-9 gives a quantitative evaluation of the effect of an excitatory impulse on the membrane potential of a motoneuron simultaneously subjected to a stimuli of an inhibitory nerve fiber.

Motoneurons are actually subjected to a continuous bombardment by impulses which expose their membranes to both inhibitory and excitatory transmitter substances. They discharge impulses only when the excitatory synaptic activity is momentarily so dominating that the depolarization exceeds the threshold value causing a short circuit of the membrane. The synapses are distributed fairly uniformly over the soma and dendrites of the motoneuron, but at some areas of the motoneuron the threshold for a depolarization is actually lower than in other areas. Here excitatory synapses could initiate an impulse despite a considerable inhibitory bombardment of areas remote from this focus. As a result of the longer latency for inhibitory impulses they are most effective when they precede the excitatory volley.

Functionally there are only two types of nerve cells, excitatory and inhibitory. Any one class of nerve cells operates by the same chemical transmitter substance at all its synapses, and the substance always has the same synaptic action on all nerve cells connected: an excitation or inhibition. All inhibitory nerve cells are short axon neurons lying in the gray matter, while all transmission pathways, including the peripheral afferent and efferent pathways, are formed by the axons of excitatory neurons. Many short axon interneurons also belong to the class of excitatory neurons (Eccles, 1957).

Fig. 4-9 With a microelectrode (M.) implanted in a motoneuron the changes in the cell's internal electric potential can be displayed on an oscilloscope (O.). When the excitatory nerve fiber is stimulated, an EPSP is evoked, and the threshold for producing a spike is eventually reached (dotted line at top). Stimulation of the inhibitory nerve fiber gives rise to an IPSP (dotted line at bottom). Inhibition of a spike discharge is an electrical subtraction process. The IPSP widens the gap between the cell's internal potential and the firing threshold (here about −55 mv). Thus, if the motoneuron is simultaneously subjected to both excitatory and inhibitory stimulation, the IPSP is subtracted from the EPSP, and no spike occurs (full line). (Modified from Eccles, 1965.)

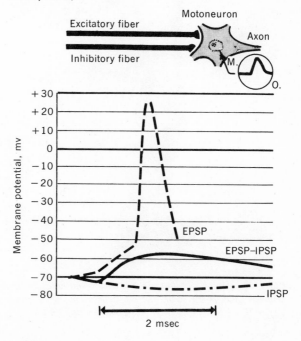

Fig. 4-10   Motor endplate. In the motor endplate, the knoblike terminations of the telo-dendron of a motor axon fit into depressions of a multinucleated mass of sarcoplasm. (After Willy Schwartz, from Elias and Pauly, 1961.)

TRANSMISSION OF IMPULSES FROM NERVE AXON TO SKELETAL MUSCLE: THE MOTOR UNIT

As the axon of the motoneuron approaches the muscle fiber, it loses its medullary sheath. The axis cylinder branches, and the terminals make close contact with the sarcoplasm of the muscle, forming *motor endplates* (Fig. 4-10). Each motoneuron supplies from about five (eye muscles) up to several thousand muscle fibers (limb muscles). The motoneuron and the muscle fibers it supplies function as a *motor unit*, since an impulse in the axon will activate all the fibers almost simultaneously. The gain in muscular tension with the activation of one motor unit will depend on the number of muscle fibers included in the unit. The fewer there are, the finer can be the grading of the contraction. The fibers in a motor unit may be scattered and intermingled with fibers of other units, and they can be spread over an approximately circular region with an average diameter of 5 mm (Buchtal et al., 1957). Single muscle fibers may be innervated by more than one motoneuron, but the majority of the muscle fibers possess only a single endplate which is supplied by a single nerve fiber. The terminal branches of the axon end near the middle of the muscle fiber.

Figure 4-11 presents a schematic drawing of an endplate. The surface membranes of the nerve fiber and muscle fiber are comparable with the pre- and postsynaptic membranes, respectively, of a central synapse. In the synaptic ending there are minute vesicles in great number, similar to those found in the synaptic knobs in the CNS. They contain acetylcholine, and this chemical substance is responsible for the depolarization of the postjunctional membrane.

At rest a miniature endplate potential can be recorded and is probably produced by ejections of a few quanta of ACh from vesicles that spontaneously burst on the surface of the membrane. The arrival of an impulse from the motoneuron to the nerve terminal causes an almost synchronous ejection of many quanta of ACh. The released ACh diffuses rapidly across the narrow junctional gap and attaches to the specific receptors

Fig. 4-11 Schematic drawing of endplate: ax., axoplasm with its mitochondria; my., myelin sheath; tel., teloglia; sarc., sarcoplasm with its mitochondria; m.n., muscle nuclei. (From R. Couteaux, in G. H. Bourne (ed.), "The Structure and Function of Muscle," vol. 1, Academic Press Inc., New York, 1960.)

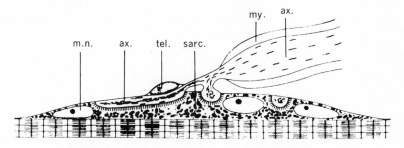

on the membrane, producing an endplate potential. When it reaches a certain critical magnitude, it depolarizes the surface membrane of the muscle fiber and evokes a propagated muscle action potential, which travels simultaneously to both ends of the muscle fiber. The electrochemical events are actually very similar to the ones described for the excitatory processes in the CNS. The ACh effect, drastically increasing the permeability of the postjunctional membrane to sodium ions and later on to potassium ions, is very short-lived, for in the junctional gap there is also the ACh-destroying enzyme, cholinesterase. Thus the membrane permeability can again be reduced, and the membrane potential is gradually restored to its resting level. The membrane of the muscle fiber is now sensitive to a subsequent release of transmitter substance.

It should be recalled that within the CNS the algebraic sum of the electrochemical effects of impulses arriving at excitatory and inhibitory synaptic knobs in contact with a neuron determines whether it becomes excited. In the neuromuscular function there is, however, only *one transmitter, and its effect is excitatory*. The quanta of ACh released is usually more than adequate to excite the muscle fiber. It should be emphasized that the only way to promote relaxation of the muscle fibers is to decrease or stop the discharge of the responsible motoneurons.

The motor unit also obeys the all-or-none law, i.e., an effective stimulus of the motoneuron causes a maximal muscle action potential, and the muscle fibers react "to the best of their ability." However, the strength developed will depend on the length of the muscle fibers, temperature, oxygen supply, and the frequency of stimuli. These factors will be discussed later. The action potentials can be recorded with needle electrodes or surface electrodes attached to the skin over the active muscle, and the obtained *electromyogram* (EMG) reveals the sum of the motor unit activity.

Summary   An impulse started in a motoneuron propagates in the axon and is transmitted in the motor endplate by ACh to the muscle fibers supplied by the terminal branches of the nerve. The ACh affects the muscle membrane, giving rise to an ion flux reversing the resting membrane potential. A muscle action potential propagates over the membrane and, in a manner still unknown, initiates the mechanical-chemical mechanisms which cause the myosin and actin in the sarcolemma of the muscle to react (Chap. 3). The developed muscle tension will depend on (1) the number of motor units activated and (2) the frequency with which they are stimulated. These two fundamental variables will be discussed later in this chapter.

REFLEX ACTIVITY AND SOME OF THE BASIC FUNCTIONS OF PROPRIOCEPTORS

Synapses

It has been pointed out that various nerves, either directly or through interneuronal relays, expose the motoneurons to a continuous bombardment of transmitter substances.

There is a minimum of two neurons in a reflex chain: the *afferent* (receptor) neuron and *efferent* (effector) neuron. The nutrient cells of the afferent neurons are in the dorsal root ganglion (or cranial equivalent), and they convey information from cutaneous, muscular, or special senses. The cell body of the efferent nerve of the motoneuron is located in the ventral horn (or motor cranial nucleus). The afferent fibers entering the dorsal root of the spinal cord may thus end monosynaptically around motoneurons, but generally several or many connecting interneurons intervene between the afferent and efferent neurons.

Most neural control of skeletal muscles is actually reflex in nature. The excitability of the motoneurons is increased or decreased, depending on the algebraic sum of the excitatory and inhibitory activity in the synaptic knobs. If the muscles are the slaves of the motoneurons, the motoneurons are slaves of spinal and supraspinal mechanisms. To illustrate the reflex nature even of voluntary movements, it may be pointed out that most of the pyramidal tract fibers from the cerebral cortex motor area evidently terminate on interneurons. Two systems of interneurons are involved, one at the cortical level and one at the spinal stage of pyramidal tract innervation. "It can be stated that by establishing synaptic connections with interneurons rather than with motoneurons the pyramidal tract and the other descending tracts (from the cerebellum, the red nucleus, the reticular formation) are able to operate through the coordinative mechanisms at the segmental levels of the spinal cord" (Eccles, 1957). Otherwise the only alternative of an impulse traffic in the tracts would be an excitation of the motoneuron. As it is, the intervening interneuron can give rise to excitation *or* inhibition. The descending impulses from the higher levels of the brain, including those voluntarily evoked, will necessarily be modified on the basis of coincident information from all kinds of receptors. It would also seem important that, via interneuronal circuits, information based on past experiences would permit modification of responses. Thus, the *impulse traffic in the final common path, the axons of the motoneurons, would reflect an integration of synaptic activity based on both past and present experience.*

The synapses may be created in the same segment of the spinal cord as the afferent fiber enters, or collaterals may pass up or down the spinal cord, eventually reaching the brain (Fig. 4-12). Theoretically there are innumerable pathways which an impulse can travel via nerve fibers, but some tracts are preferred, partly on an anatomic basis. One nerve may branch and terminate with several synaptic knobs on the same nerve cell and its dendrites.

In the synapse there is a delay of about 0.5 msec. The synapse transmission mechanism is more readily subject to fatigue than are the nerves. For this reason a reflex response becomes weaker during prolonged stimuli if synapses are included in the pathway for the impulses.

The more important features of the synaptic functions may be summarized as follows: if fibers from two afferent nerves ending on the same nerve cell are stimulated separately, neither causes a reflex response if the stimulus is subliminal. However, when excited, each afferent nerve liberates sufficient chemical

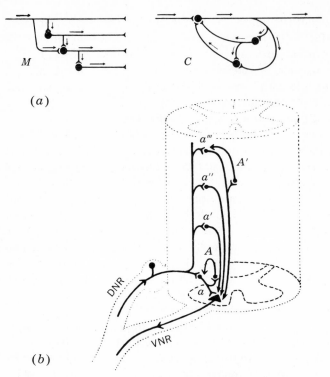

(a)

(b)

Fig. 4-12 (a) Plans of the two fundamental types of neuron circuits. M, multiple chain; C, closed chain. (After Lorente de Nó, 1938.) (b) Role of internuncial neurons in reflex action. Diagrammatic section of spinal cord. DNR = dorsal (posterior) nerve root; VNR = ventral (anterior) nerve root; a, a', a'', a''', A, A' = internuncial neurons.

    The shortest reflex arc is via DNR, a, VNR. Longer reflex arcs involving delay paths are shown via a', a'', or a'''. A, A' are "reverberators." Note how an impulse passing from DNR to VNR along a may branch to excite A, which in turn reexcites a; similarly an impulse along a''' may branch to excite A', which in turn reexcites a'''. (From Keele and Neil, 1965.)

transmitter to cause an EPSP. When both are stimulated simultaneously, the liberated quanta of transmitter substance may be enough to discharge a spike potential, and a reflex response is evoked *(spatial summation)*. The other type of summation is the *temporal summation:* individual stimuli may be ineffective, but if repeated at a high frequency a depolarization of the membrane of the connected nerve cell occurs.

    Figure 4-13*a* illustrates two other features: stimulation of the excitatory nerve *a* affects activities of muscles 1 and 2. Similarly a stimulation of *b* excites 2 and 3. Evidently the tension developed by the muscles when *a* and *b* are stimulated simultaneously will not be the algebraic sum of the tension exhibited when the afferent nerves are stimulated separately (*occlusion*). Figure 4-13*b* also presents a different possibility. A stimulus of the excitatory fiber *a* may be effective in causing a contraction of muscle 1, but it is subliminal on the motoneuron supplying muscle 2. Similarly the afferent fiber *b*, when activated separately, only evokes a spike potential in the nerve cell 3. However, a simultaneous stimulation of the fibers *a* and *b* provides an impulse strong enough to activate muscle 2. Consequently the tension developed by a stimulus of both *a* and *b* will be greater than the sum of the two reflex responses taken singly (*subliminal fringe*).

Afterdischarge is another example of spatial summation. If the axon of a motoneuron is stimulated with an electrical shock of sufficient strength, the

Fig. 4-13   Summation.

(a) *Stimulation: a; response: 1 + 2; tension: ⅔ of maximal*
Stimulation: b; response: 2 + 3; tension: ⅔ of maximal
Stimulation: a + b; response: 1 + 2 + 3; tension: maximal

(b) *Stimulation: a; response: 1; tension: ⅓ of maximal*
Stimulation: b; response: 3; tension: ⅓ of maximal
Stimulation: a + b; response: 1 + 2 + 3; tension: maximal

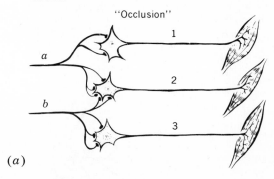

(*a*)

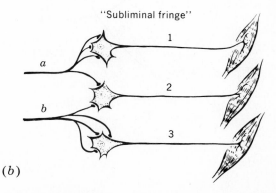

(*b*)

muscle supplied responds with a single twitch. If the stimulus is applied to the afferent nerve, the tension may remain for a much longer time because the muscle responds with repetitive twitches. Figure 4-12 explains how the motoneuron can be persistently stimulated, since volleys continuously arrive after having passed through the shorter or longer delay paths formed by the intricate maze of interneurons.

Reciprocal Inhibition (Innervation)

If a muscle such as the quadriceps muscle is stretched, impulses are evoked in the large afferent fibers from the annulospiral endings of the muscle spindle. These fibers end directly at the motoneurons of the muscle containing the excited muscle spindles and form a monosynaptic reflex arc. These long afferent fibers exert an excitatory effect. However, they also send collaterals which release the same transmitter to excite adjacent interneuron cells, which then

Fig. 4-14 The stretch reflex: If the extensor muscles are stretched, an impulse is evoked that goes from the muscle spindles in the afferent fibers, which synapse with the motoneurons of the extensor muscles, stimulating them to contraction (+). Other branches of the afferent fibers synapse with interneurons which, however, will cause an inhibition of the motoneurons of the flexor muscles, which therefore will relax (−) (reciprocal inhibition).

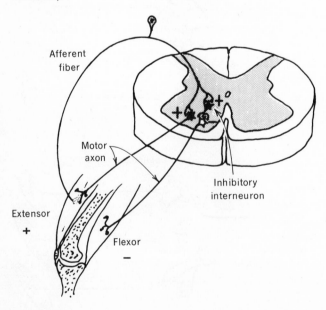

form a direct inhibitory pathway by sending their axons to the motoneurons of the antagonistic muscles, in this case the femoral biceps and semitendinosus. The effect of this stretch stimulus will be a stretch reflex, since the extensors contract and the flexors relax (Fig. 4-14). This reciprocal inhibition is an "inborn" mechanism in the CNS. Actually a reflex inhibition of antagonist muscles is an almost invariable accompaniment of reflex activation of the protagonist muscles.

Electrical or thermal stimulation of the skin of a limb can elicit inhibition as well as facilitation of motoneurons (Hagbarth, 1952, experiments on cat). Extensor muscles were inhibited from most parts of the limb but excited from a skin area localized mainly over the muscle itself. The flexor muscles were excited from most parts of the limb but inhibited from a skin area localized mainly over the antagonistic extensor muscle. Therefore, excitation and inhibition were of the reciprocal type. In 1889, Sherrington demonstrated that stimulation of a point in the motor cortex of the brain may excite flexor motoneurons and simultaneously inhibit the motoneurons to the antagonistic extensor.

Renshaw Cells

Motoneurons give off collateral branches as they traverse the spinal cord to emerge in a ventral root. They form excitatory synaptic contacts with inter-neurons located in the ventromedial region of the ventral horn. These Renshaw cells send axons which affect inhibitory synaptic connections with the same and other motoneurons of that segmental level in an overlapping and diffuse fashion (Fig. 4-15). The Renshaw cells provide a feedback, and a single volley in the axon of the motoneuron may evoke a repetitive discharge of the Renshaw cell with the consequent tendency to dampen the motoneural activity. The effect is thus a reduction in the motoneural discharge frequency. This may protect against convulsive activity and overloading of the muscles. The Renshaw cells may also be excited by afferent impulses from muscles and skin.

In accordance with the fact that every neuron is releasing only one type of trans-mitter, either an excitor or an inhibitor, the substance delivered from the branch of the motoneuron contacting the Renshaw cell is ACh. As mentioned earlier, this substance is also responsible for the stimulating effect at the neuromuscular junction. Incidentally, ACh is the only transmitter substance identified (within the CNS) in the presynaptic knobs stimulating the Renshaw cells.

Figure 4-15 also illustrates how an inhibition may be inhibited. Many motoneurons are normally subjected to a steady background inhibition by the activity of inhibitory nerve cells, eventually driven from higher levels of the CNS. Renshaw cells may form synapses with such inhibitory neurons. When these Renshaw cells are stimulated they will inhibit or depress the activity of the inhibitory neurons and release the motoneurons from inhibition. The effect will eventually be a "recurrent facilitation" of the motoneurons.

These are examples of postsynaptic inhibition and how it can be suppressed. There is evidence that a *presynaptic inhibition* may occur. Some inhibitory cells form synapses with the terminals of other nerve cells' axons, and their transmitter substance

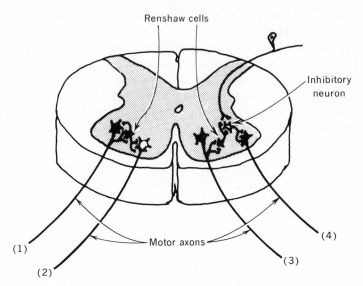

Fig. 4-15 Some motoneurons give off branches, called recurrent collaterals, before they leave the gray matter of the spinal cord. These collaterals synapse with inhibitory interneurons named Renshaw cells. They synapse in turn with the same or other motoneurons. To the left in the figure the activated motoneuron (1) stimulates the Renshaw cell, which then inhibits both motoneurons (1) and (2). To the right we have an example of inhibition of inhibition: an impulse in the afferent nerve stimulates the inhibitory neuron, and therefore, the motoneuron (4) will become inhibited; if the motoneuron (3) is now stimulated, it excites via its recurrent collaterals the Renshaw cell, which in turn inhibits the inhibitory neuron, and as a consequence, the motoneuron (4) will be released from the inhibition and may more easily respond to excitatory impulses.

reduces the amplitude of nerve impulses traveling along the contacted fiber. The result is a liberation of less than the normal amount of transmitter at the synapse and a reduced effect on the postsynaptic membrane. Such a presynaptic inhibition is often found to be acting on terminals of afferent fibers from the skin.

Renshaw cells are thus subjected to both excitatory and inhibitory inputs. The main spinal excitatory input comes through the recurrent collaterals, whereas impulses mediated through afferent nerves are predominantly inhibitory. Descending impulses from supraspinal structures (e.g., cerebellum, brain stem) can inhibit, facilitate, or excite the Renshaw cells. In many cases these pathways converge on excitatory or inhibitory interneurons forming synapses with the Renshaw cells.

The Gamma Motor System

The axons of the motoneurons so far discussed are of the so-called alpha type (12- to 20-μ diameter), and supply the skeletal muscles (extrafusal fibers). In the

ventral horn of the spinal gray matter there is also a different type of motoneuron giving rise to axons which are thinner, and they are of the gamma (γ) type.
These gamma fibers supply only the intrafusal muscle fibers of the muscle
spindles. This proprioceptor sense organ is shown diagrammatically in Fig. 4-16.
It is an elongated fusiform capsule lying parallel to the skeletal muscle fibers.
Its central part consists of a noncontractile nuclear bag or nuclear chain. On
either side some intrafusal muscle fibers are attached, and the other ends of
these fibers terminate either at a tendon or in the endomysium of extrafusal
muscle fibers.

Activation of the gamma fibers causes the intrafusal muscle fibers to contract, and the nuclear part between the polar regions becomes stretched.
Within the bag is the receptor of the annulospiral nerve ending. If the stretch
is strong enough, the receptor will react with an action potential in the afferent

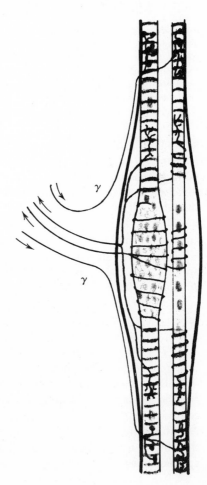

*Fig. 4-16 Diagrammatic illustration of a muscle
spindle. In the middle it has a nonstriated region
supplied by so-called annulospiral afferent nerve
fibers. There is a second type of afferent nerve fibers,
called flower spray endings, mainly localized at the
ends of the muscle spindle. The polar region of the
muscle spindle with intrafusal muscle fibers is
supplied by gamma (γ) motor efferents. The muscle
spindles are arranged parallel to the extrafusal
muscle fibers.*

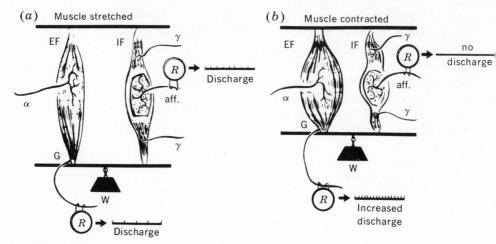

Fig. 4-17 Schematic illustration of the activity in afferent fibers from muscle spindle (aff.) and Golgi tendon organ (G, lower record) when the muscle is (a) stretched and (b) contracted. EF = extrafusal muscle fibers, IF = intrafusal fibers innervated by gamma fibers. R = recording instrument with its electroneurogram. The muscle spindle is stimulated by stretch, but there is eventually a pause in its discharge during muscle contraction, during which there is an accelerated rate of discharge from the tendon organ.

nerve fiber propagating toward the spinal cord. The impulse is monosynaptically transmitted to alpha neurons of the same muscle, providing an excitation (Fig. 4-14). When the muscle contracts, the ends of the muscle spindle come closer together, and the stretch on the nuclear bag is reduced. The stimulus of the receptor will be less or will cease completely (Fig. 4-17). As mentioned above, the afferent impulses, via an interneuron, also cause an inhibitory effect on the alpha neurons of the antagonists (Fig. 4-14).

A passive stretch of the muscle will also cause the muscle spindle to fire, and by reflex the same muscle responds by contracting, thus tending to counteract the stretching force. Tapping the quadriceps tendon, as is routinely done in medical examinations, stretches the muscle and its contained muscle spindle, causing a jerk of the lower leg, "knee reflex." The explanation for this is the one presented above.

The reflex can be evoked (1) by stretch of the muscle and (2) by increased activity in the gamma fiber.

Regarding the latter, it should be emphasized that the activation of the gamma fibers per se does not cause any increase in the muscular tension. When contracting, the intrafusal muscle fibers do not provide any noticeable tension. An intact afferent nerve from the muscle spindle is necessary to induce a contraction as a response to gamma activity or stretch. Adaptation in the muscle spindle takes place very slowly; i.e., the discharge in the afferent nerve fiber continues for as long as the muscle is stretched, though the frequency of the discharge gradually declines. A stimulus which increases slowly in intensity produces a lower frequency of impulses than a stimulus which rises very rapidly to the same level. Thus, a pull on a muscle of a cat with a given force attained within 1 sec produces an afferent impulse frequency of more than 100/sec, but a slower increase in stretch, until the same force is applied, will within 6 sec give a peak volley of about 40 impulses/sec (Granit, 1962).

It may thus be concluded that the muscle has an excellent instrument to measure its length (or more precisely the difference in length between the parallel extrafusal and intrafusal muscle fibers), the extent of a mechanical stimulation, and the rate with which the stretch is applied. Furthermore, the spindle control enormously increases the number of functions that a slightly activated alpha pool of motoneurons can perform (Granit, 1955, 1962).

The structure and function of the muscle spindles are much more complicated than revealed by the discussion presented. Figure 4-16 shows schematically *two types of intrafusal muscle fibers:* to the left the type which has a nuclear bag in the equatorial region; to the right the type with a structure called nuclear chain. The mechanical properties of the nuclear bag are different from those of the nuclear chain. In the polar regions the fibers are striated and contractile.

There are *two types of gamma efferents* with different motor endings in the intrafusal fibers: in the plate and in the trail endings. It is not settled whether each type of intrafusal muscle fiber, bag or chain, receives each type of motor ending, or whether the plate fibers terminate on bag fibers and the trail fibers go to chain fibers (Granit, 1966). The effect on the innervated fibers is somewhat different depending on which type of γ fiber is activated.

Lastly, there are *two types of afferent nerve fibers:* The annulospiral endings wrap around the nuclear fibers at the equatorial regions. These primary endings have large afferent fibers. The secondary endings, or flower spray endings, with smaller afferent fibers, lie mainly on the chain fibers, but may have a few sprays on the nuclear bag fibers at the polar regions. Morphologically these two types of sensory endings are clearly differentiated. Also functionally they are different. The primary endings have a lower threshold than do the secondary endings, and their larger afferent fibers conduct an impulse more rapidly than do the smaller ones.

It is necessary to differentiate the muscle spindles into two distinct systems, separately innervated. Some of the motor fibers increase the sensitivity of the muscle spindle to the dynamic component of stretch. They are called dynamic motor fibers and they terminate as trails. Other motor fibers mainly increase the response of the receptor to static stretch. They are called static motor fibers and terminate as plates. The primary endings are affected by both dynamic and static stretch. However, the secondary endings are activated by the static fibers but very little by activity in dynamic fibers.

When activated, the primary endings generally excite monosynaptically the motoneurons supplying their own and synergistic muscles, but they inhibit those of the antagonistic muscles. This is the "classical" function of the muscle spindles. The effect of impulses from secondary endings is more complicated and partly unknown. They may excite flexor motoneurons irrespective of whether the secondary endings themselves lie in flexor or in extensor muscles. Afferent fibers of both types of endings may be effective in inhibition via interneurons and in presynaptic inhibition of other primary afferent fibers to diminish their synaptic action.

The muscle spindles are structurally and functionally complex, and much remains to be explained about their role in coordination of movements. On a subconscious level they provide the CNS with information about the state of the muscle, both with regard to its length (via primary and secondary afferent fibers) and to the velocity at which it is being stretched (via primary endings). The muscle spindles are of the utmost importance for the execution of well-coordinated movements (Hagbarth, 1964; Granit, 1966).

Neural Control of Muscle Tonus

By an asynchronous discharge of impulses in the motoneurons, the muscles are often slightly contracted even at rest. This state of muscle tonus is chiefly maintained by gamma fiber activity causing contraction of the intrafusal fibers; the sensory nerve endings of the muscle spindle are discharging, and the afferent nerve fibers activate the alpha motoneurons. This mechanism provides an excellent state of preparedness for the muscle to respond to impulses arriving at its motoneuron. Furthermore, the slightest change in the length of the muscle will promptly affect the spindle discharge. The more or less continuous gamma activity makes it highly prepared to readjust and restore the tension on the nuclear fibers if the extrafusal muscle fibers shorten, and the muscle spindle is again ready to register variation in tension and length of the muscle.

A destruction of either the afferent nerve, the posterior spinal root (as in tabes dorsalis), the motoneuron cell body (poliomyelitis), a section of the ventral root, or an efferent muscle nerve will abolish the muscle tone for reasons that should be evident from the discussion above.

The Golgi Tendon Organs

These organs are located in the tendons of the muscles and in series with them. Both passive stretch and active contraction of the muscle will therefore pull the tendon receptor (Fig. 4-17). When stimulated, the afferent nerve fibers cause a reflex inhibition of their corresponding muscle elicited via interneurons. Excitatory interneurons are also stimulated, but they form synapses on the motoneurons to the antagonistic muscles. The threshold for a stimulus of the muscle spindle receptor is lower than for the Golgi tendon organs, and therefore, the reflex response upon a stretch of a muscle is principally first an excitation of its motoneurons and then eventually an inhibition. It should also be emphasized that the reflex arc of the gamma loop is monosynaptic, but the Golgi tendon organ stimulates via interneurons, and this delays the impulse. The tendon receptors not only inhibit the alpha motoneuron of the same muscle, but also the gamma motoneuron. This is a wise arrangement if the elicited reflex is to effectively prevent the development of a dangerous tension in muscle and tendon, since otherwise the gamma system would evoke an excitatory stimulus of the alpha motoneuron. The Golgi tendon reflex also causes a smooth retardation of a movement.

Joint Receptors

In the ligaments and the capsules of the joints there are different kinds of receptors (Ruffini end organs and receptors of the Pacinian and Golgi type). Some of these proprioceptors are specialized to respond to movements of the joint; others show an impulse discharge that varies with the exact position of the joint but are less sensitive to movements. The afferent nerve fiber's synapse in the spinal cord and other neurons can transmit impulses to cerebellum, thalamus, and sensory cortex respectively. These impulses are of great importance for the coordination of movements and for our information concerning the exact position of the joint. The ability, with the eyes shut, to touch one's nose with a finger tip or the left knee with the right heel is mainly based on information originating in the joint receptors. (The impulses from muscle spindles do not reach conscious levels in the brain.)

Supraspinal Control of Motoneurons

The muscle tone can be modified by both facilitatory and inhibitory impulses which constantly travel from the brain via descending tracts (Fig. 4-18). Mention should be made of (1) the pyramidal tract from the upper motoneurons (in the cortex) and (2) the extrapyramidal tracts from basal ganglia, mesencephalic and pontine reticular formation, vestibular nucleus, and cerebellum. Transection, at the lower part of the midbrain, gives prominence to the excitatory impulses to extensor muscles. The result is the decerebrated rigidity. In such an animal preparation, bending of the head downward evokes a violent activity of both alpha and gamma fibers to the soleus muscle. When attempting to flex the joint, the stretch of the extensor is strongly opposed. If the force is maintained, however, the muscle will suddenly give way (Sherrington's lengthening reaction). The hypersensitive stretch reflex is then inhibited by the high-threshold Golgi tendon organs. A spinal transection provides less muscle tone than normal.

Descending nerve fibers driving the gamma fibers have been traced up to the *reticular formation*. This poorly defined part of the brain stem (mainly midbrain, pons, and the tegmentum of the medulla) is formed by many nuclei scattered throughout the central part of the brain stem. It is also characterized by an interlacing network of nerve fibers. It sends and relays excitatory and facilitatory extrapyramidal reticulospinal fibers to the spinal neurons. A stimulus of the *facilitatory reticular neurons* increases the rate of discharge of the gamma efferents. Conversely, the activated *inhibitory reticular neurons* reduce or annul the gamma activity. Similar effects can be elicited from the hypothalamus. Movements induced either by stimulus in the motor cortex or pyramidal tract

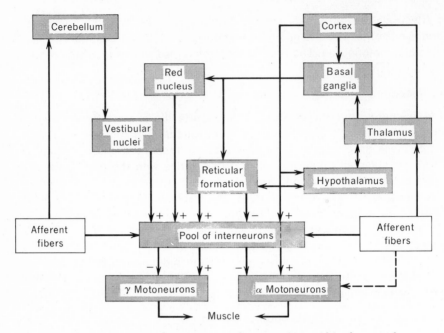

Fig. 4-18 Schematic illustration of some of the connections within the central nervous system essential for coordinated muscular contractions. (For simplicity most of the connections to and from the cerebellum are excluded.) The spinal motoneurons and interneurons are exposed to excitatory (+) as well as inhibitory (−) impulses from various levels of the CNS. The γ and α systems are, in a way, independent of each other, but are normally linked together via the interneurons. The afferent fibers come from muscle spindles, Golgi tendon organs, joint receptors, and cutaneous receptors. They constitute feedback channels used in the integration of motor activity.

or by reflex mechanisms can be abolished by impulses from the inhibitory reticular nerves. The net effect of the descending impulse traffic normally is a facilitation of spinal neurons, especially those innervating the extensor (anti-gravity) muscles.

The reciprocal innervation is the determinant for the effect on motoneu-rons irrespective of the origin of the impulse traffic. The alpha and gamma sys-tems are largely linked together, coexcited, and coinhibited, often with the gamma reflexes being predominant (Granit, 1955). A brief description of some other reflexes will stress this point. The stretch reflex has been discussed at length. If pain receptors in the skin are stimulated (by a needle prick in the sole of the foot, for instance), the limb will withdraw from the source of pain even before any sensation of pain has been experienced. This response is executed by means of excitatory connections in the spinal cord between the afferent fibers whose nerve endings were stimulated and the motoneuron of the flexor muscles of the extremity. Several interneurons are included in the reflex arc. Some of the intercalated interneurons will have an inhibitory influence on the motoneurons of the extensors of the extremity. This is an example of an ipsilateral flexor reflex.

If the nocuous stimulation is strong enough, the extensor muscle of the opposite limb of an animal is contracted and its flexor muscles reciprocally inhibited almost simultaneously with the development of flexion on the ipsilateral limb. This crossed-extensor reaction (or contralateral extensor reflex) will further assist in removing the foot from the irritating needle. A stimulus of the hind limb may spread to the forelimbs via long nerve collaterals within the spinal cord.

Summary   It has been emphasized that the gamma system is very important for the execution of coordinated movements with the muscle spindles as a servo-mechanism. Many familiar activities may be initiated over the gamma efferents, and the alpha motoneurons are reflexly "forced" to follow, stimulating a certain number of motor units to contract at a certain frequency. The muscle spindles sense whether the tension developed by the extrafusal muscle fibers is the appropriate one and they can, via their afferent fibers, modify the impulse output of the alpha motoneurons according to the demands. If, for instance, the muscle tension is not enough to overcome a resistance, the muscle spindle is still stretched, since the muscle fibers at its polar regions are contracting in response to the gamma efferents. The receptor will therefore continue to bombard the alpha motoneuron with excitatory impulses until the developed muscle force makes a shortening possible, thereby reducing the stretch on the muscle spindle.

In unfamiliar activities the motor cortex becomes more important in the direct control of movements, and the pyramidal system and alpha motoneurons are initiating the muscular contractions. This is particularly true when we are thinking about how we should best use the muscle. With practice the extrapyramidal system can take over with the gamma motoneurons as "conductors" and triggers. Actually complex coordination cannot be mastered until certain basic movement patterns have been brought to the automatism of conditioned reflexes. As a relay station the cerebellum is of extreme importance.

Posture

The upright position is maintained by muscular activity against the force of gravity. In the erect posture the line of gravity runs in the midline through (1) the mastoid processes (in front of the atlanto-occipital articulation); (2) a point just in front of the shoulder joints; (3) the hip joints (or just behind); (4) a point just in front of the center of the knee joints; and (5) a point (3 to 7 cm) in front of the ankle joints (Basmajian, 1967). None of the joints engaged in the erect position are moved to the extreme of their mobility, and the body therefore does not "hang" in the ligaments or capsules of the joints. However, when loads are supported so that traction is exerted across a joint, ligaments, and not muscles, normally maintain the integrity of the joint. For instance, the muscles that could support the arches in the foot are generally inactive in standing at rest. With a subject seated upright with his arms hanging in the relaxed neutral position and with heavy downward pull applied to an arm, the muscles that cross the shoulder joint and the elbow joint are not active to prevent dislocation of these joints (Basmajian, 1967). The center of gravity of the head and trunk is very close to the supporting column of bones, so that the antigravity muscles are only very slightly loaded. Basmajian points out that among mammals man has the most economical antigravity mechanisms once the upright posture is attained. (A quadruped, when standing with his joints partly flexed, has a much more wasteful antigravity machinery, and a direct comparison of the function of antigravity muscles of man and animals may therefore be misleading.) A very important function of the so-called antigravity muscles of man is to produce the powerful movements necessary for the changes from a lying to a sitting or standing position and to provide a firm foundation for the variety of muscular activities of everyday life.

The stretch reflex is the basic reflex in postural control. The muscles antagonizing the pull of gravity are stretched, and thereby the muscle spindles located in the muscles are also stretched. Afferent impulses are evoked, and the muscle contracts so that the pull of gravity is counterbalanced. Since the intrafusal muscle fibers of the spindle can be activated from higher centers via the gamma fibers, its receptor may be more or less prone to respond to a stretch. The hypothalamus is probably one important relay center, and a feeling of happiness, alertness, or attention may increase the gamma activity, while unhappiness, drowsiness, or lack of attention may decrease the activity in gamma fibers. In this way the very noticeable relation between an individual's mood and his posture may be explained.

Some muscle fibers, represented in the antigravity muscles, are innervated by small alpha neurons. These motoneurons have a relatively low threshold, a long after-hyperpolarization, and a slow conduction rate. They readily respond tonically on stimuli; i.e., they discharge for long durations, and adaptation is negligible. The muscles supplied by these nerves are predominantly the slow, often called red, tonic muscles. In contrast, the fast, pale or white, phasic muscles are innervated by large alpha motoneurons which have a high threshold and show a discharge only at the onset of a stretch of the corresponding muscle (adaptation is rapid). In man, a mixture of slow fibers (richer in sarcoplasm) and fast fibers is found in most muscles, but the slowly contracting fibers are more suitable for maintaining posture. They are the dominating fibers in the soleus, for instance. The fast muscle fibers, which contract more rapidly, are therefore more differentiated to perform rapid phasic movements and are abundant in the gastrocnemius muscle (Granit, 1962).

The tonic, slow muscles are more affected by the gamma loop than are the fast muscles. The recurrent collaterals within the spinal cord exert an effective inhibition of the tonic motoneurons, whereas many of the phasic muscles lack this particular feedback control. This powerful recurrent inhibition on the tonic motoneurons will, together with the long after-hyperpolarization, hold back their rate of discharge. The afferent discharge from the muscle spindles increases in direct proportion to the degree of extension of the muscle, but the efferent rate of discharge from the motoneurons is fairly constant. A stronger afferent impulse traffic recruits more motoneurons rather than increasing the rate of discharge.

The small motoneurons innervate relatively few, slow muscle fibers in contrast to the large motoneurons, which contact many fast fibers. This explains why a stimulus of a small motoneuron innervating a cat gastrocnemius may produce a tension of about 5 g, but a stimulus of a large motoneuron may increase the tension by 120 g. In studies on cats and chickens it has been shown that fast muscle fibers are more adapted to anaerobic function: these muscles have a higher content of glycogen and a higher activity of some glycolytic enzymes than slow muscles. Slow muscles have a higher ribonucleic acid content and a higher rate of incorporation of amino acids into the muscle than do the fast muscles. This may indicate a higher turnover of proteins in the slow muscle, perhaps linked together with the mobility of protein-bound Ca^{++} ions in the sarcoplasmatic reticulum of the muscle cell. They also have a higher content of myoglobin than the fast muscle and a higher level of oxidative enzymes, and the density of the capillary network seems more developed around the slow muscle fiber. In other words, the slow muscle fibers are more adapted to function aerobically.

There is evidence that there is a neural influence operating in the development of this differentiation in metabolism and contraction behavior of fast and slow muscle fibers (Gutmann and Syrevý, 1967; Guth, 1968).

Summary The slow muscles innervated by the small alpha motoneurons are especially suitable as antigravity muscles. They can respond readily and for a long duration to impulses set up in the muscle spindles which are stimulated by the pull of gravity.

In addition to the gamma system, many other reflexes contribute to the integrated activity regulating the normal posture. On the ventral horn, nerve fibers converge from dorsal nerve roots and from all levels of the brain and spinal cord. Impulses from the eye, vestibular apparatus, and sole of the foot are especially important for the modification of the activity of the alpha motoneurons. The cerebellum and reticular formation constitute important relay stations and links between the alpha and gamma systems, and the cerebral cortex exhibits an overall control (partly inhibitory).

The free normal posture is characterized by a "postural sway," so that the center of gravity varies with respect to its projection on the ground with a frequency of 5 to 6 "cycles"/min. With the eyes closed (or in the dark) the swaying is more pronounced. The muscle spindles are actually pulled upon irregularly, and their rate of discharge is therefore highly irregular. Electromyographic studies reveal that the antigravity muscles may be activated with a frequency of 5 to 20/sec. The alternating activity-inactivity in the motor units involved and the postural sway will prevent fatigue and facilitate the blood flow through the muscles; it also assists the venous return. It should be emphasized that fatigue in the well-balanced, standing individual is usually not due to muscular fatigue, but is more likely caused by an inappropriate distribution of the blood. The energy output in the erect position is only slightly elevated, say from 0.25 liter/min in the supine to 0.30 to 0.35 liter/min, but the cardiac output and stroke volume are decreased and the heart rate elevated. In free positions where the center of gravity of limbs and/or trunk is shifted from a balanced position, activity in counteracting muscle groups must compensate; this increases the load on the muscles. The slow muscles respond with tetanic contractions even if the discharge frequency in their motoneurons is low; hence the blood flow may be obstructed, causing local fatigue. It is an interesting finding that during activity with forward flexion of the spinal column, there is a marked muscular activity until flexion is extreme, at which time the ligamentary structures assume the load and discharge from the trunk muscle ceases (Floyd and Silver, 1955; Basmajian, 1967).

PROPERTIES OF THE MUSCLE AND MUSCULAR CONTRACTION

The Motor Units

The alpha motoneuron and the muscle fibers which it innervates are the motor unit. As discussed above, the number of muscle fibers in a unit varies from about 5 to about 2,000, and a single muscle fiber may have a polyneuronal innervation. When a nerve impulse reaches the motor endplate, ACh is liberated and the membrane of the muscle fibers is depolarized. This evokes an action potential which propagates along the muscle fiber at a speed of about 5 m/sec. The whole length of each muscle fiber and all the fibers of the motor unit are therefore brought into action almost instantly, even if the unit is acted upon by a nerve fiber at only one point in its length. The main feature of the muscle action potential is very similar to the action potential of the nerve. It is generated and propagated by basically the same mechanism. A single, adequately strong, stimulus of the motor nerve gives rise to a twitch of the innervated muscle. After a short latent period the tension developed by the muscle increases and thereafter it decreases again. The muscle action potential is completed during the early phase of the contraction (Fig. 4-19a). If the motor nerve of the muscle is stimulated repeatedly and the second impulse reaches the muscle before it has relaxed after the first stimulus, it contracts again. Since the second twitch starts from a higher tension level, the tension resulting from the two stimuli will be considerably greater than that from a single stimulus of the same strength (summation). At high rates of stimulation (about 50/sec) the muscle does not relax before the next contraction, and the muscle fibers are in full tetanus; i.e., there is a complete mechanical fusion of the contractions. The tension developed by a muscle in tetanus may be 4 to 5 times greater than that exerted during a single twitch. At lower rates of stimulation the mechanical fusion is incomplete and the developed tension is not maximal (Fig. 4-19b). Even in a tetanus the muscle action potential develops in response to each stimulus, revealing the rate at which the nerve is being stimulated (Fig. 4-19c). It has been observed that the

Fig. 4-19   Action potentials (e) and mechanical changes (m) in skeletal muscle in response to stimulation of its motor nerve (from mammalian nerve-muscle preparation).

(a) A single stimulus gives rise to a single twitch. The diphasic action potential is completed in the early part of the contraction phase. Each mark on the abscissa denotes 0.05 sec.

(b) Time (abscissa) is not marked; the ordinate gives tension in kilograms. The lower curve, partly dotted, shows one response of the muscle to a single maximal stimulation to its nerve. Curves A to D show the responses to repetitive, maximal stimuli: A at 19, B at 24, C at 35, and D at 115 stimuli/sec. Note that the higher the frequency of stimulation, the greater is the tension which is developed, and it is maintained more steadily. From a state of partial or incomplete tetanus (curves A to C), the muscle goes into full tetanus (curve D).

(c) At a high frequency of stimulation (in this case 67 stimuli/sec) the muscle goes into tetanus, but the action potentials appear at the same frequency as the stimulation rate. [a and c, after Sherrington et al.: "Reflex Activity of Spinal Cord," 1932; b, after Cooper and Eccles: J. Physiol. (London): 69, 1930.]

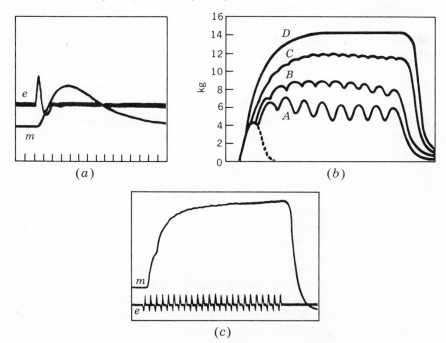

tetanic frequency necessary for fusion of the muscle response is lower for the slow muscle fiber than for the fast one.

Apparently during a single twitch, the active state of the muscle fibers is not long enough (some 30 msec) to permit the mechanical rearrangement essential for maximal tension to develop. During partial or full tetanus, the period of the active state is prolonged and thereby the tension will be increased.

Since the rate of discharge from the motoneurons may vary from very low (5 to 10 impulses/sec) to very high levels (50 or above/sec), the strength of the resulting contraction of the innervated muscle will vary correspondingly. This provides for one important mechanism for the gradation of the activity of the muscles. The second mechanism is dependent on the number of motoneurons (and motor units) activated simultaneously. Therefore, *the higher the stimulation frequency* and the *larger the number of active motor units*, the *greater will be the tension generated by the muscle.*

Figure 4-19 is based on results obtained from nerve-muscle preparations. Normally the activated motoneurons discharge asynchronously, and the muscle fibers of the different motor units are in different phases of activity. The net effect is a smooth muscle contraction, even if the individual motor unit is activated at a subtetanus frequency and therefore shows a tremulous response.

The Mechanical Work

In the activated muscle the contractile components, i.e., the myofibrils, shorten and stretch the elastic components (connective tissue, tendon). When both ends of the muscle are fixed and no movement occurs in the joint involved, the contraction is called *isometric*. If the muscle varies its length when activated, the contraction is *isotonic* (dynamic). In the latter case external work is done, and the amount of work can be calculated from the product of weight lifted $\times$ the distance. Since the distance is zero in an isometric contraction, no mechanical work is done according to physical laws. However, isometric activity demands energy and can be very fatiguing. From a physiological viewpoint, the "work" performed is definitely more related to developed force $\times$ contraction time than to force $\times$ the displacement.

An activation of muscle fibers is always associated with an *increased heat production* (Hill, 1958).

Part of this heat production actually precedes the rise in tension during isometric contraction or the shortening during isotonic activity (activation heat). During isotonic contraction, heat is also produced proportional to the degree of shortening which the muscle undergoes (shortening heat). Heat is liberated if the muscle during its relaxation lowers a previously lifted load, but not otherwise (relaxation heat). This initial heat production associated with the muscle activity is independent of the oxygen supply. After completion of the mechanical events, there is, providing that the supply of oxygen is adequate, a heat production which is about equal to the initial heat (recovery or delayed heat).

During isometric contraction, all the extra energy output is converted to heat. In dynamic exercise, at least 75 percent of the chemically available energy is transformed to heat. The chemical reactions providing the energy for contraction of the muscles are in some way controlled by the change in the length of the muscle and by the tension placed on the muscle.

The *mechanical efficiency of work* (ME), expressed in percent, is the ratio of external work performed, W, expressed in calories, to the extra energy production:

$$(\text{ME}) = \frac{W \times 100}{E - e}$$

where E = the gross caloric output and e = resting metabolic rate expressed in calories.

[It should be emphasized that for a work that is partly anaerobic, the caloric output (oxygen uptake) must be measured not only during the work period but also during the recovery period from such work (oxygen debt).] When a person exercises on a bicycle ergometer, in stair climbing, and similar performances, the mechanical efficiency is up to 20 to 25 percent; i.e., 75 to 80 percent of the energy is dissipated as heat. In isometric work and many other types of activities the mechanical efficiency is 0 percent.

The produced heat increases the temperature of the muscle. Within limits the *elevated temperature improves the performance of the muscle*. This can be explained on both chemical and physical bases. The increased tension produced by the muscle fibers upon repeated stimulation (Fig. 4-19*b*) can partly be explained by the beneficial effect of increased tissue temperature. To prevent overheating during prolonged activity of the muscle fibers, an increased local blood flow is essential as well as an increased heat conductance of the skin (Chap. 6).

Muscle Length and Speed of Contraction

The tension exerted by a stimulated muscle, either during a single twitch or during incomplete or full tetanus, is dependent on the initial muscle length. An unattached unstimulated muscle is at its *equilibrium length*, and the tension is zero. If stretched, the passive elastic tension increases as an exponential function of length, actually over a range up to 200 percent of the equilibrium length (Fig. 4-20, lower curve). Normally when attached by its tendons to the skeleton, the muscle is under slight tension, since it is moderately stretched (*resting length*). Measurements of the tension developed by the stimulated muscle show that tension is maximal when the initial length of the muscle at the time of stimulation is about 20 percent above the equilibrium length (relative length 1.2:1). The active tension falls roughly linearly at lengths below this optimum and is zero when the muscle is maximally shortened. When stretched beyond the relative length 1.2:1, the active tension produced by the stimulated muscle

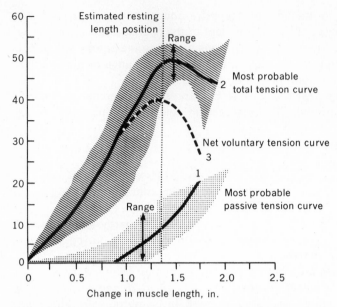

Fig. 4-20 Isometric length-tension summary for human triceps muscle. (The ordinate gives the maximal tension in pounds.) To obtain the "net voluntary tension curve" the values from the "passive tension curve" were subtracted from the "total tension curve." For further explanation, see text. (After University of California, "Fundamental studies of human locomotion and other information relating to design of artificial limbs," 2, 1947.) (See Ruch and Fulton, 1960.)

becomes progressively smaller and is zero when the muscle is elongated to about twice its resting length. This failure to yield tension when overstretched can be explained on the sliding hypothesis of muscular contraction. When the actin and myosin filaments cease to overlap, the cross-linkages between them cannot be formed upon stimulation, no sliding movement can occur, and therefore no active tension develops (Huxley, 1962). Figure 4-20 shows experimental data from studies in which (1) the tension produced by passive stretching of the triceps muscle in man (lowest curve) and (2) the tension developed during maximal voluntary effort were measured (upper curve). The explanation of why less tension is exerted when the muscle shortens is not known. The tension also decreases as the speed of shortening increases (see below). The velocity in a muscle contraction is maximal with zero load. With a load which the muscle just fails to lift, the velocity is zero, and the maximal isometric tension develops.

Striated muscles can actually, within a second, shorten at a rate up to 10 times their length. In the intact body anatomic limitations of the joints usually restrict the lengthening and shortening of the muscles, and the permitted range is normally within 0.7 to 1.2:1 (for some muscles up to 1.4:1) of the equilibrium length. The maximal tension can therefore usually be exerted when the muscle is maximally stretched, and as it shortens, the tension produced decreases. Since the skeletal muscles exert their effect on external resistance via levers, the geometric arrangement of the bony levers must be included in an analysis of the optimal work positions and the most effective utilization of the forces of muscle contractions. Figure 4-21 illustrates a position in which the flexor of the arm holding a weight must produce a force that is about 10 times greater than the force of gravity acting on the load. The optimal lever arm for the biceps muscle is obtained with the arm flexed with an angle of about 90° in the elbow joint. An extension or further flexion of the arm from the right angle causes a decrease in the lever arm of the biceps. If a constant force acts on the forearm against the pull of the biceps, this muscle must increase its exerted tension to balance the resistance as the lever arm decreases in length. If the tendon of biceps were closer to the hand, this would certainly favor its ability to flex the forearm against resistance but at the expense of speed of movement.

Summary In any analysis of the most favorable position, if maximal force is to be exerted on an external resistance, consideration must be given to the following facts: (1) The maximal tension which any muscle fiber can develop depends on the relative length of the muscle fiber at the time of contraction. It has a maximum at a relative length of about 1.2:1 and decreases at lower and higher lengths. (2) The lever arms in the body, through which the muscle tensions are transformed into pulls, pushes, etc., alter with changing positions of the movable joints.

With regard to terminology there can be no confusion about the term *isometric* muscle contraction. However, since the lever arm usually alters during a movement in a joint, it is very seldom that a muscular contraction is purely isotonic (= constant tension). Even if the external load is kept constant, the

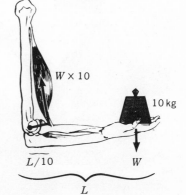

Fig. 4-21   In order to hold a weight of 10 kg in the hand in the position shown in the figure, the arm flexor must exert a force 10 times greater (or about 100 kg), since the lever of the weight (L) is 10 times the length of the lever of the arm flexors.

force developed by the muscle varies as the lever arms become shorter or longer. Therefore it is more correct to use the term *dynamic exercise* than *isotonic exercise* when there is a movement in joints involved.

In functional muscle activities several anatomically different muscles collaborate. The parts of the muscle group that act in synergism may change with position of the limb. Consequently it is very difficult to predict from theoretical considerations the most efficient work position which will produce the greatest strength.

Asmussen and coworkers (Molbech, 1963; Asmussen et al., 1965) have studied *maximal isometric and dynamic contractions at different speeds.* In one set of experiments the movement was a pull of a handle at shoulder height. The hand was moved from the starting position (fully stretched and supinated arm) in a horizontal line to a position about 40 cm closer to the shoulder (Fig. 4-22). Consequently this movement was accomplished through a flexion in the elbow and an extension of the shoulder joint (*dynamic concentric contraction*). The handle was furnished with electric strain gauges so that the exerted force could be measured and recorded. The resistance to the pull could be varied by having the handle attached to the shaft of a piston moving in a cylinder containing oil. The oil was forced through adjustable openings from the front of the piston

Fig. 4-22 A dynameter for measuring isometric as well as dynamic excentric and concentric strength. Resistance in the dynamic strength test can be varied; therefore the speed of shortening and lengthening also varies. With a strong motor pulling the handgrip to the right, the excentric contraction is obtained. In the figure the subject is halfway through his dynamic effort. (By courtesy of Asmussen.)

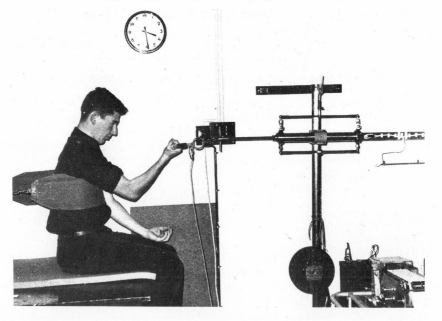

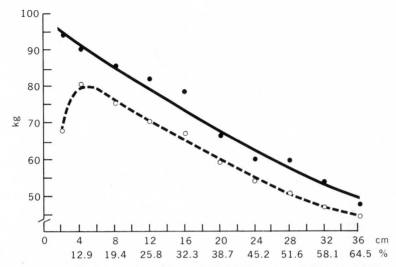

Fig. 4-23   Maximum horizontal pull (kg) with the arm in different positions. Abscissa shows movement of the hand from outstretched position in centimeters and in percent (lower line) of total arm-length. Upper curve: isometric measurements. Lower curve: dynamic measurements, speed 15 percent arm-length per second. Data on one subject. (Asmussen et al., 1965.)

to the back of the piston via a bypass. The pull of the handle was maximal throughout the contraction, but the speed of movement could be varied with the resistance in the bypass. In corresponding experiments, a motor pulled the handle away from the subjects who tried to resist it (*dynamic excentric contraction*). The maximal isometric strength was measured in the various positions through which the dynamic movement passed, i.e., in 10 positions between 2 and 36 cm from start (arm stretched). The results are summarized in Figs. 4-23 and 4-24. As expected from experiments with isolated muscles, the dynamic strength exerted during shortening of the muscles was always less than the isometric strength in the same positions. The strength decreased as the muscle length became shorter (Fig. 4-23). In maximal concentric contraction the developed strength was lower with faster movement (Fig. 4-24). On the other hand, the maximal dynamic strength during lengthening of the contracted muscles always surpassed the isometric strength, and to a greater degree as the speed of movement increased (Fig. 4-24). There was a high correlation between isometric and dynamic strength at a given velocity. In these experiments the relationship was independent of the subjects' athletic prowess.

Electromyographic studies (Lippold, 1952; Bigland and Lippold, 1954a) show the following: (1) When a voluntary isometric contraction of a muscle is made, the electrical activity, measured by integrating the action potentials from surface electrodes, bears a linear relation to the tension that is being exerted. (2) At constant velocity of shortening or lengthening, the electrical activity in the muscle is directly proportional to the tension. However, the slope of this correla-

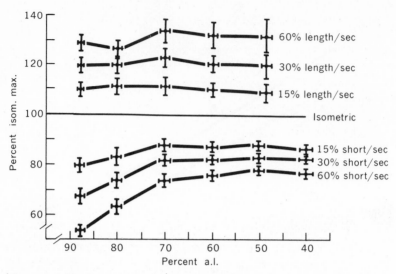

Fig. 4-24   Maximum pulls with arm at different speeds of movement. Isometric strength at any given arm-length set to 100. Three lower curves: during shortening; three upper curves: during lengthening. Abscissa: movement of hand from out-stretched position, in percent of arm-length (a.l.). Velocities in percent arm-length per second. Data from 18 subjects. Two times standard error is drawn as horizontal and vertical bars through the average points. (Asmussen et al., 1965.)

tion falls off during lengthening (Fig. 4-25). This means that the degree of muscle excitation required to produce a given force of contraction is smaller when the active muscle is forcibly stretched than it is when the muscle shortens at the same velocity. (3) At constant tension the electrical activity increases linearly with velocity of shortening but remains almost independent of speed when the muscle is being lengthened.

Roughly, the oxygen consumption of the active muscles should reflect the number of active motor units and their frequency of excitation, i.e., the electrical activity displayed. As a matter of fact, the oxygen uptake increases linearly with the work load in many types of activities. A comparison of the oxygen uptake at a given rate of work, in exercise where the muscle groups work concentrically ("positive work") and excentrically ("negative work"), respectively, reveals that the negative work demands much less oxygen (Fig. 4-26) (Asmussen, 1953; Bigland and Lippold, 1954a). These results fit with the electromyographic studies and the force-velocity characteristic of a muscle.

From studies on individuals of different body size, age, and sex Asmussen et al. (1965) and Lambert (1965) concluded that the correlation r between symmetrical muscle groups (right and left) is quite high ($r = 0.8$). Between flexors and extensors of the same extremity it is also fairly high, but between muscles from different parts of the body, the correlation is rather low ($r =$ about 0.4

or less). Therefore, they conclude, the *general muscle strength should not be evaluated from measurements in one single muscle group*, e.g., the finger flexors in a handgrip, but from application of a battery of selected, well-standardized muscle tests (Hettinger, 1961). Since the correlation between dynamic and isometric strength was high (r = about 0.8), the maximal dynamic strength can be roughly predicted from the simpler measurements of the maximal isometric strength of the same muscles, provided that the subject is not particularly well trained in one or other type of exercise (see Chap. 11).

"Regulation of Strength"

In a completely relaxed muscle no motor units are active, and the muscle is electrically silent. As a result of the elasticity of the myofibrils and fibrous tissues, there is, however, a certain tension (tonus) even in the relaxed muscle. As discussed above, the *gradation of a muscle contraction is brought about by varying the number of active motor units and their frequency of excitation.* Electromyographic studies demonstrate that differentiated, reproducible innervation patterns occur during different movements. During voluntary contraction when the exerted tension reaches a certain value, a particular motor unit starts to fire and continues firing until the tension again drops below the threshold level. The

Fig. 4-25 The relation between integrated electrical activity and tension in the human calf muscles. Recording from surface electrodes. Shortening at constant velocity (above) and lengthening at the same constant velocity (below). Each point is the mean of the first 10 observations on one subject. Tension represents weight lifted and is approximately $\frac{1}{10}$ of the tension calculated in the tendon. (Bigland and Lippold, 1954a.)

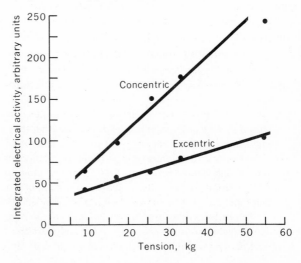

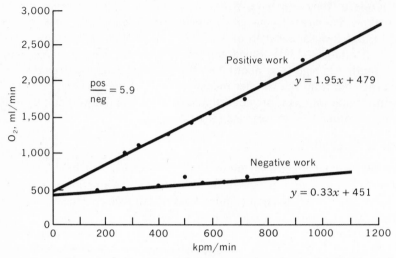

Fig. 4-26 Oxygen uptake in positive (upper curve) and negative (lower curve) work consisting of riding a bicycle on a motor-driven treadmill, uphill in positive and downhill in negative work (with the movements of the pedals reversed). The work load was measured as the product of weight of subject plus bicycle times the vertical distance this weight was lifted or lowered. Oxygen uptake at zero load was measured during free-wheeling. Rate of pedaling = 45 rpm. Note that the cost of positive work was on an average 5.9 times higher than that of negative work. (From Asmussen, 1953.)

same unit always starts firing at about the same tension level. Most motor units which are activated at higher tensions (above 25 percent of the maximal voluntary contraction) start to fire abruptly at relatively high frequencies. They vary their discharge rates very little in the course of a contraction until the contraction is almost maximal. Those units active at low tensions usually have a lower starting frequency and show a greater frequency range than those active only at higher tensions. According to Bigland and Lippold (1954b), the *gradation of contraction in the muscles is brought about mainly by motor-unit recruitment,* except at very low and high contraction strengths.

As mentioned, in a given contraction pattern motor units are usually recruited in roughly the same order, and this holds true whether they are activated by voluntary effort or through proprioceptive reflex arcs. However, this recruitment order may differ depending on the velocity of the contraction. A rapid contraction may be initiated by some motor units setting up only a few discharges (EMG) and then other units take over at a constant rate of discharge. Some motor units, probably consisting of fast muscle fibers, may only be active during rapid and almost maximal efforts (Grimby and Hannerz, 1968).

As previously discussed, motoneurons are exposed to both excitatory and inhibitory impulses from various levels in the CNS and from the muscle spindles of the innervated muscle, as well as from the Golgi tendon organs of the muscle tendons and the Renshaw cells. In the normal control of locomotion and posture, the muscle activity can be initiated by the gamma loop, by a simultaneous activation of the gamma and alpha motoneurons, or without assistance from the gamma system.

In below-elbow amputees who had the distal tendon of the biceps brachii separated from its lateral attachment, the force that could be developed in maximal voluntary contractions at various lengths of the muscle was invariably greater with the elbow flexed 90° than with the elbow extended (Blaschke et al., 1952). Apparently impulses from structures in or around the elbow joint could inhibit the motoneurons innervating biceps brachii if the elbow was extended.

Ralston (1957) points out that in the body an already shortened muscle cannot be activated as fully as an artificially stimulated muscle because the alpha motoneuron excitability is reduced as a result of the lack of facilitation via the spindles. Neither is the stretched muscle able to produce as much tension as an artificially stimulated one because of the inhibition of some of the alpha motoneurons via Golgi tendon organs and tendon afferents.

In isolated muscles the maximal strength is proportional to the cross-sectional area of the muscle. Roughly, this is also true for the muscles *in situ*. The striated muscles can exert a tension of about 4 to 6 kg for each square centimeter of their cross section. The strength per unit cross-sectional area is, according to Ikai and Fukunaga (1968), almost the same in male and female individuals, regardless of age. They did not find any significant difference in ordinary and trained adults. As mentioned, the relative length of the muscle, speed of contraction, type of contraction (isometric, dynamic, excentric or concentric, proprioceptive afferent inflow, etc.) can modify the motoneuron activity and the muscle's ability to develop strength when subjected to this neural activity.

If a muscle contracts against a force that lengthens the muscle, the stimulus of the low-threshold muscle spindles would be maximal, and the motoneurons would be subjected to facilitation. As a matter of fact, the highest tension can be developed in an excentric contraction (Fig. 4-24). Whether the electrical activity within a muscle can reach higher values in a maximal excentric contraction than in a concentric or isometric contraction is not known. It is, however, noteworthy that when a given tension is produced, the electrical activity is markedly less in an excentric contraction than in a concentric contraction (Fig. 4-25). Thus, other factors apparently compensate for an eventual variation in the afferent impulses from the muscle spindles.

It should be emphasized that the afferent discharge from the muscle spindles increases in direct proportion to the degree of extension, but there is no part of the range of muscle length where one muscle spindle is particularly sensitive (Granit, 1962).

The inhibitory feedback from the Renshaw cells and Golgi tendon organs to the motoneurons cannot be ignored in a discussion of the neuromuscular "control" of muscle strength.

Ikai and Steinhaus (1961) conducted experiments in which the subjects made a maximal arm flexion every minute during a 30-min period. They found that the firing of a 22-caliber gun 2 to 10 sec before a pull, a shout, various drugs, or hypnosis could significantly modify the exerted maximal strength. Thus, the performance was distinctly higher after the gunshot than before. Shouting, hypnosis, epinephrine, or amphetamine also tended to improve performance as compared with the controls. The positive effect on strength was noticeable on untrained subjects, but slight or absent on well-trained athletes, e.g., weight lifters. Ikai and Steinhaus cite Pavlov: ". . . any unusual sensory experience or excitement may inhibit inhibitions." They emphasize that their findings ". . . support the thesis that in every voluntarily executed, all-out maximal effort, psychological rather than physiological factors determine the limits of performance. Because such psychological factors are readily modified, the implications of this position gravely challenge all estimates of fitness and training effects based on testing programs that involve measures of all-out or maximal performance."

It is a well-known phenomenon that an individual in a stress situation can perform better than otherwise; he becomes exceptionally powerful. In controlled experiments it is established that epinephrine and norepinephrine increase both excitability and contractility above normal values, but the mechanism for such effects is not clear. During stress the production of these hormones is increased.

Merton (1954) reports that a maximal voluntary effort exerted by the adductor pollicis developed the same tension as a maximal tetanus artificially excited via its motor nerve in the wrist.

Ikai et al. (1967), on the other hand, report experiments with the same muscle group involved, showing that the maximal strength caused by an electrical stimulation of the peripheral nerve to the muscle gave a tension that was about 30 percent greater than the strength developed during maximal voluntary effort (isometric contraction, electrical stimulation of the ulnar nerve in the elbow region with 50 to 60 volts and 50 cps).

In our opinion there is overwhelming evidence to show that a voluntary maximal muscle effort in most situations and with unconditioned subjects does not engage all the motor units of the active muscle at tetanus frequency. There exist effective inhibitions of varying degree on some motoneurons, depending on supraspinal and proprioceptor activity. In a specific situation, e.g., an emergency, and perhaps as an effect of training, inhibition decreases (or facilitation increases), and the muscle mass can become more completely utilized in the contraction. We remind the reader about the rather massive inhibitory interaction between interneuronal paths in the spinal cord and the many inhibitory or excitatory impulses that may descend from higher levels of the CNS. Mobilization of extra muscular strength in critical moments may be elicited from excitatory areas of the reticular formation. (Consult Figs. 4-15 and 4-18 for possible mechanisms.) In most sport events there is an important technique factor involved. Studies have revealed that the skilled weight lifter initiates the critical maneuver with a slight, hardly visual stretch of the muscles engaged in the lifting. Concomitant afferent impulses from the muscle spindles, stimulated by the stretch, should facilitate the motoneurons by way of reflex. [During a rapid movement the interneurons mediating the inhibitory impulses from the Golgi tendon organs to the motoneurons may actually be inhibited (Granit, 1966, p. 273).] Many athletes shout in the critical stage of an effort, and in the light of the experiments by Ikai and Steinhaus, this may increase the motoneuron activity. A training in technique may empirically teach the athlete how to take advantage of facilitating stimuli and reflex mechanisms and to avoid inhibitions which reduce the performance. A proper understanding of the underlying mechanisms is very important and useful, not only for sport activities, but also in physical therapy.

The thesis that central factors are of decisive importance for the development of strength is also supported by the following observations. The strength can increase without a proportional hypertrophy of the muscles (McMorris and Elkins, 1954; Rasch and Morehouse, 1957). When striving for muscle strength for a particular activity, the best training is that activity itself. The gain in strength when "unfamiliar" procedures are employed is comparatively modest, even if the trained muscles are activated thereby (Rasch and Morehouse, 1957; Petersen et al., 1961; Hansen, 1967) (Chap. 11).

The explanation for these results could be that a gain in strength after a training program is not only due to changes in the muscle tissue but also to a modification of the impulse traffic reaching the motoneurons. In a different procedure, receptors, nerves, and synapses which are not "trained" are engaged. Furthermore, a slight change in a movement may reduce the load on one muscle and increase it on another.

Basmajian (1967) found in his studies of the elbow flexors (the biceps brachii, the brachialis, and the brachioradialis) that there is a wide range of individual patterns of activity in these muscles during flexion and extension of the forearm. These muscles also differ in their flexor activity in the three positions of the forearm: prone, semiprone, and supine (Fig. 4-27). The speed of a movement also influences the activity pattern. Therefore, a training of the elbow flexors with the forearm in a prone position may be quite ineffective for a performance done with a supine forearm.

Asmussen and Heebøll-Nielsen (1955) pointed out that functions of importance for physical performance, which depend on maximal muscular exertion, increase with height at a much greater rate than predicted for boys seven to seventeen years of age. Their results pointed to the fact that an increasing ability to mobilize and coordinate the muscles of the body was the reason for the rapidly increasing physical capability in schoolboys.

Summary   Muscle strength is a very complicated function, and the number of muscle fibers available is only part of the story. It is not surprising that even in well-standardized measurements of muscle strength, the standard deviation of the results obtained in repeated tests on the same subject is of the order of ± 10 percent or even higher (Rodahl et al., 1965). Furthermore, the correlation

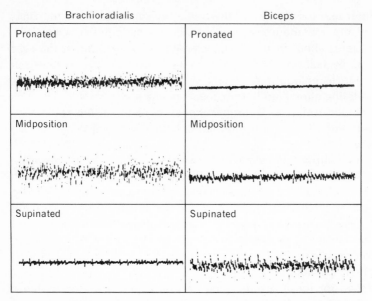

Fig. 4-27   Electromyograms recorded from brachioradialis and biceps brachii during maximal isometric contraction (flexion) in three different positions of the forearm. Note the high electrical activity in brachioradialis and the "silence" in biceps when the forearm is pronated. In the supinated position, the biceps is very active but the brachioradialis contributes much less than in the pronated position. A training of the flexor muscles of the arm with the forearm pronated cannot be so effective if the aim is to improve the ability to perform with the arm supinated, since this position partly engages different muscles. (By courtesy of Carlsöö.)

between isometric muscle strength recorded in different parts of the body is rather low. The practical application of measurements of muscle strength is therefore doubtful.

Every athlete has experienced how his performance may vary from day to day; it may be a question of a more or less successful inhibition of the inhibition on a subconscious level!

Coordination

Under certain circumstances a movement is the result of contraction in only one or two muscles (Basmajian, 1967). In more complex (and most) movements, groups of muscles are cooperating. Hubbard (1960) points out that the key to skilled movement is the timing of the period of muscular tension development in a movement cycle so that approximately full utilization of internal tension under essentially isometric coordination is possible. In fast movements (ballistic movements) at least a spurt of activity in the agonist produces momentum and kinetic energy in the segment, and then it relaxes as the limb proceeds by its own momentum. By reciprocal inhibition the antagonist relaxes completely except perhaps at the end of a movement or when the movement is stopped by the limits of the joint or an external force. Also, in slow movements in some activities discrete bursts of neural activity are observed in agonists and antagonists to produce muscular impulses, which alternately act independently to accelerate and decelerate the segment (Hubbard, 1960). The feedback loops from the proprioceptors are probably responsible for a periodicity and modulation of the motoneural activity. In fast movements proprioceptor and visual stimuli are probably not relayed in time to correct a misdirected movement but in time to make adjustments in succeeding movements. In slow movements a continuous close control is possible.

It has been noted (Carlsöö and Johansson, 1962) that when one falls to the ground on an outstretched hand there is a strong activity in all the muscles which surround the elbow joint, even before the hand touches the floor. This reflex can efficiently protect the involved joint.

Hubbard (1960) also points out that skilled pitching, shot-putting, and discus-throwing are excellent examples of developing momenta serially to stretch agonists successively. In terms of efficient production, the important factor is to develop tension under conditions as like the isometric (or even excentric) as possible and to maintain this condition as long as possible. This can be accomplished by moving the proximal segment ahead of the distal segment so that the agonist develops tension while lengthening or remaining at the same length as long as possible. The difference between a good discus-thrower and a poor one is that the poor one uncoils as he spins across the circle, and the good one stays coiled until set to throw.

During a strong effort in a particular muscle there is a high incidence of activity (in a predictable pattern) in far removed muscles of the same limb and trunk musculature, and this is especially the case in children (Janda, quoted from Basmajian, 1967, p. 93).

Learning: Use and Disuse

When unfamiliar and complicated movements are performed, they are performed clumsily and with difficulty. With proper practice they become smooth and easy. However, we are far from understanding how this is accounted for in physiological terms, how new connections are formed during learning, etc. The movements of the newborn are characterized by uncoordinated movements of large parts of the body, but gradually, coordinated reflexes develop (postural reflexes, tonic neck reflexes, righting reflexes, walking, etc.). The extrapyramidal and pyramidal motor centers and tracts develop, so that the movements become more complex. Nevertheless, when beginning school, the child's extrapyramidal centers dominate the movements, and specific practical tasks directed toward external objects cause the activity in play, imitations, etc. The child usually fails if he performs more complicated "artificial" movements. It is *during puberty that the pyramidal system first attains full functional maturity* and the child becomes apt to develop the fine coordinated movements necessary for writing, etc., based on the integration of nervous activity from various levels of the CNS and the impact from all peripheral receptors. The changing body dimensions during adolescence necessitate a continuous modification in the interpretation of impulses exchanged between muscles and CNS to secure a correct innervation pattern for given tasks. Apparently there are definite anatomic and physiological limitations to the complex movements that can be performed during early adolescence.

The CNS is, like other tissues, characterized by a plasticity in its structural, biochemical, and functional properties. Eccles (1964) believes that repeated exercise of a junction increases the potency of the normal excitatory process responsible for transmission, leaving an enduring trace of increased synaptic efficiency.

Postsynaptic inhibition is far more powerful and prolonged in higher centers than it is in the spinal cord. Learning involves selectivity of response, and presumably inhibition would be significantly concerned in the regression of irrelevant responses. Furthermore, Eccles points out that excitatory synapses on pyramidal cells are concentrated on the dendrites and particularly on the dendrite spines. Possibly the dendrite spines are especially concerned in learning phenomena, e.g., they hypertrophy as a consequence of intense activation, which may greatly enhance the enduring changes produced in these cells by intensive stimulation. In his opinion use possibly gives increased function by enhancing the manufacture and availability of transmitter substance, but enlargement of synaptic knobs, and even the sprouting of new knobs, would seem more probable devices for developing an increased synaptic action that may persist throughout a lifetime. Unique events may be remembered for a lifetime, and reinforcement may occur by the replaying of the specific space-time patterns each time the memory is recalled.

With regard to the idea that functional strengthening of connections occurs by use alone with the transmitter mechanisms as an important factor, Sharpless (1964) has a different opinion. He emphasizes that the excitatory process at neural junctions adjusts itself to the level of input, tending to become less effective with excessive use and more effective with disuse. The result would be that the level of activity of excitable elements, averaged over relatively short periods of time, tends to remain constant. There is evidence that the information required in the learning process is encoded in the molecular structure of ribonucleic acid (RNA) of a special type in macromolecules derived from the nucleus of nerve cells (Hydén and Egyhazi, 1962). Rats were trained 45 min per day for 3 to 4 days to walk a light wire to obtain food. In the trained group a special type of RNA was formed in the neurons which were presumed to have been engaged in the wire balance walk.

There is further evidence reported in support of the phenomenon that the injection of brain-derived fractions prepared from trained donors results in significant behavioral changes in the injected animals, shortening the time required to learn a given task, i.e., it is possible to transfer information among animals via brain extracts (Byrne and Samuel, 1966).

After damage to peripheral nerve fibers, the axis cylinder and myelin sheath distal to the site of injury degenerate. A regeneration can take place and is facilitated by the presence of uninterrupted endoneural tubes. The regenerating axons grow into the peripheral endoneural tubes at a rate of several millimeters per day. Chemical factors released from the degenerated nerve fibers stimulate and guide the collateral growth of the axons. The remyelination closely follows, and the anatomic recovery is completed within 1 year. However, the functional recovery takes longer. Many of the sprouting central axons may get lost, not reaching the peripheral sheath; other axons will establish faulty connections. A touch fiber originally supplying one receptor may connect with a touch receptor in a different region or may even connect with a temperature receptor. The effect will be reduced sensitivity, false localization, or complete misinterpretation of a stimulus of the receptor, e.g., touch giving a sensation of heat. Prolonged practice may be necessary for a return of the finer and discriminative sensibility. A functional regeneration of damaged nerves within the CNS will not take place.

The reinnervation of the muscle by axonal sprouts can reestablish the old pattern of innervation, but if the collateral sprouts happen to be derived from inappropriate parent fibers, transient or permanent functional disorder may be the consequence. To some extent, eventual inappropriate neuromuscular connections can be mastered by a reeducation and, also to some extent, by a reorganization in the CNS. Nerve fibers in the CNS can be stimulated to grow and achieve new functional connections.

A muscle deprived of its neural supply becomes paralyzed. The motor endplates disappear and the muscle shrinks. The denervated muscle becomes hypersensitive to ACh, and the chemoreceptive area expands so that the entire surface of the muscle fiber becomes sensitive. It also becomes sensitive to norepinephrine. In the reinnervated muscle a new motor endplate develops, and the chemosensitivity shrinks to this zone. A denervated muscle can be functionally restored after as long a time as 2 to 3 years, provided that it again gets its neural contact with the CNS.

The structural and metabolic changes in a denervated muscle are not the same as in a muscle that has been inactivated by long-lasting block of nervous impulses (Gutmann et al., 1961). With an intact nerve supply the muscle is much less affected by inactivity. Observations suggest that substances are supplied to the muscle by the innervating nerve. This "trophic function" of the nerve cell is apparently essential for the muscle's metabolism and protein syntheses. It may be an axonal migration of proteins and nucleic acids. The continuous subthreshold release of ACh at the motor endplate may also exert an effect on the muscle. (For details see Guth, 1968.)

The adaptation of muscle fibers to training involves considerable metabolic and morphological changes, and there are also, as discussed above, changes in nerve cells involved in motor activity, particularly in the RNA synthesis. There may be a functional connection on a biochemical level so that the muscle's adaptivity might be dependent on a trophic effect of the innervating nerve cell, and not only of the impulse traffic as such.

A stimulation of a denervated muscle, even a very intensive one, will not lead to a hyper-trophy as noticed in a normal animal subjected to training (Gutmann et al., 1961).

Eccles et al. (1962) studied the effect of cross-anastomosis of motor nerves in kittens: i.e., the nerves supplying antagonistic muscles, such as the lateral gastrocnemius and plantaris muscle, were sectioned and transposed. Six to ten months after the cross-union was made, intracellular recordings in the motoneurons revealed a significant increase in the monosynaptic action of the afferent fibers from the in-series synergists of the plantaris muscle (flexor digitorum brevis) in the lateral gastrocnemius motoneuron that had reinnervated the plantaris muscle. They also found evidence that there could be a regression of some of the central synaptic connections of the afferent fibers involved in the cross-union which, as a consequence of their changed peripheral origin, had now become functionally inappropriate. They conclude that it is highly probable that plastic changes may ensue when motoneurons and muscle afferent functions are changed by cross-union operations.

Missiuro and Kozlowski (1963) have shown that transplantation of a knee flexor of the hind leg of a rabbit to the distal tendon of a severed extensor modifies the coordinat-ing relations reflected in electromyograms. The transplanted flexor begins to act as a knee extensor. The first electromyographic manifestations that the transplanted muscle is taking over the new function sometimes become evident within a few days after the transplantation and gradually become stronger and more firmly established. Over a period of 4 months after the operation the transplanted muscle periodically displays a tendency to revert to its original function. On retransplantation the muscle immediately resumes the original function. The authors' explanation for this is that the nervous sys-tem is capable of functional adaptation of the transplanted muscle and corresponding nerve centers to a new situation.

Buller et al. (1960) describe experiments on kittens or cats in which the nerves to a slow muscle (e.g., soleus) and a fast one (e.g., flexor digitorum longus) were divided and cross-sutured. When a nerve from fast or phasic motoneurons had been made to inner-vate a slow muscle, the muscle was transformed to a fast muscle, even in the adult cat. Likewise, slow tonic motoneurons converted fast muscles to slow. Apparently the differ-ence of the speed of the contractile material and specific biochemical properties depend in one way or the other, presently unknown, on the nerve by which the muscle is innervated (Gutmann and Syrevý, 1967).

Muscular Fatigue

Fatigue is a very complex conception, especially since heavy exercise loads respiration and circulation, as well as the neuromuscular function. Fatigue with-out preceding exercise is not uncommon. The disposition to subdue the feeling of fatigue is individually very different.

Central or peripheral fatigue?   Merton (1954), in his studies on voluntary strength and fatigue, found that when strength failed in adductor pollicis, electrical stimulation of the motor nerve to the muscle could not restore the strength. Since, even in extreme fatigue, the action potentials evoked by nerve stimulation were not significantly diminished, he concluded that neuromuscular block could not be important in the fatigue of volitional tetanus. Further evidence that the fatigue is peripheral in nature, at least in experiments lasting only for minutes, was provided by the fact that recovery from fatigue did not take place if the circulation to the muscle was arrested.

In contrast we have to consider the experiments reported by Ikai et al. (1967) (also discussed earlier in this chapter). Their subjects made a voluntary maximal contraction once a second, altogether 100 times. After every fifth contraction an electrical stimulus was applied to the nerve of the muscle. In both conditions there was a decrease in developed force with time, but the decrease was more pronounced in the voluntary effort than in the electrically induced contraction. (The maximal developed force was at the end about 40 percent of the initial value for the voluntary contractions, but only reduced to 66 percent when the muscle was stimulated electrically. The initial force was in the order of 12 kp in the voluntary contraction and 16 kp with the nerve stimulated artificially.) Ikai et al. conclude that the decrease in muscular strength in the course of fatigue is based on two components, a peripheral one and a central one, involving the CNS.

It should be emphasized that the electrical stimulation of a nerve leads to a synchronized activation of all muscle fibers innervated, but under normal conditions the impulses to the motor units come asynchronously. At a high frequence of stimulation it may not make any difference, since there is then a complete or almost complete fusion of the mechanical contraction waves (Fig. 4-19). With an electrical stimulation of the peripheral nerve, afferent fibers are also activated with consequences which are difficult to evaluate.

In the subjects of Ikai et al., the energy yield was apparently not taxed to maximum in voluntary efforts; the changes in the intra- and extracellular compartments of the muscle during the "natural" fatigue did not impede an increased performance when the muscle was stimulated artificially. However, a muscle that is stimulated repeatedly will gradually show a prolonged latency period and the rise of tension becomes slower and smaller; the muscle relaxes more gradually and incompletely, and may finally cease to respond to a stimulus, irrespective of whether we are dealing with a normal or an artificial one.

This complicated question is by no means settled. From extensive studies of the literature, Ernst Simonson (personal communication) concludes that the fatigue developing in maximal voluntary muscular effort is, to a large degree, located in the CNS; he excludes transmission fatigue at the neuromuscular junction as a plausible weak point.

In this section we shall now mainly limit the discussion to the fatigue localized in the exercising muscles.

Blood supply　For its contraction the muscle consumes energy. Metabolites are formed, and oxygen, if available, is used up, with a concomitant production of carbon dioxide and water. A restoration of the internal equilibrium necessitates an adequate blood supply. During contraction the active muscle swells and becomes hard. In maximal static contractions of the quadriceps femoris in humans the pressure within the muscle can be several hundred millimeters of mercury. Since the peak arterial blood pressure at rest is some 120 mm Hg and during exercise below 200 mm Hg in most cases, the blood flow through the active muscle will be partly or completely blocked. The time for maintained tension in a maximal static effort is only a few seconds. The active period can be prolonged with lower tension developed and the contraction time varies exponentially with the relative tension (Fig. 4-28) (Grose, 1958; Rohmert, 1960; Kogi and Hakamada, 1962). The subject's motivation and his ability to endure the unpleasant feeling from his muscles are of importance for the contraction time. The highest isometric work that can be maintained for long periods of time is about 10 to 15 percent of the maximal tension (Rohmert, 1960, 1961; Kogi and Hakamada, 1962).

With the blood flow occluded just before and during a vigorous contraction, there is no difference in the initial tension developed as compared with the controls. At lower tensions an occlusion of the local blood flow reduces the maximal isometric contraction time. The difference appears at tension when the intramuscular pressure falls below systolic blood pressure, which partially reestab-

Fig. 4-28   Maximal work time plotted versus force in isometric muscular contraction against a load expressed in percentage of maximal isometric strength in the same type of work. Average of results obtained in studies on different muscle groups in 21 subjects. Note that a 50 percent load can be maintained for just 1 min, but as long as the muscular force is less than 15 percent of the maximal force (dashed line), the contraction may be maintained almost "indefinitely." The subjects experienced discomfort and aches in the working muscle some time before they were compelled to terminate the effort because of muscular exhaustion. (From Rohmert, 1960.)

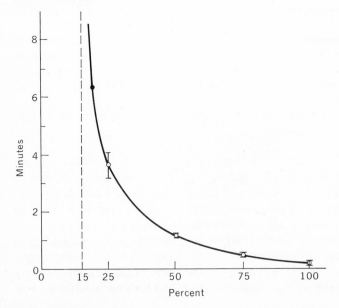

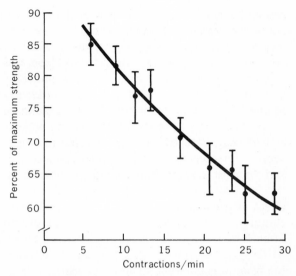

Fig. 4-29   Percent of maximum isometric strength that can be maintained in a steady state during rhythmic contractions. Points are averages for finger muscles, hand muscles, arm muscles, and leg muscles, combined. Vertical lines denote ± standard error. (Molbech, 1963.)

lishes the blood flow (Merton, 1954; Royce, 1958). Metabolites and carbon dioxide can be washed away from the muscle, and the oxidative restoration of the energy-producing mechanism is reestablished. Dynamic contractions also periodically hinder the passage of blood, partly or totally. The work load, in relation to the duration of the contraction periods, and the intervals between the periods of contraction determine the length of time the work can be endured. In exercises including frequent dynamic concentric contractions, the energy output for a given tension is relatively high. According to Asmussen, this type of exercise can probably be performed for long periods of time only if the developed strength does not exceed 10 to 20 percent of the maximal isometric strength.

Figure 4-29 shows results from experiments in which the subjects performed rhythmic maximal isometric contractions on a dynamometer in pace with a metronome (Molbech, 1963). Gradually the tensions decreased because of fatigue, but they finally leveled off at a value that could be maintained for a long time. With 10 contractions/min, about 80 percent of the maximal isometric strength could be applied without impairment. With 30 contractions/min the maximal load was reduced to 60 percent. The values seemed to be independent of the size of the activated muscle group.

Apparently the ability of the muscle fibers to maintain a high tension and the individual's subjective feeling of fatigue are highly dependent on the blood

flow through the muscle. In very short spells of work, ATP and creatine phosphate can yield energy and the oxygen present in the muscle (bound to the
myoglobin) also makes an energy delivery from aerobic processes possible. A
prolonged activity period with reduced blood flow may cause the oxygen need
to exceed the oxygen supply, and the anaerobic processes must contribute
markedly to the energy yield. The impaired blood flow not only limits the oxygen
supply but also the removal of metabolites and heat. Exactly which factor limits
the performance is not known. It could be an accumulation of lactic acid, of H^+,
and/or heat. With appropriately spaced pauses, the blood flow can secure the
supply of oxygen and energy-rich compounds and wash out the produced substances, and the work can proceed aerobically for long periods of time.

Effect of prolonged exercise   In heavy exercise prolonged for hours the work
output during maximal efforts becomes gradually decreased (Saltin, 1964).
After 1 hr rest, a work load that normally could be tolerated for 6 min had to
be terminated after about 4 min due to exhaustion. The peak lactate level in
the blood was correspondingly decreased. It is believed that the limiting factor
must be sought at the cellular level in the exercising skeletal muscles, and could
be anything from a change in the properties of the membrane of muscle fibers,
a disturbed ATP-ADP "machine," etc., to a depletion of the glycogen stores or a
reduced capacity to neutralize the metabolites produced.

Nöcker (1964) points out that prolonged exercise to exhaustion decreases the
potassium concentration within the active muscle cells, e.g., from 635 to 460 mg/100 ml
in rats. An increase in the hydrogen ion concentration increases the permeability of the
cell membrane. The coupled $Na^+ - K^+$ pump may be less efficient in prolonged activity
of the muscles. Since the potassium-sodium balance is of the utmost importance for the
excitability and the recovery of the muscle fibers, it is reasonable to assume that the
muscle's decreased ability to contract can be linked to a *disturbed ion balance*, eventually,
with a hyperpolarization of the cell membrane. There is also a possibility of modifying
the afferent impulses from a muscle subjected to prolonged severe exercise with an
increased inhibition of the motoneurons as a consequence. In emergency situations this
inhibition can, however, eventually be inhibited. A direct stimulus of the fatigued muscle
(prolonged work) has increased the force of contraction in some experiments.

There are *characteristic changes of the EMG* in muscle fatigue, indicating a change
in both the impulse traffic in the motor nerve and the muscle reaction to the discharge.
The amplitude increases and the rhythm slows down. A grouping and synchronization
of the discharges appears which, at least partly, can be attributed to a decrease of the
proprioceptive afferent impulses from muscle spindles, as shown by Kogi and Hakamada
(1962). These authors found that the quotient of the electrically integrated amplitude of
slower components divided by that of the faster components increased gradually and
steadily in fatigue experiments of isometric-isotonic contractions of various strength.
The appearance of a high "slow wave" ratio was significantly related to the onset of a
local fatigue sensation, to the feeling of pain, and to the subject's incapability of maintaining the intended tension.

According to Nöcker (1964), the potassium balance in the exercising skeletal muscles can be markedly disturbed without affecting the normal ion concentration in the intra- and extracellular fluid of the heart muscle cells. This finding can be taken as another evidence for the attractive hypothesis that a normal heart cannot be exercised to the limit of its capacity, since the skeletal muscles fail to work and force the subject to stop working before the heart has reached its maximal limit.

"Fatigue" in joints   Basmajian (1967) subjected a sitting person's upper limbs to a strong downward pull. The muscles crossing the shoulder joints and the elbow joints were electrically quiescent, but the subject felt local fatigue. Basmajian thinks that this fatigue originates from the painful feeling of tension in the articular capsule and ligaments.

Motivation   With the tools now available the conductor of physiological exercise experiments does not have to rely exclusively on the subject's subjective statement for an evaluation of the relative load placed on the muscles, the circulation, or other functions. The production of lactic acid, the pH of the blood, EMG, heart rate, blood sugar level, body temperature, etc., may be used as an indication of how the subject should feel. In healthy subjects, young and old, untrained and well-trained, the complaint after a maximal effort is, in most cases, that the stiff muscles refused to obey the will any longer; if they are well motivated they may, however, proceed for some time.

Recovery   Muscle fatigue is a reversible phenomenon, provided that proper periods of rest or mild exercise are inserted. Actually, light exercise with other muscle groups than those fatigued can promote the recovery. Hormones, e.g., epinephrine, and drugs can temporarily increase the strength exerted by a fatigued muscle.

Sore Muscles

If an untrained individual performs vigorous exercise, the engaged muscles may become very painful and hard. The symptoms usually appear after about 12 hr, become more severe during the next day, and fade gradually away, so that the muscles are symptom-free after 4 to 6 days. The pain is probably caused by injuries, especially on connective tissues within the muscle and its attachment to the tendon. Secondarily, histamine and other substances are produced, which may cause edema and pain. The repair of the damaged tissue results in a stronger muscle much less susceptible to further injuries, even if the subsequent exercise is much more severe. A sore muscle is more likely to develop in the beginner if the muscle works excentrically, even if the metabolic rate is much lower than during concentric contractions; the exerted tension may, however, reach much higher values in excentric work.

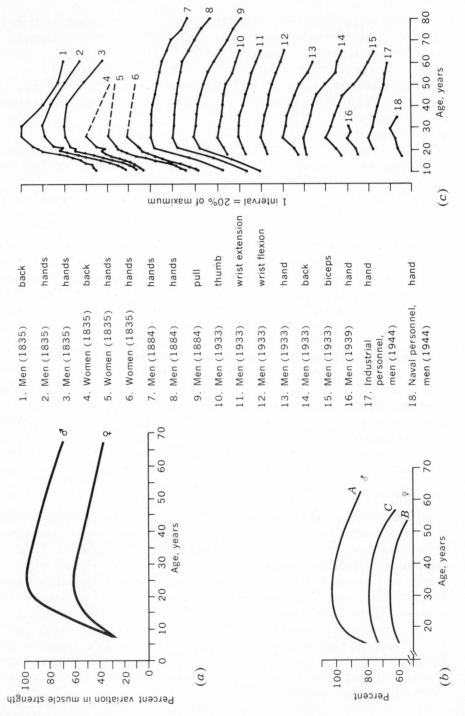

1. Men (1835)	back	
2. Men (1835)	hands	
3. Men (1835)	hands	
4. Women (1835)	back	
5. Women (1835)	hands	
6. Women (1835)	hands	
7. Men (1884)	hands	
8. Men (1884)	hands	
9. Men (1884)	pull	
10. Men (1933)	thumb	
11. Men (1933)	wrist extension	
12. Men (1933)	wrist flexion	
13. Men (1933)	hand	
14. Men (1933)	back	
15. Men (1933)	biceps	
16. Men (1939)	hand	
17. Industrial personnel, men (1944)	hand	
18. Naval personnel, men (1944)	hand	

(c) 1 interval = 20% of maximum — Age, years

(a) Percent variation in muscle strength — Age, years

(b) Percent — Age, years

The localized, sustained, and painful *cramp* that occasionally can throw a muscle into vigorous involuntary contraction, e.g., after prolonged monotonous exercise or even during sleep, cannot be explained at present. A stretch of the muscle usually relieves it from the cramp, probably due to inhibitory impulses from the stimulated Golgi tendon organs (Sherrington's lengthening reaction).

MUSCLE STRENGTH: SEX AND AGE

Muscle strength depends on many factors, some of which are discussed in the preceding pages. In tests of muscle strength, the day-to-day variation is usually of the order of $\pm$ 10 to 20 percent. The correlation between the strength of different muscle groups in the same individual is, as mentioned above, low, moderate, or fairly high, depending on which muscle groups are compared. Figure 4-30 presents data from several studies on average muscle strength for different muscle groups in female and male subjects of different ages. The maximal strength is reached between the ages of twenty and thirty, after which it decreases gradually, so that the strength of the sixty-five-year-old is approximately 80 percent of that attained between the ages of twenty and thirty (Fisher and Birren, 1947; Hettinger, 1961). In both sexes the rate of decline with age in the strength of the leg and trunk muscles is greater than in the strength of arm muscles. It should be emphasized that training of muscle strength and therefore the degree of engagement of muscle synergists in the daily routine will highly influence the results.

For adult women the strength of any muscle group is lower than for men of the same age. On an average, the muscle strength of women is about two-thirds that of men.

A difference in body dimensions must necessarily be considered when evaluating the variation in strength with sex and age. With strength proportional to the transverse sectional area of the muscle, the isometric strength should vary with the individual's height in the second power, providing there is a geometrical similarity between the individuals of different body size (Asmussen, 1962). If the height of a child increases by a factor of 1.5 (e.g., from 120 to 180 cm), the developed strength during a maximal pull or push should increase by a factor of $1.5^2 = 2.25$. In measurements of strength as torque, the distance over

Fig. 4-30   Variation in maximal isometric strength for various muscle groups with age.
(a) Changes in strength with age in men and women.
(b) Curve A is for men and curve B for women. The data in curve C are derived from curve B but corrected for the sex difference in body height with the assumption that muscular strength is related to H^2 where H = body height. The average strength for the 22-year-old man = 100 percent. (By courtesy of Asmussen.)
(c) Data compiled by Fisher and Birren (1947). Figures in parentheses indicate year in which study was made.

which the muscle can shorten, being proportional to the height, must also be considered. Therefore, in the given example the torque should increase by a factor of $1.5^3 = 3.375$. (See Chap. 10.)

In Fig. 4-30b, the strength in women has been corrected for body size, and thereby, the difference in strength between women and men has been reduced, so that adult women in average attain about 80 percent of the men's value.

Fig. 4-31   Maximal isometric strength in kilograms (or kiloponds) of some muscle groups in girls and boys 7 to 17 years of age (note the double logarithmic plot). (From Asmussen et al., 1959.)

(a) Strength of the trunk muscles is modified by the sex of the subjects regardless of height.
(b) Strength of the upper extremity muscles is greater for boys than for girls regardless of height; at heights 150 to 160 cm the boys were about 13 years old.
(c) Strength of the muscles of the lower extremities of boys and girls is not appreciably different when strength is related to height.

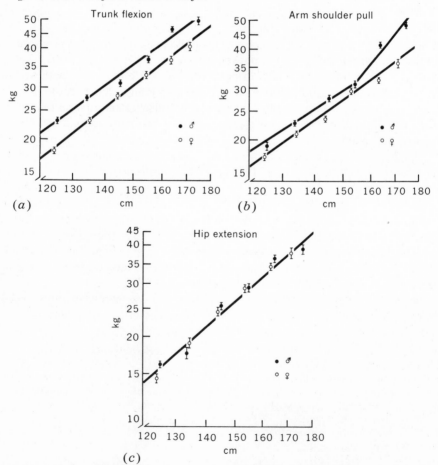

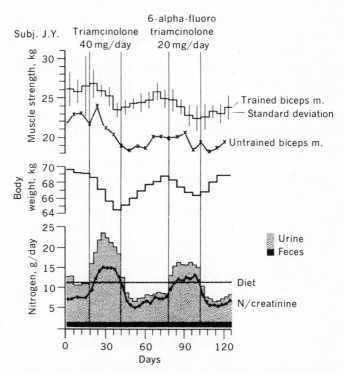

Fig. 4-32   Effect of corticosteroids on muscle strength, body weight, and nitrogen balance in a normal young man. Corticosteroid administration causes an increase in urinary nitrogen elimination, a loss of body weight, and a loss of muscle strength in both the trained and the untrained biceps muscle. (Rodahl et al., 1965.)

However, the body composition and geometric proportions are different in both sexes, and the strength per square centimeter of muscle tissue may be the same in a woman and a man, age and degree of training being the same. The skeleton in a woman of a given height is more slender than in a man, and the ratio muscle mass to the lean body mass is smaller in a woman.

Asmussen et al. (1959, 1962) report from studies on children seven to seventeen years of age that when muscle strength is plotted in relation to body height on a double logarithmic scale, the data usually fit straight lines. However, for some muscle groups, especially the arm muscles, the slope of this line increases suddenly in boys at a height of about 155 cm, i.e., at about the age of thirteen. A similar tendency was not observed in girls. This sudden increase in strength in boys may be of a hormonal origin. There was no difference in the strength of the leg muscles between girls and boys when differences in body height were eliminated, but in trunk and arm muscles the boys were considerably stronger than the girls, and this difference appeared as early as age seven. These results are presented in Fig. 4-31.

In performances of a more functional character, e.g., sprinting and jumping, muscle strength is of importance. In such events girls below about twelve years of age perform on a par, or slightly under par when compared with boys. Between the ages of twelve and eighteen the boys show an improvement in results that is much greater than predicted from the increase in body size. In girls a comparable improvement does not take place (Asmussen, 1962).

Asmussen concludes that in children age affects muscle strength by (1) increased size of the anatomic dimensions, (2) the results of aging itself (one extra year of age increases the strength by 5 to 10 percent of the average strength of the same height group, and this gain may be attributed to the maturation of the CNS), and (3) the developing of the sexual maturity of the child (probably the male sexual hormones are of special importance for this effect). As a matter of fact, between the ages of six and twenty, about one-third of the increase in body height occurs, but during the same period of time, four-fifths of the development of strength takes place.

Even with correction for body size, age, and sex, large individual differences in muscle strength are observed, and a standard deviation from a mean value of ± 15 to 20 percent must be accepted as a normal finding.

EFFECT OF ANABOLIC AND ANTIANABOLIC AGENTS

It is a common clinical experience that prolonged administration of corticosteroids causes weakness and wasting of muscles. In carefully controlled metabolic balance studies, in which the effects of 20 to 40 mg triamcinolone (Squibb, Kenacort) per day were examined on muscle strength, nitrogen balance, and body weight in healthy young men, it was found that corticosteroid administration caused an increase in urinary nitrogen elimination, a loss of body weight, and a loss of muscle strength (Fig. 4-32) (Rodahl et al., 1965). By the simultaneous administration of anabolic agents, such as methyltestosterone, it was possible to partially, but not completely, counteract these changes. This indicates that the catabolic processes caused by corticosteroid treatment in man also involve the contractile proteins of the muscle.

REFERENCES

Asmussen, E.: Positive and Negative Muscular Work, *Acta Physiol. Scand.*, **28**:364, 1953.
Asmussen, E.: Muscular Performance, in K. Rodahl and S. M. Horvath (eds.), "Muscle as a Tissue," p. 161, McGraw-Hill Book Company, New York, 1962.
Asmussen, E., O. Hansen, and O. Lammert: The Relation between Isometric and Dynamic Muscle Strength in Man, *Communications from the Testing and Observations Institute of the Danish National Association for Infantile Paralysis*, no. 20, 1965.
Asmussen, E., and K. Heebøll-Nielsen: A Dimensional Analysis of Physical Performance

and Growth in Boys, *J. Appl. Physiol.*, **7**:593, 1955.

Asmussen, E., K. Heebøll-Nielsen, and S. Molbech: Description of Muscle Tests and Standard Values of Muscle Strength in Children, *Communications from the Testing and Observation Institute of the Danish National Association for Infantile Paralysis*, no. 5, 1959.

Baker, P. F.: The Nerve Axon, *Sci. Am.*, **214**(3):74, 1966.

Basmajian, J. V.: "Muscles Alive," The Williams & Wilkins Company, Baltimore, 1967.

Bigland, B., and O. C. J. Lippold: The Relation between Force, Velocity and Integrated Electrical Activity in Human Muscles, *J. Physiol. (London)*, **123**:214, 1954a.

Bigland, B., and O. C. J. Lippold: Motor Unit Activity in the Voluntary Contraction of Human Muscle, *J. Physiol. (London)*, **125**:322, 1954b.

Blaschke, A. C., H. Jampol, and C. L. Taylor: Biomechanical Considerations in Cineplasty, *J. Appl. Physiol.*, **5**:195, 1952.

Buchtal, F., C. Guld, and P. Rosenfalck: Multielectrode Study of the Territory of a Motor Unit, *Acta Physiol. Scand.*, **39**:83, 1957.

Buller, A. J., J. C. Eccles, and R. M. Eccles: Interactions between Motoneurons and Muscles in Respect of the Characteristic Speeds of Their Responses, *J. Physiol. (London)*, **150**:417, 1960.

Byrne, W. L., and D. Samuel: Behavioral Modification of Injection of Brain Extract Prepared from a Trained Donor, *Science*, **154**:418, 1966.

Carlsöö, S., and O. Johansson: Stabilization of and Lead on the Elbow Joint in Some Protective Movements: An Experimental Study, *Acta Anat.*, **48**:224, 1962.

Eccles, J. C.: "The Physiology of Nerve Cells," The Johns Hopkins Press, Baltimore, 1957.

Eccles, J. C.: "The Physiology of Synapses," Springer-Verlag OHG, Berlin, Göttingen, Heidelberg, 1964.

Eccles, J. C.: The Synapse, *Sci. Am.*, **212**(1):56, 1965.

Eccles, J. C., R. M. Eccles, C. N. Shealy, and W. D. Willis: Experiments Utilizing Monosynaptic Excitatory Action on Motoneurons for Testing Hypothesis Relating to Specificity of Neuronal Connections, *J. Neurophysiol.*, **25**:559, 1962.

Elias and Pauly: "Human Microanatomy," 2d ed., Da Vinci Publishing Co., Chicago, 1961.

Fisher, M. B., and J. E. Birren: Age and Strength, *J. Appl. Psychol.*, **31**:490, 1947.

Floyd, W. F., and P. H. S. Silver: The Function of the Erectores Spinae Muscles in Certain Movements and Postures in Man, *J. Physiol.*, **129**:184, 1955.

Granit, R.: "Receptors and Sensory Perception," Yale University Press, New Haven, Conn., 1955.

Granit, R.: Muscle Tone and Postural Regulation, in K. Rodahl and S. M. Horvath (eds.), "Muscle as a Tissue," p. 190, McGraw-Hill Book Company, New York, 1962.

Granit, R. (ed): "Muscular Afferents and Motor Control," John Wiley & Sons, Inc., New York, 1966.

Grimby, L., and J. Hannerz: Recruitment Order of Motor Units on Voluntary Contraction: Changes Induced by Proprioceptive Afferent Activity, *J. Neurol. Neurosurg. Psychiat.*, **31**:565, 1968.

Grose, J. E.: Depression of Muscle Fatigue Curves by Heat and Cold, *Res. Quart. Am. Ass. Health Phys. Educ.*, **29**:19, 1958.

Guth, L.: "Trophic" Influences of Nerve on Muscle, *Physiol. Rev.*, **48**:645, 1968.

Gutmann, E., R. Beránek, P. Hník, and J. Zelená: Physiology of Neurotrophic Relations, *Proc. 5th Nat. Congr. Czech. Physiol. Soc.*, 1961.

Gutmann, E., and I. Syrevý: Metabolic Differentiation of the Anterior and Posterior Latissimus Dorsi of the Chicken during Development, *Physiol. Bohemoslov.*, **16:**232, 1967.

Hagbarth, K.-E.: Excitatory and Inhibitory Skin Areas for Flexor and Extensor Motoneurons, *Acta Physiol. Scand.*, **26**(Suppl. 94), 1952.

Hagbarth, K.-E.: Lower Somatic Functions of the Nervous System, *Ann. Rev. Physiol.*, **26:**249, 1964.

Hansen, J. W.: Effect of Dynamic Training on the Isometric Endurance of the Elbow Flexors, *Intern. Z. Angew. Physiol.*, **23:**367, 1967.

Hettinger, Th.: "Physiology of Strength," Charles C Thomas, Publisher, Springfield, Ill., 1961.

Hill, A. V.: The Priority of the Heat Production in a Muscle Twitch, *Proc. Roy. Soc. (Biol.)*, **148:**397, 1958.

Hubbard, A. W.: Homokinetics: Muscular Function in Human Movement, in W. R. Johnson (ed.), "Science and Medicine of Exercise and Sports," Harper & Row, Publishers, Incorporated, New York, 1960.

Huxley, A. F.: Skeletal Muscle, in K. Rodahl and S. M. Horvath (eds.), "Muscle as a Tissue," p. 3, McGraw-Hill Book Company, New York, 1962.

Hydén, H., and E. Egyhazi: Nuclear RNA Changes of Nerve Cells during a Learning Experiment in Rats, *Proc. Nat. Acad. Sci. U.S.*, **48:**1366, 1962.

Ikai, M., and T. Fukunaga: Calculation of Muscle Strength per Unit Cross-sectional Area of Human Muscle by Means of Ultrasonic Measurement, *Int. Z. angew. Physiol. einschl. Arbeitsphysiol*, **26:**26, 1968.

Ikai, M., and A. H. Steinhaus: Some Factors Modifying the Expression of Human Strength, *J. Appl. Physiol.*, **16:**157, 1961.

Ikai, M., K. Yabe, and K. Ischii: Muskelkraft und Muskuläre Ermüdung bei Willkürlicher Anspannung und Elektrischer Reizung des Muskels, *Sportarzt und Sportmedizin*, **5:**197, 1967.

Keele, C. A., and E. Neil: "Samson Wright's Applied Physiology," 11th ed., Oxford University Press, Fair Lawn, N.J., 1965.

Kogi, K., and T. Hakamada: Slowing of Surface Electromyogram and Muscle Strength in Muscle Fatigue, *Rep. Inst. Sci. Labour (Tokyo)*, **60:**27, 1962.

Kuffler, S. W., C. C. Hunt, and J. P. Quilliam: Function of Medullated Small-nerve Fibers in Mammalian Ventral Roots: Efferent Muscle Spindle Innervation, *J. Neurophysiol.*, **14:**29, 1951.

Lambert, O.: The Relationship between Maximum Isometric Strength and Maximum Concentric Strength at Different Speeds, *Intern. Fed. Phys. Educ., Bull.* **35:**13, 1965.

Lippold, O. C. J.: The Relation between Integrated Action Potentials in a Human Muscle and Its Isometric Tension, *J. Physiol. (London)*, **117:**492, 1952.

Lorente de No: Analysis of Activity of Chains of Internuncial Neurons, *J. Neurophysiol.*, **1:**207, 1938.

Matthews, P. B. C.: Muscle Spindles and Their Motor Control, *Physiol. Rev.*, **44:**219, 1964.

McMorris, R. O., and E. C. Elkins: A Study of Production and Evaluation of Muscular Hypertrophy, *Arch. Phys. Med. Rehabil.*, **35:**420, 1954.

Merton, P. A.: Voluntary Strength and Fatigue, *J. Physiol. (London)*, **123:**553, 1954.

Missiuro, W., and S. Kozlowski: Investigations on Adaptive Changes in Reciprocal Innervation of Muscles, *Arch. Phys. Med. Rehabil.*, **44:**37, 1963.

Molbech, S.: Average Percentage Force at Repeated Maximal Isometric Muscle Contractions at Different Frequencies, *Communications from the Testing and Observations Institute of the Danish National Association for Infantile Paralysis*, no. 16, 1963.

Nöcker, J.: "Physiologie der Liebesübungen," Ferdinand Enke Verlag, Stuttgart, 1964.

Petersen, F. B., H. Grandal, J. W. Hansen, and N. Hvid: The Effect of Varying the Number of Muscle Contractions on Dynamic Muscle Training, *Intern. Z. Angew. Physiol.*, **18**:468, 1961.

Ralston, H. J.: Recent Advances in Neuromuscular Physiology, *Am. J. Phys. Med.*, **36**:94, 1957.

Rasch, P. J., and L. E. Morehouse: Effect of Static and Dynamic Exercise on Muscular Strength and Hypertrophy, *J. Appl. Physiol.*, **11**:29, 1957.

Rodahl, K., B. Issekutz, Jr., J. J. Blizzard, and C. H. Demos: The Effect of Anti-anabolic Steroids on Muscle Strength, Nitrogen Balance and Body Weight in Healthy Subjects, unpublished, 1965.

Rohmert, W.: Ermittung von Erholungspausen für statische Arbeit des Menschen, *Intern. Z. Angew. Physiol.*, **18**:123, 1960.

Rohmert, W.: Untersuchung statischer Haltearbeiten in achtstündigen Arbeitsversuchen, *Intern. Z. Angew. Physiol.*, **19**:35, 1961.

Royce, J.: Isometric Fatigue Curves in Human Muscle with Normal and Occluded Circulation, *Res. Quart.*, **29**:204, 1958.

Ruch, T. C., and J. F. Fulton: "Medical Physiology and Biophysics," 18th ed., W. B. Saunders Company, Philadelphia, 1960.

Saltin, B.: Aerobic Work Capacity and Circulation at Exercise in Man, *Acta Physiol. Scand.*, **62**(Suppl. 230), 1964.

Sharpless, S. K.: Reorganization of Function in the Nervous System: Use and Disuse, *Ann. Rev. Physiol.*, **20**:357, 1964.

5

Blood and Body Fluids

contents

chapter five

Blood and Body Fluids

The function of the individual cells within the body is dependent on the constancy of their internal and surrounding environment. Claude Bernard recognized that an evolution of higher forms of organisms could not have taken place without establishment of a stable *milieu interne*, its composition being guarded by regulatory mechanisms.

The muscle cell is unique in regard to its ability to increase its metabolic rate. The maintenance of a constant *milieu interne* of the cell during the transition from rest to vigorous exercise necessarily represents, at times, a tremendous challenge to the circulation. The result may be that the muscle cell must cease working or that it is forced to slow down the intensity of work as the changed composition and property of the fluid within the cell or surrounding it interferes with the various processes which are necessary for the cell to function and perform work.

BODY FLUIDS

Generally speaking, we are dealing with a large water pool of about 60 percent of the body weight in men, 50 percent in women. As adipose tissue contains very little water, the percentage of total water in the obese individual is lower than in the nonobese (Fig. 5-1). The relation between total water and fat-free body weight (lean body mass) is fairly constant; in an adult the total water is about 72 percent of the lean body mass (Hernandez-Peon, 1961).

We can divide the water space into three compartments. The intracellular fluid is separated by the cell membrane from the interstitial fluid, while the third water compartment, the intravascular fluid, is circulating within the blood vessels (Fig. 5-2). In the vascular system, only the walls of capillaries permit an exchange of materials.

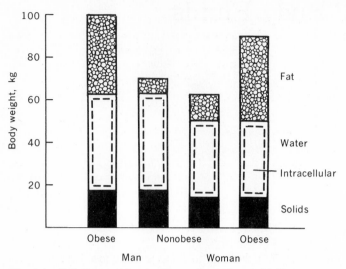

Fig. 5-1   Body composition in obese and nonobese individuals.

The fluid outside the cells, i.e., the extracellular fluid, constitutes about 30 percent of the total body water. Extracellular fluid has an ionic composition similar to that of sea water. The ions are found in approximately the same relative proportions, although the total ionic concentration of sea water is several times that of extracellular fluid. It has been postulated that this resemblance suggests that extracellular fluid was originally derived from the ancient oceans, which were more dilute than those of today (Hernandez-Peon, 1961).

The property of the capillary wall allows free passage of all substances in the blood except for the plasma proteins and blood corpuscles. On the other hand, there is a marked difference in concentration of various electrolytes between the extra- and intracellular fluid. The membrane of the resting cell acts as a "barrier," especially for positively charged ions (cations), but processes within the cells must continuously work to maintain this barrier property by throwing out some intruding ions (mainly Na^+) and by holding back others (mainly K^+). It is somewhat of a paradox that the prerequisite for homeostasis in the body is based on a "heterostasis" at the cellular level.

BLOOD

The blood and the lymph take care of the transportation of material between the different cells or tissues. The blood brings food materials from the digestive tract to the cells for catabolism or synthesis of molecules in tissue structures or for depots which are later mobilized and redistributed. Heat and the chemical products of catabolism are removed, carbon dioxide is expelled through the lungs, heat dissipated through the skin, and metabolites transported to the kidneys and liver for further processing. The blood circulation has a key position in the maintenance of a proper water balance and fluid distribution. As a carrier of hormones produced by endocrine glands and other active chemical agents, the blood can, in various ways, modify the function of cells and tissues.

Volume

The first effort to determine the blood volume in man was made on two criminals immediately after their execution. The blood vessels were flushed with a known volume of saline, and the fluid was collected and measured. However, this method has its definite limitations. The most common procedure is to inject into a vein a measured amount of a substance that does not easily escape from the blood vessels. After some time, when complete mixing has taken place, a blood sample is secured for determination of the concentration of the "tracer substance," and the subject's volume of blood can be calculated.

To illustrate the principle of current methods, the dye method will be described in some detail. After withdrawal of about 10 ml of blood from a vein (serving as a blank), an exactly measured amount of Evans blue (T-1824) is injected into the vein. This dye is bound to the proteins of the blood and escapes only slowly from the blood vessels. Then, 10 minutes' time is allowed for the dye to mix completely with the circulating blood, and a new blood sample is taken. The blood in the test tubes is permitted to clot and is then centrifuged. The concentration of the dye in the serum is measured colorimetrcially by comparing the light absorption with that of the blank and with that of a sample containing the dye in a known concentration. The more diluted the injected dye, the larger the plasma volume. By centrifugation of a blood sample, the proportion between blood-cell volume and plasma volume (hematocrit) can be determined, and hence, the total blood volume on the basis of the determined plasma volume. With a similar procedure, the volume of blood can be determined with radioactive iodinated proteins (I^{131} or I^{125}) or with radioisotope-labeled ("tagged") red cells (for example, Cr^{51}).

The individual variations in blood volume are large. Therefore, the figure presented in textbooks is nothing more than the mean value from determinations on a certain number of individuals under certain experimental conditions. The blood volume varies with the degree of training. A volume of 5 to 6 liters for men and 4 to 4.5 liters for women can be considered as normal (about 75 ml/kg body weight for men, 65 ml/kg body weight for women, and 60 ml/kg body weight for children).

Cells

The solid elements of the blood which are visible under the ordinary microscope are red corpuscles, or *erythrocytes;* white cells, or *leukocytes;* and platelets, or *thrombocytes.* When a blood sample in a tube containing an anticoagulant is centrifuged, these elements are separated from the liquid portion of the blood,

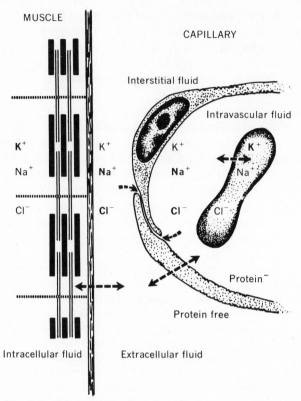

Fig. 5-2 Schematic drawing of structure of capillary wall and the exchange of material through it.

known as the *plasma*. The relative amount of plasma and corpuscles in blood is known as the *hematocrit*. For men the hematocrit is usually about 47 percent, and it is 42 percent for women and children.

The erythrocyte is a biconcave circular disk without a nucleus. It has an average diameter of 7.3 μ, and a thickness of 1 μ in the center and 2.4 μ near the edge (Fig. 5-2). The average cell count in the adult male is 5.7 million/mm³; in the female and children it is about 4.8 million. The life span of a red cell is about 4 months. Red cells are formed in the bone marrow at a speed that normally matches the rate of destruction. It can be estimated that the formation of erythrocytes proceeds at a rate of 2 million to 3 million per second. The color of the red cell (and the blood) is due to its content of hemoglobin, which is a protein (globin) united with a pigment (hematin). This pigment contains iron; each Fe^{++} atom can combine with one molecule of O_2. This combination is not a chemical oxidation but loose and reversible. In an oxygen-free medium the *hemoglobin (Hb)* is reduced, and with O_2 available, it forms *oxyhemoglobin*

(HbO_2). Another property of Hb is its affinity for CO. The Fe^{++} atom reacts with CO, and when given a choice it prefers CO (the affinity for CO is about 250 times greater than that for O_2), resulting in a proportionally decreased capacity to take up O_2. This phenomenon explains the high toxicity of CO.

In adult men the average Hb content is 15.8 g/100 ml of blood; in women it is 13.9 g/100 ml. From a statistical point of view, a normal value could be within 14.0 to 18.0 and 11.5 to 16.0 g/100 ml for men and women, respectively, with a range of 95 percent. Each gram of Hb can maximally combine with 1.34 ml of O_2. With 15.0 g Hb/100 ml of blood, the fully saturated blood can carry 20.1 ml O_2/100 ml plus some 0.3 ml O_2/100 ml dissolved in the plasma (oxygen pressure 100 mm Hg).

The concentration of Hb in blood can be determined according to two principles: (1) The blood sample is saturated with O_2, and the content of O_2 in a measured volume of blood is determined. The volume of dissolved O_2 can be calculated, and as the O_2-combining power of 1 g Hb is fixed to 1.34 ml, the concentration of Hb in the sample can easily be calculated. (2) The second method is based on the typical and specific light absorption spectra of Hb and its derivatives. A rough estimation of the Hb content of blood can be obtained from the hematocrit value, or from blood counts, with the presumption that each red corpuscle has a normal Hb content.

Plasma

The plasma occupies about 55 percent of the total blood volume. It contains about 9 percent solids and 91 percent water. The concentration of protein in the plasma is 6.4 to 8.3 g/100 ml. Three types of proteins are present: albumin (4.8 percent), globulin (various fractions, 2.3 percent), and fibrinogen (0.3 percent). If blood is permitted to clot, the fibrinogen, in the presence of thrombin, forms fibrin. The fluid "squeezed" from the clot is called *serum*.

The proteins of the blood have several important functions: fibrinogen is necessary for blood coagulation, and the globulin (especially gamma globulin) is necessary for the formation of antibodies. All fractions, but especially the albumin, have the important transport functions of carrying ionic and nonionic substances to the sites of need or elimination. The blood proteins are active in the buffer action and constitute in a way a mobile reserve store of amino acids. Finally, they play an important role in the plasma volume and tissue fluid balance.

All plasma proteins can pass through the capillary wall in small amounts, the albumin most readily. However, as most of the protein molecules are kept within the capillaries, they will exert an osmotic pressure of about 25 mm Hg. This protein or colloidosmotic pressure is an important factor in the exchange of fluid between the intravascular and interstitial spaces.

The total amount of electrolytes in the plasma is 900 mg/100 ml. The main ions are Na^+ and Cl^-. The plasma normally contains about 100 mg glucose/

100 ml. It also contains free fatty acids (FFA), amino acids, hormones, various enzymes, and about 25 different electrolytes in varying amounts.

Specific Heat

The specific heat of whole blood is 0.92 cal/g; i.e., the change in temperature of 1 g blood by 1°C will require or deliver 0.92 cal, i.e., about 0.9 kcal/liter of blood.

Buffer Action—Blood pH—CO_2 Transport

Apart from its many other functions, the blood also acts as a buffer (for a detailed discussion, see Davenport, 1958). As you may recall, a buffer solution contains an acid or base that is only slightly ionized (weak) and a highly ionized salt of the same acid or base. If we take a weak acid (HA), we have an equilibrium

$$HA \rightleftharpoons H^+ + A^-$$

If a strong acid is added to the solution, there is an increase in the hydrogen concentration, pushing the reaction to the left. As long as the buffer salt can provide A^- ions, thereby forming undissociated HA, a change in the pH of the solution is prevented, or "buffered."

According to the law of mass action,

$$H^+ = K \frac{HA}{A^-}$$

As pH $= -\log H^+$,

$$pH = pK + \log \frac{A^-}{HA}$$

The last equation is known as the Henderson-Hasselbalch equation. It is evident that with the numerical value of pK known, the pH of the solution can be estimated from the ratio of the concentration of the ionized salt to the concentration of free acid.

The pH of arterial blood is 7.40, and of mixed venous blood collected at rest, it is about 7.37. In the catabolism of the cells, CO_2 is formed, and in the case of anaerobic oxidation, lactic acid is formed. The oxidation of P and S in protein leads to the formation of phosphoric and sulfuric acid. The predominantly acid nature of the metabolites could easily explain the shift in pH as blood passes the capillary bed.

The isoelectric point of the proteins in the blood is on the acid side of the blood pH. Thus, suspended in an alkaline solution, the proteins ionize as acids and form negatively charged anions (NH_2—R—COOH $\rightleftharpoons NH_2$—R—COO$^-$ + H$^+$ where R symbolizes the protein radical; for simplicity we may write protein$^-$ for the anion). The proteins, therefore, act as hydrogen acceptors. With proteins ionized as weak acids, we have a buffer system: H protein $\rightleftharpoons$ H$^+$ + protein$^-$.

The number of ionizable groups in the protein is large, and as whole blood contains about 19 g protein/100 ml (with about 7 g/100 ml in plasma), its capacity to accept hydrogen without pH changes is considerable. In this respect, the potency of the plasma proteins is only one-sixth that of Hb.

There is another reaction of considerable physiological importance. H—HbO$_2$ is a stronger acid than reduced H—Hb; i.e., it dissociates more completely than does H—Hb. Hence, when blood is giving off O$_2$, the hydrogen-ion concentration of the blood falls, and pH rises.

Before we discuss the hydrogen exchange in the capillaries, we must draw attention to the second important buffer system of the blood. The CO$_2$ formed in the cells is dissolved and diffuses freely into the erythrocytes. Catalyzed by carbonic anhydrase, present only in the red cells, CO$_2$ forms, with water, H$_2$CO$_3$. This weak acid is dissociated as follows:

$$CO_2 + H_2O \rightleftharpoons H_2CO_3 \rightleftharpoons H^+ + HCO_3^- \tag{1}$$

The equilibrium of the first step (to the left) is such that only 0.001 of the CO$_2$ forms H$_2$CO$_3$ at equilibrium.

If we use the pH notation, the formula is approximately

$$pH = pK + \log \frac{HCO_3^-}{CO_2}$$

As the pK value and the concentration of HCO$_3^-$ and CO$_2$ can be determined, this formula can be applied to calculate the blood pH. The normal ratio of bound to free CO$_2$ is 20:1, and any change in this ratio will inevitably cause a change in the blood pH.

The equilibrium in the reactions in Eq. (1) is determined by the concentration of the various molecules and ions. If free H$^+$ can be removed from the system, the reaction should be pushed to the right. Potentially this reaction gives place for more CO$_2$ in the solution without a change in the pH. Though more H$_2$CO$_3$ will be formed, this is only an intermediate step in the formation of H$^+$ and HCO$_3^-$. In this sense, the supply of a hydrogen acceptor will actually determine the final equilibrium. Here the protein$^-$ anions can enter the picture. The CO$_2$ is produced in the tissue and diffuses into the red cell. At the same time O$_2$ is diffusing out to the tissue, where the O$_2$ concentration is lower than in the capillaries (HbO$_2$ $\rightarrow$ Hb + O$_2$). As mentioned earlier, reduced Hb is a weaker acid than HbO$_2$ (and incidentally weaker than H$_2$CO$_3$), and when reduced, it "binds" some of the H$^+$ ions dissociated according to Eq. (1) (H$_2$CO$_3$ + protein$^-$ $\rightleftharpoons$ HCO$_3^-$ + H protein). It is a happy coincidence that a willing hydrogen acceptor appears just when it is urgently needed. To a lesser extent, other protein anions contribute in a proper buffer action.

We may now give the complete picture of *the CO_2 transport from tissue to lungs:*

1   The most important reactions in the O_2 and CO_2 exchange between blood and tissue are (a) the formation of a weak acid, reduced Hb, from the stronger one, HbO_2; (b) the formation of $H^+ + HCO_3^-$ from $CO_2 + H_2O$ with the carbonic anhydrase serving as an enzyme in the intermediate formation of H_2CO_3. The interplay between (a) and (b) can in a quantitative way be illustrated as follows (Davenport, 1958). For each millimole of oxyhemoglobin reduced, about 0.7 mmole of H^+ can be taken up, and consequently 0.7 mmole CO_2 can enter the blood without there being any change in pH. If the CO_2 were derived only from fat combustion, 0.7 mole of CO_2 would be produced per mole of O_2 used [respiratory quotient (RQ) = 0.7]. This means that the formation of reduced Hb from the oxygenated HbO_2 would completely buffer the CO_2 uptake. However, at rest, 0.82 to 0.85 mole of CO_2 is formed per mole O_2 used. During heavy exercise, this figure is close to 1.00.

2   Some of the remaining CO_2 can combine directly with Hb forming carbaminohemoglobin ($CO_2 + HbNH_2 \rightleftharpoons HbNHCOOH$), and the simultaneous reduction of HbO_2 greatly favors this formation.

3   There is an increase in the volume of dissolved CO_2, so that the venous blood at rest contains about 0.3 ml $CO_2/100$ ml more than the arterial blood.

Together, these three factors explain how the CO_2 can be transported with the very small change in pH of 0.03 to the acid site.

The net result of the reaction under 1 is an increase in bicarbonate ions (HCO_3^-) in the red cells. HCO_3^- can freely diffuse across the cell membrane and enter the plasma. Actually, there is simultaneously a decrease in the concentration of the anions as undissociated HHb is formed. The consequent excess of cations within the cell and lack of positive ions in the plasma cannot be compensated by a simple diffusion of a cation (in this case K^+) out to the plasma. The cellular activity "prohibits" an exchange of metallic cations. To restore the electrochemical equilibrium (Donnan equilibrium), the Cl^- ions diffuse into the cell, thereby replacing its loss of anions. This chloride shift allows about 70 percent of the formed HCO_3^- to be transported in the plasma (Fig. 5-3).

If these reactions are understood, the events that take place in the lung capillaries should be easily comprehensible. The reactions run "backward," and in an ingenious way, the O_2 uptake and formation of the relatively strong acid HbO_2 facilitate the exclusion of CO_2. In brief, the increase in free H^+ with the dissociation of HbO_2 forces the reaction in Eq. (1), as shown earlier in this chapter, to the left. As the concentration of bicarbonate within the cell decreases, the plasma is ready to send in more HCO_3^- ions. In exchange, the plasma gets

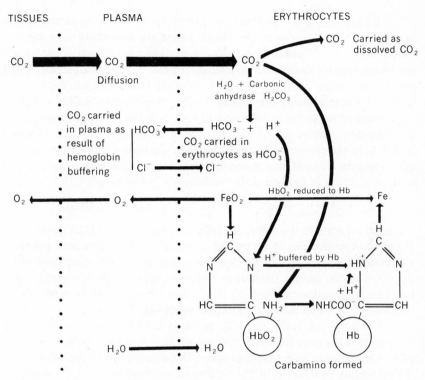

*Fig. 5-3   Schematic presentation of the processes occurring when carbon dioxide passes
from the tissues into the erythrocytes. At the bottom the effect of oxygenation and reduction
upon buffering action of the imidazole group of hemoglobin is illustrated. An increase
in the acidity of the blood drives the reaction to the right, and oxygen is given off. (A
decrease in the acidity of the solution would drive the reaction to the left, and oxygen
would be taken up by reduced hemoglobin.) In other words, the reduction of oxyhemoglobin causes the hemoglobin to become a weaker acid and to take up hydrogen ions from
the solution. (From Davenport, 1958.)*

back the Cl^- ions. At the same time, carbaminohemoglobin gives up some of
its CO_2. The degree of these reactions is dependent on the O_2 and CO_2 tension
of the alveolar air.

At rest, about 200 ml of CO_2 is produced per minute in the tissues, and the
cardiac output to transport this volume is about 5 liter/min. Thus, 100 ml of
blood transports 4 ml of this CO_2. The CO_2 content of 100 ml of arterial blood is
still about 50 ml after the delivery of CO_2 to the lungs. About 2.5 ml is dissolved
in the plasma, which is in equilibrium with a CO_2 tension of 40 mm Hg. Most of
the remaining CO_2 is in the plasma in the form of bicarbonate. As emphasized
earlier, the HCO_3^- ions are free, but they are "matched" by an equal amount
of cations, mainly Na^+. Similarly, the other buffer anions mentioned, protein$^-$
and Hb$^-$, are free ions, but for electrochemical equilibrium, free cations, mainly

K^+, are also present. These cations are often referred to as the buffer base. In other words, proteins of the blood, especially Hb, share with bicarbonate a "fixed" amount of cations, the buffer base. If stronger acids are introduced into the blood, those cations are "available" for the formation of a salt (still ionized) with the introduced anions. As weak acids are being formed, the hydrogen ions from the stronger acids are picked up (H protein, HHb, $H_2CO_3 = H_2O + CO_2$).

The buffer system composed of CO_2 and HCO_3^- is, in itself, of relatively low capacity. However, as CO_2 can be expired and actually stimulates the respiration, the physiological importance is significant. For simplicity, the buffer action may be illustrated with the following example. In heavy muscular exercise, lactic acid (HLa) is formed. After diffusion into the blood,

$$Na^+ + HCO_3^- + H^+ + La^- \rightleftharpoons Na^+ + La^- + H_2CO_3(\rightleftharpoons H_2O + CO_2)$$

the reaction goes to the right, as H_2CO_3 is a weaker acid (i.e., stronger acceptor of H^+) than lactic acid. The increase in free CO_2 and decrease in pH of the blood stimulate to a hyperventilation, thereby decreasing the CO_2 content of the blood and body. This causes the buffer base to decrease. As the lactic acid is later oxidized or transformed into glycogen, the blood becomes more alkaline, but this tendency is opposed by a withholding of CO.

The powerful buffer capacity of the blood has been emphasized, and the various proteins play an important role in this function. As the protein content of many tissues is high (average about 15 percent), we should expect that such tissues could contribute to the acid-base equilibrium. In fact, it appears that about five times as much acid is neutralized by other tissues as by the blood. The body of an average man might neutralize one equivalent (in other words, 1 liter of a 1.0 N solution) of a strong acid, such as HCl, before the blood pH would fall below 7.0, close to the lowest pH value compatible with life (Hitchcock, 1960).

The buffer capacity of blood (and tissues) must be considered as a "first aid" service not capable of maintaining a constant acid-base equilibrium for long periods. *The kidney function* stands as a final guard over the body pH: (1) Excess acids can be eliminated via the kidneys in the form of weak organic acids or as ammonium salts. (2) The tubular epithelium can produce NH_3, which diffuses into the tubular fluid, where it might form NH_4^+. This formation and excretion of NH_4^+ permits an excretion of anions, and an equivalent amount of metallic cations can be replaced. (3) The third mechanism that can acidify the urine is a variation in the ratio $(HPO_4^{--})/(H_2PO_4^-)$ according to the equation

$$pH = pK + \log \frac{HPO_4^{--}}{H_2PO_4^-}$$

The formation and excretion of acidic phosphate, as well as the other two processes, assist the conservation of sodium by the body. On the other hand, excess alkali may be eliminated with the urine as bicarbonate or basic phosphate.

Viscosity

The viscosity, or resistance to flow, of blood is mainly dependent on the plasma proteins and the cell content. The higher the hematocrit value, the higher the viscosity. Blood with a normal hematocrit of 45 percent is 2.1 times as viscose as water. With an increase in hematocrit to 55 percent, the viscosity is 2.6 times higher than that of water. A polycythemia that occurs after acclimatization to high altitude will evidently influence the work of the heart, as the resistance of the blood to flow will vary with its viscosity. At 0°C, the viscosity of blood is 2.5 times as great as at 37°C, which is an important factor in reducing the circulation in tissues exposed to cold, as in frostbite (Burton, 1965).

Blood has an "anomalous viscosity." The red cells tend to accumulate in the axis of the blood vessels, which leaves a zone near the wall relatively free from cells. This axial accumulation of the erythrocytes may result in small side branches of a blood vessel containing a volume of red cells that is considerably less than that for mixed blood (effect of "plasma skimming"). Another effect is a lower viscosity of the blood than expected. The axial accumulation of cells is more pronounced with an increase in velocity of the blood. However, in the physiological range of flow, the axial accumulation is complete, and therefore, the effective viscosity is constant (i.e., blood behaves as if it were a Newtonian fluid). There is a third factor that influences the effective viscosity of blood. In the very narrow vessels (arterioles and capillaries) the blood behaves as if the viscosity were reduced (Fåhraeus-Lindquist effect), which also contributes to reducing the load on the heart (Fig. 5-4). This is of great importance during heavy muscular activity (Burton, 1965).

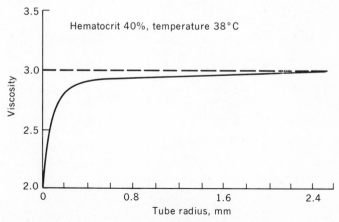

Fig. 5-4   Viscosity of erythrocyte suspension as a function of tube radius. (From Haynes & Rodbard, 1962, in "Blood Vessels and Lymphatics," ed. by D. I. Abramson, Academic Press Inc., New York.)

REFERENCES

Burton, A. C.: Hemodynamics and the Physics of the Circulation, in T. C. Ruch and
     H. D. Patton, "Physiology and Biophysics," pp. 523–542, W. B. Saunders Company,
     Philadelphia, 1965.
Davenport, H. W.: "The ABC of Acid-Base Chemistry," The University of Chicago Press,
     Chicago, 1958.
Hernandez-Peon, R.: Physiology in Body Fluids, in D. S. Dittmer (ed.), "Blood and Other
     Body Fluids," Federation of American Societies for Experimental Biology, Wash-
     ington, D.C., 1961.
Hitchcock, D. I.: Physical Chemistry of Blood, in T. C. Ruch and J. F. Fulton (eds.),
     "Medical Physiology and Biophysics," pp. 529–551, W. B. Saunders Company, Phila-
     delphia, 1960.

6

Circulation

contents

Circulation

HEART

It is assumed that the reader is familiar with the anatomy and basic physiology of the heart (Fig. 6-1a).

Cardiac Muscle

Under a microscope the heart muscle shows transverse striation essentially similar to that of the skeletal muscle, but the nuclei are placed centrally within the cell; the fibers give off branches which anastomose with adjacent fibers, giving an impression of a syncytial continuity of the muscle fibers. However, electron microscope studies give a different and clearer picture of the structure. Apparently the heart muscle is subdivided into numerous cell territories by cell boundaries fanning transversely across the branches of the heart muscle tissue (Sjöstrand and Andersson-Cedergren, 1960). These boundaries are associated with the intercalated disks of the heart, which form wavy "membranes" about 100 Å apart (Fig. 6-1b). Because of this "cutting" of the muscle fibers, the heart muscle does not represent a true syncytium. The role of the intercalated disks in the propagation of the stimulus through the heart muscle tissue is not understood at present.

    The conductive system of the heart and the effect of the refractory period after an excitation give the heart its rhythmical activity, unceasing from early embryonic life until death.

Blood Flow

Normally the blood vessels in the heart do not form anastomoses if the diameter is larger than 40 μ (the size of small arterioles). Therefore, in experiments on animals, a sudden occlusion of an artery will be followed by an infarction of the tissue supplied by that artery. On the other hand, a gradual narrowing and final

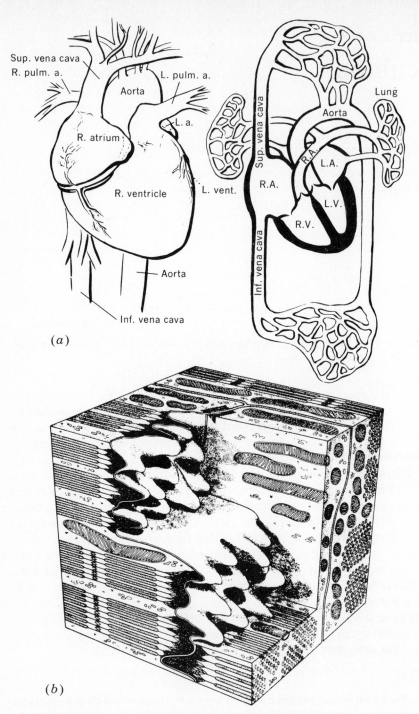

(b)

Fig. 6-1   (a) Schematic drawing of the heart and circulation. (b) Three-dimensional presentation of intercalated disk of mouse cardiac muscle. (Reproduced from Sjö-strand and Andersson-Cedergren, 1960.)

occlusion of a coronary artery will promote the development of a very rich net-work of anastomotic vessels. Experiments on dogs with coronary obstruction (Eckstein, 1957) have shown that effective collateral vessels develop more pro-fusely if the dogs are exercised. Eckstein found that the greater the coronary narrowing, the greater the development of collateral circulation. This is in agree-ment with the clinical experience that development of collateral circulation does occur in cardiac patients. The capillary network in the heart is extraordinarily rich, with at least one capillary going to each muscle fiber.

The contraction of the heart muscle interferes mechanically with the blood flow through the heart. In the left ventricle of the dog at rest there is a complete stop during the isometric contraction, and a backflow is actually established, but a peak flow is reached when the systolic pressure is at maximum, during early ejection phase. A second peak flow is reached during early diastole, after which it drops gradually to about 70 per-cent of maximum just at the end of diastole (Gregg and Green, 1940) (Fig. 6-2). In the right heart, where pressure is lower during systole, the fluctuations are not so pronounced as in the left ventricle. On the average, the volume flow of blood through the coronary vessels during diastole is about 2.5 times larger than during systole; during exercise there is an increased coronary inflow per heartbeat. The oxygen uptake rate is reported to be higher during systole than diastole (Gregg and Coffman, 1962).

Oxygen deficiency has a strong dilating effect on the arterioles of the heart. Excesses of CO_2, H^+, and lactic acid have a milder vasodilating effect. Epinephrine has some dilating influence on the vessels, but this is at least partly a secondary effect, caused by metabolites liberated by the increased metabolism induced by epinephrine. Norepinephrine has a similar, but weaker, effect. No evidence is reported for a reflex effect on coronary blood flow.

Pressures during a Cardiac Cycle

Figure 6-2 describes the pressure variations in the left heart chambers as well as in the aorta. The systole of the left ventricle starts with an isometric contrac-tion, for the mitral valves close rapidly as the pressure in the ventricle exceeds that of the atrium (phase within the vertical lines). Within about 0.05 sec, the ventricular pressure is brought up to and above the level in the aorta, and the aortic valves open. The isotonic contraction increases the pressure further. The peripheral resistance does not permit the same volume of blood to escape from the aorta as is ejected into it. Part of this volume is "stored" in the distended aorta ("windkessel vessel"). Then, as the pressure falls in the ventricle during the relaxation of the muscle, the aortic valves close, and the elastic property of the aortic wall can propel the stored blood out into the arterial tree. The intermittent energy outbursts of the heart would give an intermittent flow if the vessels were rigid tubes. Part of the potential energy is, however, taken up by the elastic arterial wall and then released during diastole of the heart, keeping the hydraulic energy level close to the heart continuously high. During the systole, blood has returned to the large veins close to the heart and the atrium. There

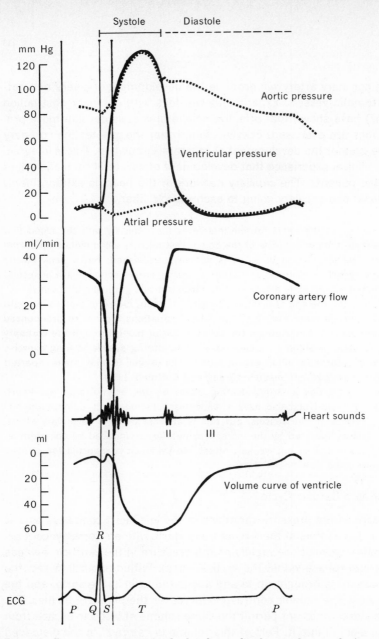

Fig. 6-2 Pressure variations in the left heart chambers and aorta during the cardiac cycle. See text for details. [The heart sounds can be objectively analyzed under various pathological conditions by phonocardiography. Closure of the atrioventricular valves contributes largely to the occurrence of the first sound (I), but so do vibrations in the tissues and turbulence of the blood flow; closure of the aortic valves is of prime importance for the occurrence of the second heart sound (II); vibrations of the chamber walls caused by movement of blood into the relaxed ventricle produce the third sound (III).] (Modified in part from Wiggers, "Circulatory Dynamics," Grune and Stratton, New York, 1952.)

is, perhaps, a passive lengthening of the atrium as the ventricle contracts, which may facilitate the filling of the atrium. During diastole there is a period of rapid filling of the ventricle after the opening of the mitral valves. The period of diastasis follows, during which the filling is much less rapid. The next cycle begins with the atrial contraction, which more or less empties the atrium.

With a heart rate of 75, the time for diastole is about 0.48 sec and for systole 0.32 sec (40 percent of the heart cycle); with a heart rate of 150, the periods are 0.19 and 0.21 sec (52 percent), respectively. It should be noted that an increase in heart rate occurs mainly at the expense of the length of the diastole and its period of diastasis. As the ventricle may eventually offer resistance to filling at the end of diastole, the pressure gradient between atrium and ventricle drops. An increase in heart rate can, in such a case, improve the filling of the heart and increase the output.

The physical events in the right heart are essentially similar to those just described, but the ventricular and pulmonary artery pressures during systole are about one-fifth of those in the left heart.

In the *aorta*, the pressure at rest varies between 120 mm Hg during systole and 80 mm Hg at the end of diastole. In the *pulmonary artery* the values are 25 and 7, respectively. It should be emphasized that the sphygmomanometer cuff technique does not always give the same value as direct measurement or measurement via a catheter in the artery. This holds true particularly during exercise.

Innervation of the Heart

Many *parasympathetic* nerve fibers from the vagus terminate in the region of the pacemaker, i.e., the sinoatrial node. When stimulated, they deliver acetylcholine, causing a slowing of the heart rate (inhibition). *Sympathetic* nerves are the efferent fibers from the upper part of the paravertebral ganglia which join the cardiac sympathetic nerves. They end in a dense network within the heart muscle. The effect upon stimulation is (1) an increase (acceleration) of the heart rate and (2) an increase in contractile force of the muscle fibers. Both epinephrine and norepinephrine cause a more rapid increase in the systolic pressure and a more complete emptying of the heart, i.e., the diastolic volume of the heart becomes smaller. The vagus does not seem to influence the contractility of the heart muscle. Via this automatic innervation of the heart, the inherent rate of beating can be highly modified; the healthy heart in a young, fully grown individual can cover a range from about 40 at rest to 200 beats/min during heavy exercise.

HEMODYNAMICS

We shall present in this summary a few definitions and also call attention to some of the physical laws governing the behavior of fluids, and especially blood in motion. For a more complete review, the reader should consult, for example, Burton, 1965; and Rodbard, 1962.

Definitions:

Heart rate (HR) is the number of ventricular beats per minute as counted from records of the electrocardiogram or blood pressure curves. The heart rate can also easily be determined by auscultation with a stethoscope or by palpation over the heart, both during rest and exercise.

Pulse rate is actually the frequency of pressure waves (waves per minute) propagated along the peripheral arteries, such as the carotid or radial arteries. In normal, healthy individuals, pulse rate and heart rate are identical, but this is not necessarily the case in patients with arrhythmias. In such cases, the output of blood by some beats may be too small to give rise to a detectable pulse wave.

Cardiac output is the volume of blood ejected into the main artery by *each* ventricle, usually expressed as liters per minute ($\dot{Q}$). With small fluctuations, the cardiac outputs of the right and left ventricles are identical. The cardiac output divided by the estimated surface area gives the "cardiac index," which relates the cardiac output to the body size. (For determinations, see below.)

Stroke volume (SV) is the volume of blood ejected into the main artery by each ventricular beat. The stroke volume is usually calculated by dividing the cardiac output by the heart rate ($\dot{Q}/HR$).

Oxygen uptake is the volume of oxygen (at 0°C, 760 mm Hg, dry = STPD*) extracted from the inspired air, usually expressed as liters per minute ($\dot{V}_{O_2}$). If the oxygen content of the body remains constant during the period of determination, the oxygen uptake equals the volume of oxygen utilized in the metabolic oxidation of foodstuffs. One liter of oxygen, then, corresponds to 4.7 to 5.05 kcal of energy liberation.

Arteriovenous oxygen difference is usually expressing the difference in oxygen content between the blood entering and leaving the pulmonary capillaries (a-$\bar{v}O_2$ diff.). Usually the oxygen content of the mixed venous blood ($C\bar{v}_{O_2}$) is determined in blood withdrawn from a long thin tube (catheter) introduced into a cubital vein and then passed through the right atrium and ventricle into the pulmonary artery. The arterial oxygen content (Ca_{O_2}) is analyzed in blood samples taken from a systemic artery, usually the femoral, brachial, or radial artery.

The relationship between the functions discussed so far can be summarized as follows:

Cardiac output ($\dot{Q}$) = stroke volume (SV) × heart rate (HR)

Oxygen uptake ($\dot{V}_{O_2}$) = cardiac output ($\dot{Q}$)
$$\times \text{ arteriovenous } O_2 \text{ difference } (a\text{-}\bar{v}O_2 \text{ diff.})$$

* STPD = standard temperature and pressure, dry.

Methods for Determination of Cardiac Output

1   Direct Fick method: Catheters are inserted into a systemic artery and the pulmonary artery. Blood samples are drawn simultaneously and analyzed for content of oxygen. At the same time, the oxygen uptake in the lungs is measured. Let us assume that 250 ml of oxygen has been taken up over a 1-min period. If the $a\text{-}\bar{v}O_2$ difference was 50 ml per liter of blood, 5 liters of blood must have passed through the pulmonary capillary bed (lesser circulation), or $250/50 = 5.0$.

2   Dye method (Stewart principle): The volume of a fluid can be calculated by adding a tracing substance, such as dye or radioactive isotope, in a known quantity and then measuring the concentration of the added substance, after complete mixing of the substance and fluid.

The formula for the calculation of volume (V) is $V = I/C$ where I is the quantity of substance added and C is the concentration of the substance. The Stewart principle for determination of the volume flow in a tubular system is as follows: It is assumed that all the fluid passes one or several single tubes with good conditions for mixing. By knowing the quantity of dye injected before the mixing point(s), and by determining the mean concentration of dye in the tubular system and the time from first appearance of dye until concentration is again zero, the flow per unit time can be calculated. This principle can actually be applied for determination of the cardiac output. The dye (or substance) is injected into a vein, preferably as close to the heart as possible; the mixture is assumed to be complete in the lungs and heart; and the concentration of dye can be determined on blood samples collected from a systemic artery. A continuous withdrawal of blood into a series of test tubes or through the cuvette of a colorimeter is necessary to establish the time course for the changes in concentration of dye. Recirculation of blood occurs before all the dye has passed the sampling artery, and this must be considered. However, if good curves are obtained when plotting dye concentration against the time for sampling, the time when concentration would have been back to zero can be obtained by extrapolation (Fig. 6-3). The formula for the calculation of cardiac output is then $\dot{Q} = I/C \times t$ where I is the quantity of dye injected, C is the mean concentration of dye, and t is the time it takes the dye to pass through the artery.

Mechanical Work and Pressure

Chemical energy is transformed into mechanical (external work) energy plus heat by the contraction of the heart muscle. The mechanical efficiency of the heart, i.e., the external work divided by the total energy exchange, is rather low, or only some 10 percent at rest.

The external work is calculated from data on force $\times$ distance moved, or for a fluid, pressure $\times$ volume moved. The pressure can create kinetic energy and a flow of the fluid. The higher the velocity of the fluid, the higher the kinetic energy (related to velocity squared). On the other hand, it may be concluded that in the part of a vessel where the velocity is highest, the fluid pressure against the wall is smallest, since total hydraulic energy = pressure energy + kinetic energy (Bernoulli's principle). If the

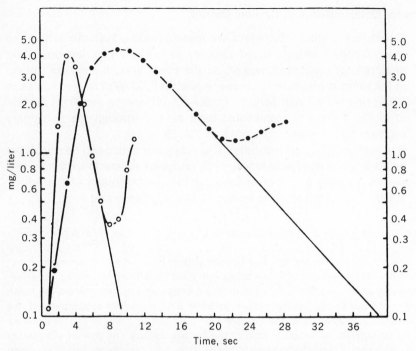

Fig. 6-3 Dye concentration curves. ● *= rest experiment; cardiac output 4.7 liters/* *min.* ○ *= work experiment, 1260 kpm/min; cardiac output 21.9 liters/min. (From* *Asmussen and Nielsen, 1953.)*

pressure is measured in a vessel with a catheter or tube connected to a manometer with the opening against the flow, the kinetic energy of flow is reconverted to pressure (*end pressure*) and the total energy is measured. With only a side hole in the catheter the *side* or *lateral pressure* is measured, and it will be less than the end pressure by an amount equivalent to the kinetic energy of flow, according to the formula given above. It can be calculated that at rest the kinetic factor in the aorta flow is only about 3 percent of the total work of the heart, but during exercise with a cardiac output five times the resting level, it may be an important part of the total work of the heart, possibly as much as 30 percent. This means that a simultaneous measurement of end pressure and side pressure in the aorta may show a difference of some 75 mm Hg, the end pressure being highest (see discussion below) (Burton, 1965). In the other arteries and in the smaller vessels, the kinetic energy factor is normally negligible, at least at rest. It should be pointed out that the sphygmomanometer cuff method for the measurement of blood pressure includes the end pressure.

Hydrostatic Pressure

The pressure in a vessel is created by the continuous bombardment by the molecules of the fluid against the inner surface of the vessel. The pressure is equal at all points lying in the same horizontal plane in a static liquid. The hydro-static pressure in a fluid at rest, under the influence of gravity, increases uni-formly with depth under the free surface (Pascal's law). It is evident that in the supine position the hydrostatic pressure is of similar magnitude in all parts of the body, but in the standing man it is higher (perhaps 100 mm Hg) in the vessels of the feet than in the head. The hydrostatic factor in a standing indi-vidual is modified by the action of the valves in the veins of the extremities. Arterial pressure should be measured at the horizontal level of the heart, or the value should be corrected to correspond to heart level. For practical pur-poses, measurement of the blood pressure with a sphygmomanometer cuff around the arm of a seated subject gives satisfactory values.

When a person is tilted from a supine position to an erect position with the feet down, the veins, and to some extent the other vessels below heart level, will dilate passively as a result of the hydrostatic force. Temporarily the blood is pooled, but when the vessels are filled, the flow continues unhindered by the uphill flow.

Tension

The pressure within the heart and vessels will more or less distend the walls (Fig. 6-4). The resistance to this force, producing *tension*, depends on the thick-ness of the wall and its content of elastic and collagen fibers and of active smooth-muscle fibers. The elastic tissue can balance the pressure without any energy output, but the contraction of the muscles requires a continuous expenditure of energy. The tension T is roughly proportional to the pressure difference P on the inside and outside of the wall (transmural pressure) and the radius r of the vessel, or $T = aPr$, where a is a constant (Laplace's law). Therefore, the same pressure applied in a small vessel does not produce the same tension as within a vessel with a larger radius (Fig. 6-4). The very thin membrane of the capillary, essential for the exchange of materials, can withstand the capillary blood pressure because its radius is so small.

The same law may, with some reservations, be applied to the heart. With a given tension developed by the heart muscle, a lower pressure is produced if the heart is dilated (increased in size, i.e., radii of curvature). If the diameter of the heart is doubled, the tension per unit length of ventricular wall must be about twice as great to produce the same pressure. The energy cost, i.e., the oxygen uptake, of the heart is related to the tension that must be developed and the time this tension is maintained. Therefore, an increase in the size of the heart increases the load on the heart.

As pointed out by Burton (1965), an increase in heart rate also increases tension work, since the time for rest (no tension developed) is relatively shorter. He also empha-sizes that an evaluation of the load on the heart must be based on the tension-time integral of the heart muscle, and that the external work, pressure × volume flow, only

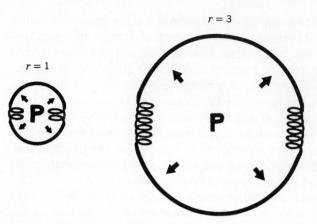

*Fig. 6-4   The vascular tension at a given transmural pressure,
P, is roughly proportional to the radius of the vessel, r. (This
is a simplification neglecting the thickness and properties of
the wall.) Thus, if the radius is tripled, three times more tension
will be required to maintain the same pressure.*

reflects a small fraction of this load (see Mechanical Work and Pressure). If the external
work is calculated, the integral of pressure with respect to flow should be used, not the
mean pressure $\times$ flow. When using the aortic pressure instead of the ventricular, one
must add the kinetic energy factor (end pressure); this is especially important during
exercise.

It should be borne in mind that the heart muscle follows the same law
as the skeletal muscle, i.e., its ability to produce tension increases with length.
By applying Laplace's law, one finds that the heart has to pay more to maintain
a given pressure if a reduction of muscular strength is compensated for by a
dilatation (increase in length of the fibers).

Flow and Resistance

Since the difference in the hydraulic energy of a fluid in two parts of a system
cannot be resisted by the fluid, it flows with a rate proportional to the energy
difference or pressure head. The resistance to flow of blood results from the
inner friction or viscosity of the blood. There is a cohesive force between the
blood and the wall of the vessel which "retards" the flow of the molecule layers
close to the wall. The nearer the center of the vessel, the higher the speed of
each lamina of fluid, and this "friction" phenomenon results in a maximal speed
in the very center of the vessel. The hydraulic energy provided the blood by the
contracting heart muscle is gradually spent and transformed to heat. Fluid flow
can be *laminar* (streamlined), i.e., each lamina of liquid slips over adjacent
laminae without mixture or interchange of fluid. At a critical velocity (expressed

by Reynolds number) the laminae of fluid move irregularly and start to mix with each other; the flow becomes *turbulent*. During turbulence the energy loss is larger and for a given pressure gradient the flow is lower (Fig. 6-5). The velocity of flow in the ventricles of the heart and aorta is normally turbulent during the early phase of contraction. This turbulence produces vibrations, causing high-pitched sounds which can easily be heard by listening over the heart (Fig. 6-2). If the velocity is abnormal, as in mitral stenosis, the heart sounds are abnormal. Similarly, a shunt between aorta and the pulmonary artery (open ductus Botalli) can give rise to a turbulent flow in that shunt vessel. In measurement of the blood pressure by applying a measured pressure over the brachial artery (with the sphygmomanometer cuff), the peak blood pressure suddenly overcomes the resistance of the gradually reduced compression. The blood flows with a high velocity through the narrowed artery, and a turbulence gives rise to sounds detectable by the stethoscope.

The radius of a tube is a deciding factor in the flow through it; the resistance to flow is a function of its radius to the fourth power. A decrease to half the radius, other things being equal, will actually decrease the flow to one-sixteenth of the original value. A dilatation of 10 percent increases the blood flow about 50 percent compared to the flow through the same vessel constricted. The variation in activity of the smooth muscles of the vessels gives a very sensitive instrument for the control of blood flow.

Fig. 6-5   In turbulent flow, the energy loss is greater and the rate of flow slower than in laminar flow at a given pressure gradient. Length of arrows indicates rate of flow. (From Haynes and Rodbard, 1962.)

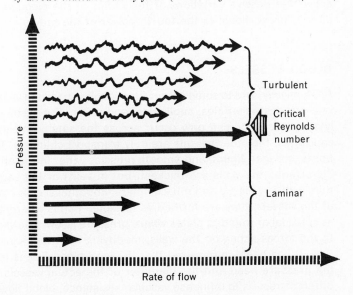

Flow in the blood vessels is normally laminar except close to the heart. The laminar blood flow (F) is actually proportional to the driving force (ΔP), inverse to the viscosity (η), proportional to the radius of the vessel (r), and inverse to the length of the vessel (ℓ):

$$F = \Delta P \times \frac{\pi}{8} \times \frac{1}{\eta} \times \frac{r^4}{\ell} \qquad \text{(Poiseuille-Hagen formula)} \qquad (1)$$

The resistance to laminar flow (R) is proportional to pressure gradient per rate of flow. This simple formula can help to evaluate the resistance offered to the blood flow in the vascular bed. If the blood pressure in the right atrium is zero, an increase in mean aortic blood pressure (BP_{mean}) at constant cardiac output $(\dot{Q})$ gives evidence of an increased peripheral resistance to flow (as $R \times \dot{Q} = BP_{mean}$).

The complex composition of blood and the distensible blood vessels complicate the situation in some ways. If a given pressure gradient ("driving pressure") is maintained between artery and vein, the flow is not, as expected, constant at varying pressure levels. At low pressures the flow becomes zero, even though there is still a considerable driving pressure. The pressure at which the vessels close completely is called the critical closing pressure. The decisive factor here is the transmural pressure. When the pressure outside the inner layer of the wall (tissue pressure and active muscular contraction in the vessel wall) exceeds the intravascular pressure, a collapsible vessel collapses ("artificially" this happens in the regular blood pressure measurement commented on earlier). At pressures well above the critical closing pressure, the flow is essentially proportional to the driving pressure.

Summary The pressure-flow curve obtained in studies of the hemodynamics of circulation is not linear. The effect of driving pressure can be modified by the transmural pressure, especially in the arterioles and capillaries, and at low intravascular pressure. The resistance to flow can be profoundly altered by active and passive variations of the radius of the blood vessels, the flow being directly proportional to the fourth power of the radius (r^4).

BLOOD VESSELS

Anatomically, and to some extent functionally, the blood vessels may be classified into arteries, arterioles, capillaries, and veins. The exchange of water, electrolytes, gases, etc., can only occur across the semipermeable membrane of the capillaries; the other vessels are only transport channels. The amount of elastic fibers, collagen fibers, and smooth muscles varies in the different vessels; in the big arteries the walls are thickest, but in veins of the same size the walls are thin, with few elastic and muscle fibers. Endothelial cells cover the inner surface of the vessels; they are formed as flattened plates where pressure is high, and as cuboidal or rounded plates where pressure is low. The vascular system adapts to the forces acting on the walls, modifying their mechanical property.

It has been emphasized that the local blood flow is mainly determined by the pressure head and the diameter of the actual vessels. The capacity of the different vessels to influence vascular resistance, blood flow, and blood distribu-

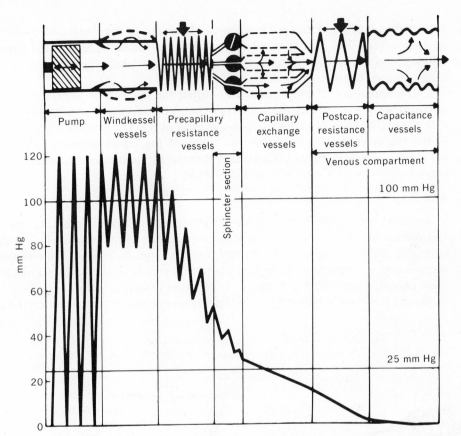

Fig. 6-6   A schematic illustration of the functionally different, consecutive sections of the vascular bed, related to the blood pressure fall along the circuit (ordinate blood pressure in mm Hg). Note the marked pressure drop and absorption of the pulse amplitudes in the precapillary resistance vessels. The line at 25 mm Hg represents the plasma protein osmotic pressure (colloid-osmotic pressure). (By courtesy of Folkow.)

tion is, in a functional way, described by the following subdivision: windkessel vessels (main arteries), resistance vessels (arterioles are the most important, but capillary and postcapillary sections are also of significance), precapillary sphincters (determining the functioning capillary surface area), shunt vessels (e.g., arteriovenous anastomoses), capacitance vessels (veins), and exchange vessels (capillaries) (Folkow, 1960). Most of this terminology is illustrated in Fig. 6-6.

Arteries

The arteries serve as windkessels, or pressure tanks, during the ejection of blood from the heart, and the elastic tissue tends to recoil when stretched. This property enables it to store and release energy generated by the heart and convert an intermittent flow to a continuous flow. The resistance to flow in the large arteries (and veins) is very small. The velocity of flow is high, but the diameter of the vessels is large.

The expansion of the arterial wall during the ejection of blood causes a pressure wave that travels along the peripheral blood vessels at a speed of 5 to 9 m/sec (the velocity of the blood in the aorta at rest is 0.5 m/sec). The more elastic the arterial wall, the lower the speed of the pulse wave. In practical application, the frequency of this wave is counted as the pulse rate, since it can easily be felt over the radial or carotid arteries.

Arterioles

In arterioles the peripheral resistance is high, causing a marked pressure drop. The velocity is still high because the total cross section of the arterioles is not so large. The individual vessels are narrow, from 0.1 mm to 60 μ. In the wall are transversely oriented smooth-muscle fibers that can be activated by stimuli of their nerves, transmitter substance, or by other chemical factors. Figure 6-7, A, B, and C, shows schematically various degrees of constriction of an arteriole

Fig. 6-7   A schematic representation of the structural pattern of the capillary bed. The distribution of smooth muscle is indicated in the vessel wall; see text for discussion. AVA = arteriovenous anastomose. (Modified from Elias and Pauly, 1961.)

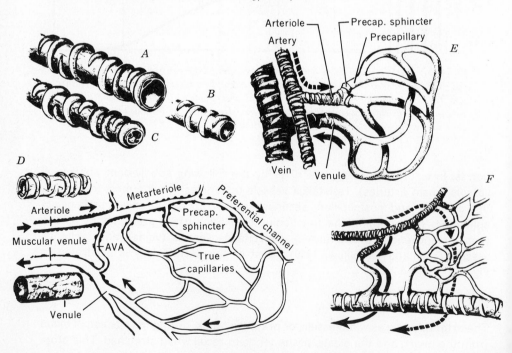

caused by contraction of the smooth-muscle fibers. It should be stressed that the vessels do not have muscles that can actively dilate them.

Figure 6-8 illustrates how the arterioles (and capillaries) are arranged in parallel-coupled circuits between the arteries and the veins. They can effectively alter the total resistance against the outflow of blood from the arteries and thereby the arterial blood pressure and work of the heart; they have a decisive

Fig. 6-8   Schematic drawing showing how the arterioles and capillaries are arranged in parallel-coupled circuits between the arteries (top) and the veins. The cardiac output may be increased fivefold when changing from rest to strenuous exercise. The figures indicate the relative distribution of the blood to the various organs at rest (lower scale) and during exercise (upper scale). During exercise the circulating blood is primarily diverted to the muscles. The area of the black squares is proportional to the minute volume of blood flow.

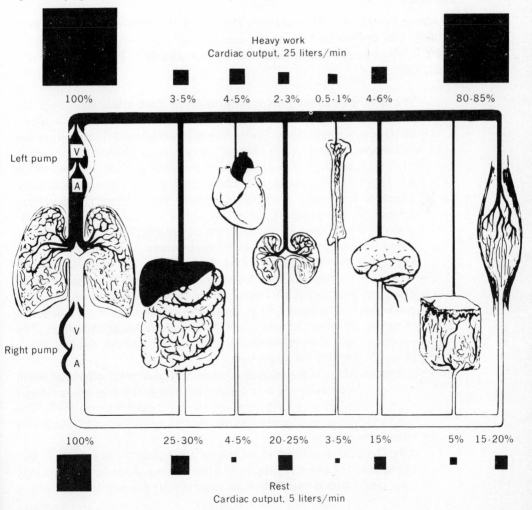

influence on the distribution of blood flow to the various organs. An increase in caliber of the arterioles in a muscle will decrease the resistance in that area and hence increase the flow. If this local vasodilatation is not compensated for by a vasoconstriction in another area or an increase in cardiac output, the arterial blood pressure will inevitably fall. The arterioles are arranged in a series of parallel channels joining the capillary bed (and veins) with the arterial side, effectively "regulating" the blood flow through the organs and tissues. The pressure is still high when the blood enters the arterioles, but then drops to 30 to 40 mm Hg (the pressure is somewhat higher in the kidneys). The systolic and diastolic pressure variations will usually disappear before the blood reaches the capillaries (Fig. 6-6).

Capillaries

The microarchitecture of the capillary network is different in different organs (Fig. 6-7). The precise pattern of the capillary arrangement is not completely known. It may be assumed to be as follows (see Zweifach, 1950, 1961; Lutz and Fulton, 1962):

1   The arterioles branch off into terminal or final arterioles. There the smooth-muscle fibers are arranged in circular or spiral alignment and constitute a single layer. The terminal arteriole ends in a capillary-like channel or "thoroughfare channel" (preferential channel), from which the capillaries branch (Fig. 6-7,D). In the proximal wall portions of the thoroughfare channel is a discontinuous coat of thin muscle cells. There is frequently a group of muscle fibers forming precapillary sphincters, especially at the origin of the capillary. The preferential channel finally loses its muscle coat completely, but can be differen-tiated from adjoining capillaries by the presence of a somewhat thicker supporting connective-tissue coat. The true capillaries are thin-walled endothelial tubes with no muscle or fibrous fibers. They are kept in place by fine connective-tissue fibers. In skeletal muscles the true capillaries are at least 8 to 10 times as numerous as the preferential channels. In other tissues there are relatively fewer capillaries. The capillaries and the preferential channels drain into the venules, which in turn drain into the larger veins.

2   Another pattern of the microcirculation may be a net of arterioles which divide into anastomosing true capillaries (thin endothelial tubes without muscle cells) (Fig. 6-7,E). Precapillary sphincters can be seen. The capillaries converge into venules. No vessels resembling a thoroughfare channel are found.

3   Arteriovenous anastomoses or shunts are typical for the vascular bed in the skin but are also present in other tissues. Anatomically this may be a vessel with a relatively thick muscle coat running from an artery

or arteriole to a vein or venule (Fig. 6-7,F). The diameter can, in the latter case, be as small as 15 to 20 μ. Similar shunts may be located proximally to the terminal arterioles or form a direct continuation from one of the branches of an arteriole to the venous circulation. In both cases, this shunt provides a good possibility of maintaining blood flow and low resistance, but the capillary vascular bed is bypassed (dark arrows in Fig. 6-7,F).

Summary   The circulation of blood from arteries to veins can take various routes via thoroughfare channels, true capillaries, and arteriovenous or arteriolovenous shunts. The pressure gradient and the activity of the smooth-muscle cells in the walls or the precapillary sphincters decide which route the blood flow will take.

Capillary Structure and Transport Mechanisms

Let us now take a closer microscopic and ultramicroscopic look at the architecture of the capillary. Its length is less than 1 mm (down to 0.4 mm) and its inner diameter ranges from 5 to 20 μ; therefore, there is sometimes hardly room for an erythrocyte (diameter 7.2 μ), which may temporarily stop the flow and be squeezed out of shape as it is forced through. The capillary wall is made of very thin endothelial cell plates of protein and lipid, with a rather homogeneous cytoplasm and long granular nucleus. The wall has no muscle cells or other contractable elements but has elastic properties.

On the other side of the endothelium is a fine mesh of connective tissue and a thin layer of a mucopolysaccharide substance ("basement membrane"). The space between the cells is filled with a cement, possibly a calcium proteinate. Earlier findings of the existence of pores or channels penetrating the capillary wall are questioned. They may, however, be too small to be detected by the electron microscope.

The most important processes for the exchange of substances across the capillary membrane are filtration and diffusion. Isotope studies have shown that gases, water, and molecules in water and lipid can rapidly diffuse back and forth in the direction of the concentration gradients through the capillary wall. Other materials can, however, be transported actively through the capillary wall against the gradient. Numerous indentations of the endothelial plates can be seen, as well as small vesicles ("ejection capsules") in the inside of the wall. A similar transporting process (pinocytosis) has been suggested for other cells.

Filtration and Osmosis

Figure 6-9 shows the driving forces in a process so well outlined by Starling in 1896. Within the capillary the hydrostatic pressure (h.p.) is normally well above the pressure in the interstitial fluid. As the plasma proteins cannot pass through

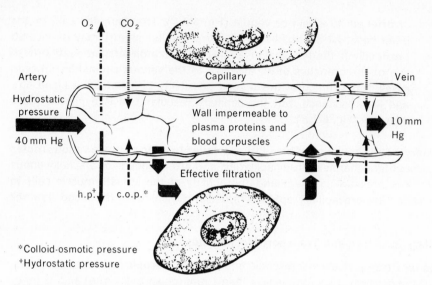

Fig. 6-9   Driving force in capillaries. For details, see text.

the capillary wall (which is a statement more didactic than true; within 1 to 2 days the entire quantity of plasma protein may be exchanged), they will exert a colloid-osmotic pressure much higher than that existing outside the capillary. For simplicity, the pressures outside the capillary are disregarded (or corrected for) in Fig. 6-9. The gradual change in colloid-osmotic pressure as water escapes or returns to the vessel is also omitted.

When entering the capillary, the net hydrostatic pressure may be 40 mm Hg, and this pressure drops gradually to 10 mm Hg on the venous side of the capillary. The net colloid-osmotic pressure is about 25 mm Hg (see Fig. 6-6). The net filtration pressure is then 40 − 25 mm Hg "proximally," giving an outflow of fluid, but it is 10 − 25 "distally," giving a pressure differential of 15 mm Hg. This means that fluid is sucked back with a force of 15 mm Hg. It is easy to see how small variations in pressures can profoundly change the exchange of fluid. Three examples are given: (1) If the arterial pressure falls, for instance during blood loss, there is also a reduced capillary pressure. Thereby the balance between the hydrostatic pressure and protein osmotic pressure becomes disturbed, with an increased absorption of fluid from the interstitial fluid as a consequence. (2) A decrease in concentration of plasma protein (prolonged starvation or loss due to kidney disease causing albuminuria) lowers the plasma protein osmotic pressure and the return of fluid from the tissue. An increase in the extravascular fluid follows (edema). (3) A rise of the venous pressure increases the mean capillary pressure and hence the outward passage of fluid. Furthermore, the reabsorption of fluid at the venous side of the capillary is decreased (it may be 15 − 25 = −10 mm Hg). It is easy to understand how a failure of the left ventricle of the heart can cause pulmonary edema. When the right ventricle eventually becomes incompensated by the increased load, the edema "moves" to the systemic circulation area. The liver and dependent parts of the body, which are most affected by the hydrostatic pressure, are sensitive to an elevation of pressure in the systemic veins.

It was mentioned that many capillaries are "too small" for the red cells. Consequently the red cell becomes temporarily squeezed in the bulging capillary, with no plasma separating the red cell from the protein-poor interstitial fluid. The red cell sucks up water, and the loss of oxygen and uptake of carbon dioxide are also enhanced. The deformation of the red cell, eventually down to a 3 μ diameter also promotes the exchange of various substances through its membrane. When returning to the vein or lung capillary, the red cell will release the excess water to the surrounding plasma (Hansen, 1961).

When metabolites are formed in active cells, the osmotic pressure outside the capillaries might temporarily increase, and water is withheld.

Diffusion–Filtration

The importance of a net filtration of fluid across the capillary membrane, as illustrated in Fig. 6-9, has probably been overemphasized in the past. It is now believed that diffusion accounts for the major exchange of fluid. It should be noted that diffusion can take any direction regardless of how filtration moves the fluid. It is important to keep in mind that these two processes are separate.

The transport of CO_2 from the tissue to the capillary starts immediately as the blood reaches the thin membrane tube and continues as long as there is a pressure gradient for CO_2. In the same way, oxygen diffuses from the capillary throughout its length, irrespective of the direction of the net water flow across the membrane.

Veins

The collecting venules have a supporting coat of connective tissue and irregularly spaced smooth-muscle cells (Fig. 6-7, E and F). In the distal part of the venules a well-defined muscle layer is developed. Also, the larger veins have muscles in their walls which can constrict the lumen of the vessel. Valves are frequent in the veins of the extremities of the body.

The veins within a skeletal muscle will be compressed mechanically during the muscular contraction. Blood is squeezed from the veins, and because of the higher resistance on the capillary side and the design of the valves, the blood return to the heart is facilitated by this "muscle pump." During muscular relaxation the venous pressure drops, and therefore blood may flow from the superficial veins, anastomosing the deep veins. The blood flow to a working muscle is higher during the period of relaxation than during contraction. The venous pressure has its minimum at the time of relaxation, which increases the arteriovenous pressure difference and thus the perfusion.

The beneficial effect of leg muscle activity on circulation when the person is in the erect position is illustrated by a drop in venous pressure of 60 mm Hg or more, which has been observed in the vein of the foot when the leg muscles are active. The vasomotor reflexes activated when changing position are discussed later in this chapter.

The resistance to flow in the larger veins is very small, with a pressure gradient between the foot and the groin of only 6 mm Hg in the horizontal position (Ochsner et al., 1951). Haddy et al. (1962) have analyzed the resistance to flow in the vessels of the limb. The fraction caused by veins larger than 0.5 mm was about 10 percent of the total under normal conditions.

Bazett (1950) has calculated that the resistance to flow in the vessels distal to capillaries (postcapillary resistance) was 0.26 "unit" as compared with 0.67 unit for the precapillary resistance. The venous resistance was thus calculated to be about 25 percent of the total. During exercise where cardiac output was assumed to increase three times, he calculated the two components of resistance to be 0.10 and 0.33 unit, respectively. The total resistance was lower than at rest, and the decrease in venous resistance may be due partly to the presence of a muscular pump. The venous fraction was still some 23 percent of the total resistance. These data are included to emphasize the venous side as an important factor in the circulation, not merely a transport channel.

It has been calculated (since direct measurements are presently impossible) that the venous systems normally contain some 65 to 70 percent of the total blood volume. The veins are therefore often referred to as capacitance vessels (Fig. 6-6). The pulmonary veins may account for about 15 percent of the total blood volume. (At rest the blood in the capillaries is estimated to be only 5 percent of the total; the spleen plays only a minor role in man as a blood or red-cell depot.)

Vascularization of Skeletal Muscles

Nerves and vessels enter the muscle at a neuromuscular hilus often located at half-length of the muscle. The artery enters the muscle substance and branches freely in its course along the perimysium (Fig. 6-10). By anastomoses, a primary arterial network is established. Finer arteries arise and create a secondary network infiltrating the muscle tissue. The smallest arteries and terminal arterioles branch off, usually transversely to the long axis of the muscle fibers and at fairly regular intervals of 1 mm. The arterioles then supply the capillary network oriented parallel to the individual muscle fibers, but also form frequent transverse linkages over or under the intervening fibers, thereby forming a delicate oblong mesh. The veins have valves (from a caliber of 40 μ or larger) directing the blood flow toward the heart, and follow the course of the arterioles and arteries. The capillary network at the motor endplate is especially well developed.

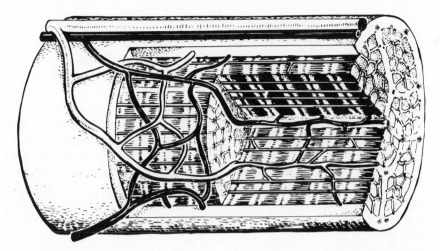

Fig. 6-10   Vascularization of skeletal muscle. Capillaries run parallel to the fibrils which make up the fiber.

The muscles have a rich capillary supply. In his classical studies Krogh (1929) calculated that there are up to 1,350 capillaries/mm² in the cross section of a horse gastrocnemius. In smaller animals, this figure may rise to 4,000 capillaries/mm². In a large man with a muscle mass of 50 kg and 2,000 capillaries/mm², the total length of the capillaries (each being less than 1 mm) would be 100,000 km (62,000 miles), or reach 2.5 times around the earth at the equator. The total surface area is estimated to be 6,300 m². When passing through the capillary bed, 1 mm³ of blood with its content of 5 million red cells comes in contact with 6,000 cm² of capillary surface. Krogh has also shown that the number of capillaries containing circulating blood varies. At rest, 100 capillaries/mm² of cross section of the skeletal muscle may be open, but during exercise the number may be 3,000 open capillaries/mm².

The capillary density in skeletal muscles in man is not exactly known. It is probably less than the 3,000/mm² cross section as calculated by Krogh (1929). There may be about 500 capillaries/mm² open during heavy muscular activity (Folkow, personal communication). The total surface area of the capillary bed is then about 250 m² in an individual with 35 kg of muscles. With an average capillary diameter of 8 μ the area of the capillaries would be 2.5 percent of the total tissue area; the distance from a capillary to the most remote cell would be about 20 to 25 μ if the capillaries are evenly distributed. The precapillary sphincters can close the inlet to the capillary, and at rest the number of open capillaries is markedly reduced. The opening or closing is supposed to be operated by local chemical factors of a hypoxic or metabolic nature, and only to a smaller degree by nerve activity (see below) (Kjellmer, 1965). In a given degree of metabolic activity in the tissue the number of working capillaries may be fairly constant, but the individual capillary is intermittently open or closed.

It is still uncertain which type of capillary system is predominant in the muscle (Fig. 6-7); whether it is the thoroughfare channel (preferential channel) system or a system in which capillaries form a parallel-coupled circuit between terminal arterioles and venules. There are probably no real arteriovenous or arteriolovenous shunts.

The abundance of capillaries in the muscles provides good facilities for the supplying of oxygen and nutritive materials to the cells "bathing" in the interstitial fluid, and for the removal of products from metabolic activity. Krogh (1929) calculated that with 100 open capillaries/mm^2, a gradient in oxygen tension of 12 mm Hg from a capillary to the most distant point would be sufficient during resting conditions to provide those distant parts with oxygen by diffusion. With all capillaries open a lower oxygen gradient would be enough to drive a diffusion; the mitochondria are functioning with an oxygen pressure of 1 mm Hg or even less.

REGULATION OF CIRCULATION AT REST

This could be a very confusing discussion, for the theories on "regulation of circulation" are almost equal to the number of physiologists working in the field. Moreover, there is no unanimous agreement on the terminology of *regulation* and *control*. From a demonstration of an effect on heart rate or vasomotor tone of a stimulation of some part of an animal, it may be tempting to construct a hypothesis concerning regulatory mechanisms. On the contrary, the noticed effect may represent an interference with or a disturbance of the normal function and by no means a reflection of a regulatory mechanism.

It should be remembered that by following nerve fibers one can travel from one cell to almost any other cell within an animal's body. The anatomic basis for a regulation, or interference, in the function of cells and organs is therefore definitely present. Hormones and other chemical substances (including anesthetic agents) can exert a purposeful effect on some cells but, by accident, disturb the function of others.

The evolution of Homo sapiens has passed through many stages. Nerve tracts and organs have probably changed their functions, but remnants of once very essential functions and mechanisms may still remain, normally suppressed or transiently noticeable only in specific situations. Similar obsolete mechanisms may eventually be revived under special conditions. If so, this is most likely to occur when points in the central nervous system or nerves are artificially stimulated, which may rouse centra or tracts from "sleep" or inhibition. The effect thus displayed cannot, however, be considered as a mechanism of regulation. In research, artificial situations and tools are essential in studying the cells, organs, and even the intact organism. The most critical and crucial part of a study is the interpretation and application of results obtained. With these difficulties and pitfalls in mind, we shall now discuss the regulation of circulation at rest and during exercise.

Any change in cellular activity should be met by a corresponding variation in local blood flow through the capillary bed. If an individual cell, in one way or another, could control its environment by varying the blood supply in balance with the actual nutritional demand, that cell would benefit. But other cells or tissues might suffer if some cells selfishly take more than their share, leaving others with little or nothing. Hence coordinative mechanisms are essential if the distribution of blood is to be balanced properly. An active regulatory mechanism ensures that more active and less active cells, as well as more susceptible and less susceptible organs, are supplied according to their need and to the capacity of the whole circulation.

Arterial Blood Pressure and Vasomotor Tone

The blood pressure in the aorta is maintained by an integration of the following factors: (1) cardiac output, (2) peripheral resistance, (3) elasticity of the main arteries, (4) viscosity of the blood, and (5) blood volume. Evidently a regulation of the arterial blood pressure can use factors 1 and 2, since tools for 3 to 5 are normally not at its disposal for rapid modifications.

The local blood flow is mainly determined by the pressure head and the diameter of the actual vessels. The smooth muscles of the arterioles and veins in many regions continuously receive nerve impulses that keep the lumen of the vessels more or less constricted. This vasomotor tone is provided by the sympathetic vasoconstrictor fibers driven from the vasomotor area in the medulla oblongata. The transmitter substance is norepinephrine. The heart and brain receive few vasomotor fibers; the supply to the abdominal organs (by splanchnic nerves) and skin is very rich. The muscles have an intermediate position. The vasomotor tone at rest can be demonstrated by section or blocking of the sympathetic nerve fibers in an animal. The arterioles dilate, and the arterial blood pressure falls. The effect of such an inhibition of vasomotor tone is very marked in the skin, but less pronounced in skeletal muscles. From a basal blood flow of 3 to 5 ml/(100 ml) (min) of muscle tissue, a doubling of the flow may occur. During exercise it can, however, amount to some 50 ml/min or more (Folkow, 1960). Folkow points out that the smooth muscles in some blood vessels may exhibit spontaneous rhythmic contractions creating a basal vascular tone. He suggests that intravascular pressure may be a stimulating factor. Some of the smooth muscle fibers, predominantly localized in the most narrow vessels, may actually serve as stretch receptors and pacemakers and thereby as triggers for the neighboring cells. Thus a wave of depolarization and contraction is propagated in the proximal direction (Folkow, 1967). The vaso- motor tone is important in keeping arterial blood pressure and cardiac output on an economical level. The splanchnic area could contain the whole blood volume after maximal dilatation of the vessels. Also, the vascular bed of skin and muscles has a similarly large capacity. Fainting (vasovagal syncope) may be the result of a central inhibition of the vasomotor efferent impulses.

The cardiac output can certainly not exceed the venous return. A constriction of the postcapillary vessels (the capacitance vessels) with their large content of blood will increase the blood flow toward the heart, making an increase in cardiac output possible. A decrease (inhibition) in the vasomotor tone, which actually causes a relaxation of the smooth muscles in the vessel wall and therefore a vasodilatation, can be obtained principally in three ways: (1) By an effect on the smooth muscles by chemical substances liberated locally from neighboring cells or delivered from the blood. More or less effective as dilating agents are hypoxia, lowered pH, an excess of CO_2 and lactic acid, adenosine compounds, an increase in extracellular potassium, P, or hyperosmolarity (heat can exert a similar effect on the skin vessel, but only to a small degree in the skeletal muscles). It should be noted that smooth muscles in the precapillary vessels in the skeletal muscles are effectively relaxed by metabolites. A constriction of veins induced by sympathetic nerve activity is, on the other hand, well maintained even at extensive metabolite accumulation (Kjellmer, 1965; Folkow, 1967). Vasodilatation (2) by a decreased discharge in the sympathetic vasomotor nerves and (3) by liberation of acetylcholine from the nerve endings of active sympathetic vasodilator fibers (cholinergic effect which is blocked by atropine). The final common path of those nerve fibers starts in the lateral spinal horns, i.e., they have from here on a similar anatomy to the sympathetic nerves. However, these sympathetic cholinergic vasodilator fibers are distributed only to the vessels of the skeletal muscles and, possibly, also to the coronary vessel (Folkow, 1960).

Summary   The peripheral resistance to blood flow is determined by the vasomotor tone. The degree of contraction of the smooth muscles in the arterioles is of special importance for the local blood flow as well as the total resistance. In some tissues (e.g., muscles) this tone is probably partly spontaneous, the smooth muscle contractions (myogenic activity) being triggered by mechanical stretch induced by the intravascular pressure, but a sympathetic vasomotor tone is superimposed. In other regions (e.g., skin) this sympathetic vasomotor tone is definitely dominating. A vasodilatation can occur after a local inhibition of the sympathetic effect or the myogenic activity (by metabolites, heat, activity in the sympathetic vasodilator fibers) or by central inhibition of the sympathetic impulse traffic.

The Heart and the Effect of Nerve Impulses

The heart has its own pacemaker, initiating about 70 impulses/min if left alone. Both sympathetic and parasympathetic nerve impulses can modify the heart rate. The parasympathetic activity from a cardioinhibitory center via the vagus nerve (and acetylcholine) causes a slowing of the heart rate (bradycardia), and the sympathetic cardiac nerves (norepinephrine) can produce an increased heart rate (tachycardia). In a resting subject, a blocking of the vagus nerve will cause the heart to beat faster, indicating a predominating parasympathetic tone. It is not presently known whether the very slow heart rate, characterizing athletes trained for endurance, is affected by a further increase in vagus tone. The sympathetic nerves can increase the contractile force of the heart muscle fibers, but sympathetic nervous control of the vasomotor tone in the heart vessels is probably insignificant. Sympathetic vasodilator fibers may, however, operate within the heart muscle. The blood vessels of the heart dilate willingly when affected by hypoxia and metabolites.

Control and Effects Exerted by the Central Nervous System (CNS)

As we have now analyzed the tools available for a variation and redistribution of blood flow within the body, the actual control and regulating mechanisms may be discussed.

Important and essential nuclei are located in the medulla oblongata. The *medullary vasomotor area* can be divided into a *vasoconstrictor center* and a *vasodepressor area*, the latter operating through inhibition of the sympathetic vasoconstrictor outflow (Peiss, 1962). Efferent nerve fibers extend to sympathetic connector cells in the lateral horns of the spinal cord (thoracic region and first two segments of the lumbar region). The neurogenic vasomotor tone of the blood vessels originates essentially in the vasomotor area, and a continuous, somewhat rhythmical discharge can be detected in the nerve cells. This discharge is probably caused by the influence of the chemical composition of the interstitial fluid that bathes the cells. The spontaneous sympathetic vasomotor activity can then be modified by impulses from the vasodepressor area (inhibitory) or from higher levels of the CNS (inhibitory or facilitating) (Fig. 6-11). Whether the inhibition operates directly on the tonically active cell bodies of the vasoconstrictor center or is accomplished at the preganglionic cell bodies in the spinal cord is not known.

The vasodepressor area is essentially a relay station without spontaneous activity, but it is activated by afferent impulses, especially from the baroreceptors in the aortic and carotid bodies.

On higher levels of the CNS there are areas, especially in the cerebral cortex and diencephalon, from which cardiovascular reactions can be elicited. Although these higher centers do not contribute to the continuous vasomotor tone, many adjustments are initiated primarily from the brain above the level of the medullary centers. Of special interest are the *sympathetic cholinergic vasodilator fibers* (Folkow, 1960, 1967; Uvnäs, 1960). They can be traced from the motor cortex and followed through the anterior hypothalamus and mesencephalon (relay stations). The nerve fibers bypass the vasomotor center and run to the lateral spinal horns and the sympathetic final common path. When stimulated, they can be activated in synergism with vasoconstrictor fibers. The combined effect is then a vasodilatation of precapillary resistance vessels in

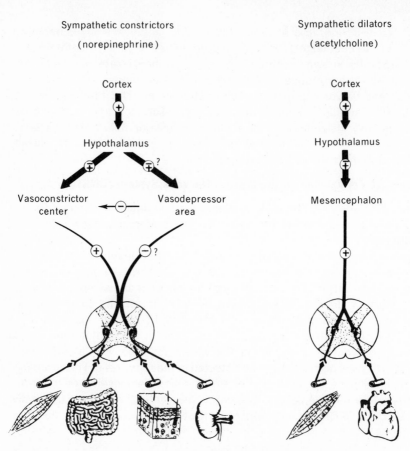

Fig. 6-11 *Central vasomotor integration. The inhibition on the vasoconstrictor activity exerted from the vasodepressor area may operate directly on the center or eventually at the spinal level. The vessels in skeletal muscles and the gastrointestinal tract are, at rest, easier to involve in a constrictor activity ("low-threshold areas") than are the cutaneous and kidney vessels (Folkow, 1967). For detailed description, see text.*

the skeletal muscles and vasoconstriction of the vessels of the abdominal organs and skin. "With very little shift in arterial pressure this automatic activation pattern leads to a remarkable and instantaneous redistribution of cardiac output to favor skeletal muscles" (Folkow, 1960). Simultaneously, accelerator fibers to the heart may also be stimulated, and the medulla of the suprarenals may give rise to a liberation of epinephrine. This hormone dilates the resistance vessels of the skeletal muscles and excites the smooth muscles of the capacitance vessels; norepinephrine strongly contracts both resistance and capacitance vessels (Folkow, 1960).

Similar reaction patterns are characteristic for emergency conditions, such as fear, and can be elicited by electrical stimulation of the hypothalamus (Abrahams et al., 1960; Folkow, 1960; Uvnäs, 1960) and even by cutaneous stimulation (Abrahams et al., 1960).

Rushmer (1965) found in his experiments on unanesthetized dogs that the initial cardiovascular response to treadmill exercise could be precisely duplicated by electrical stimulation in the diencephalon, and also by focusing the attention of the trained dog on the switch that turns on the motor-driven treadmill. The further adjustments and regulation of the cardiovascular system during exercise will be discussed below.

Figure 6-11 describes the central vasomotor integration in a schematic way (Peiss, 1962).

Mechanoreceptors in Systemic Arteries

The afferent input is far from being completely mapped out. Important afferent fibers come from mechanoreceptors in blood vessels and in the heart. The systemic arterial receptors are located in the tissue of the carotid sinus, aortic arch, right subclavian artery, and common carotid artery (Heymans and Neil, 1958; Neil, 1960). Mechanical deformation of the walls of the vessels is the normal stimulus of the receptors, and they respond to the rate of rise of blood pressure as well as amplitude of pulse pressure.

The receptors are actually stretch receptors, and pressure as such is not the adequate stimulus. (The commonly used name "baroreceptors" is therefore somewhat misleading.) However, an increase in the intravascular pressure does expand the vessel wall and stretch the receptors. They respond with a discharge transmitted to the CNS. If the wall of the vessel, where the stretch receptors are located, becomes less distensible, due to increased activity of smooth muscles in the wall or a progressive structural change (in a hypertensive patient or aging person), a given pressure would induce less deformation of the receptors and a reduced impulse output (Peterson, 1967).

Pulsatile pressure about a given mean blood pressure is more effective than a steady mean pressure to set up an impulse traffic in the afferent nerves (sinus branch of the glossopharyngeal nerve and afferent fibers of the vagus). The threshold at which a stimulus is effective varies for the different receptors. A recording of activity in the sinus nerve normally reveals a continuous discharge and a variation in the impulse traffic with each pulse beat.

The baroreceptors can report a fall as well as a rise in blood pressure to the cardiovascular centers, primarily the medullary vasomotor area. At rest, the baroreceptors exert a restraining influence on the cardiovascular system,

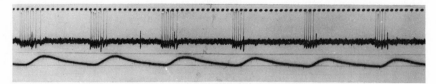

Fig. 6-12   Action potentials from mechanoreceptors in the carotid sinus recorded from the sinus nerve in the cat. Dots at the top give the time, 50 cps; the bottom line shows the blood pressure recorded in the carotid artery with calibration lines at 100 and 150 mm Hg respectively. Note that on the left side, when pressure is high, there are 9 to 10 large "spikes" per heartbeat. When the pressure is lower, there are only 5 spikes. (The spikes occur at the pressure rise, but there was some delay in the pressure recording.)

causing a reflex bradycardia and reflex inhibition of the medullary vasomotor center (Fig. 6-12).

Posture

The physiological interplay of the various factors involved in the maintenance of an adequate arterial blood pressure can be elucidated by the following experiment:

On a tilting table, a subject is tilted from supine to a head-up position (about a 60° angle to the horizontal). Due to the force of gravity, blood is pooled in the parts of the body below the heart level. Thus the venous return to the heart is temporarily reduced. Consequently, the cardiac output decreases and so does the arterial blood pressure. The strain exerted on the baroreceptors is reduced, and fewer impulses are transmitted from them to the CNS. The impulse output from the parasympathetic cardioinhibitory center is diminished (which results in an increase in heart rate); the vasodepressor area becomes inhibited, and from the adrenergic sympathetic centers there is an increased impulse traffic (the effect is a vasoconstriction in resistance vessels and capacitance vessels, especially in the splanchnic area, and an increase in heart rate). Thus, the peripheral resistance becomes higher, the cardiac output can be restored to an adequate level, and the arterial blood pressure can increase. The variation in heart rate in a subject passively tilted to different body positions is illustrated by Fig. 6-13 (Asmussen et al., 1939). If blood pressure cuffs are placed around the upper parts of the thighs and a pressure of about 200 mm Hg is applied when the subject is in a horizontal or head-down position, the heart rate response to the head-up position is less pronounced (Fig. 6-13). The hydrostatic forces are acting on a shorter "column," as the blood is prevented from circulating to the legs. If the pressure within the cuffs is suddenly released, the fall in arterial blood pressure may be very pronounced, and eventually the subject faints as the blood, and thereby the oxygen supply to the brain, becomes inadequate. The capacity of the legs to retain blood has increased, for its arterioles

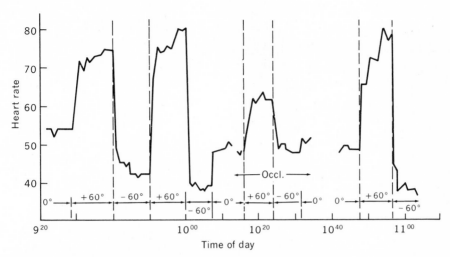

Fig. 6-13   Variation in heart rate (ordinate) in a subject passively tilted to different body positions. +60° = tilting to a head-up position; −60° = head-down position; Occl. = experiments in which inflated blood pressure cuffs were placed around the upper parts of the thighs. For further details, see text. (From Asmussen et al., 1939.)

and capillaries dilate as anaerobic metabolites accumulate during the period of circulatory occlusion. Tilting of the subject to a head-down position will quickly restore the circulation and consciousness as the legs are drained. Figure 6-14 shows how the heart rate can be lowered some 10 beats/min if the subject in a head-up position (on a tilting table) contracts his leg muscles. The massaging effect of the repeatedly contracting muscles on the capillaries and veins enhances the venous return to the heart, and the heart rate is lowered. Most likely the nerves from the cardiovascular mechanoreceptors form an important link in the reflex chain.

The beneficial effect of the muscle pump on the venous return should certainly be stressed for people who work in a fixed sitting or standing position. Bandaged legs can partly reduce the hydrostatic shift of fluid to the legs in the upright position, and thereby the circulation is facilitated (Lundgren, 1946; Arenander, 1960). A sudden standstill after prolonged exercise, particularly in a hot environment, can cause fainting, as the blood pools in the dilated vessels in exercised legs and in the skin. The unexpected fall of the tall soldier during a military parade can also be explained by an inappropriate distribution of blood.

There are two important factors counteracting an edema formation in the legs in erect position: (1) A pressure-induced facilitation of the rate of myogenic contractions of the smooth muscles of the arterioles and precapillary sphincters, decreasing the surface area of the capillary bed available for filtration; (2) an effective reabsorption of fluid via the red cells squeezed in the narrow capil-

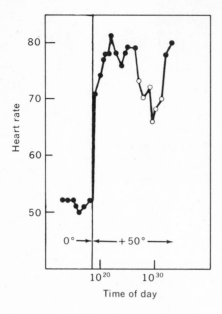

Fig. 6-14   Effect of voluntary contraction of leg muscles on heart rate during passive standing. ● = *passive standing;* ○ = *voluntary contraction of leg muscles.   (From Asmussen et al., 1939.)*

laries. The close contact between red cells and tissue fluid will promote an uptake of water in the capillaries as well as the exchange of gases, as discussed earlier in this chapter.

Summary   A change in body positions will inevitably affect the circulation as long as the individual stays under the influence of the pull of gravity. A head-up position will primarily increase the blood volume in the legs and decrease the central blood volume and cardiac output. Secondary variations in arterial blood pressure are reported from the mechanoreceptors in some arteries to inform the cardiovascular centers in the brain. The activity in sympathetic and parasympathetic nerves varies by reciprocal innervation in such a way that the arterial blood pressure and cardiac output return to a level fairly close to the one typical for the individual in a supine position. The nerves from the mechanoreceptors were appropriately called "buffer nerves" by Samson Wright.

Other Receptors

In the walls of the pulmonary artery there are *mechanoreceptors* with reflex effects on the systemic circulation and heart similar to those caused by the systemic arterial baroreceptors. A third group of mechanoreceptors are located in the walls of the atria and ventricles of the heart. When stimulated, they cause by reflex a vasodilatation, bradycardia, and systemic hypotension, so that the load on the heart diminishes (Neil, 1960). In the heart, at least in the left atrium, are stretch receptors that serve as one sensory mechanism in a reflex regulation of blood volume by control of urine output (Henry and

Pearce, 1956). The filling volume of the cardiac atria, related to the circulating or thoracic blood volume, is likely to be the appropriate stimulus; a variation in the production of antidiuretic hormone from the hypophysis is likely to be the tool.

The *chemoreceptors* in the carotid and aortic bodies, stimulated by low oxygen tension in the circulating blood, not only influence the pulmonary ventilation, but also indirectly the circulation. In artificially ventilated dogs, stimulus of the chemoreceptors by hypoxia causes an increase in sympathetic vasoconstrictor discharge, a bradycardia, and decrease in cardiac output (Daly and Scott, 1958; Daly, 1964). In a spontaneously breathing animal, however, a hypoxia causes a tachycardia, probably secondary to the hyperventilation. Consequently, it can be concluded that the tachycardia of systemic hypoxia does not have a chemoreceptor reflex origin.

In this context it is of interest that one of the primary *adjustments which takes place during a dive* is circulatory in nature (Scholander et al., 1962; Irving, 1963). Most animals, including man, display a diving bradycardia, which usually develops gradually. Thus during the dive the frequency may drop to one-half of normal in some species and to one-tenth in others. As a rule the bradycardia stays with the animal whether it exercises or not during the dive. Blood pressure is maintained at a normal or even an elevated level as a result of peripheral vasoconstriction. In the diving animal the reduction and selective redistribution of the circulation can save oxygen for vital organs such as the brain and heart. In man the heart rate increases immediately at the start of a dive and slows down very markedly as the dive progresses, even if the diver exercises vigorously (Scholander et al., 1962; Olsen et al., 1962; Irving, 1963). Prompt return to normal sinus rhythm occurs with the first breath. (Some fascinating aspects of the physiological response to diving in ducks are discussed by Folkow, 1968.)

This bradycardia during diving is not a result of apnea only, since it is reinforced by water submersion. Skin receptors, wetting of the nose, and asphyxial reflexes may contribute to the development of the bradycardia. The onset of the decline in heart rate is very rapid. However, it may be that the chemoreceptors gradually contribute to the bradycardia, by a progressive hypoxic drive, since the drop in alveolar oxygen tension is very marked, at least during exercise combined with breath-holding (P.-O. Åstrand, 1960). In any case, there is a striking similarity in the circulatory response of isolated chemoreceptors in dogs to a stimulus of hypoxia and the effect induced by diving on the circulation in the intact animal.

Phylogenetically, the chemoreceptors may be a very ancient receptor mechanism of importance for diving animals, but they received a new function when the animal left the water for good.

Afferent impulses to the cardiovascular centers from receptors in the skin, muscles, and joints have been suggested but have not as yet been shown to exist.

Any damage to superficial tissues causes a regional vasodilatation response. The afferent fibers of pain receptors branch, and impulses pass along those branches to arterioles, causing a dilatation (antidromic activation or *axon reflex*).

REGULATION OF CIRCULATION DURING EXERCISE

The conclusions of the preceding discussions are as follows: The blood flow in the precapillary vessels of the muscles is "regulated" locally, for the most part, by the level of metabolism in the muscle cells. The actual nutritional demand is the determining factor and will be superimposed on any nervous influence; the vessels then become functionally sympathectomized. The blood flow in the splanchnic area is, on the other hand, under control of the CNS. The flow is varied so that the systemic arterial blood pressure is maintained at an adequate level to supply brain, heart, and some other vital organs with blood. In this case, neural vasoconstrictor activity can be superimposed upon the local dilator control. A stimulation of the vasoconstrictor fibers can increase the resistance in the vessels of the gastrointestinal tract about 20 times (Folkow, 1960). The vessels of the skin also subserve centrally controlled mechanisms. Impulses from the vasomotor area are important, but the final control is probably exerted by the temperature-regulating centers, at least at rest or during submaximal exercise.

At rest the kidneys receive about 25 percent of the cardiac output. There are no tonic impulses from the CNS to renal blood vessels, but electrical stimulation of the renal nerves causes intense renal vessel constriction with associated changes in blood flow (down to 250 ml/min) and in excretion of water and electrolytes (Pappenheimer, 1960). Exercise, postural changes, and circulatory stress in general may cause profound alterations in renal function, mediated through the hemodynamic effects of renal nerves.

This summary gives the main tools available for the circulatory system during exercise. The approximate blood distribution to the various organs is illustrated by Fig. 6-8.

The changes in heart function and circulation, from the moment muscular exercise begins (or even before), are initiated from the brain levels above the medullary centers (probably the cerebral cortex and diencephalon). By a reciprocal innervation there is a simultaneous increase in the sympathetic activity and decrease in the parasympathetic impulse traffic. The skeletal muscles may receive an increased share of the cardiac output because of their innervation with sympathetic cholinergic vasodilator fibers. The activity in the sympathetic adrenergic vasoconstrictor fibers reduces the blood flow to skin and splanchnic area (Fig. 6-15). The balance in the redistribution of blood flow can be such that there is no (or only a mild) decrease in the systolic blood pressure during the initial period of exercise, despite a marked dilatation of the resistance vessels in the muscles (Holmgren, 1956). Constriction of the veins (capacitance vessels), pumping action of the working muscles, and forced respiratory movements assist the venous return to the heart. Since the heart escapes from the vagal inhibition, and sympathetic impulses can increase the force and frequency of the muscle contractions, the heart gains capacity to take care of the increased inflow of blood, and if necessary, pumps it out against an elevated resistance. The volume of blood in the heart and lungs actually increases, at least during exercise in the supine position (Mitchell et al., 1958a; Braunwald and Kelly, 1960).

In the working muscles the effect of the increased metabolism is a local change in pH and in composition of the interstitial fluid, causing an opening of

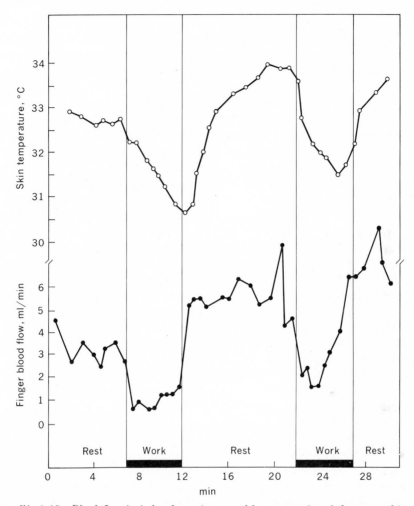

Fig 6-15 Blood flow in index finger (measured by means of a plethysmographic method) and skin temperature on the same finger at rest and during two periods of work on a bicycle ergometer (1080 kpm/min, or 180 watts). (From Christensen and Nielsen, 1942.)

capillaries and those arterioles not already dilated by the sympathetic vasodilator activity. For several reasons this local control of blood flow is the most important factor in securing an efficient blood supply to the working muscles. It appears that the mechanisms controlling distribution of open capillaries in the muscle are more responsive to muscular exercise than to hypoxia (Otis, 1963). This difference in response can be explained, since during exercise the potassium ions released from the intracellular space reach such a high extracellular concentration that they can account for a major part of the vascular dilatation accompanying muscular activity (Kjellmer, 1965).

Furthermore, exercise leads to hyperosmolarity by release of particles (metabolites) from the striated muscle fibers into interstitial fluid space. This change in the environment of the smooth muscles of the blood vessels may inhibit vascular tone (Mellander and Johansson, 1968).

Skinner and Powell (1967) point out that the concentration of oxygen in the tissue can substantially influence the reaction of the vascular bed to potassium. Oxygen-deficient blood strengthens the vasodilating effect of potassium at all levels of potassium concentration in the tissue.

An eventual neurogenic vasodilatation in skeletal muscles affects the resistance vessels almost exclusively, but it is not restricted to working muscles.

The capacitance vessels are rather sensitive to constrictive influence. Actually, at a given low impulse traffic in the sympathetic vasoconstrictor fibers, the constrictor response from the capacitance vessels is relatively much more pronounced than from the resistance vessels (Mellander, 1960; Kjellmer, 1965; Folkow, 1967).

In experiments on man Bevegård and Shepherd (1967) have observed an increase in tension of the venous walls in both exercising and nonexercising limbs. This increase in tension persists throughout the exercise and is proportional to the severity of the work. They emphasize that this venoconstriction is not released by metabolites produced in the working muscles.

At least in experiments on animals under various conditions, the activity in vasodilator fibers is of short duration (Bevegård and Shepherd, 1967). It makes physiological sense to interpret the initial vasodilatation in muscles as a state of general preparedness; arterioles and thoroughfare channels are flooded by a rapid bloodstream just at the doors of the capillaries. In the metabolically active areas the capillary sphincters open, and the nutritive vessels can immediately be perfused by blood (part of an emergency reaction). After perhaps some 10 sec, the sympathetic vasodilator activity ceases. In the active muscles the vessels remain open in proportion to the metabolic rate, but in the resting muscles the arterioles constrict by the now-dominating activity in sympathetic vasoconstrictor fibers. Whether the vasodilator fibers fire effectively when rate of work is varied, new muscle groups are engaged, etc., is not known. The metabolic control of the blood flow in muscles is illustrated by the finding that the oxygen content of venous blood from a resting extremity falls as low as that of one which is exercising (Donald et al., 1954; Mitchell et al., 1958b; Carlson and Pernow, 1959).

It should be recalled that epinephrine, liberated from the activated suprarenal medulla, excites the smooth muscles of the capacitance vessels, but dilates the resistance vessels of skeletal muscles. Norepinephrine from the sympathetic vasoconstrictor fibers contracts muscles of both capacitance and resistance vessels (the threshold being different).

As the exercise proceeds, the blood vessels of the skin, especially the arteriovenous anastomoses, dilate, so that the produced heat can be transported to the surface of the body. The heavier the exercise and the higher the environmental temperature, the more pronounced is this secondary vasodilatation in the skin. Indirectly impulses in sympathetic nerve fibers are partly behind this dilatation, and the temperature-regulating center in the hypothalamus is guiding the impulse traffic. These nerve fibers stimulate the sweat glands to sweat production by acetylcholine, but the glands also deliver an enzyme acting on proteins in the tissue fluid, and a substance with vasodilator effect is formed, identical with or related to bradykinin (Barcroft, 1960). The local skin temperature also affects the lumen of the vessels. The dilatation of the vascular bed in the skin is apparently not combined with a further increase in cardiac output. In a few subjects studied by Williams et al. (1962), the cardiac output was essentially the same when they performed exercise in hot and cold environment.

During work, the integrated effect of neural and chemical factors (including hormones) gives a cardiac output that may be markedly higher and with quite a different distribution than at rest (Fig. 6-8).

Most investigators report an elevated *arterial systolic* and *pulse pressure* during exercise (Fig. 6-22), and the effect of such increased pressure on the mechanoreceptors in the arterial walls has been discussed. There are, however, important aspects to consider when interpreting blood pressure curves and the effects of the pressure on mechanoreceptors.

A simultaneous measurement of intra-arterial blood pressure in a peripheral artery and in the aorta during exercise gives a significantly higher systolic end pressure in the peripheral artery, but the mean and diastolic pressures are about the same as in the aorta (open-ended catheter directed against the flow axis) (Kreeker and Wood, 1955; Holmgren, 1956; Marx et al., 1967). The systolic blood pressure in a peripheral artery is higher in a resting than in a working limb (P.-O. Åstrand et al., 1965). The progressive increase in systolic pressure (and pulse pressure) along an artery is at least in part due to a distortion in the transmission because of summation of the centrifuge wave and the reflected waves from the periphery. The importance of the wave reflection increases when the peripheral resistance is high, as is the case in a resting limb.

Marx et al. (1967) report that the pressure measured in the aorta with a catheter with side holes gives a significantly lower pressure during exercise than does measurement with an open-ended catheter directed upstream. The reader is reminded of the discussion earlier in this chapter. The total energy of the blood is the sum of the kinetic energy and the pressure energy. The side-hole catheter is only measuring the pressure energy (side pressure), but the catheter with the opening directed upstream also includes the kinetic energy. Since the kinetic energy factor of the blood in the aorta is high during exercise with a pronounced increase in cardiac output, the two catheters should give quite different pressure readings. Marx et al. have convincingly shown that this is the case.

It can be calculated (1) that the blood pressure measured in a peripheral artery during exercise does not give a true picture of the *systolic* blood pressure in the aorta; (2) a measurement in the aorta against the flow (end pressure) gives an evaluation of the pressure-load on the heart, but (3) it does not reflect the side pressure and strain on the vessel walls. The mechanoreceptors cannot sense the kinetic energy factor of the passing blood. Lateral distending pressure should therefore be used for an evaluation of the strain on the mechanoreceptors (data on mean pressure are of no value in this connection).

Even if the pressure variation in the walls of the aorta and the carotid sinus during heavy exercise has been overestimated in the past, it is still an irrefutable fact that the blood pressure variations synchronously with heart rate occur more rapidly and frequently during exercise than at rest. If the stiffness of the vessel walls were the same in the two conditions (which may not be the case), the mechanoreceptors must be far more activated in the exercising individual. (With everything else being constant a heart rate of 180 must give *at least* three times more activity in the mechanoreceptors per unit of time compared to resting conditions with a heart rate of 60.)

As emphasized above, the mechanoreceptors are lying in the vessel walls and they convert distortion of the wall into nerve impulses. A sympathetic vasoconstrictor activity involving the smooth muscles of the vessel wall makes it stiffer. Under such conditions a given intravascular pulsatile pressure gives less deformation of the wall than when the smooth muscles are less active; therefore the discharges from the mechanoreceptors are reduced (Peterson, 1967).

At least during submaximal exercise, the mechanoreceptor mechanism continues to oppose the rise in heart rate and blood pressure through a negative feedback. By creating subatmospheric pressures in a Plexiglas box enclosing the neck, researchers have found that blood pressure, heart rate, and cardiac output were significantly reduced at rest as well as during exercise (with cardiac output elevated up to 17 liters/min) (Bevegård and Shepherd, 1964, 1967).

If the *cardiac output* $(\dot{Q})$ during exercise is four times the resting value, giving a 25 percent increase in arterial mean pressure (P_{mean}), it follows that the resistance to flow (R) is reduced to more than one-third of what it was at rest, since $\dot{Q} \times R = P_{\text{mean}}$. The lowered resistance is, however, due to the dilated vascular bed in the working muscle, and the vessels in the abdomen are still obeying signals in the sympathetic vasoconstrictor fibers (Wade et al., 1956). It should be recalled that if nonmuscular tissues receive 80 percent of the cardiac output at rest but only 20 percent during heavy exercise, their blood flow would in both cases be some 4 liters/min. The distribution of this flow of 4 liters/min is probably different at rest than during exercise (Fig. 6-8). Thus, the actual flow through the splanchnic bed is reported to diminish during exercise (Chapman et al., 1948; Bucht et al., 1953; Wade et al., 1956; Bishop et al., 1957; Rowell et al., 1964), but the cerebral blood flow is increased (Scheinberg et al., 1954), as is the skin circulation during prolonged muscular activity.

The linear increase in heart rate and, within a wide range, cardiac output with the increase in oxygen uptake suggests a "neat" regulatory mechanism with the metabolic activity as an important guide. Whether or not the heart muscle has a capacity to pump out more blood than it actually does during the most severe exercise but is prevented from doing so by an inhibition elicited from mechanoreceptors, for instance, is still an open question.

The importance of impulses from higher levels of the CNS to the first circulatory adjustments to exercise is emphasized by the experimental findings that some of the effects can be elicited even before the exercise starts (Rushmer, 1961). On the other hand, for the muscular blood flow the neural control of the vessels of the muscles prob-ably has minor practical importance. Barcroft and Swan (1953) have shown that the hyperemia which follows standard exercise is essentially the same in a normal and in a sympathectomized muscle. This fits with the previous statement that local factors are superimposed on any nervous influence on the vascular bed of the muscles.

The cardiac nerves are not of essential importance for the ability to perform heavy exercise. Donald et al. (1964) found that dogs with chronic cardiac denervation showed an almost unchanged capacity for work as measured by oxygen uptake or time for a race over a given distance.

Summary   We can schematically describe the circulatory response to exercise by considering four stages:

1   At rest the skeletal muscles receive only some 15 percent of the minute blood flow, and their arterioles are constricted by a continuous vaso-constrictor activity and some sort of a spontaneous vascular tone. Few capillaries are open, but the individual capillaries open and close alter-natively. The heart rate is kept down by a parasympathetic outflow via the vagal nerve.

2   When or even before exercise begins, there is an inhibition of the parasympathetic activity and an increased sympathetic impulse traffic. The heart escapes from its inhibition and beats faster and with in-creased force. Impulses from the higher levels of the CNS, transmitted by sympathetic cholinergic vasodilator fibers, dilate arterioles in the muscles thereby increasing their blood flow. On the other hand, sym-pathetic adrenergic vasoconstrictor fibers act on the vessels of abdomi-nal organs and skin so that a decreasing share of the cardiac output flows through those tissues. The veins become constricted by activity in the constrictor fibers. This constriction of veins together with the pumping action of the working muscles and the forced respiratory movements facilitate the blood return to the heart, making an increased cardiac output possible.

3   The appropriate adjustment of the circulation occurs. In the working muscles the increased metabolism and potassium ions cause changes in the environment which locally dilate arterioles and open capillaries. The sympathetic vasodilator fibers are probably inactive or without effect, and in the resting muscles the arterioles constrict. Hormones contribute in the constriction of vessels in nonactive areas.

4   For temperature balance within the body, the produced heat is trans-ported to the skin, since skin vessels become dilated.

The resistance vessels, particularly the precapillary sphincters in muscles, are dominated by local vasodilative factors, while the capacitance vessels are more sensitive to the constrictive influence. Therefore, the

blood flow to the muscle and the distribution of blood within it is determined by the metabolic requirements, and pooling of blood in the active
muscle is prevented by nervous activity (Kjellmer, 1965).

The raised capillary pressure leads to a net outward filtration of fluid, the
flow of which is facilitated by a simultaneous increase of the capillary area. The
capillary permeability does not change during exercise (Kjellmer, 1965).

CARDIAC OUTPUT AND TRANSPORTATION OF OXYGEN

Efficiency of the Heart

Studies on cardiac output and the energy output and work efficiency of the heart have
shown that (1) a given stroke volume can be ejected with a minimum of myocardial shortening if the contraction starts at a larger volume; (2) energy losses in the form of friction
and tension developed within the heart wall are also at a minimum in a dilated heart;
(3) the stretched muscle fiber can, within limits, provide a higher tension than the un-
stretched one; (4) loss of energy is larger when the contraction occurs rapidly, i.e., with
heart rate high as compared with a slower contraction time; (5) on the other hand, the
greater the volume of the heart, the higher is the tension of the myocardial fibers necessary to sustain a particular intraventricular pressure (as a consequence of Laplace's
law, as discussed earlier in this chapter). The energy need for a contraction is closely
related to the tension that has to be developed.
     There are a few reports on the energy output and mechanical efficiency of the
heart muscle when heart rate and stroke volume are altered. In the isolated heart it has
been found that an increase in cardiac output caused by an increased stroke volume
markedly improved the efficiency of the heart work (blood pressure and heart rate were
kept constant) (Evans, 1918). This means that the increase in oxygen uptake was not
proportional to the increased performance.
     If the cardiac output was maintained unaltered, variations in heart rate were
followed by an almost linear variation in oxygen uptake of the heart; the high heart rate
was associated with a low mechanical efficiency.
     The *individual with a high capacity for oxygen transport* because of natural endow-
ment and/or training is characterized by a large stroke volume and a slow heart rate.
Cardiac output and systolic/diastolic blood pressures at given work loads or at rest are
not noticeably different from normal. Of the factors listed above, the first four tend to
act in the individual's favor as far as each single heart beat is concerned, while the fifth
factor acts against him. However, as relatively few contractions are performed per
minute, the total energy cost to maintain a given work level (flow × pressure) may be
relatively low and the efficiency high. At maximal work levels, the person with the large
diastolic heart volume and stroke volume may have as high a heart rate as the individual
with a small diastolic filling. In that situation, factors 1 to 4 are still favoring the fit man, but
factor 5 tends to decrease the efficiency when performance per unit time is considered.
     It should be borne in mind that at a given maximal heart rate, the heart which
has the capacity to provide the largest stroke volume can attain the highest cardiac

output. One condition is a good diastolic filling, which inevitably leads to the drawback of increased tension required to produce the pressure.

A rich capillary network in the myocardium would provide the capacity to meet the demand of an increased metabolism. For the exercising skeletal muscles, however, it is essential that the increased demand for blood flow through the myocardium does not consume too many of the extra liters of blood that the heart eventually can manage to pump out. No data are available which would permit an analysis of this problem.

From a general viewpoint it is considered an advantage if a given level of cardiac output can be maintained with a low heart rate, i.e., a large stroke volume. The reason why a training with "overload" apparently is necessary to improve the efficiency of the heart is presently unknown.

For the *cardiac patient* the situation is different if a large diastolic filling is combined with a small stroke volume and a high heart rate. The dilatation of the heart muscle gives improved ability for the muscle fibers to produce tension as long as they are stretched, as in factor 3. Factors 1 and 2 help in keeping the efficiency high. Factors 4 and especially 5 tend to reduce the efficiency. Undoubtedly the heart has to pay for its compensating of the basically reduced myocardial strength. Sooner or later there will be a critical equation for energy requirement and energy supply of the heart, and the vascular bed in the myocardium is the key in that formula. At rest the capillary blood flow may cover the need, but during even mild exercise, there arises a discrepancy, and the patient is forced to stop because of symptoms such as angina pectoris, which is probably a pain caused by hypoxia in the myocardium. It should be emphasized that with a high heart rate, the tension time, i.e., the time during which the heart muscle is contracted per minute, is prolonged, and the blood flow through the heart muscle is therefore reduced (Fig. 6-2).

Summary A large dilated heart may provide tension when the muscle fibers are stretched, but if tension decreases markedly as the muscle shortens, the frequency of contractions must be high. A high heart rate lowers the mechanical efficiency of the heart and increases the oxygen uptake for a given cardiac output. An impaired circulation in the myocardium can further complicate the situation. The difference in fitness of the athlete's large heart and that of the cardiac patient is evident.

Venous Blood Return

(See Guyton, 1963.) The cardiac output cannot exceed the flow of blood returning to the heart. In the supine position the hydrostatic pressure on the venous side of most of the capillaries is about 10 mm Hg, but the pressure gradient to the right atrium is increased by the negative intrathoracic pressure. At rest this pressure is about 5 mm Hg less than the ambient barometric pressure because the elastic tissue of the lungs is expanded to the size of the thoracic cavity and the recoil of the tissue exerts a tension on the thin-walled vessels within the thorax. The inspiration increases this pulling force, and blood is sucked into the thorax. At the same time, the abdominal veins are compressed as the diaphragm contracts. The variations in intrathoracic and intra-abdominal pressures with breathing will significantly enhance the venous return because a

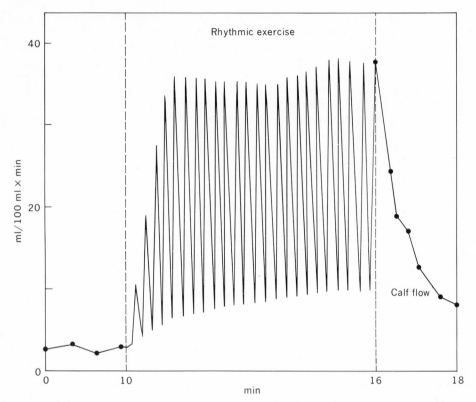

Fig. 6-16   Schematic representation of changes in blood flow through muscles of calf during strong rhythmic contractions. During contraction the blood within the muscles is emptied toward the heart but at the same time the inflow is greatly reduced. (From Barcroft and Swan, 1953.)

backflow in the veins is hindered by the capillary resistance and venous valves. During an expiratory effort with the glottis closed (Valsalva's maneuver) the intrathoracic pressure increases, impairing both the venous return and the cardiac output; this is usually the case in weight lifting. Hyperventilation followed by Valsalva's maneuver may cause fainting (for explanation, see Chap. 7).

A second important factor improving the venous return is dynamic activity of the skeletal muscles (Figs. 6-14, 6-16). As pointed out earlier in this chapter, the muscle pump is very effective in propelling the blood toward the heart. When the leg muscles contract rhythmically, there is a decrease in the blood volume of the legs, indicating the emptying of blood; and blood can thus be driven from a segment against a resistance of 90 mm Hg (Barcroft and Swan, 1953). The action of the muscle pump is especially important in the erect position. If exercise is started (walking), the pressure in the veins of a foot can decrease from 100 mm Hg in passive standing to about 20 mm Hg. The increase in venous return to the heart as muscular work starts will promote the possibility of an immediate increase in cardiac output. The bigger the muscle mass involved, the more pronounced is this effect. As pointed out above, when the exercise stops, the blood stays temporarily in the dilated vascular bed, and the decrease in venous return to the heart may cause such a drop in cardiac output and arterial blood pressure that fainting occurs. This transient pressure drop is more likely to occur if the skin vessels are dilated to secure heat elimination and the person remains motionless in a standing position.

At rest the venous system (capacitance vessels) contains about 65 to 70 percent of the total blood volume. By constriction of the venules and veins, blood is emptied toward the heart. Vasomotor activity in the splanchnic area, the skin, and the lungs is probably of special importance in this connection. Variations in blood volume by blood loss, dehydration, or prolonged inactivity may influence the filling of the heart.

To complete the picture, it should be emphasized that the normal heart has the capacity to pump all the blood that is returned to the heart into the arteries. Increases in heart rate and force of contraction can thus keep the atrial pressures low even during heavy exercise. A failure of the left or right ventricle would cause a rise in the central venous pressure and severe disturbances in the capillary fluid exchange (congestive heart failure).

Summary   The venous return to the heart is determined by the balance between the filling pressure and the distensibility of the heart, i.e., the intraventricular pressure minus the intrathoracic pressure. The filling is enhanced by (1) the variation in intrathoracic and intra-abdominal pressures during the respiratory cycle, (2) the effect of the muscle pump during muscular movements, and (3) a vasoconstriction in the postcapillary vessels. Changes in body position will, at least temporarily, affect the volume of blood in central veins.

Cardiac Output and Oxygen Uptake

At rest in the supine position, the cardiac output is 4 to 6 liters/min with an extraction of 40 to 50 ml O_2/liter of blood and a total oxygen uptake of 0.2 to 0.3 liter/min. When a subject strapped to a tilting table is tilted from the horizontal to the feet-down position, the cardiac output may fall from 5 to 4 liters/min. This is due to the previously discussed venous pooling. Stroke volume is reduced and heart rate is usually increased. Activation of the muscle pump propels the blood toward the heart, and the heart rate may even decrease as stroke volume increases (Fig. 6-14). In the passive feet-down position the oxygen uptake is unchanged, and hence the arteriovenous O_2 difference is increased.

During exercise the cardiac output increases with the increase in oxygen uptake, but not linearly, if a range from rest value up to maximal is considered (Fig. 6-17). On 11 women and 12 men, well-trained, twenty to thirty years of age, the cardiac output, oxygen uptake, heart rate, and oxygen content of arterial

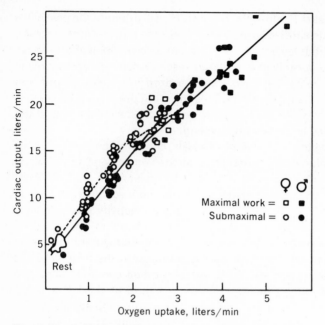

Fig. 6-17   Individual values on cardiac output in relation to oxygen uptake at rest, during submaximal, and during maximal exercise on 23 subjects sitting on a bicycle ergometer. Regression lines (broken lines for women) were calculated for experiments where the oxygen uptake was (1) below 70 percent and (2) above 70 percent of the individual's maximum. (From P.-O. Åstrand et al., 1964.)

blood were determined at rest sitting on a bicycle ergometer and on four to five different work loads up to the maximum that could be maintained for 4 to 6 min. Dye dilution technique with indocyanine green was used for determination of cardiac output. Two to three measurements were done at each load, the first after about 5 min of exercise (except for maximal load). The mean values were used to calculate cardiac output, stroke volume, and $a\text{-}\bar{v}O_2$ difference (P.-O. Åstrand et al., 1964). The physically very fit man can increase his oxygen uptake from 0.25 to 5.00 liters/min or more when working on a bicycle ergometer or treadmill. This increase is, let us assume, met by an increase in heart rate from 50 to 200, and in stroke volume from 100 to 150 ml, which means that during maximal exercise the cardiac output has increased from 5.0 to 30.0 liters/min, or six times the resting level. Since the oxygen uptake increased 20 times, the $a\text{-}\bar{v}O_2$ difference must have changed from 50 to 165 ml/liter of blood, or 3.3 times the resting level (and 3.3 × 6 is close to 20). This better utilization of the oxygen transported by the blood is reached principally in two ways: (1) The blood flow is redistributed during exercise so that skeletal muscles with their pronounced ability to extract oxygen may receive 80 to 85 percent of the cardiac output, as compared with some 15 percent at rest. (2) The oxygen dissociation curve is shifted so that more oxyhemoglobin is reduced than normally at a given pressure for oxygen. This "Bohr effect" is easier to understand by studying Fig. 6-18. In the working muscles the temperature may exceed 40°C and the pH may be lower than 7.0, so there is really not a fixed relation between oxygen tension and oxygen saturation of the hemoglobin.

At normal pH (7.40), CO_2 tension (40 mm Hg), and blood temperature (37°C), the blood keeps about 33 percent HbO_2 at an oxygen tension of 20 $mm\ Hg$. At a pH of 7.2 and temperature of 39°C, the percentage of HbO_2 at the same oxygen tension is reduced to about 17, which means that 1 liter of blood with an O_2 capacity of 200 ml can deliver 26 ml

Fig. 6-18   Effects of CO_2, pH, temperature (°C), and CO on the oxygen dissociation curve of the blood. (From various sources; see Ruch and Patten, p. 765, 1965.)

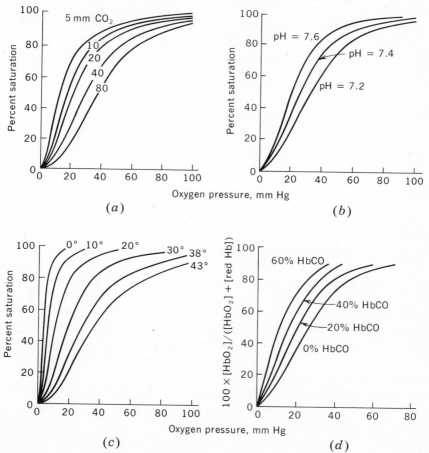

more oxygen without changes in the pressure gradient for oxygen between capillary and muscle cell. With 24 liters of blood circulating the working muscles per minute, this extra oxygen delivery amounts to about 0.6 liter/min or 12 percent of the total uptake, in the example given above. With pH 7.0 and muscle temperature 40°C the extra oxygen uptake would, however, not be further increased because the oxygen content of the arterial blood would be affected negatively by the low pH and high temperature.

With carbon monoxide present carboxyhemoglobin (HbCO) is formed. Such a conversion effects the oxyhemoglobin dissociation curve with a shift to the left. The effect of carbon monoxide on oxygen transport is therefore twofold: it reduces the amount of hemoglobin available for oxygen transport, and it interferes with the unloading of oxygen in the tissues. In the example just discussed an unloading to 35 percent saturation occurred at an oxygen tension of 20 mm Hg at normal pH and temperature. With 10 percent HbCO present in the blood the same unloading to 35 percent demands an oxygen tension of about 14 mm Hg, which may be critical for many cells remote from the capillaries (Roughton, 1964). For smokers the effect of CO on the hemoglobin becomes a real handicap during exercise.

The shift in the dissociation curve is a result of the heat production by the working cell and the formation of CO_2 and lactic acid during the heavy exercise. The effect of CO_2 in releasing O_2 from the blood is actually twofold: CO_2 lowers the pH of the blood, and by combining with hemoglobin, reduces its affinity for O_2. (During exercise the role of carbamine-bound CO_2 is probably less than at rest; Roughton, 1964.)

By the two mechanisms, based on (1) a regulation of the circulation and (2) an inherent characteristic of hemoglobin, the oxygen uptake can be elevated 20 times, but the cardiac output has to increase to only 30 liters/min, not $20 \times 5 = 100$ liters/min.

Oxygen Content of Arterial and Mixed Venous Blood

During exercise there is a *hemoconcentration* of the blood, which is partly explained by a withdrawal of fluid to the active muscle cells and by the interstitial fluid (receiving the metabolites produced in the cells). Hence the osmotic pressure is highest within and close to the working cells. The raised capillary pressure and surface area also lead to an increased outward filtration. In the experiments on 23 subjects (P.-O. Åstrand et al., 1964) the oxygen capacity of the arterial blood was about 10 percent higher during maximal exercise than at rest. The actual oxygen content of the blood drawn from the brachial artery was 3 percent higher during heavy exercise than at rest. This hemoconcentration makes the blood more viscous, but it also increases the transportation capacity per liter of blood for both oxygen and carbon dioxide.

There is evidently a discrepancy between the increase in oxygen binding capacity of the blood and the extra oxygen actually taken up during strenuous exercise. In other words there is a slight reduction in saturation of the arterial blood during maximal exercise, despite a normal or even elevated oxygen tension

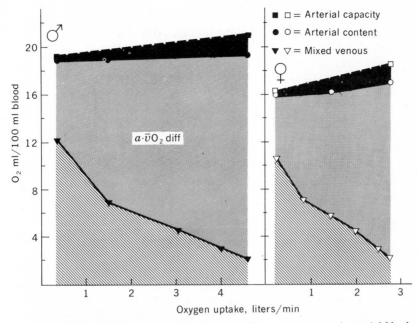

Fig. 6-19   Oxygen-binding capacity and measured oxygen content of arterial blood; calculated oxygen content of mixed venous blood at rest and during work up to maximum on bicycle ergometer. Mean values for five female (right part) and five male subjects (left), 20 to 30 years of age, and with high maximal aerobic power. (From data presented by P.-O. Åstrand et al., 1964.) During maximal work the arterial saturation is about 92 percent as compared with 97 to 98 percent at rest, and the venous oxygen content is very low and similar for women and men.

in the lung alveoli. The arterial pH may, however, be below 7.2 and the blood temperature markedly elevated, and therefore the shift in the oxygen dissociation curve to the right from 97 toward 90 percent saturation (Figs. 6-18, 6-19) is noticeable even at high oxygen tensions. This shift is a disadvantage in the lungs, but the overall effect is, as discussed above, an improved oxygen delivery due to the advantage at the tissue level, both in active and nonactive areas.

The increased extraction of oxygen from the arterial blood as exercise becomes heavier is illustrated in Fig. 6-19. During maximal exercise the venous blood leaving the muscles has a very low oxygen content. In this study the calculated oxygen content in *mixed venous blood* averaged about 2.0 ml/100 ml of blood for both women and men. At rest the blood flow through most tissues is luxurious as far as the oxygen need is concerned, since other functions determine the flow distribution (e.g., through the vascular bed in kidneys, intestines, and skin; for the oxygen supply to the kidneys, about 50 ml of blood per min would be enough, but at rest the actual flow exceeds 1 liter/min). As emphasized in previous discussions the blood flow, during exercise, is redistributed with the primary object of supplying metabolically active tissues with oxygen and removing the produced carbon dioxide (Fig. 6-8).

Saltin et al. (1968) report the oxygen content in the femoral vein during maximal running on the treadmill to be 1.4 vol percent (average of four subjects). The oxygen tension was 12 mm Hg and the pH 7.09. They noted that the difference in oxygen content between femoral venous and mixed venous blood (containing about 2 vol percent of oxygen) was small, particularly after training.

It is reasonable to assume that some correlation should exist between the oxygen content of the arterial blood and the cardiac output at a given oxygen uptake. For some unknown reason, women have about 10 percent lower concentration of Hb in the blood than men. In the mentioned study the cardiac output required to transport 1.0 liter of oxygen was 9.0 liters for women (O_2 content in arterial blood: 16.7 ml/100 ml), and 8.0 liters for men (O_2 content: 19.2 ml/100 ml) during submaximal work with an oxygen uptake of 1.5 liters/min. During maximal exercise the figures were 7.0 and 6.0 liters, respectively. The cardiac output in males is more effective in its oxygen-transporting function than in women, and this difference can be explained by the Hb content of the blood (P.-O. Åstrand et al., 1964).

In recent experiments, the relationship between oxygen uptake during maximal exercise and the oxygen content of arterial blood has been further analyzed. The subjects were exposed to acute hypoxia by reducing the ambient pressure of the inspired air to simulate an altitude of 4,000 meters ($P_{bar} = 460$ mm Hg). The cardiac output attained during submaximal exercise was higher at high altitude than at sea level, but during maximal effort no difference in cardiac output was observed. The oxygen uptake during maximal work was, however, reduced in proportion to the decrease in oxygen content of the arterial blood, or to about 70 percent of what it was at sea level (Stenberg et al., 1966).

The *myocardial oxygen extraction* is reported to be approximately the same at rest, during exercise (oxygen uptake was increased two- to threefold over the resting value), and during recovery in normal healthy subjects (Messer et al., 1962). The arteriovenous oxygen difference is as high as 16 to 17 vol percent at rest. During exercise it may increase to 18 to 19 ml/100 ml blood flow (oxygen content of arterial blood 20 vol percent or higher).

Stroke Volume

Factors affecting the stroke volume are (1) the venous return to the heart and (2) the distensibility of the ventricles. The degree of diastolic filling has an anatomic limitation (children-adults, women-men), but within a range various factors, some of which were discussed above, affect the stretching of the muscle fibers. The final factors determining the stroke volume are (3) the force of contraction in relation to (4) the pressure in the artery (aorta or pulmonary artery).

The heart adjusts itself to changing conditions by an inherent self-regulatory mechanism. Starling, using his famous lung-heart preparations, found that the normal heart tended to empty itself almost completely. It was distended to a greater diastolic volume in response to either a greater venous return or an increase of the arterial pressure. In the latter case there was a transient decrease in stroke volume, but as the force of contraction increased with a greater initial length of the muscle fibers, the stroke volume and cardiac output became normal. By stimulation of the sympathetic cardiac nerves, the contraction force increased, and the arterial resistance could be overcome despite a greater extent of myocardial shortening.

The results from Starling's studies of the isolated heart were also considered applicable in the intact animal. Most earlier textbooks concluded that the diastolic volume of the heart was smaller at rest but increased during exercise, when the venous return increased and the arterial pressure was elevated. The same end-systolic volume could be maintained or was dependent on the strength of the heart muscle. The well-trained person was characterized by a small residual volume of blood in the heart after systole at rest as well as during exercise. The net effect was a substantial increase in stroke volume during exercise, according to these standard tests.

Then followed a novel approach which was in complete disagreement with Starling's heart law. From x-ray studies at rest and during exercise, it was concluded that the end-diastolic volume of the heart was fairly constant during rest and exercise (Reindell, 1943; Kjellberg et al., 1949). The systolic emptying of the ventricles was limited at rest, but during muscular activity the stroke volume increased, resulting in a smaller residual volume. Sympathetic nerve impulses and catecholamines were thought to increase the contraction force essential for the more complete emptying of the ventricle, actually against an elevated arterial blood pressure.

Earlier experimental data on cardiac output and stroke volume were obtained with various indirect methods using nitrous oxide (Krogh and Lindhard, 1912), acetylene (Christensen, 1931), and by other methods. Exercise was performed on a treadmill or in a sitting position on a bicycle ergometer, and the control values were usually secured on the subject in the same position at rest. The typical finding was an increase in stroke volume in transition from rest to work but a tendency for the stroke volume to remain unchanged with further increase in work load. In some studies no increase in stroke volume was found. The data thus obtained did, to some extent, confirm Starling's law, but in essential parts they refuted it.

With newly developed methods, especially the direct Fick method and dye dilution technique and the refined methods worked out by Rushmer and coworkers (1960, 1965), the situation has become somewhat clearer. Using dogs, Rushmer et al. measured the heart rate, the rate of change of ventricular dimensions, the power developed by part of the myocardium, the stroke work and work per unit time by that sample of myocardium, and the rate of change of ventricular pressure. When the dogs stood up, the diastolic diameter abruptly diminished and remained reduced. During exercise on a treadmill, the stroke change in diameter was essentially unaltered or sometimes increased through greater systolic ejection. An increase in diastolic dimensions rarely occurred, except in experiments in which elastic circumference and length gauges were employed. Elevated effective filling pressure was occasionally, but not consistently, recorded. The consistent finding in 100 exercise studies on more than 40 dogs was (1) an increase in heart rate and (2) an increase in systolic pressure. These responses were noticed from the very beginning of exercise. With the exercise there was also an increased power of the contraction, increased work per stroke, and increased work per unit time.

In conclusion, the experiments by Rushmer and his group suggested that in these animals the increase in cardiac output was attained primarily through tachycardia, with little contribution by increased stroke volume.

The present concept on the stroke volume in man during exercise adhered to by most cardiophysiologists can be summarized as follows: Work performed in the supine or erect position does not significantly increase the stroke volume from the value measured at rest in the supine position (Wade and Bishop, 1962; Bevegård, 1962; Bevegård and Shepherd, 1967).

When the position is changed from supine to standing or sitting, there is a diminution in end-diastolic size of the heart and decrease in stroke volume, as discussed above. If muscular work is then performed, the stroke volume increases to approximately the same size as obtained in the recumbent position. The discrepancy between earlier and recent studies is in most cases not due to inconsistent results, but must be blamed on erroneous conclusions and bold extrapolations of the data to conditions not examined. Certainly Starling's law holds true for the isolated heart-lung preparation, but when the heart is functioning in the intact animal, other mechanisms are superimposed.

The importance of the *central blood volume* for the stroke volume was demonstrated in 1939 by Asmussen and Christensen. Subjects were exercising with the arms in the sitting position. In some experiments the subjects laid down with their legs elevated for about 10 min before the exercise started. The circulation to the legs was then arrested by pressure cuffs around the thighs. When assuming a sitting position following this procedure there was approximately 600 ml of blood less in the legs compared to sitting without occlusion of the blood flow to the legs. It was noted that the cardiac output was about 30 percent higher when the legs were "blood free," i.e., the central blood volume was high, as compared to the experiments with blood pooling in the legs. The high cardiac output was due to a high stroke volume, for the heart rate was actually lower than in exercise with reduced central blood volume (and low cardiac output).

In Fig. 6-20 the individual data on stroke volume are plotted, expressed in percentage of the maximum reached. The maximal oxygen uptake of the subject is similarly set to 100 percent, and the submaximal loads are defined in percent of this maximum. At rest, the stroke volume is for most subjects between 50 and 70 percent (mean 63 percent) of the maximum measured during exercise. Hence, during exercise in the sitting position there is a definite increase in stroke volume with an "optimum" reached when oxygen uptake exceeds 40 percent of maximal aerobic power. The heart rate at that load is 110 to 120. Methodological errors and biological variations result in a variation in calculated stroke volume of ± 4 percent as oxygen uptake further increases. There is no tendency toward a decrease in stroke volume at maximal work. Of the 23 subjects, 12 reached their maximal stroke volume at the peak load. No significant correlation was found between maximal heart rate and eventual decrease in stroke

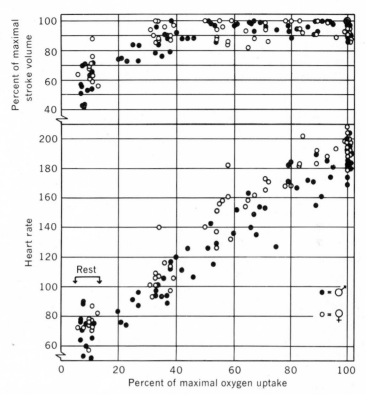

Fig. 6-20   Stroke volume in percent of the individual's maximum, and heart rate at rest and during exercise. The oxygen uptake on the abscissa is expressed in percent of the subject's maximum. Circled dot at "100 percent" represents 11 of the 23 subjects. Measurements were made with the subjects in the sitting position (same subjects as in Fig. 6-17).

volume during the maximal exercise. In our opinion, this rules out the hypothesis that a high heart rate (about 200) during exercise should interfere with the filling of the heart.

In this discussion of the variation in stroke volume it should be emphasized that in experiments on animals reflexes that evoke tachycardia also evoke simultaneously, either directly or indirectly, an increase in contractility of the heart muscle, so that stroke power and stroke work increase (Sarnoff and Mitchell, 1962).

Braunwald et al. (1967) observed in experiments on man that an increase in heart rate increases the velocity of the shortening of the myocardial fibers. They emphasize that the exercise tachycardia as such contributes to an improvement of the contractile state of the myocardium. They conclude that "the normal cardiac response to exercise involves the integrated effects on the myocardium of simple tachycardia, sympathetic stimulation, and the operation of the Frank-Starling mechanism. During submaximal levels of exertion, cardiac output can rise even when one or two of these influences are blocked. However, during maximal levels of exercise, the ventricular myocardium requires all three influences in order to sustain a level of activity sufficient to satisfy the greatly augmented oxygen requirements of the exercising skeletal muscles."

Heart Rate

In many types of work, the increase in heart rate is linear with the increase in work load. There are exceptions and those exceptions are perhaps more frequent among untrained subjects. When the subject performs very heavy exercise, the $a\text{-}\bar{v}O_2$ difference may increase so that the oxygen uptake increases relatively more than the cardiac output. The evaluation, from submaximal work loads, of an individual's maximal oxygen uptake or capacity to perform work is, in most test procedures, based on a registration of the heart rate during steady state and then an extrapolation to a fixed heart rate or to an assumed maximal heart rate (see Chap. 11). There are many pitfalls in this method. Here, the following should be noted: (1) The standard deviation for maximal heart rate during exercise is ± 10 beats/min. Hence, for twenty-five-year-old individuals, women or men, the maximal heart rate for 5 out of 95 subjects may be below 175 or above 215, since the maximum is about 195 on an average. (2) There is a gradual decline in maximal heart rate with age, so that the ten-year-old attains 210, the sixty-five-year-old only about 165 beats/min (Fig. 6-21) (Robinson, 1938; P.-O. Åstrand, 1952; I. Åstrand, 1960; Hollmann, 1963).

Fig. 6-21   The decline in maximal heart rate with age, and heart rate during a submaximal work load. Mean values from studies on 350 subjects. The standard deviation in maximal heart rate is about ±10 beats/min in all age groups. (From Åstrand and Christensen, 1964.)

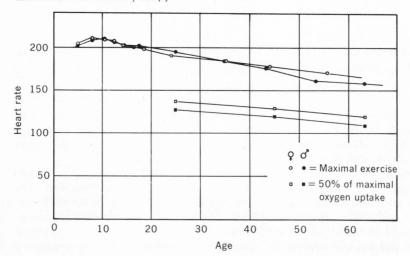

When 50 percent of the maximal aerobic power is used, the heart rate in the twenty-five-year-old man is about 130, but the same relative work load and feeling of strain is experienced at a heart rate of 110 for the sixty-five-year-old man (Fig. 6-21) (I. Åstrand, 1960). For women, the 50 percent oxygen uptake is attained at a heart rate of about 140 beats/min at the age of twenty-five.

Prolonged exercise in a hot environment causes a higher heart rate than exercise at a low room temperature. Emotional factors, nervousness, and apprehension may also affect the heart rate at rest and during work of light and moderate intensity. During repeated maximal exercise the heart rate is, however, remarkably similar under various conditions, with a standard deviation of ±3 beats per min (P.-O. Åstrand and Saltin, 1961a).

The heart rate at a given oxygen uptake is higher when the work is performed with the arms than with the legs (Christensen, 1931; P.-O. Åstrand et al., 1965; Stenberg et al., 1967.) Static (isometric) exercise also increases the heart rate above the value expected from work load. The mechanism for these differences in heart-rate response to exercise is not understood. However, the elevated heart rate is usually accompanied by a decreased stroke volume. It is known that a variation in heart rate at a given oxygen uptake at rest and during submaximal exercise often produces a change in stroke volume, so that the cardiac output is maintained at an appropriate level. (This information is based on studies in patients with artificial pacemakers or irregular heart rate, and in subjects submitted to various drugs influencing the heart rate.) (I. Åstrand et al., 1963; Bevegård and Shepherd, 1967; Braunwald et al., 1967.)

The regulation of the circulation in exercise is probably guided primarily by factors sensitive to an adequate cardiac output. Heart rate and stroke volume are the variables, and the stroke volume is more likely to be directly influenced by such factors as venous return or peripheral vascular resistance.

Blood Pressure

(Some methodological aspects of measurements of blood pressure have been discussed earlier in this chapter.) The side pressures (lateral pressures) in the aorta during systole and diastole are reported not to be markedly different during exercise compared with the resting condition (Marx et al., 1967). The aortic (arterial) pressures, also including the kinetic energy factor (end pressure), increase linearly with the increase in oxygen uptake as the exercise becomes heavier.

As a result of the vasodilatation in the vascular bed in the working muscles, the peripheral resistance to blood flow is reduced during exercise, but the elevation in cardiac output causes the blood pressure to rise. The arterial pressure obtained in a peripheral artery at rest, 120 mm Hg in systole and 80 mm Hg in diastole, may exceed 175 and 110 mm Hg, respectively, during exercise (Fig. 6-22).

It should be noted that the arterial blood pressure is significantly higher in arm exercise than in leg work (Fig. 6-22). The high blood pressure at a given

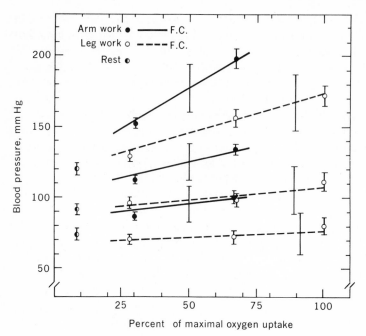

Fig. 6-22   Effect of exercise on blood pressure (end pressure). Regression lines of arterial systolic, mean, and diastolic blood pressures, respectively, in relation to oxygen uptake (in percent of the maximum) during arm and leg exercise in the sitting position for 13 subjects. F.C. = femoral artery catheter. The figure summarizes data from 23 submaximal and 13 maximal work loads with arm work (cranking) and 44 and 13 experiments, respectively, with leg work. The vertical heavy lines represent ±1 SD (standard deviation) around the regression line. The dots and thin lines represent the mean ±1 SEM (standard error of the mean) for three groups of values at different levels of oxygen consumption. (The systolic pressure measured in a peripheral artery is higher than in the aorta, but the mean and diastolic pressures are similar in the two vessels.) (From P.-O. Åstrand et al., 1965.)

cardiac output, when the work is performed by the arms, induces an increased stroke work of the heart. Therefore, for untrained individuals or for cardiac patients, it may be hazardous to work hard with the arms, e.g., shovel snow, dig in the garden, or carry heavy trunks. The relatively high blood pressure in exercise with small muscle groups is probably due to a vasoconstriction in the inactive muscles. The larger the activated muscle groups, the more pronounced is the dilatation of the resistance vessels. The lower peripheral resistance is reflected in a lower blood pressure (since $\dot{Q} \times R = P_{mean}$).

We have a similar situation with a considerable ventricular load in isometric work. Donald et al. (1967) describe a powerful cardiovascular reflex causing an unexpectedly high rise in blood pressure in response to sustained contractions above 15 percent of maximal voluntary force. The pressor response appears to be largely independent of the muscle bulk involved in the contraction, provided the relative tension is constant. They point out that relatively moderate and localized isometric work can cause a far higher pressure component than noticed in dynamic work; this "may well be dangerous to a person with a compromised heart or impaired integrity of the arterial wall."

I. Åstrand et al. (1968) studied carpenters using a hammer to nail at different heights. When they were hammering into the ceiling, their heart rate and intra-arterially measured blood pressures were significantly higher than when hammering at bench level. They were also higher than during leg exercise on a bicycle ergometer with a similar level of oxygen uptake (see Chap. 13).

Reindell et al. (1960) report that when comparing the arterial blood pressure response to exercise in subjects of different ages, the older men had consistently higher systolic and diastolic pressures than the younger ones. At rest the twenty-five-year-olds averaged 125/75 mm Hg and during exercise at a rate of 100 watts (oxygen uptake about 1.5 liters/min) the pressures were 160 and 80 mm Hg in systole and diastole, respectively. For the fifty-five-year-old group the increase was from 140/85 at rest up to 180/90 mm Hg, the work load being the same. Similar results are reported by Hollmann (1963).

Type of Exercise

The cardiac output at a given oxygen uptake during submaximal exercise is consistently 1 to 2 liters less in the erect position than when the subject is recumbent; the heart rate is about the same (Reeves et al., 1961; Bevegård, 1962; Wade and Bishop, 1962). The compensation for the lower cardiac output must, by definition, be an increased $a\text{-}\bar{v}O_2$ difference when erect. It should be emphasized that the maximal oxygen uptake during cycling in the supine position is about 15 percent lower than during work in the sitting position on the bicycle ergometer (Fig. 6-23) (Åstrand and Saltin, 1961b; Stenberg et al., 1967). Similarly, the cardiac output is somewhat lower during maximal exercise with the legs in the supine position. Combined arm and leg work in the sitting or supine position reveals the same values for maximal oxygen uptake, heart rate, and cardiac output as work in the sitting position with only the leg muscles (Fig. 6-23) (P.-O. Åstrand and Saltin, 1961b; Stenberg et al., 1967).

Heart Volume

The size of the heart can be visualized by means of roentgenograms, and its volume can be computed by application of empirical formulas. A high correlation has been established between heart volume and various parameters, such as blood volume, total amount of hemoglobin, and stroke volume in healthy younger individuals (Kjellberg et al., 1949; Sjöstrand, 1953; Reindell et al., 1960; Mellerowicz, 1962). The difference between the athlete's heart and the dilated heart of the cardiac patient is not only the configuration of the heart, but also

the disproportion between heart size and maximal aerobic power, total hemoglobin content, etc. Figure 6-24a shows how a group of 30 young girls have a
calculated heart volume that in many cases is much larger than expected for
their body size. The girls were some of the best Swedish swimmers and were not
very likely to suffer from any heart disease. When the heart volume is related
to the girls' maximal aerobic power (Fig. 6-24b), the findings make functional
sense. The girl with the highest oxygen uptake and largest heart was actually
also the best swimmer; she was second in the 400-m free-style in the Olympic
Games, 1960 (P.-O. Åstrand et al., 1963).

On the average the heart volume calculated from roentgenograms has been
found to be largest (above 900 ml) for well-trained athletes engaged in events
calling for endurance (bicyclists, canoeists, cross-country skiers, long-distance
runners). The average for middle-distance runners, swimmers, and soccer and
tennis players was between 800 and 900 ml, and for boxers, fencers, gymnasts,
jumpers, sprinters, throwers, and untrained controls, below 800 ml (Reindell et
al., 1960; Mellerowicz, 1962).

Figure 6-25a gives the relationship between maximal stroke volume and
calculated heart volume. If both volumes were closely related to the body size
of the subject, we should expect the exponent b in the equation (of the logarithmic values) $y = a \times x^b$ to be 1.0. In fact, it is 0.87, which is not significantly different from 1.00. For the plotting of maximal cardiac output per minute

*Fig. 6-23   Oxygen uptake (to the left) and cardiac output; comparison between the highest
individual values attained in the sitting (abscissa) and supine (ordinate) position for arm
work (x), leg work (○), and combined arm and leg work (●). Line of identity and lines corresponding to 10 percent deviation are drawn. Symbols with arrows give the mean of the different
groups. Note that leg work in the supine position did not bring oxygen uptake and cardiac
output to a maximum, but when the arm muscles were also exercised, the oxygen uptake and
cardiac output did increase to the same level as in leg work, or in combined arm + leg work
in the sitting position. (From Stenberg et al., 1967.)*

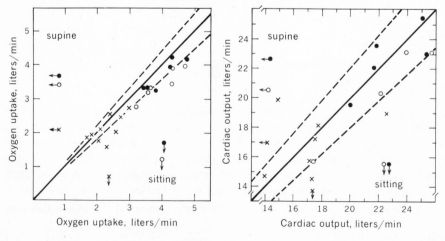

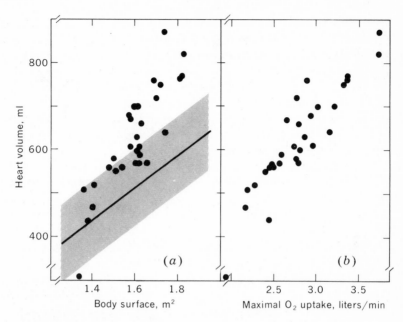

Fig. 6-24   Relationship between heart volume and (a) calculated body surface area and (b) maximal O₂ uptake in 30 young well-trained girl swimmers. Shadowed area gives the 95 percent range for "normal" girls. (Modified from P.-O. Åstrand et al., 1963.)

and heart volume the value for b should be 0.667, and it is actually 0.76 (Fig. 6-25b). However, for the relationship between maximal oxygen uptake and heart volume (stroke volume) (Fig. 6-25c), the b value is 1.14 or significantly different from the expected value of 0.667. Apparently in the examined group the cardiac output is more efficient with regard to the oxygen transportation in the individuals with larger hearts. The most probable explanation is the higher Hb concentration for the male subjects. A cardiac patient with dilated heart would most likely have his value for heart size on the lower-right side of the regression lines in Fig. 6-25. The aged individual also has a large heart in relation to his maximal aerobic power.

Age

At a given work load or oxygen uptake the old individual attains, on the average, the same heart rate as the younger one (I. Åstrand, 1960; Hollmann, 1963; Strandell, 1964). Strandell (1964) found that the cardiac output at a given work load was about 2 liters/min lower in the sixty- to eighty-year-old men at any level of oxygen uptake compared with the young ones. Stroke volume was also significantly lower for the older men (about 20 percent).

As already mentioned, the heart rate reached during maximal exercise decreases with age. The value typical for the ten-year-old girl or boy is 210, for

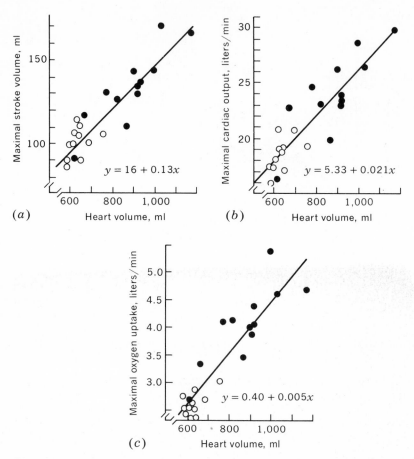

Fig. 6-25   Relationship between heart volume and (a) maximal stroke volume, (b) maximal cardiac output, and (c) maximal oxygen uptake (same subjects as in Fig. 6-17). (From P.-O. Åstrand et al., 1964.) A further discussion of body dimensions is presented in Chap. 10.

the twenty-five-year-old 195, and for the fifty-year-old 175 beats/min (Fig. 6-21). Therefore the decrease in circulatory capacity in the old individual is more marked than predicted from heart rate, stroke volume, and cardiac output observed during submaximal exercise if the norms are the same as when evaluating young individuals. The old man has a larger heart volume, calculated from roentgenograms taken in supine position, than does the young man; blood volume and total amount of hemoglobin are not different in young and old men (Strandell, 1964; Grimby and Saltin, 1966). These findings should be related to the decrease in maximal stroke volume, cardiac output, and aerobic work capacity in old men (see Fig. 10-6).

Whether the decrease in maximal heart rate with age is a consequence of an arteriosclerosis in the vessels of the heart is not known. The oxygen cost of a cardiac performance involving a high heart rate is great, and therefore the load on the heart is reduced by a lowered ceiling for heart rate. The lower heart rate during maximal exercise in the old individual is probably not a direct response to hypoxia, since breathing pure oxygen instead of room air during the exercise does not further elevate the heart rate (I. Åstrand et al., 1959).

During exercise the arterial blood pressure (in systemic as well as pulmonary arteries) is higher in the old than in the young man (Reindell et al., 1960; Hollmann, 1963; Strandell, 1964).

Training and Cardiac Output

It is an old observation that the heart rate at rest, during standard exercise, as well as during recovery from such work, is lowered with a training of the oxygen-transporting system. The few studies published on the cardiac output during standard exercise repeated during a course of training indicate that cardiac output is maintained at the same level (Musshoff et al., 1959; Wade and Bishop, 1962). The reduction in heart rate should then mirror an increase in stroke volume. However, more data are necessary before any definite conclusions are drawn on the circulatory response to exercise (see Chap. 12).

Top athletes in endurance events are characterized by a very high maximal oxygen uptake (maximal aerobic power). Their maximal values for circulatory parameters must therefore be high compared to less athletic individuals. Ekblom and Hermansen (1968) have collected data obtained during maximal work on the treadmill in athletes, using the dye dilution technique. Some data are presented in Table 6-1. Both an intensive training and superb natural endowments contribute to the remarkable circulatory capacity for oxygen transportation in these subjects.

TABLE 6-1

| Subject | \multicolumn{5}{Maximal values} |
	$\dot{V}_{O_2}$, liters/min	Cardiac output, liters/min	Heart rate	Stroke volume, ml	$a\text{-}\bar{v}O_2$ difference, ml/100 ml
G. P.	6.00	39.8	188	212	15.1
C. R.	5.77	37.8	188	201	15.3
A. H.	5.60	34.4	189	182	16.3
C. S.	5.50	36.2	198	183	15.2
B. T.	5.64	38.0	193	197	14.8
L. R.	6.24	42.3	206	205	14.8
Mean	5.79	38.1	194	197	15.3

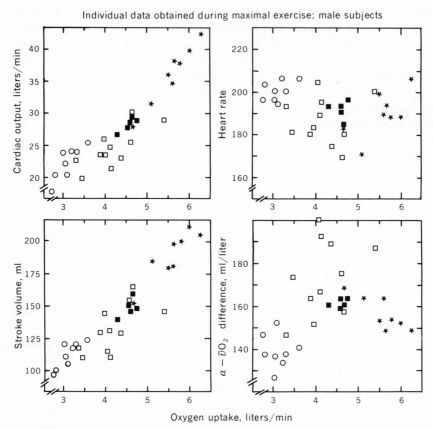

Individual data obtained during maximal exercise; male subjects

Fig. 6-26   *Cardiac output, heart rate, stroke volume, and arteriovenous oxygen differ-*
ence during maximal exercise in relation to maximal oxygen uptake in top athletes who
were very successful in endurance events (stars), well-trained but less successful athletes
(filled squares), and 25-year-old habitually sedentary subjects (unfilled circles). (From
Ekblom, 1969.) Included are also maximal values on the male subjects presented in Fig.
6-17 (unfilled squares).

Figure 6-26 summarizes data on 32 subjects with maximal aerobic power
ranging from 2.8 to 6.2 liters of oxygen/min. This figure shows a clear relationship between maximal cardiac output and oxygen uptake. It also shows that it
is the stroke volume which to a large extent determines the maximal cardiac
output.

SUMMARY AND CONCLUDING REMARKS

Figure 6-27 presents mean values of some functions studied at rest and during
submaximal and maximal exercise, on young fairly well-trained individuals.

Some sex difference in cardiac output (and necessarily arteriovenous oxygen difference) can be observed at a given oxygen uptake. The most pronounced difference between the sexes is the smaller stroke volume and higher heart rate during exercise of a given severity for the women compared with men. Actually, a similar difference in stroke volume and heart rate is usually observed when comparing individuals of the same age but with a low and high performance capacity, respectively.

Fig. 6-27   The figure is based on average values from measurements on 11 women and 12 men, all of them relatively well trained and working on a bicycle ergometer in the sitting position (P.-O. Åstrand et al., 1964). The individual data are presented in Figs. 6-17 and 6-20. (Since the abscissa gives the oxygen uptake in absolute values, the calculated mean curves can be misleading. The less-fit subjects have both a low maximal oxygen uptake and low stroke volume. Those with a high capacity for oxygen uptake also have a larger stroke volume. A man with a maximal aerobic power of 5 liters/min eventually attains maximal stroke volume first at a work load giving an oxygen uptake of 2 liters/min. The one with a maximal oxygen uptake of 3.5 liters/min reaches his plateau for stroke volume when the oxygen uptake exceeds 1.3 liters/min.)

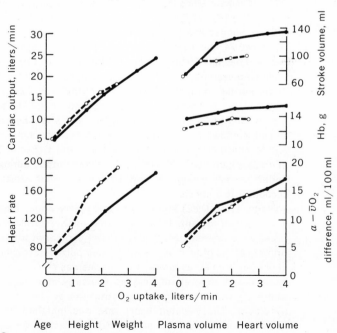

	Age	Height	Weight	Plasma volume	Heart volume
○ ♀	21 yrs	169 cm	62.7 kg	2.87 liters	640 ml
● ♂	24 yrs	179 cm	74.7 kg	3.70 liters	880 ml

Cardiac output, heart rate, and stroke volume at rest and during exercise can be modified by various factors. A few examples should be given:

1   As discussed above, the cardiac output at a given oxygen uptake is 1 to 2 liters/ min higher in the supine position than in the upright position, the heart rate being the same. In maximal leg exercise on a bicycle ergometer in the recumbent position, the heart rate, oxygen uptake, and cardiac output are, however, lower than in maximal exercise in the upright position. With the arm muscles engaged in exercise simultaneously with the legs, the oxygen uptake, heart rate, and cardiac output are not significantly increased above the values obtained in maximal erect leg exercise (Fig. 6-23). The feeling of strain is more related to the load per square centimeter of activated muscle than to the actual oxygen uptake and cardiac output. A work load of 2100 kpm*/min could be performed for 3 min in leg exercise, but for as long as 6 min if the arms shared the burden performing 600 and the legs 1500 (= 2100 kpm/min). The oxygen uptake, heart rate, and most likely the cardiac output were not different in the two experiments (P.-O. Åstrand and Saltin, 1961b).

2   Acute exposure to hypoxia elevates the cardiac output and heart rate during submaximal exercise, oxygen uptake being unchanged (Asmussen and Nielsen, 1955b; Stenberg et al., 1966). During maximal exercise, however, the cardiac output and heart rate are not different at normal barometric pressure or at 462 mm Hg. The maximal oxygen uptake is reduced at low oxygen tension in the inspired air, since the oxygen content of arterial blood is reduced (Stenberg et al., 1966) (Fig. 17-5).

3   After prolonged heavy exercise for 2 to 3 hr or exposure to high environmental temperature, the water loss may be 4 to 5 percent of the body weight. During submaximal standard exercise performed about 1 hr after exposure to heat or prolonged exercise, the heart rate was higher and the stroke volume lower than under normal conditions. The cardiac output was slightly reduced. During maximal exercise no systematic difference was noted in oxygen uptake, cardiac output, and heart rate attained before and after the exposure to heat or prolonged exercise, but the time for which the maximal test could be maintained was significantly reduced in the second effort (Saltin, 1964). Figure 6-28 summarizes experimental findings on cardiac output during maximal leg exercise under normal conditions and in the different situations discussed under sections 1 to 3 above.

4   Williams et al. (1962) studied the circulatory and metabolic reactions to work in a hot environment. They found that neither cardiac output nor $a-\bar{v}O_2$ differences were altered significantly in heat compared with normal conditions. The primary effect of heat on the circulatory parameters concerned in the transport of oxygen from lungs to working muscles was manifested in heart rate. This was increased at all levels of work up to nearly maximal levels of oxygen uptake, where, however, maximal heart rate was similar in hot and in comfortable conditions. The elevated heart rate was matched by a decrease in stroke volume, since the cardiac output was unaffected. During submaximal exercise

---

* kpm = kilopond meter; 1 kilopond is the force acting on the mass of 1 kilo at normal acceleration of gravity.

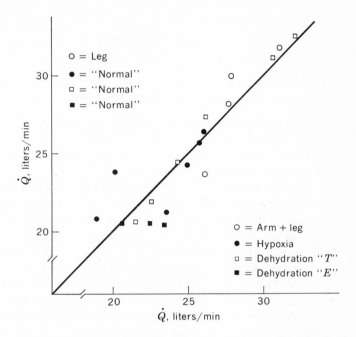

Fig. 6-28   Relation between maximal cardiac output during leg
exercise (bicycle ergometer) under normal conditions, plotted on
ordinate, and values during other forms of exercise and under conditions of hypoxia or dehydration (abscissa). The hypoxia was
attained by simulating an altitude of 4,000 m. The dehydration represented water loss to about 4 percent of body weight. Dehydration
"T" was achieved in a hot room; dehydration "E" resulted from water
loss during prolonged heavy exercise. Measurements were made after
1½ hours' rest at normal temperature. (Data from Saltin, 1964;
Stenberg et al., 1966, 1967.)

in the heat, part of the cardiac output was probably diverted away from working muscles to the skin (say 1 to 2 liters/min). Therefore, the muscles may
suffer a higher degree of anaerobic metabolism when exercise is performed
in a hot room. The somewhat lowered oxygen uptake and the appearance of
lactates in the blood at lower levels of work in hot conditions, noted by Williams
et al., are in agreement with this assumption.

5   Grimby and Nilsson (1963) studied the effect of pyrogen-induced fever on circulatory functions at rest and during exercise. During the flush phase the
cardiac output was higher than under normal conditions at rest and during
light to moderate exercise in the supine position. The stroke volume was not
significantly altered at rest and at light work but was somewhat reduced during
heavier exercise in the flush phase compared with normal conditions. The
response to maximal exercise in the two conditions was the same with regard
to cardiac output and oxygen uptake; the heart rate in the maximal test was a
few beats higher in fever than in normal condition.

6   Emotion may influence the heart rate markedly at rest and during submaximal work, but the circulatory parameters are usually not affected by psychological factors when the work becomes heavy.

Ingestion of meals and smoking may affect cardiac output and heart rate. No information is available on the effect when maximal exercise is performed.

This discussion of the oxygen-transporting system during muscular activity reveals that a regulation of the circulation at rest and during submaximal exercise is not guided only by the metabolic rate; various other factors may influence the circulatory response to exercise.

Of probable primary importance for the regulatory mechanisms is the relation between volume of oxygen supplied to the metabolically active tissue (cardiac output times the oxygen content of arterial blood) and the oxygen demand of the tissue. Within limits, other demands can be met by compensatory mechanisms; e.g., if the stroke volume is reduced, the heart rate may increase so that an adequate cardiac output is still maintained.

During maximal exercise, however, the cardiac output, oxygen uptake, and heart rate are remarkably fixed to values typical for the individual even if the performance is made under adverse conditions. In this situation apparently all circulatory functions of decisive importance for a maximal oxygen supply to working muscles are actually devoted to this task. Irrespective of environment, external and internal (within limits) maximal vasoconstriction occurs in the blood vessels of the viscera and skin, so that practically the entire cardiac output is diverted to the vigorously working muscles. Maximal exercise involving large muscle groups creates an emergency reaction in the circulatory adjustment which favors the exercising muscles, including the heart, at the expense of all other tissues with the exception of the central nervous system.

REFERENCES

Abrahams, V. C., S. M. Hilton, and A. Zbrozyna: Active Muscle Vasodilation Produced by Stimulation of the Brain Stem: Its Significance in the Defense Reaction, *J. Physiol. (London)*, **154**:491, 1960.

Arenander, E.: Hemodynamic Effects of Varicose Veins and Results of Radical Surgery, *Acta Chir. Scand.* (Suppl. 260):1, 1960.

Asmussen, E., and E. H. Christensen: Einfluss der Blutverteilung auf den Kreislauf bei körperlicher Arbeit, *Skand. Arch. Physiol.*, **82**:185, 1939.

Asmussen, E., E. H. Christensen, and M. Nielsen: Pulsfrequenz und Körperstellung, *Skand. Arch. Physiol.*, **81**:190, 1939.

Asmussen, E., and M. Nielsen: The Cardiac Output in Rest and Work Determined Simultaneously by the Acetylene and the Dye Injection Methods, *Acta Physiol. Scand.*, **27**:217, 1953.

Asmussen, E., and M. Nielsen: Cardiac Output during Muscular Work and Its Regulation, *Physiol. Rev.*, **35**:778, 1955a.

Asmussen, E., and M. Nielsen: Cardiac Output in Rest and Work at Low and High Oxygen Pressures, *Acta Physiol. Scand.*, **35**:73, 1955b.

Åstrand, I.: Aerobic Work Capacity in Men and Women with Special Reference to Age, *Acta Physiol. Scand.*, **49**(Suppl. 169), 1960.

Åstrand, I., P.-O. Åstrand, and K. Rodahl: Maximal Heart Rate during Work in Older Men, *J. Appl. Physiol.*, **14**:562, 1959.

Åstrand, I., T. E. Cuddy, J. Landegren, R. O. Malmborg, and B. Saltin: Hemodynamic Response to Exercise during Atrial Flutter and Sinus Rhythm, *Acta Med. Scand.*, **173**:121, 1963.

Astrand, I., A. Guharay, and J. Wahren: Circulatory Response to Arm Exercise with Different Arm Positions, *J. Appl. Physiol.*, **25**:528, 1968.

Åstrand, P.-O.: "Experimental Studies of Physical Working Capacity in Relation to Sex and Age," Munksgaard, Copenhagen, 1952.

Åstrand, P.-O.: Breath Holding during and after Muscular Exercise, *J. Appl. Physiol.*, **15**:220, 1960.

Åstrand, P.-O., and B. Saltin: Oxygen Uptake during the First Minutes of Heavy Muscular Exercise, *J. Appl. Physiol.*, **16**:971, 1961a.

Åstrand, P.-O., and B. Saltin: Maximal Oxygen Uptake and Heart Rate in Various Types of Muscular Activity, *J. Appl. Physiol.*, **16**:977, 1961b.

Åstrand, P.-O., L. Engström, B. O. Eriksson, P. Karlberg, I. Nylander, B. Saltin, and C. Thorén: Girl Swimmers, *Acta Paediat.* (Suppl. 147), 1963.

Åstrand, P.-O., T. E. Cuddy, B. Saltin, and J. Stenberg: Cardiac Output during Submaximal and Maximal Work, *J. Appl. Physiol.*, **19**:268, 1964.

Åstrand, P.-O., and E. H. Christensen: Aerobic Work Capacity, in F. Dickens, E. Neil, and W. F. Widdas (eds.), p. 295, "Oxygen in the Animal Organism," Pergamon Press, New York, 1964.

Åstrand, P.-O., B. Ekblom, R. Messin, B. Saltin, and J. Stenberg: Intra-arterial Blood Pressure during Exercise with Different Muscle Groups, *J. Appl. Physiol.*, **20**:253, 1965.

Barcroft, H.: Sympathetic Control of Vessels in the Hand and Forearm Skin, *Physiol. Rev.*, **40**(Suppl. 4):81, 1960.

Barcroft, H., and H. J. C. Swan: Sympathetic Control of Human Blood Vessels, Edward Arnold (Publishers) Ltd., London, 1953.

Bazett, H. C.: A Consideration of the Venous Circulation, in B. W. Zweifach and E. Shorr (eds.), "Factors Regulating Blood Pressure," p. 53, Josiah Macy Jr. Foundation, New York, 1950.

Bevegård, S.: Studies on the Regulation of the Circulation in Man, *Acta Physiol. Scand.*, **57**(Suppl. 200), 1962.

Bevegård, S., and J. T. Shepherd: Circulatory Effects of Stimulating the Carotid Artery Stretch Receptors in Man at Rest and during Exercise, *Clin. Res.*, **12**:335, 1964.

Bevegård, B. S., and J. T. Shepherd: Regulation of the Circulation during Exercise in Man, *Physiol. Rev.*, **47**:178, 1967.

Bishop, J. M., K. W. Harold, S. W. Taylor, and P. N. Wormald: Changes in Arterial-hepatic Venous Oxygen Content Difference during and after Supine Leg Exercise, *J. Physiol. (London)*, **137**:309, 1957.

Braunwald, E., and E. R. Kelly: The Effect of Exercise on Central Blood Volume in Man, *J. Clin. Invest.*, **39**:413, 1960.

Braunwald, E., E. H. Sonnenblick, J. Ross, Jr., G. Glick, and S. E. Epstein: An Analysis of the
Cardiac Response to Exercise, *Circulation Res.*, **20** and **21**:44, 1967.

Bucht, H., J. Ele, H. Eliasch, A. Holmgren, B. Josephson, and L. Werkö: The Effect of
Exercise in the Recumbent Position on the Renal Circulation and Sodium Excretion in Normal Individuals, *Acta Physiol. Scand.*, **28**:95, 1953.

Burton, A. C.: Hemodynamics and the Physics of the Circulation, in T. C. Ruch and
H. D. Patten (eds.), "Physiology and Biophysics," pp. 523–542, W. B. Saunders
Company, Philadelphia, 1965.

Burton, A. C.: "Physiology and Biophysics of the Circulation," The Year Book Medical
Publishers, Inc., Chicago, 1965.

Carlson, L., and B. Pernow: Oxygen Utilization and Lactic Acid Formation in the Legs at
Rest and during Exercise in Normal Subjects and in Patients with Arteriosclerosis
Obliterans, *Acta Med. Scand.*, **164**:39, 1959.

Chapman, C. B., A. Henschel, J. Minckler, A. Forsgren, and A. Keys: The Effect of Exercise
on Renal Plasma Flow in Normal Male Subjects: *J. Clin. Invest.*, **27**:639, 1948.

Christensen, E. H.: Beiträge zur Physiologie schwerer körperlicher Arbeit. Minutenvolumen
und Schlagvolumen des Herzens während schwerer körperlicher Arbeit, *Arbeitsphysiol.* **4**:453, 470, 1931.

Christensen, E. H., and M. Nielsen: Investigation of the Circulation in the Skin at
Beginning of Muscular Work, *Acta Physiol. Scand.*, **4**:162, 1942.

Daly, M. de B.: Reflex Circulatory and Respiratory Responses to Hypoxia, in F. Dickens
and E. Neil (eds.), "Oxygen in the Animal Organism," p. 267, Pergamon Press,
New York, 1964.

Daly, M. de B., and M. J. Scott: The Effect of Stimulation of the Carotid Body Chemoreceptors on Heart Rate in the Dog, *J. Physiol. (London)*, **144**:148, 1958.

Donald, D. E., S. E. Milburn, and J. T. Shepherd: Effect of Cardiac Denervation on the
Maximal Capacity for Exercise in the Racing Greyhound, *J. Appl. Physiol.*, **19**:849,
1964.

Donald, K. W., J. K. Bishop, and O. L. Wade: A Study of Minute to Minute Changes of
Arteriovenous Oxygen Content Difference, Oxygen Uptake and Cardiac Output
and Rate of Achievement of a Steady State during Exercise in Rheumatic Heart
Disease, *J. Clin. Invest.*, **33**:1946, 1954.

Donald, K. W., A. R. Lind, G. W. McNicol, P. W. Humphreys, S. H. Taylor, and H. P.
Staunton: Cardiovascular Responses to Sustained (Static) Contractions, *Circulation Res.*, **20** and **21**:15, 1967.

Eckstein, R. W.: Effect of Exercise and Coronary Artery Narrowing on Coronary Collateral
Circulation, *Circulation Res.*, **5**:230, 1957.

Ekblom, B.: Effect of Physical Training on Oxygen Transport System in Man, *Acta Physiol.
Scand.* (Suppl. 328), 1969.

Ekblom, B., and L. Hermansen: Cardiac Output in Athletes, *J. Appl. Physiol.* **25**:619,
1968.

Evans, C. L.: The Velocity Factor in Cardiac Work, *J. Physiol.*, **52**:6, 1918.

Folkow, B.: Range of Control of the Cardiovascular System by the Central Nervous System, *Physiol. Rev.*, **40**(Suppl. 4):93, 1960.

Folkow, B.: Homeostasis of Peripheral Circulation, in "Les Concepts de Claude Bernard
sur le Milieu Intérieur," p. 165, Masson et Cie, Libraires de l'Académie de Médecine, Paris, 1967.

Folkow, B.: Circulatory Adaptations to Diving in Ducks, *Proc. 24th Congress Intern.*

Union Physiol. Sci., **6**:23, 1968.

Gregg, D. E., and J. D. Coffman: Coronary Circulation, in D. I. Abramson (ed.), "Blood Vessels and Lymphatics," chap. 9, p. 269, Academic Press Inc., New York, 1962.

Gregg, D. E., and H. D. Green: Registration and Interpretation of Normal Phasic Inflow into Left Coronary Artery by Improved Differential Manometric Method, *Am. J. Physiol.*, **130**:114, 1940.

Grimby, G., and N. J. Nilsson: Cardiac Output during Exercise in Pyrogen-induced Fever, *Scand. J. Clin. Lab. Invest.*, **15**(Suppl. 69):44, 1963.

Grimby, G., and B. Saltin: Physiological Analysis of Physically Well-trained Middle-aged and Old Athletes, *Acta Physiol. Scand.*, **179**:513, 1966.

Guyton, A. C.: "Circulatory Physiology: Cardiac Output and Its Regulation," W. B. Saunders Company, Philadelphia, 1963.

Haddy, F. J., L. A. Sopirstein, and R. R. Sonnenschein: Arterial and Arteriolar Systems: Biophysical Principles and Physiology, in D. I. Abramson (ed.), "Blood Vessels and Lymphatics," chap. 2, p. 61, Academic Press Inc., New York, 1962.

Hansen, T.: Osmotic Pressure Effect of the Red Cells: Possible Physiological Significance, *Nature*, **190**:504, 1961.

Henry, J. P., and J. W. Pearce: The Possible Role of Cardiac Atrial Stretch Receptors in the Induction of Change in Urine Flow, *J. Physiol. (London)*, **131**:572, 1956.

Heymans, C., and E. Neil: "Reflexogenic Areas of the Cardiovascular System," Churchill, London, 1958.

Hollmann, H.: "Höchst- und Dauerleistungsfähigheit des Sportlers," Johann Ambrosius Barth, Munich, 1963.

Holmgren, A.: Circulatory Changes during Muscular Work in Man, *Scand. J. Clin. Lab. Invest.* (Suppl. 24), 1956.

Irving, L.: Bradycardia in Human Divers, *J. Appl. Physiol.*, **18**:489, 1963.

Kjellberg, S. R., U. Rudhe, and T. Sjöstrand: The Amount of Hemoglobin (Blood Volume) in Relation to the Pulse Rate and Heart Volume during Work, *Acta Physiol. Scand.*, **19**:152, 1949.

Kjellmer, I.: Studies on Exercise Hyperemia, *Acta Physiol. Scand.*, **64**(Suppl. 244), 1965.

Kreeker, E. J., and E. H. Wood: Comparison of Simultaneously Recorded Central and Peripheral Arterial Pressure Pulses during Rest, Exercise and Tilted Position in Man, *Circulation Res.*, **3**:623, 1955.

Krogh, A.: "The Anatomy and Physiology of Capillaries," rev. ed., Yale University Press, New Haven, Conn., 1929.

Krogh, A., and J. Lindhard: Measurements of the Blood Flow through the Lungs of Man, *Skand. Arch. Physiol.*, **27**:100, 1912.

Lundgren, N.: The Physiological Effects of Time Schedule Work on Lumber Workers, *Acta Physiol. Scand.*, **13**(Suppl. 41), 1946.

Lutz, B. R., and G. P. Fulton: Structural Basis of the Microcirculation, in D. I. Abramson (ed.), "Blood Vessels and Lymphatics," chap. 5, p. 137, Academic Press Inc., New York, 1962.

Marx, H. J., L. B. Rowell, R. D. Conn, R. A. Bruce, and F. Kusumi: Maintenance of Aortic Pressure and Total Peripheral Resistance during Exercise in Heat, *J. Appl. Physiol.*, **22**:519, 1967.

Mellander, S., Comparative Studies on the Adrenergic Neurohormonal Control of Resistance and Capacitance Blood Vessels in the Cat, *Acta Physiol. Scand.*, **50**(Suppl. 176), 1960.

Mellander, S., and B. Johansson: Control of Resistance, Exchange, and Capacitance Functions in the Peripheral Circulation, *Pharm. Rev.*, **20**:117, 1968.

Mellerowicz, H.: "Ergometrie," Urban & Schwarzenberg, Munich, 1962.

Messer, J. V., R. J. Wagman, H. J. Levine, W. A. Neill, N. Krasmow, and R. Gorlin: Patterns of Human Myocardial Oxygen Extraction during Rest and Exercise, *J. Clin. Invest.*, **41**:725, 1962.

Mitchell, J. H., B. J. Sproule, and C. B. Chapman: Factors Influencing Respiration during Heavy Exercise, *J. Clin. Invest.*, **37**:1693, 1958a.

Mitchell, J. H., B. J. Sproule, and C. B. Chapman: The Physiological Meaning of the Maximal Intake Test, *J. Clin. Invest.*, **37**:538, 1958b.

Musshoff, K., H. Reindell, and H. Klepzig: Stroke Volume, Arteriovenous Difference, Cardiac Output and Physical Working Capacity and Their Relationship to Heart Volume, *Acta Cardiol. Brux.*, **14**:427, 1959.

Neil, E.: Afferent Impulse Activity in Cardiovascular Receptor Fibers, *Physiol. Rev.*, **40**(Suppl. 4):201, 1960.

Ochsner, A., Jr., R. Colp, Jr., and G. E. Burch: Normal Blood Pressure in the Superficial Venous System of Man at Rest in the Supine Position, *Circulation*, **3**:674, 1951.

Olsen, C. R., D. D. Fanestil, and P. F. Scholander: Some Effects of Breath Holding and Apneic Underwater Diving on Cardiac Rhythm in Man, *J. Appl. Physiol.*, **17**:461, 1962.

Otis, A. B.: The Control of Respiratory Gas Exchange between Blood and Tissues, in D. J. C. Cunningham and B. B. Lloyd (eds.), "The Regulation of Human Respiration," p. 111, Blackwell Scientific Publications, Ltd., Oxford, 1963.

Pappenheimer, J. R.: Central Control of Renal Circulation, *Physiol. Rev.*, **40**(Suppl. 4):35, 1960.

Peiss, C. N.: Sympathetic Innervation of Arterial Tree, in D. I. Abramson (ed.), "Blood Vessels and Lymphatics," chap. 4, p. 96, Academic Press Inc., New York, 1962.

Peterson, L. H.: Cardiovascular Control and Regulation, in "Les Concepts de Claude Bernard sur le Milieu Intérieur," p. 191, Masson et Cie, Libraires de l'Académie de Médecine, Paris, 1967.

Reeves, J. T., R. F. Grover, S. G. Blount, Jr., and G. F. Filley: Cardiac Output Responses to Standing and Treadmill Walking, *J. Appl. Physiol.*, **16**:283, 1961.

Reindell, H.: Über den Kreislauf der Trainierten. Über die Restblutmenge des Herzens und über die besondere Bedeutung röntgenologischer (kymographischer) hämodynamische Beobachtungen in Ruhe und nach Belastung, *Arch. Kreislaufforsch.*, **12**:265, 1943.

Reindell, H., H. Klepzig, H. Steim, K. Musshoff, H. Roskamm, and E. Schildge: "Herz Kreislaufkrankheiten und Sport," Johann Ambrosius Barth, Munich, 1960.

Robinson, S.: Experimental Studies of Physical Fitness in Relation to Age, *Arbeitsphysiol.*, **10**:251, 1938.

Rodbard, S.: Arterial and Arteriolar Systems: Biophysical Principles and Physiology, in D. I. Abramson (ed.), "Blood Vessels and Lymphatics," chap. 2, p. 31, Academic Press Inc., New York, 1962.

Roughton, F. J. W.: Transport of Oxygen and Carbon Dioxide, in W. O. Fenn and H. Rahn (eds.), "Handbook of Physiology: Respiration," vol. 1, p. 767, American Physiology Society, Washington, D.C., 1964.

Rowell, L. B., J. R. Blackmon, and R. A. Bruce: Indocyanine Green Clearance and Estimated Hepatic Blood Flow during Mild to Maximal Exercise in Upright Man, *J. Clin. Invest.*, **43**:1677, 1964.

Ruch, T. C., and H. D. Patten: "Physiology and Biophysics," W. B. Saunders Company, Philadelphia, 1965.

Rushmer, R. F.: Regulation of the Heart's Function, *Circulation*, **21**:744, 1960.

Rushmer, R. F.: Control of Cardiac Output, in T. C. Ruch and H. D. Patten (eds.), "Physiology and Biophysics," p. 644, W. B. Saunders Company, Philadelphia, 1965.

Saltin, B.: Aerobic Work Capacity and Circulation of Exercise in Man, *Acta Physiol. Scand.*, **62**(Suppl. 230), 1964.

Saltin, B., G. Blomqvist, J. H. Mitchell, R. L. Johnson, Jr., K. Wildenthal, and C. B. Chapman: Response to Submaximal and Maximal Exercise after Bedrest and Training, *Circulation* **38** (Suppl. 7), 1968.

Sarnoff, S. J., and J. H. Mitchell: The Control of the Function of the Heart, in W. F. Hamilton and P. Dow (eds.), vol. 1, p. 489, American Physiological Society, Washington, D.C., 1962.

Scheinberg, P., L. I. Blackburn, M. Rich, and M. Saslaw: Effects of Vigorous Physical Exercise on Cerebral Circulation and Metabolism, *Am. J. Med.*, **16**:549, 1954.

Scholander, P. F., H. T. Hammel, H. LeMessurier, E. Hemmingsen, and W. Garey: Circulatory Adjustment in Pearl Divers, *J. Appl. Physiol.*, **17**:184, 1962.

Sjöstrand, F. S., and E. Andersson-Cedergren: Intercolated Discs of Heart Muscle, in G. H. Bourne (ed.), "Structure and Function of Muscle," vol. 1, chap. 12, p. 421, Academic Press Inc., New York, 1960.

Sjöstrand, T.: Volume and Distribution of Blood and Their Significance in Regulating Circulation, *Physiol. Rev.*, **33**:202, 1953.

Skinner, N. S., Jr., and W. J. Powell: Regulation of Skeletal Muscle Blood Flow during Exercise, *Circulation Res.*, **20** and **21**:59, 1967.

Stenberg, J., P.-O. Åstrand, B. Ekblom, J. Royce, and B. Saltin: Hemodynamic Response to Work with Different Muscle Groups in Sitting and Supine, *J. Appl. Physiol.*, **22**:61, 1967.

Stenberg, J., B. Ekblom, and R. Messin: Hemodynamic Response to Work at Simulated Altitude, 4,000 m, *J. Appl. Physiol.*, **21**:1589, 1966.

Strandell, T.: Circulatory Studies on Healthy Old Men, *Acta Med. Scand.*, **175**(Suppl. 414), 1964.

Uvnäs, B.: Sympathetic Vasodilator System and Blood Flow, *Physiol. Rev.*, **40**(Suppl. 4): 69, 1960.

Wade, O. L., and J. M. Bishop: "Cardiac Output and Regional Blood Flow," Blackwell Scientific Publications, Ltd., Oxford, 1962.

Wade, O. L., B. Combes, A. W. Childs, H. O. Wheeler, A. Cournand, and S. E. Bradley: The Effect of Exercise on the Splanchnic Blood Flow and Splanchnic Blood Volume in Normal Man, *Clin. Sci.*, **15**:457, 1956.

Williams, C. G., G. A. G. Bredell, C. H. Wyndham, N. B. Strydom, J. F. Morrison, J. Peter, P. W. Fleming, and J. S. Ward: Circulatory and Metabolic Reactions to Work in Heat, *J. Appl. Physiol.*, **17**:625, 1962.

Zweifach, B. W.: Basic Mechanisms in Peripheral Vascular Homeostasis, in "Factors Regulating Blood Pressure," Transactions of the Third Conference, 1949, New York, Josiah Macy Jr. Foundation, New York, 1950.

Zweifach, B. W.: "Functional Behaviour of the Microcirculation," Charles C Thomas, Publisher, Springfield, Ill., 1961.

7

Respiration

contents

chapter seven

Respiration

MAIN FUNCTION

The living cell uses oxygen for its metabolism, in the course of which carbon dioxide is produced. Thus, the concentration of oxygen within the cell is lowered, and oxygen will tend to diffuse toward the place of combustion. Similarly, the carbon dioxide produced will tend to diffuse away. The exchange of O_2 and CO_2 is dependent on the distance the molecules have to travel and the pressure gradient. In the single-cell organism the surface can effectively be utilized for the respiratory exchange. By diffusion, oxygen can easily reach every point within the cell, and CO_2 can be eliminated. The calculations of Krogh (1941) indicated that when metabolism is fairly high, diffusion can only provide sufficient oxygen to organisms with a diameter of 1 mm or less. A spherical organism with a 1-cm radius having an oxygen uptake of 100 ml/(kg) (hr) would need an external oxygen pressure of 25 atm to secure the oxygen supply to its center by diffusion. In multicelled animals the problem of gas transports is solved in various ways; e.g., numerous airways, like tracheae in insects, or specialized organs, like lungs or gills, are developed, exposing an enlarged respiratory surface to the external medium to effect exchange of oxygen and carbon dioxide. With the gas exchanger located distant from the sites of metabolism, two transport systems carry O_2 and CO_2 between the respiratory surface and the cells in which the metabolism proceeds: a circulatory system supplies the blood, and a respiratory system supplies the air to the lungs. The spongelike structure of the gas-exchange area provides an enormous contact surface between air and blood. This air-tissue-blood interface of an average adult human lung is estimated to be some 70 to 90 m^2 with a thickness of the tissues varying from 0.2 μ to several μ (average some 0.7 μ). Fresh air and "new" blood must continuously be supplied to the many million gas exchange units, since the store of oxygen in the body is very limited and the central nervous system and heart muscle do not tolerate any lag in the supply of oxygen.

In this chapter we will discuss the exchange of oxygen and carbon dioxide between ambient air and blood. For a comprehensive discussion of the many aspects of respiration the reader is referred to the "Handbook of Physiology" (Fenn and Rahn, eds., 1964–65) and other books (e.g., Cherniack and Cherniack, 1961; Bates and Christie, 1964; Comroe, 1965; Cotes, 1965).

The ideal gas exchanger should (1) provide *a large contact area* between the air and blood with a very thin membrane separating the two media, since diffusion is directly proportional to the area but inversely related to the thickness of the membrane. The membrane should cause a minimal resistance to gas flow. (2) The inspired air must become *saturated with water vapor and heated* to tissue temperature to *protect the delicate membranes from injury;* any particles and agents in the air which may be harmful should be removed during the passage through the airways, and if introduced, be expelled. (3) Variations in oxygen and carbon dioxide in the blood leaving the lungs should only vary within small limits, and therefore, *the distribution of gas and blood in the many exchange units should be closely matched.* (4) *The gas exchange and therefore the perfusion of the units must be proportional to the uptake of oxygen and production of carbon dioxide by the cells.* This demands some sort of regulative mechanism linking the needs of the distant cells with the external respiration. The anatomic location of airways and lungs inevitably causes respiration to be influenced by movements of the trunk. Speaking and singing modify the respiration.

Point (4) of the factors listed above is of special interest in this connection, since muscular exercise may increase the oxygen uptake of the body some 20 times the resting level with a similar rise in CO_2 production. The respiration also plays an important role in the *maintenance of the pH of the blood at normal levels.* Hydrogen ions cannot be exchanged between air and blood, but the acid-base equilibrium of the blood (and tissue) is closely associated with the CO_2 content and pressure in the blood (see Chap. 5). During heavy exercise, lactic acid, an acid stronger than H_2CO_3, is produced in the working muscles and more CO_2 is formed, which stimulates respiration. Elimination of CO_2 during hyperventilation will reduce the effect of lactic acid on the blood pH.

Before we discuss respiration during exercise, a summary of the basic respiratory function will be presented.

ANATOMY AND HISTOLOGY

Airways

Figure 7-1 illustrates the respiratory tree and its subdivision into finer airways until they finally terminate in numerous blind pouches, the alveoli. Figure 7-2 presents schematically the general architecture of the airways. Via nose or mouth the inspired gases pass through the pharynx, larynx, and trachea into the bron-

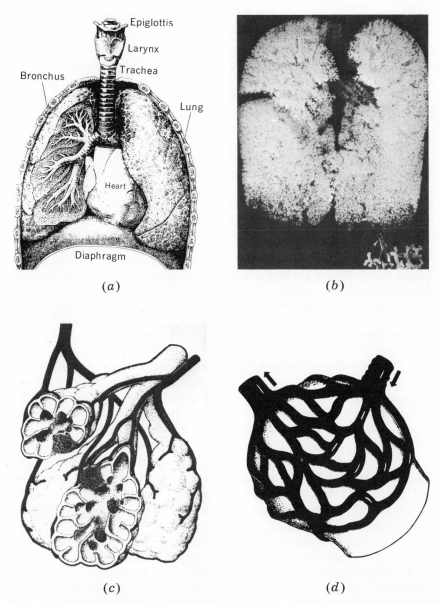

Fig. 7-1   (a) *Principal organs of breathing. In this drawing the ribs, the large arteries from the heart, and part of one lung have been cut away.*
(b) *A cast of the complete air spaces of the lung, showing the millions of air sacs at the end of the bronchioles. Inset: the terminal portion of a bronchiole magnified. (From C. M. Fletcher, BBC Publication 3 s. 1963.)*
(c) *Schematic drawing of a bronchiole and its air sacs and alveoli.*
(d) *An alveolus embraced by the capillary network.*

chial tree. The airways branch by regular dichotomy and the two daughter branches in turn become parent branches, etc. They terminate in the *alveoli*, polyhedral or cuplike outpouchings of the finer airways. Their diameter is up to 300 μ. The surface of the alveolus is not smooth but is corrugated by the capillaries and by various subcellular structures, like nuclei, bulging into the alveolus (Fig. 7-3). The alveoli share their walls with their neighbors and form the spongelike texture of the lungs (Fig. 7-1).

In the adult man about 23 such generations of branches can be traced as the airways subdivide into the periphery of the lung and distribute them among the numerous respiratory units. The first 16 generations roughly constitute the "conductive zone" where practically no gas exchange occurs between blood and air. Then follow the "transitory and respiratory zones": generations 17 to 19 of branching form the respiratory bronchioles with a diameter of about 1 mm, which subdivide to produce about 1 million alveolar ducts. The last of a short series of alveolar ducts terminates in rotundate enclosures: the alveolar ducts with a diameter of about 400 μ. The cylindrical surface of the respiratory bronchioles bears a smaller number of variously spaced alveoli, but the alveolar ducts and sacs are fully alveolated. Therefore, they lack proper walls but open out on all sides into alveoli, some 300 million altogether in the adult (Fig. 7-1). The respiratory and transitory zones, including the alveoli, amount to about 90 percent of the lung volume; about 65 percent of the air in the lungs is in the actual alveoli at three-fourths of the maximal inflation of the lungs (Weibel, 1964).

The bronchi and bronchioles that are wider than about 1 mm have a discontinuous cartilaginous support in the wall. Muscle fibers in circular or crisscrossing bundles are incorporated into a complex connective tissue framework of collagenous reticular and elastic fibers. The inner surface is covered with a ciliated epithelium, which is usually simple. Goblet cells occur singly or in groups between the epithelial cells and produce a secretion. In the finest bronchioles the mucus-secreting elements become sparse and finally absent. They are lacking cartilages, and the ciliated cells also disappear gradually. Muscle fibers as well as elastic, collagenous, and reticular fibers provide the supporting latticework of the interalveolar septum and a framework for the entrance of the alveoli and alveolar sacs and ducts.

In the fetus the alveoli are atelectatic, but the first influx of air in the newborn child provides a force that stretches the original cuboidal epithelium lining of the alveoli into an extremely thin layer of squamous cells. The lung of the newborn is not fully developed. The airways have subdivided into only some 17 generations of branchings and the number of alveoli is less than one-tenth of that found in the adult. As time goes on additional branches are added and many more alveoli are formed as new ramifications grow out, so that before the age of ten years the adult number is reached. Whether or not strenuous physical efforts may serve to add more branches and alveoli to the mature lung is not known (see below). The relative alveolar volume decreases with age.

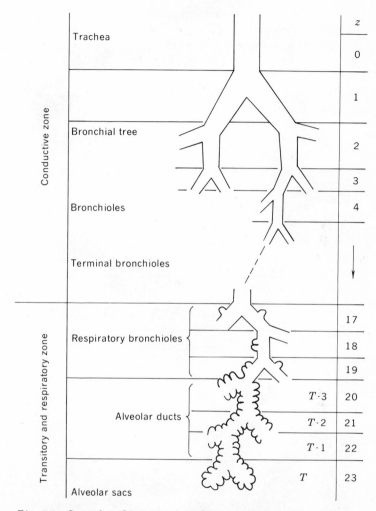

Fig. 7-2 General architecture of conductive and transitory airways. z designates the order of generations of branching, T, the terminal generation. A more detailed discussion appears in the text. (Weibel, 1963.)

Despite the latticework fibers supporting the alveolar walls, the air-tissue-blood interface would provide special problems due to the surface tension which is created. This tension tends to decrease the surface wall to a minimum so that the alveoli may collapse. The alveoli are, however, lined with an insoluble surface film of lipoprotein, about 50 Å thick. It is produced in the alveoli and keeps the alveoli open and free from transudate from the blood by lowering the surface tension. This function is especially important as the volume decreases, e.g., during forced expiration, which would otherwise empty the small alveoli. During quiet breathing there may actually be a reduced effect

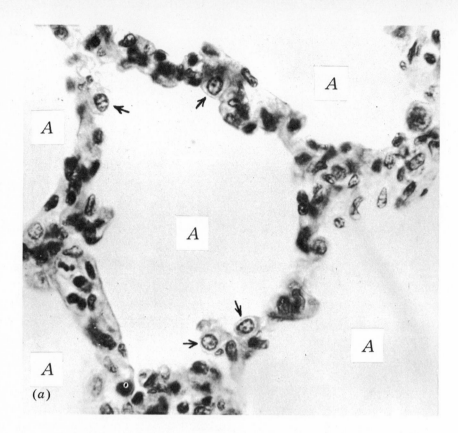

(a)

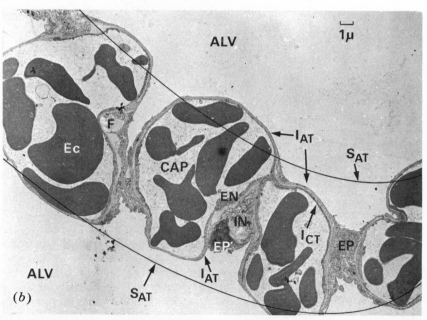

(b)

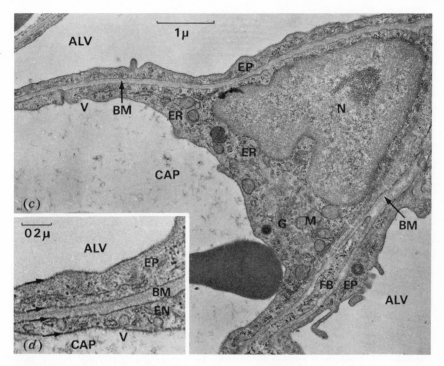

Fig. 7-3 (a) *Histologic section through one complete alveolar outline (center) and portions of four adjacent alveoli (A); magnification ×650. (Modified from Krahl, 1964.)*
(b, c, d) *Electron micrographs of interalveolar septum of rat lung in cross section showing the barrier composed of alveolar epithelial (EP), capillary endothelial cells (EN), and some interstitial elements. (From Weibel, 1964.)*
(b) *Magnification ×5,500. Note the difference between the estimated alveolar surface (S$_{AT}$) and the real "corrugated" air-tissue interface (I$_{AT}$). I$_{CT}$ = tissue-blood interface; CAP = capillaries; IN = interstitium; F = fibrous elements; Ec = erythrocytes.*
(c) *Magnification ×23,500. BM = basal membranes; N = nucleus; EP = endoplasmic reticulum; V = pinocytotic vesicle.*
(d) *Magnification ×59,000 of the thin portion of air-blood barrier with four membranes.*

of the surface film and an occasional collapse of alveolar units. Forced inflation of the lungs may cause more material to be provided for the lining film and open the alveoli. A yawn or deep breath may exert such a beneficial function. The surface film thus stabilizes the small alveolar spaces and enables the lung to retain air at low inflation pressure (Pattle, 1965).

Blood Vessels

The pulmonary arteries enter the lungs with the stem bronchi and provide arterial partners to the airways as they subdivide toward the respiratory zones of the lungs. The arterioles follow the bronchioles, alveolar ducts, and sacs, and provide short twigs to capillary networks enveloping the alveoli surrounding the

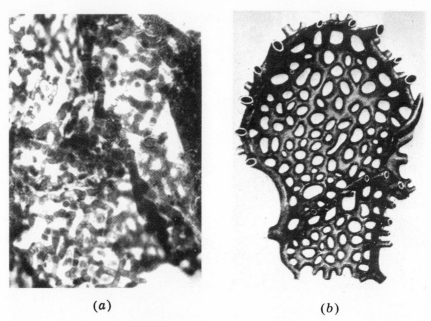

(a)   (b)

Fig. 7-4   (a) Network of capillaries in alveolar walls. Magnification ×375. (From Miller, 1947.)
(b) Blood-filled capillary network in interalveolar septum of human lung. Larger, dark vessel to the right (arteriole) gives off short precapillaries which open at once into pulmonary capillaries. Magnification ×650. (From Krahl, 1964.)

particular airway terminal and to any other alveoli in the immediate vicinity. Each alveolus may be covered by a capillary network consisting of almost 2,000 segments; the capillary networks in the lungs are the richest in the body and are more or less continuous throughout large parts of the lungs (Fig. 7-4). The air-blood "barrier" is formed by the continuous alveolar epithelial and capillary endothelial cells with a tenuous interstitium with fibrous elements in between the two cell layers. The thickness of the "barrier" can vary from about 0.2 μ to several μ, and the variations are caused by various structures scattered throughout the continuous cell layers (Fig. 7-3). It is evident that the capillary networks are so arranged that each vessel is usually in contact with two neighboring alveoli.

The airways, pulmonary vessels, and lymphoid tissue are supplied by vessels derived from the systemic circulation (bronchial arteries).

Nerves

Vagal efferent fibers go to the smooth muscles of the bronchial tree as far as the terminations of the alveolar ducts and sacs, and to the bronchial mucous glands.

Nerve impulses stimulate the smooth muscles to contract and activate the glands. *Vagal afferent fibers* carry impulses from special stretch receptors scattered in lungs and pleura. Sympathetic fibers act as bronchodilators. There are many free nerve endings in the bronchial walls and pleura, but their function is not known.

"AIR CONDITION"

The inspired air may be cold or hot, dry or moist, but due to the rich blood supply of the mucous membranes of the nose, the mouth, and the pharynx, the air temperature becomes adjusted to body temperature and also moistened. In man exposed to the cold air in the Arctic, or the hot air in the Tropics, the inspired air is about 37°C by the time it reaches the pharynx. As a matter of fact, in experimental animals exposed to $-100°C$ and up to $+500°C$ the air temperature was warmed or cooled during its passage through the upper respiratory tree and attained body temperature in the tracheobronchial tree (Moritz et al., 1945). The mouth and pharynx can perform these air-conditioning functions as effectively as the nose and pharynx. Air saturated with water vapor at 37°C has a $P_{H_2O} = 47$ mm Hg; the content of water is then 43.9 g/m^3. At low temperatures the water content in the air is low. Even if saturated, the air at 0°C only contains 4.85 g H_2O/m^3. In a normal climate about 10 percent of the total heat loss of the body at rest or during work takes place through the respiratory tracts by the air conditioning of the inspired air. At -15 to$-20°C$ the percentage would be about 25 (see Chap. 15). The respiratory tract serves as a regenerative system: the heating and humidifying of inspired air cool the mucosa. But during expiration, some of the heat and water are recovered by the mucosa from the passing alveolar air. Body heat and water are conserved (Cole, 1954). On a very cold day this condensation of water vapor may result in excessive accumulation of water in the nostrils, leading to a runny nose! A cross-country skier breathing 100 liters/min of air at $-20°C$ must, in 1 hr, add about 250 ml of water to this air. Not all of this water volume is expired, however, thanks to the regenerative system.

FILTRATION AND CLEANSING MECHANISMS

If living in a city, we may inhale billions of particles of foreign matter every day. Particles larger than about 10 μ are effectively removed from the inspired air in the nose, where they are trapped by the hair or the moist mucous membranes. Those particles which escape these obstacles usually settle on the walls of the trachea, the bronchi, and the bronchioles. Therefore only a few and very small particles are likely to reach the alveoli, and this part of the lung is practically sterile.

As mentioned above, the epithelium of the airways within the lungs consists of *ciliated cells*. In the conductive zone each cell carries up to 300 cilia about 6 to 7 μ in length, at the free cell surface. The cilia of many thousands of cells beat in an organized whiplike fashion in strokes, like oars of a boat, with a rapid upward propulsive stroke followed by a slower recovery downward stroke. This goes on continuously day and night. The cilia are covered by a continuous surface of watery mucus. By the ciliar activity this fluid carpet with all the entrapped particles moves towards the larynx at a speed of well over 1 cm/min. This mucus is either expectorated or swallowed. The ciliary escalator is remarkably resistant to noxious influences. However, cigarette smoke has a deleterious effect on the ciliar function. They slow down or stop their beating when exposed to the smoke.

From time to time we may sneeze or cough, and with the explosive blast this causes (the air moves with a speed actually approaching the speed of sound) foreign particles may be expelled.

MECHANICS OF BREATHING

Pleurae

The lungs increase and decrease their volume with the reciprocating movement of the bellowslike pump, the thorax. The thoracic cavity is covered by the very thin parietal pleura, and the lungs by the pulmonary (visceral) pleura. These very thin membranes of single layers of flat epithelial cells on fibrous connective sheets continue uninterrupted from one pleural surface to the other across the pulmonary hilus. The two pleurae surfaces are held close together with a thin fluid film in between, providing smooth lubricated surfaces. If the thorax is opened so that the atmospheric pressure prevails in the intrapleural space, the elastic recoil of the lungs causes them to collapse, and the chest expands a little, since a retractive force of the lungs is normally counterbalanced by an outward spring of the chest cage (pneumothorax). Such an injury, of course, makes the lung involved incapable of any respiratory function. Normally, however, the pleurae are in close, but friction-free contact with each other. Any volume changes in the thorax are completely transmitted to the lungs. The two pleural surfaces may be compared with two flat sheets of glass placed face to face with a thin layer of water between the two opposing surfaces. While the two sheets of glass can easily be slid back and forth, it requires a great force to remove the two sheets away from one another by forces acting perpendicular to the glass surfaces.

It would appear plausible that gas and fluid from the blood might collect in the intrapleural space due to the discrepancy in the size of the thoracic cavity and the lungs: the opposing forces of the lung and chest wall tend to separate the two pleurae with a pressure of a few centimeters of water lower than atmospheric pressure. Gas is, however, absorbed because the sum of the gas tensions in venous blood (and pleural liquid) is less than arterial (and atmospheric) pressure. Water from the space is effectively absorbed, since the colloid osmotic pressure of the plasma proteins in the pulmonary capillaries easily matches the slightly lower hydrostatic pressure in the pulmonary vessels.

Respiratory Muscles

Figure 7-5 shows the contours of the thorax and lungs at the end of an expiration and an inspiration, respectively. During quiet breathing at rest, the diaphragm is the principal muscle driving the inspiratory pump: the abdominal muscles relax, the abdomen protrudes, the thoracic volume increases, and the lungs expand. The contraction of the diaphragm causes its dome to descend some 1.5 cm, and the intraabdominal pressure increases. This movement is actually of the same order of magnitude both in the so-called costal and in the diaphragmatic type of breathing. During deep breathing, the vertical movements of the diaphragm may exceed 10 cm. The external intercostal muscles assist in the inspiration, expecially during exercise. The fibers slope obliquely downward and forward from the caudal margin of one rib to the cranial margin of the rib

Fig. 7-5 Diagram of frontal section of thorax, based on roentgenograms, showing changes in size of thorax and position of heart and diaphragm with respiration. To the left: inspiration is designated with light stippling, and expiration with heavy stippling. (From Braus, 1956.)

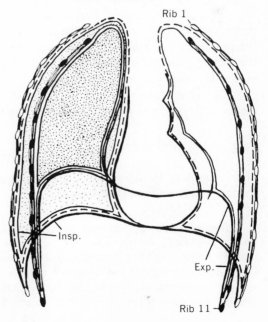

below. The lower insertion is located more distant from the center of rotation than the upper one. When the fibers contract, the force exerted by the muscle is equal at both insertions, but the longer leverage of the lower rib gives a torque that raises rather than lowers the upper rib. The net effect is a lifting of the ribs when the external intercostal muscles contract. The elevation of the ribs, rotating around the axis of their necks, increases the dimensions of the rib cage in both transverse and dorsoventral directions, similar to that of the handle of a bucket when lifted. When the inspiratory muscles, during quiet breathing, relax, the elastic recoil forces in the lung tissue, thoracic wall, and abdomen restore the chest to the resting position without any help of the expiratory muscles. During exercise or forced breathing at rest, with a ventilation exceeding 2 or 3 times the resting value, these recoil pressures are supplemented by activity of the expiratory muscles. The internal intercostal fibers have a direction opposite to those of the external intercostal muscles. Therefore the function of the inner layer of the intercostal muscles is to facilitate expiration. The muscles of the abdominal wall are essentially expiratory muscles, but do not become engaged forcefully until the pulmonary ventilation reaches high levels.

At a ventilation exceeding 50 liters/min and especially at extremely high ventilation, accessory muscles may assist. The sternocleidomastoids and scaleni are the most important ones during forced inspiration. When the athlete grasps for support after an exhausting spurt, this posture may facilitate the action of the respiratory muscles.

The activity of the inspiratory muscles increases progressively throughout inspiration and they actually continue to contract while being stretched during the early part of expiration. Part of the work done during inspiration is "stored" in the elastic structures of the system and is then available to supply part of the power for the expiration. If an expiration decreases the lung volume below the resting level of the system, the chest wall recoils outwardly causing a passive inspiration back to the resting volume.

During exercise the inspiratory and expiratory muscles are activated reciprocally, especially the expiratory ones in the last part of expiration.

Total Resistance to Breathing

The respiratory muscles work mainly against an airway resistance and a pulmonary tissue and chest wall resistance.

The work done against inert forces to accelerate tissues and gases is in this connection negligible. Most of the tissue resistance is offered by elastic forces. But the collagen fibers, providing the supporting framework for the delicate structures of the lungs, also contribute to the resistance when a volume change occurs. Thanks to the soft, yielding tissues of the lungs, the resistance is low. Of the total pulmonary resistance only about 20 percent is a tissue resistance and 80 percent is airway resistance.

At high flow velocities, as during heavy exercise, the air flow is turbulent in the trachea and the main bronchi, giving a high flow resistance. Due to the large total cross area of the finest air tubes, the air flow in this region is low and therefore laminar.

At rest the oxygen cost of the breathing is only a small fraction of the total resting energy turnover. It has been estimated to be about 0.5 to 1.0 ml/liter of moved air. With a pulmonary ventilation of 6 liters/min the oxygen uptake of the respiratory muscles would be up to 6 ml, compared to a total resting oxygen uptake of the body as a whole of about 250 to 300 ml. With the high pulmonary ventilation during heavy exercise the energy cost per liter ventilation becomes progressively greater, and the oxygen cost of breathing may be up to 10 percent of the total oxygen uptake. The air resistance when breathing through the nose is 2 to 3 times greater than that obtained by breathing through the mouth. It is therefore natural for the athlete to breathe through his mouth when performing heavy exercise, since this reduces the airflow resistance.

Summary   When the respiratory muscles are relaxed (resting volume) the chest wall is retracted by the elastic recoil of the lungs. The beginning of an inspiration is assisted by the recoil of the chest wall but the lung tissue is further stretched. During deeper inspiration there is a retractive force from both the chest wall and the lungs. Part of the energy provided by the inspiratory muscle, mainly the diaphragm and the external intercostal muscles, is "stored" in the elastic structures and is utilized during the expiration. An expiration below the resting volume increases the outward recoil of the chest wall. In any volume position the lungs and the chest cage behave like opposing springs, and at the resting volume the forces exerted exactly counterbalance each other. The work done by the respiratory muscles is mainly devoted to doing elastic work and to overcoming the airway resistance.

VOLUME CHANGES

Terminology and Methods for the Determination of "Static" Volumes

Figure 7-6 should be consulted for the terminology. When the respiratory muscles are relaxed, there is still air left in the lungs. This air volume is the *functional residual capacity* (FRC). A forced maximal expiration brings the volume down to the *residual volume* (RV) by expiration of the *expiratory reserve volume*. (Actually, the limit of a maximal expiration is not only the capacity of the expiratory muscles to compress the thoracic cage. Many small airways become occluded during the forced expiration, and the lungs, with trapped gas, are also compressed.) A maximal inspiration from the functional residual capacity adds the *inspiratory capacity* and the gas volume contained in the lungs is then the *total lung capacity* (TLC). The maximal volume of gas that can be expelled from the lungs following a maximal inspiration is called the *vital capacity* (VC). It fol-

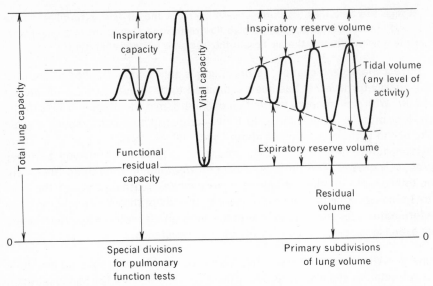

Fig. 7-6   Diagram of lung volumes and capacities. (From Pappenheimer et al., 1950.)

lows that vital capacity plus the residual volume constitute the total lung capacity. The volume of gas moved during each respiratory cycle is the *tidal volume* (V_T).

The vital capacity and its subdivisions are commonly measured with the help of a spirometer. With the subject connected to the spirometer via a wide-bore tube, any change in lung volume is reflected in a volume displacement in the spirometer. A calibration factor translates this displacement recorded on a kymograph into liters.

The functional residual capacity can be measured with the closed-circuit methods (*gas-dilution method*). A closed spirometer contains a small, known amount of helium (or hydrogen). After a normal expiration the subject is connected to the spirometer and rebreathes from the system. The expired carbon dioxide is absorbed by soda lime. Oxygen is added to the circuit at a rate to keep the volume at the end of expiration at a constant level. This refilling can be adjusted automatically. The concentration of the indicator gas falls in the spirometer and rises in the lungs. The final concentration is a simple function of the added gas volume, i.e., the functional residual capacity. The principle for this method is clarified by Fig. 7-7. The concentration of the indicator is analyzed continuously, for example with a katharometer, and a constant reading for about 2 min indicates a complete mixing. If the subject then performs a maximal expiration, followed by a maximal inspiration to total lung capacity, the recordings permit calculation of residual volume and the subdivisions discussed above (Fig. 7-6). Normally about 5 min of rebreathing is enough for complete mixing of the indicator gas within spirometer-lungs, but in patients with an impaired lung function up to 20 min of rebreathing may be necessary. The reason for using helium or hydrogen as indicator gas is that these gases are only absorbed by the lung tissues and blood to a negligible degree.

If He_1 and He_2 are the initial and final concentrations, respectively, of helium and V_s is the volume of gas in the spirometer to the point of the subject's mouth, the functional residual volume, V_{FRC}, can be calculated from the formula

$$V_s \times He_1 = (V_{FRC} + V_s)He_2 \quad \text{or} \quad V_{FRC} = \frac{V_s(He_1 - He_2)}{He_2}$$

The lung volumes are expressed at BTPS, i.e., gas volume at body temperature, ambient pressure (P_B), and saturated with water vapor ($P_{H_2O} = 47$ mm Hg) and therefore the gas volumes recorded by the kymograph must be recalculated. For a spirometer temperature of $t°C$ with a water pressure of P_{H_2O} we have

$$FRC = V_{FRC} \times \frac{310}{273 + t} \times \frac{P_B - P_{H_2O}}{P_B - 47} \text{ liters (BTPS)}$$

If the subject is connected with the spirometer after a maximal expiration and then rebreathes deeply three times before being disconnected, after a maximal expiration, the residual volume can be directly determined by the application of the same formula.

For a determination of the functional residual capacity the open-circuit *gas washout method* can also be applied. After a normal expiration, the subject starts to inhale pure

Fig. 7-7   A spirometer with a measured volume of gas (V_s) contains helium in a small, analyzed concentration (He_1). After a normal expiration (lung volume = functional residual capacity = V_{FRC}), the subject rebreathes from the spirometer until a homogeneous gas mixture is attained with a new and lower helium concentration (He_2) due to its dilution with the air in the lungs. V_{FRC} can now be calculated (see text).

$$V_{FRC} = V_s (He_1 - He_2)/He_2$$

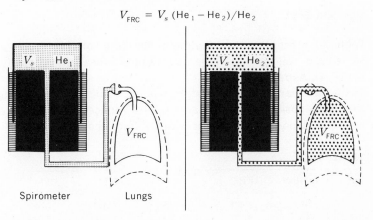

Spirometer   Lungs

oxygen. As time goes, the nitrogen in the lungs (about 79 percent at the beginning) will be flushed out. The exhaled air is collected, and its concentration of nitrogen is followed continuously. The oxygen breathing continues until the N_2 concentration is reduced to 2 percent. This normally takes less than 7 min. The volume of nitrogen flushed from the lungs can be calculated from the volume of expired air and its concentration of nitrogen. Principally this nitrogen was in the functional residual capacity, and the volume with its 79 percent N_2 required to contain this amount of exhaled nitrogen can easily be calculated. However, a correction must be made, since some nitrogen has come out of solution in the body fluid as a result of a reduction in the nitrogen pressure in the alveoli.

A third method to measure all lung volumes is by the use of a *body plethysmograph with pressure recorders*. The subject sits in a closed chamber. After a normal expiration, a shutter blocks off the airway at the mouth, and if the subject makes an expiratory effort against the shutter with his glottis open, the air in the thorax becomes compressed and the air in the plethysmograph becomes expanded. If the pressure changes in the chamber air are measured, the thoracic volume change can be calculated. A simultaneous recording of the pressure changes in the alveoli (= in the mouth) permits an application of Boyle's law to calculate the intrathoracic volume. In principle

$$P \times V = (P + \Delta P) \times (V - \Delta V) \qquad V = \frac{P\Delta V}{\Delta P}$$

where P is alveolar pressure, V is the thoracic gas volume, ΔP is the change in pressure during expiration against the shutter, ΔV is the change in volume due to compression of the thorax by the respiratory muscles during this expiration. (The formula is simplified since the product $\Delta P \Delta V$ may be neglected.) (For references, see DuBois, 1964.)

The complete picture of the lung volume is then easily obtained by measuring the expiratory reserve volume and the inspiratory capacity by spirometry, as described above.

As was briefly discussed, there are several methods available for fairly accurate measurements of the absolute gas volumes and air spaces in the airways. The results obtained by the gas dilution, gas wash out or by the body plethysmography methods are closely comparable and reproducible with a coefficient of variation of roughly ± 5 percent.

Age and Sex

Table 7-1 presents data on some of the lung volumes in liters obtained from some fairly well-trained students (physical education), about twenty-five years old (P.-O. Åstrand, 1952).

These data were measured in the standing position. During tilting from standing to supine position, the TLC and VC are reduced 5 to 10 percent, because of a shift of blood to the thoracic cavity from the lower part of the body. This

TABLE 7-1

Sex	Number	FRC	VC	RV	TLC
♀	51	2.60	4.25	1.15	5.40
♂	45	3.40	5.70	1.50	7.20

illustrates the effect of gravity on the blood distribution within the body (see Chap. 6).

VC, RV, and TLC are related to body size and vary approximately as the cube of a linear dimension, such as body height, up to the age of twenty-five. In other words, these volumes in children are of a size that could be expected from theoretical considerations (see Chap. 10).

The individual dimensions are, however, not exclusively decisive for the size of the lung volumes. In women the lung volumes are about 10 percent smaller than for men of the same age and size. For the average person the lung volumes are up to 20 percent smaller than the values listed in Table 7-1. Training during adolescence will eventually increase the VC and TLC. After the age of about thirty, the residual volume and functional residual capacity increase and the vital capacity decreases. The ratio of RV/TLC $\times$ 100 in the young individual is about 20 percent, but for the fifty- to sixty-year-old individual this ratio increases to about 40 percent, an increase which can be accounted for almost entirely by changes in lung elasticity with age (Turner et al., 1968).

Athletes have similar or slightly higher values for VC and TLC compared to the data in Table 7-1. (Highest value for VC recorded by us in Stockholm so far is 8.1 liters for a cross-country skier with a maximal oxygen uptake of 5.9 liters/ min and many Olympic gold medals!)

The vital capacity has previously been proposed as one method to assess physical work capacity. In a group of about 190 individuals, seven to thirty years of age, a significant correlation was found between vital capacity and maximal oxygen uptake (Fig. 7-8). A closer examination of the individual figures reveals, however, that individuals with a vital capacity of about 4 liters/min may have a maximal oxygen uptake from about 2.0 to 3.5 liters/min. From this and similar studies it is evident that vital capacities of 6.0 liters may be associated with oxygen uptake capacities varying from about 3.5 to 5.5 liters/min. This example serves to show that one function may appear closely related to another if the data are derived from persons of greatly different size. However, the scattering of the data may still be considerable and sufficiently large to make any prediction of an individual's maximal oxygen uptake from such parameters as vital capacity rather unreliable. The conclusion may be drawn, however, that an oxygen uptake of 4.0 liters/min or more does require a vital capacity of at least 4.5 liters.

The measurement of the vital capacity as part of a larger test battery may yield valuable information, especially concerning the distensibility of the respiratory system. Certain pathological conditions are associated with a reduced vital capacity.

"Dynamic" Volumes

Dynamic spirometry, i.e., the determination of ventilatory capacity per unit time, is also used to assess an individual's respiratory function. The subject breathes into a low resistance spirometer, and its displacements are recorded with the aid of a kymograph.

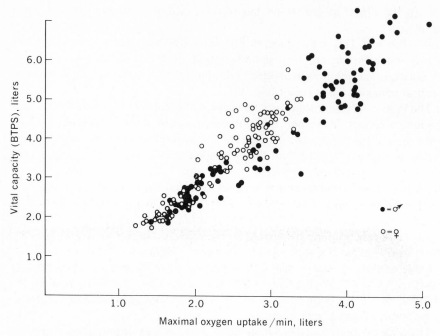

Fig. 7-8   Individual data on vital capacity measured in standing position in relation to maximal oxygen uptake during running or cycling in 190 subjects from seven to thirty years of age. (From P.-O. Åstrand, 1952.)

For the determination of *forced expiratory volume* (FEV), the subject first takes a deep breath and inspires maximally. The subject then exhales as force-fully and completely as he can. In this way it is determined how much of his vital capacity he can exhale in the course of 1 sec ($FEV_{1.0}$), and this volume is expressed in percent of the individual's entire vital capacity. A normal figure for a twenty-five-year-old individual is about 80 percent. The maximal flow is limited by the rate by which the muscles are able to transform chemical energy into mechanical energy and also by a rising flow resistance. Thus, $FEV_{1.0}$ is reduced in persons who have any airway obstructions.

An evaluation of the mechanical properties of the lungs and the chest wall can also be made by determining the *maximal voluntary ventilation* (MVV) (also referred to as maximal breathing capacity). The subject is asked to breathe as rapidly and as deeply as he can during a given time interval, usually 15 sec. The individual differences in MVV are large. In the case of healthy twenty-five-year-old men the mean value is about 140 liters/min, with a range from 100 to 180 liters/min. For women the normal values range from about 70 to 120 liters. The pulmonary ventilation during maximal work is somewhat lower than that obtained during the determination of MVV.

Since the volume is also affected by the breathing frequency, it may be advisable to have the subject maintain a fixed respiratory rate such as 40 respirations per minute (MVV_{40}), especially in the case of longitudinal studies. The respiratory volume may be recorded with the aid of a spirometer, or the expiratory air may be collected in a Douglas bag. The volume of air is then expressed in liters per minute (BTPS). Since the result depends to a great extent on the complete cooperation of the subject, it is essential that every effort be made to encourage the subject to exert himself to make a maximal effort. MVV depends, among other factors, on the body size of the individual, the forces of the respiratory muscles, the mechanical properties of the thoracic wall and lungs, and on the airway resistance. The measurement of MVV is therefore a measure of the overall capacity of the breathing apparatus to pump air. The maximal air flow during short periods of peak flow during expiration may reach values up to 400 liters/min. One of the limiting factors is the rising air-flow resistance in the tracheobronchial tree, which becomes progressively compressed as the intrathoracic pressure increases during the expiratory effort. Forced expiration tends to collapse the walls of the intrathoracic airways.

COMPLIANCE

The lungs and the thorax are partly made of elastic tissue. During inspiration these tissues are stretched. Due to the elastic nature of these tissues, they return to their resting position as soon as the inspiratory muscles are relaxed. The more rigid these tissues are, the greater muscular force must be applied in order to achieve a given change in volume. The relation between force and stretch or between pressure and volume can be measured. Thus a measure is obtained of the tissue's elastic resistance to distension, or its so-called compliance. With the aid of a balloon placed in the intrathoracic esophagus, the pressure may be measured at the end of a normal expiration and again after the subject has inhaled a known volume of gas. These measurements may be repeated at different volume changes. The volume change in liters produced by a unit of pressure change in centimeters H_2O gives the lung compliance. If a pressure change of 5 cm H_2O produces a change in lung volume of 1 liter, the lung compliance is 1.0 liters/5 cm H_2O, or 0.2 liters/cm H_2O, which is the normal value at quiet breathing. With a respiratory depth of about 0.5 liter the pressure variations in overcoming the resistance are, in consequence, a few cm of water. At lung volumes closer to maximal inspiration, or maximal expiration, a greater pressure is required for a given volume change, i.e., the compliance is reduced. If, due to pathological changes, such as interstitial or pleural fibrosis, the lungs are more rigid and less distensible, the compliance is also reduced, and the respiratory work is increased.

In the foregoing the principle for the measurement of the compliance of the *lungs* has been discussed. It is also of interest to assess the compliance of the thoracic cage. This can be estimated by measuring the compliance of the respiratory system as a whole and then subtracting the compliance of the lungs alone. The chest wall compliance decreases markedly with age.

AIRWAY RESISTANCE

In addition to overcoming the elastic resistance of the respiratory system, part of the energy of the respiratory muscles has to be applied to overcome two types of nonelastic resistance: a tissue viscous resistance due to friction, and a resistance to the movement of air in the air passages. This airway resistance may be doubled by bronchial smooth muscle contraction or reduced to half the normal resistance by bronchodilatation. The airway resistance may also be increased by mucous edema or by intraluminal secretion. The factors causing this bronchoconstriction may be local or they may be a reflex response to inhaled fine, inert particles, smoke, dust, noxious gases, or to the action of the parasympathetic system. The effect of the sympathetic system and epinephrine on bronchial tone is to dilate the airways. The increased sympathicus-tonus during muscular effort thus tends to lower the airway resistance.

In this connection it should be pointed out that inhalation of the smoke from a cigarette within seconds causes a two to threefold rise in airway resistance which may last 10 to 30 min (Comroe, 1966b). At rest this increased airway resistance is not noticeable. In order to give rise to subjective symptoms of distress the airway resistance has to be increased 4 to 5 times the normal value. The causative factor is not nicotine but particles which have a smaller diameter than 1 μ and which affect the sensory receptors in the airway path. However, during muscular effort, with its increased demand on pulmonary ventilation, the effect of tobacco smoking becomes apparent. The chronic effect of tobacco smoking is an increased secretion in the respiratory tract and a narrowing of the air passages. FEV as well as MVV may be reduced. It is a common observation that athletes involved in events requiring endurance never smoke. This may be explained by the fact that cigarette smoking reduces the respiratory function and increases the amount of carboxyhemoglobin. The latter reduces the oxygen transporting capacity of the blood. A sprinter, shot-putter, or diver may be unaffected by cigarette smoking, however, since the requirement for aerobic power in these athletic events may be insignificant.

PULMONARY VENTILATION AT REST AND DURING WORK

The pulmonary ventilation is the mass movement of gas in and out of the lungs. The pulmonary ventilation is mainly regulated so as to provide the gaseous exchange required for the aerobic energy metabolism. Some gaseous exchange does take place through the skin, and some gas is lost in the urine and other secretions, but the volume of gas thus exchanged is negligible.

Methods

The gas volumes are usually measured very accurately with a water-filled spirometer. Room air (or a mixture of gases) is inhaled through a respiratory valve. The expired air is either collected in a bag (Douglas bag) or collected directly in the spirometer.

The volume may also be measured by other means, with the aid of a gas meter or flow meter, for instance. The latter is constructed on the principle that the pressure gradient along a rigid tube of uniform cross section is linearly related to the flow of a gas or fluid, as long as the flow is linear. A flow resistance, which may be a fine mesh screen of a dimension which will not affect respiration, is inserted in the tube. With a differential pressure manometer the pressure difference across the resistance can be measured when gas is flowing. The volume of gas flow is obtained by graphic or electric integration. (For literature references concerning methods see Mead and Milic-Emili, **1964.)**

The amount of inhaled and exhaled air is usually not exactly equal, since the volume of inspired oxygen in most situations is larger than the volume of carbon dioxide expired. Pulmonary ventilation usually means the volume of air which is *exhaled* per minute. It is exceedingly important that the mouthpiece, respiratory valve, tubes, and stopcocks are so constructed that they cause a minimum of increased airway resistance during heavy physical exertion. Thus, corrugated external breathing tubes should not be used, since these may give rise to turbulence. The diameter of the tubes and all openings should be about 30 mm or *wider* (see Appendix).

Pulmonary Ventilation during Exercise

Figure 7-9 shows how ventilation ($\dot{V}_E$) increases during increasing work loads up to the maximal level. From a resting value of about 6.0 liters/min the ventilation increases to 100, 150, and in extreme cases, to 200 liters/min (Saltin & P.-O. Åstrand, 1967) (BTPS = gas volume at normal body temperature and ambient barometric pressure, saturated with water vapor). The increase is semilinear, with a relatively greater increase at the heavier work loads. (The explanation for this is discussed in connection with the discussion of the regulation of respiration.)

Figure 7-10 presents data on maximal pulmonary ventilation for about 225 subjects, from four to about thirty years of age, collected during maximal running for about 5 min. A positive correlation exists between maximal $\dot{V}_E$ and $\dot{V}_{O_2}$, but it is evident that maximal ventilation cannot be used for prediction of maximal oxygen uptake. Maximal pulmonary ventilation is actually not a well-defined parameter. Figure 7-9 illustrates a marked increase in ventilation during heavy exercise without any further increase in oxygen uptake. A well-motivated subject may continue to exercise at very high work loads despite strain (as judged from blood lactic acid concentration), and he attains a high ventilation; the less motivated one just quits exercise when still submaximal in a physiological sense.

If pulmonary ventilation is expressed in relation to the magnitude of oxygen uptake, it is 20 to 25 liters/liter O_2 at rest and during moderately heavy work, but increases to 30 to 35 liters/liter O_2 during maximal work. In children under

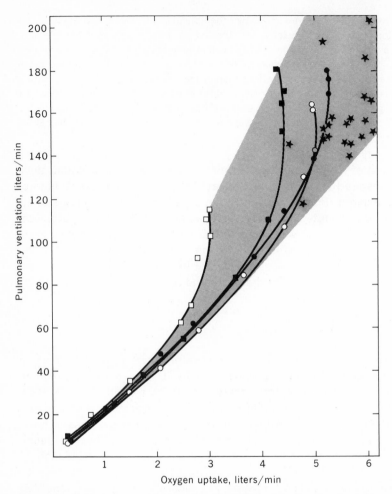

Fig. 7-9 Pulmonary ventilation at rest and during exercise (running or cycling). Four individual curves are presented. Several work loads gave the same maximal oxygen uptake. Work time from 2 to 6 min. Stars denote individual values for top athletes measured when maximal oxygen uptake was attained. (Data from Saltin and P.-O. Åstrand, 1967.) Individuals with maximal oxygen uptake of 3 liters/min or higher usually fall within the shadowed area. Note the wide scattering at high oxygen uptakes.

ten years of age the values are about 30 liters during light work and up to 40 liters/liter O_2 uptake during maximal work (Fig. 7-10).

Figure 7-11 presents mean values for maximal pulmonary ventilation during exercise (running or cycling) in different age groups. The lower ventilation in the older individuals is associated with a reduced maximal oxygen uptake (Fig. 9-12).

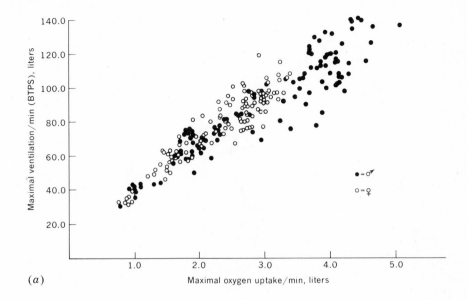

(a)

Maximal oxygen uptake/min, liters

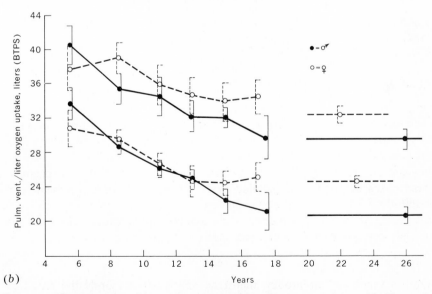

(b)

Years

Fig. 7-10   *Data on 225 subjects from four to about thirty years of age.* (a) *Maximal pulmonary ventilation in relation to maximal oxygen uptake measured during running on a motor-driven treadmill for about 5 min.*
(b) *Average values of ventilation per liter oxygen uptake in relation to age. The upper curves show maximal values (attained during running); and the lower ones, submaximal values during running or cycling, with an oxygen uptake which was 60 to 70 percent of the subject's maximal aerobic power [same subjects as in (a)]. Vertical lines denote ± 2 SEM. (From P.-O. Åstrand, 1952.)*

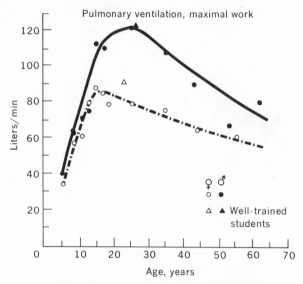

Fig. 7-11   Pulmonary ventilation measured after about 5 min exercise with a work load that brought the oxygen uptake to the individual's maximum. Mean values on 350 women and men and about 80 well-trained subjects; exercise on motor-driven treadmill or bicycle ergometer. (Mainly based on data from P.-O. Åstrand, 1952; I. Åstrand, 1960.)

Dead Space

Only a part of the inhaled volume of air reaches the alveoles where the gaseous exchange can take place. This part is known as "the effective tidal volume" (V_A). Part of the inspired tidal volume (V_T) occupies the conducting airways. This part is called "dead space" volume (V_D) because it does not take part in the gaseous exchange between alveolar air and blood. During expiration this dead space component is exhaled first. It has a composition similar to moist inspired air. Then comes the alveolar component, which has a relatively high concentration of carbon dioxide and a low oxygen concentration. The total expired gas is therefore a mixture of dead space and alveolar gas, or

$$V_T = V_A + V_D$$

From a functional standpoint this dead space is not merely the result of the anatomic features of the respiratory tract. In addition to the air volume which remains stagnant in the conductive airways, some air eventually reaches alveoles that are not at all or are poorly perfused by capillary blood. This reduces the gaseous exchange. In patients suffering from pulmonary disease, an unfavorable relationship between ventilation and perfusion may increase the physiological dead space.

The volume of the dead space may be estimated with the aid of Bohr's formula, which is based on the fact that the expired volume of oxygen at each respiration $(V_T \times F_{E_{O_2}}{}^*)$ is equal to the sum of the volume of oxygen contained in the dead space compartment $(V_D \times F_{I_{O_2}})$ and the volume of oxygen coming from the alveolar air $(V_A \times F_{A_{O_2}})$. We therefore arrive at the following formula:

$$V_T \times F_{E_{O_2}} = V_D \times F_{I_{O_2}} + V_A \times F_{A_{O_2}}$$

Since $V_A = V_T - V_D$, the formula may be simplified as follows:

$$V_D = V_T \frac{F_{E_{O_2}} - F_{A_{O_2}}}{F_{I_{O_2}} - F_{A_{O_2}}}$$

If the oxygen content of the inspired air is 21 percent, the oxygen content of the expired air 16 percent, the oxygen content of the alveolar air is 14 percent, and the depth of respiration, V_T, is 500 ml:

$$V_D = 500 \frac{16 - 14}{21 - 14} = 143 \text{ ml}$$

The same calculation can be made on the basis of CO_2.

With a depth of respiration (tidal volume) of 500 ml at rest, the dead space constitutes approximately 150 ml. The rest of the tidal volume reaches the alveoles. It should be noted, however, that the first portion of the inhaled air is, in reality, the respiratory air which remained in the dead-space compartment from the previous respiration. The "fresh" air which is pulled down into the alveoles is diluted into a relatively large volume, i.e., the functional residual capacity. The variations in the gas concentration are therefore relatively small in the alveoles during rest and normal breathing.

Due to methodological difficulties, it is not easy to measure the exact behavior of the dead space during exercise. Asmussen and Nielsen (1956) estimated that with a tidal volume of 3 liters the dead space was 300 to 350 ml, whereas Bargeton (1967) concluded from his data that the increase in dead space with increasing tidal volume is very slight and can be taken as a constant for moderate changes in tidal volume. Relatively speaking, the dead space is reduced with increasing tidal volume. If, at a ventilation of 6.0 liter/min, the respiratory frequency is 10, and the dead space 0.15 liter, the alveolar ventilation is

$$6.0 - 0.15 \times 10 = 4.5 \text{ liters/min}$$

If the respiratory rate, on the other hand, is 20, and the gross ventilation and dead space are assumed to be unchanged, the alveolar ventilation is only

$$6.0 - 0.15 \times 20 = 3.0 \text{ liters/min}$$

* F = Fraction of oxygen in the expired, inspired, and alveolar air respectively.

Animals which depend on evaporative heat loss from the respiratory tract for their temperature regulation avoid hyperventilation of the alveoles thanks to a high respiratory rate and a low alveolar ventilation (panting).

From the pulmonary ventilation data which are given in Fig. 7-9, part of the volume does not participate in the gas exchange. The method of concealment, often described in adventure stories, by hiding submerged in water and breathing through a snorkel represents a considerable complication of the gas exchange. The tube (snorkel) represents an extension of the respiratory dead space, and the tidal volume has to be increased by an amount equal to the volume of the tube if the alveolar ventilation is to be maintained unchanged. The breathing may therefore become very laborious.

A second complication of this diving is the increased load on the inspiratory muscles. Within the lungs there is the same pressure as at the water surface, i.e., atmospheric pressure. The outside of the thorax is, however, subjected to atmospheric pressure *plus* the pressure of the column of water above the diver. At a depth of 1.0 m this extra pressure will be 0.1 atm, or about 76 mm Hg. The highest pressure the inspiratory muscles can overcome is just above 70 mm Hg, and therefore, a depth of 1.0 meter would be the maximum that can be tolerated even if the problem of extra dead space could be solved by a system of valves.

Tidal Volume–Respiratory Frequency

By definition, the pulmonary ventilation equals the frequency of breathing multiplied by the mean expired tidal volume, or

$$\dot{V}_E = f \times \bar{V}_T$$

At rest the respiratory frequency is between 10 and 20. Inspiration occupies less than half the total cycle, the rise in flow being more abrupt than the fall. During physical work of low intensity, it is primarily the tidal volume that is increased. This may, in many types of exercise, amount to 50 percent of the vital capacity when the work load is moderately heavy or heavy (Fig. 7-12). The respiratory frequency is also increased, especially in the case of heavy work. Children about five years of age may have a respiratory frequency of about 70 at maximal work, twelve-year-old children about 55, and twenty-five-year-old individuals 40 to 45 (Fig. 7-12) (P.-O. Åstrand, 1952).

The increase in tidal volume is brought about through the utilization of both the inspiratory reserve volume and the expiratory reserve volume (see Fig. 7-6). Inspiration and expiration become more equal in both time and pattern. Naturally the vital capacity limits the tidal volume, but rarely more than 50 percent of the vital capacity is utilized.

Similarly the upper limit for the respiratory frequency is determined by the rate at which the neuromuscular system can generate alternating movements. Studies have indicated that an individual spontaneously balances the depth

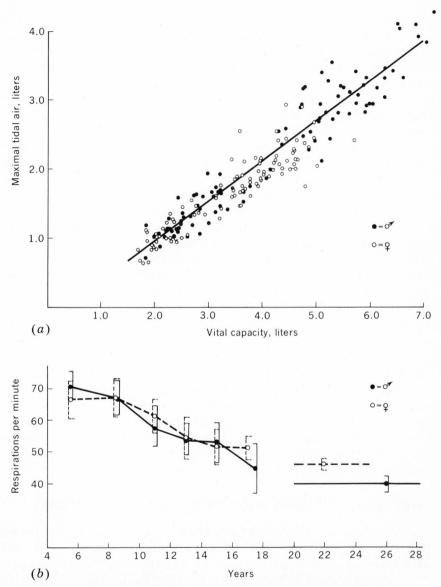

(a)

(b)

Vital capacity, liters

Fig. 7-12 (a) *Highest tidal volume measured during running at submaximal and maximal speed (work time about 5 min) related to the individual's vital capacity measured in standing position. Altogether 190 subjects from seven to thirty years of age. On an average 50 to 55 percent of the vital capacity is used as maximal tidal air.*
(b) *Respiratory frequency during running at a speed that brings the oxygen uptake up to maximum; average values (± 2 SEM) for 225 subjects from four to thirty years of age. (From P.-O. Åstrand, 1952.)*

of respiration and respiratory frequency in such a way that a certain ventilation takes place at optimal efficiency, i.e., with the utilization of a minimum of energy by the respiratory muscles (Milic-Emili et al., 1960). The greater the pulmonary ventilation, the narrower the range of respiratory frequencies appears to be, yielding minimal energy expenditure (Otis, 1964). In athletic performances it is therefore advisable to allow the athlete to assume the respiratory pattern which seems natural for him. In many types of physical work the respiratory frequency tends to become fixed to the work rhythm. Needless to say this certainly holds for crawl swimming, but it also holds for such activities as bicycle riding, sculling, and running. Therefore the ventilatory pattern is not exclusively guided by demand for minimal energy expenditure of the respiratory muscles (Flandrois et al., 1961). This is, however, natural since the thoracic cage is also highly affected by muscles other than the true respiratory ones.

In swimming, instruction in the breathing technique may be necessary, but in other cases the respiratory pattern should be allowed to follow its natural pattern. It should be noted that work on the bicycle ergometer with a pedaling frequency of, for instance, 50 usually gives a respiratory frequency related to this pedaling frequency. It often increases stepwise from 12.5 to 16.6, 25.0, 33.0, up to 50 at maximal work. It is important to keep this in mind if one is studying the mechanics and regulation of breathing during work, using only the bicycle ergometer as a means of providing the work load.

Summary   The respiratory frequency at which the respiratory work is minimal increases progressively with increasing ventilation. The respiratory frequency at each level of ventilation usually corresponds to the frequency which is spontaneously chosen by the subject. How this regulation is brought about is not known. Probably various receptors, including muscle spindles and the gamma system, also play an important role in the adjustment of the respiratory frequency to the work rhythm.

Respiratory Work (Respiration as a Limiting Factor in Physical Work)

The work of the respiratory muscles consists primarily of overcoming the elastic resistance and the flow-resistive forces. A precise determination of the mechanical efficiency of breathing is not simple, and data in the literature range from a few percent to about 25 percent (Milic-Emili and Petit, 1960). At rest the respiratory muscles require from 0.5 to 1.0 ml O_2/liter of ventilation. With increasing ventilation the oxygen cost per unit ventilation becomes progressively greater. It has been estimated that the respiratory muscles during heavy work may tax as much as 10 percent of the total oxygen uptake (Liljestrand, 1918; Nielsen, 1936; Otis, 1964).

A question of considerable importance is whether or not hyperventilation may limit the oxygen uptake capacity. The answer is probably negative for the following reasons: (1) After the maximal oxygen uptake is reached, it is still possible for the subject to continue to work at a higher work load because of the anaerobic processes. At the same time the pulmonary ventilation is markedly increased, without any distinct ceiling being reached (Fig. 7-9). (2) At an extremely heavy work load which can only be tolerated for a few minutes at the most, the pulmonary ventilation is greater than at a somewhat lower but still maximal load which may be tolerated for about 6 min. The oxygen uptake is nevertheless the same in both cases (P.-O. Åstrand and Saltin, 1961). (3) At maximal work it is possible voluntarily to increase the ventilation further, showing that the ability of the respiratory muscles to ventilate the lungs evidently is not exhausted during spontaneous respiration. (4) At heavy work loads the alveolar oxygen tension increases and the carbon dioxide tension decreases, which indicates an effective gas exchange in the lungs. The oxygen tension of the arterial blood is maintained or only slightly reduced. However, an exact analysis of the tension of the different blood gases during extremely heavy work has not as yet been done (Fig. 7-13).

These considerations refer to bicycling and running. The situation appears to be comparable for these two types of work.

During running on the treadmill, a group of about 40 male students attained a ventilation of 111 liters/min with an oxygen uptake of 4.04 liters/min. During maximal work on the bicycle ergometer, the group's values were 116 and 4.03 liters/min respectively. In a group of about 40 girls the pulmonary ventilation was 90 and 88 liters/min respectively (P.-O. Åstrand, 1952). In other types of work the situation may be different, as in strenuous work with the arms, which may hamper free respiration. The same may be true during swimming. In a group of 30 female elite swimmers the pulmonary ventilation was 35.5 liters/liter O_2 uptake during maximal loads on the bicycle ergometer but only 27.5 liters during swimming. No measurements were made concerning the partial pressures of the different blood gases, but it is conceivable that a reduction of the oxygen tension, and thereby the oxygen saturation, occurred. This may partly explain the 8 percent lower maximal oxygen uptake observed during swimming, as compared with the maximal oxygen uptake on the bicycle ergometer (P.-O. Åstrand et al., 1963).

To some extent the pulmonary ventilation may be a limiting factor even though the maximal capacity of the respiratory muscles is not fully taxed. Thus, if the energy demand of the respiratory muscles, in order to increase the pulmonary ventilation, necessitates such a marked increase in the oxygen consumption that all of the achieved increase in oxygen content of the alveolar air (and increase in oxygen content of the arterial blood) is entirely utilized by the respiratory muscles themselves, then none of this extra oxygen will benefit the rest of the working muscles of the body. In other words, an increase in pulmonary ventilation beyond a certain point would not be physiologically useful, since all of the additional oxygen thus gained would be required for the work of breathing (Otis, 1964). It is even conceivable that the oxygen utilization by the respiratory muscles may become so great that the oxygen supply to other tissues is reduced. However, such a critical limit is probably not reached in normal individuals. It is likely that the blood flow in the vessels of the respiratory muscles is maximal even at a ventilation

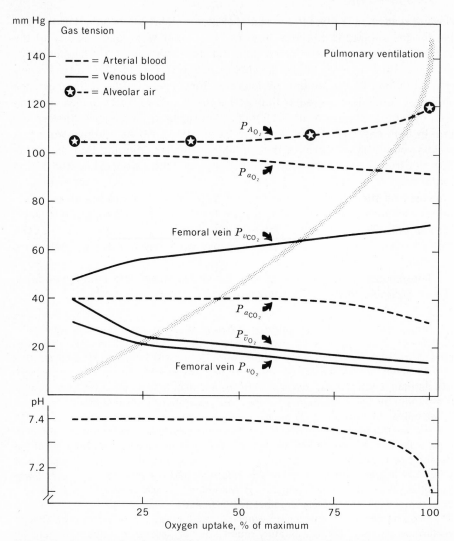

Fig. 7-13 Oxygen and carbon dioxide tensions in blood and alveolar air at rest and during various levels of work up to and exceeding the load necessary to reach the individual's maximal oxygen uptake (= 100 percent). At bottom, arterial pH. Curves are based on data from different authors and unpublished studies. (The line denoting pulmonary ventilation refers to an individual with maximal aerobic power of 4.0 liters/min if the figures on the ordinate are valid.)

below the maximal ceiling and that the oxygen content of the blood is more or less completely extracted.

A further increase in ventilation beyond this point is probably met by anaerobic
processes. A point in favor of this view is the fact that the oxygen uptake reaches a distinct plateau during extremely heavy work, even if the work load, and thereby also the
pulmonary ventilation, is further increased (Fig. 7-9). Even though the full capacity of the
respiratory muscles may not be utilized during heavy work, the maximal force which
these muscles may develop is limited by the rate at which chemical energy can be transformed into mechanical energy.

DIFFUSION IN LUNG TISSUES, GAS PRESSURES

The role of respiration is to provide the gaseous exchange between the blood
and the ambient air. This is accomplished by the flowing of blood through capillaries of extremely small caliber which are located only a few microns from the
alveolar air, which is a derivate of the ambient air (Fig. 7-3). The gas exchange
between the capillary blood and the alveolar air takes place by the process of
diffusion (for details, see Forster, 1964).

This diffusion takes place as a movement of gas molecules from a region
of higher to one of lower chemical activity. The partial pressure of the gas is a
measure of this activity. The normal pressure of oxygen, carbon dioxide, and
nitrogen in atmospheric air ($P_{Bar} = 760$ mm Hg), in alveolar air, and in mixed
venous blood and arterial blood at rest is given in Fig. 7-14. (If, for instance, the
oxygen concentration in the alveolar air is 15 percent of the dry gas, its partial
pressure is

$$P_{O_2} = {}^{15}\!/\!_{100}(760 - 47) = 107 \text{ mm Hg}$$

since the partial pressure of water vapor is 47 mm Hg at 37°C.)

Blood flow through a tissue is not always determined by the metabolic
activity in the tissue in question. Thus, the oxygen uptake of tissues such as the
kidneys and skin is small compared to the magnitude of the blood flow through
these tissues. For this reason the partial pressure of oxygen in the venous blood
remains high and the CO_2 pressure relatively low. Comparatively more O_2 is utilized in the muscle, and here a greater amount of CO_2 is produced, so that the
partial pressures of these gases in the venous blood are different from those of
the above-mentioned organs.

It should be noted that the total gas pressure in venous blood is considerably lower than in arterial blood (706 mm Hg as against 760 mm Hg, see Fig.
7-14). In this way accumulation of gas in the intrapleural space in the thorax is
avoided, despite the opposed recoil of the lungs and chest wall. If gas is trapped
behind an occlusion of an airway, it becomes absorbed into the pulmonary circulation because of this subatmospheric gas pressure of venous blood.

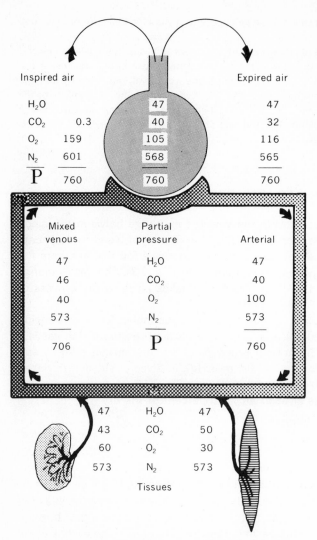

Fig. 7-14 Typical values of gas tensions in inspired air, alveolar air (encircled), expired air, and blood, at rest. Barometric pressure, 760 mm Hg; for simplicity the inspired air is considered free from water (dry). Tension of oxygen and carbon dioxide varies markedly in venous blood from different organs. In this figure gas tensions in venous blood from the kidney and muscle are presented.

Under normal conditions the diffusion processes are so rapid that the gases in the blood leaving the pulmonary capillaries are approximately in equilibrium with the gases in the alveoli. In the case of CO_2 the partial pressure is practically identical in the alveolar air and in the arterial blood (40 mm Hg). In the case of O_2, however, the partial pressure is a few millimeters higher in the alveoles than in the arterial blood (about 105 as against 100 mm Hg). The gas exchange between pulmonary air and blood is achieved entirely by the process of diffusion. No other processes, such as secretion, are involved. An analysis of the gas concentrations and pressures in the expired air at the end of the expiration gives an approximate idea of the gas pressure in the arterial blood. One may simply collect the last portion of the expiratory air volume, either at the end of a single forced expiration (Haldane-Priestley method) or from several repeated respiratory cycles (end-tidal sampling technique) for an analysis of the gaseous composition of expired air. With the aid of modern analytic and registration techniques it is also possible to follow variations in gas concentrations and pressures continuously during one or several successive respirations (for instance with the aid of a mass spectrometer). Figure 7-15 gives examples of the variations in CO_2 and O_2 pressures in the air in the trachea and in the alveoli during a single respiration. At rest the variations in the composition of the gases in the alveoli are small due to the fact that the inhaled air volume is diluted into a relatively large gas volume, the functional residual volume (FRV). During work, when the depth of respiration is increased, the variations become considerably larger.

Because of the length of the respiratory tract, the gas movement during respiration may be considered as a mass movement of gas flow. For the distribution within the small lung units, a molecular diffusion is the main determinant. Due to the small dimensions of the alveoli, a complete mixing within the alveolus probably occurs in less than 0.01 sec. The rapid equalization of the gas pressures during the passage of the blood around the alveoli is evident from Fig. 7-16. At rest, the time it takes for the blood to pass the capillary is somewhat less than 1 sec, but already after 0.1 sec the diffusion of the CO_2 has reached an equilibrium. After a further few tenths of a second, the O_2 has also reached an equilibrium. Thus during normal resting conditions the blood in the pulmonary capillaries is almost completely equilibrated with the alveolar oxygen and carbon dioxide pressures. The size of the CO_2 molecule is larger than that of the O_2 molecule, which actually slows the rate of diffusion. On the other hand, the CO_2 is about 25 times more soluble in liquids than the O_2, so that the net effect is that the CO_2 diffuses about 20 times more rapidly in aqueous liquids than does oxygen. Both CO_2 and O_2 are carried by the blood mainly in reversible chemical combinations. The hemoglobin plays an overwhelming role in this transportation. In the exchange of the respiratory gases with the blood in the lungs the primary chemical reactions of these gases occur within the red cell. The "barriers" which have to be passed are the red cell membrane, the plasma, the capillary endothelium, the basement membrane, the interstitial tissue, and the alveolar basement membrane and epithelium (Fig. 7-3). However, the process is, as shown in Fig. 7-16, very rapid. Even during heavy work, when the transit time in the capillaries may be only 0.5 sec or even less, it may be assumed that a gaseous equilibrium has been

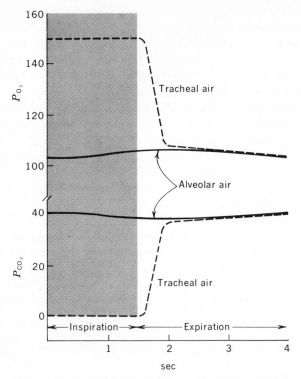

Fig. 7-15 Variations in oxygen and carbon dioxide tensions in tracheal air and alveolar air during one single breath at rest. Note the very small fluctuations in gas tensions of the alveolar air.

reached. In the narrow capillaries there is hardly any plasma between the red cell and the endothelium, which is a situation that facilitates the diffusion.

Even though the diffusion rate is sufficiently high, chemical processes are still essential if an adequate volume of gas is to pass the pulmonary membrane. Carbonic anhydrase plays an important role in the exchange of CO_2. If this were not present, the blood would have to remain in the capillaries for almost 4 min for all the CO_2 to be given off. Furthermore, when the CO_2 diffuses into the red cell and then forms protons acting on the hemoglobin to displace the O_2, the exchange of O_2 is thus affected even by the CO_2, as well as by the hemoglobin itself.

The *diffusing capacity of the lung* (D_L) is defined as the number of milliliters of a gas at STPD diffusing across the pulmonary membrane per minute and per millimeter of mercury of partial pressure difference between the alveolar air and the pulmonary capillary blood. It should be emphasized that the diffusion path includes the blood. Making use of an analogy from the field of electricity, the diffusing capacity may be compared with conductance. [Actually the word capacity may be misleading in that it varies under various conditions, especially when changing from rest to work; the term "transfer factor" has been suggested (Cotes, 1965).]

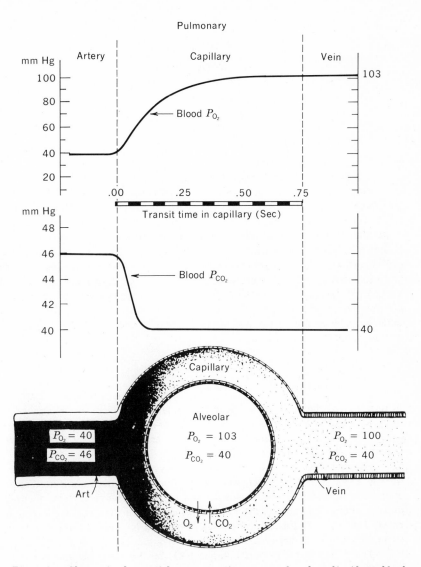

Fig. 7-16 *Change in the partial pressures of oxygen and carbon dioxide as blood passes along the pulmonary capillary. Note that already in the first part of the capillary, the blood is equilibrated with the alveolar gas. (From Cherniack and Cherniack, 1961.)*

In other words

$$\text{Diffusing capacity} = \frac{\text{gas flow}}{\text{mean driving pressure}}$$

For oxygen, the diffusing capacity is

$$D_{L_{O_2}} = \frac{\dot{V}_{O_2} \text{ ml/min}}{\bar{P}_{A_{O_2}} - \bar{P}_{c_{O_2}}}$$

where $\bar{P}_{A_{O_2}}$ = mean alveolar pressure

$\bar{P}_{c_{O_2}}$ = mean pulmonary capillary O_2 pressure

From a methodical standpoint the mean capillary P_{O_2} is difficult to determine in that the O_2 tension does not increase linearly in relation to the time during which the blood remains in the capillary (Fig. 7-16). (With a P_{O_2} of 42.5 mm Hg of the mixed venous blood and 100 mm Hg in the end-capillary blood, the mean capillary P_{O_2} would be about 84 mm Hg.) Usually CO is used for the determination of the diffusion capacity. The subject inhales about 0.3 percent CO in air during a single respiration (single-breath method, and with breath holding a certain length of time), or during a longer period of time, about 1 min (steady-state method). The capacity of the hemoglobin in the blood to bind CO is so great that the capillary P_{CO} can be assumed to be equal to zero. In this way, one difficulty in determining the mean capillary gas tension is avoided. Thus the formula may be simplified as follows:

$$D_{L_{CO}} = \frac{\dot{V}_{CO \text{ ml/min}}}{P_{A_{CO} \text{ mm Hg}}}$$

(A review of methods has been given by Forster, 1964.)

$\dot{V}_{CO}$ can be measured with great accuracy and $P_{A_{CO}}$ can be estimated even though errors may present some uncertainty in the results. $D_{L_{O_2}}$ may then be calculated by multiplying $D_{L_{CO}}$ by a factor of 1.23.

Figure 7-17 shows how the estimated oxygen diffusing capacity increases during work in a group of 20 relatively well-trained female and male subjects. From a value of about 20 ml at rest it increases relatively sharply during exercise. When the oxygen uptake exceeds about 40 percent of the maximal O_2 uptake, further increase in $D_{L_{O_2}}$ is relatively small. The mean figure for women was 46 ml, and for men about 62 ml/(min)(mm Hg) (Holmgren and P.-O. Åstrand, 1966).

The diffusing capacity is influenced (1) by the area available for gas exchange, normally the alveolar surface area, which may be some 70 to 90 m². Therefore it varies with the body size, and the difference noted between the sexes is at least partly due to a different surface area of the alveoli. It is decreased in patients with emphysema, since the destruction of the alveolar and capillary walls reduces the surface area for gas exchange. (2) The thickness of the membrane separating air from blood plays a role. In patients with interstitial or alveolar pulmonary fibrosis the thickening of the membrane decreases the $D_{L_{O_2}}$. (3) The pulmonary capillary blood volume, or rather the hemoglobin content, will influence the $D_{L_{O_2}}$. The increase in diffusion capacity observed during work is probably mainly due to an increase in the number of capillaries open (and not to an increase in the volume of capillaries already open at rest). The plateau indicated in Fig.

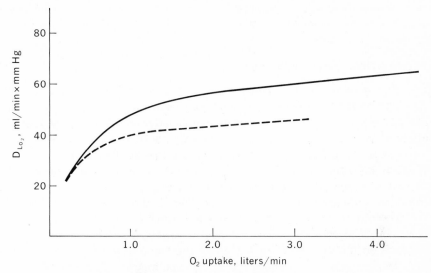

Fig. 7-17 Variation in "diffusing capacity" for oxygen (transfer of oxygen), with increasing oxygen uptake during work on a bicycle ergometer in the sitting position for 10 trained women (dotted line) and 10 trained men (full line). $D_{L_{O_2}}$ was calculated from measurements with carbon monoxide. (Modified from Holmgren and P.-O. Åstrand, 1966.)

7-17 at heavier work loads may be explained by the assumption that the capillary bed is maximally dilated even at submaximal work load. [It should be emphasized that in some studies a plateau as pronounced as that in Fig. 7-17 has not been found (Forster, 1964, pp. 861–863).]

The importance of the blood volume in the pulmonary capillaries, which at rest has been calculated to be 60 to 100 ml, is evident by the fact that $D_{L_{O_2}}$ is approximately 15 to 20 percent greater in the supine than in the sitting position, and almost 15 percent greater in the sitting than in the standing position (Bates and Pearce, 1956; Ogilvie et al., 1957). Because of a redistribution of the blood volume due to hydrostatic factors in the upright position, the perfusion is reduced, especially in the upper lobes of the lung when standing or sitting compared to the lying position (see below).

With increasing age in adults, the $D_{L_{O_2}}$ is reduced (Donevan et al., 1959). The reason for this may be found in age-related changes in all the factors referred to above.

The normal range for maximal diffusing capacity is quite wide, but as a whole it is greater the greater the individual's maximal aerobic power.

In summary, the diffusing capacity of the lung for the respiratory gases provides an index of the dimensions of the pulmonary capillary bed and the pulmonary membrane and the overall efficiency of the system in the exchange of respiratory gases. Certain technical difficulties limit the possibility for an exact estimation. Usually the $D_{L_{O_2}}$ is calculated from studies using CO. From a resting value of 20 to 30 ml O_2/min and a millimeter of mercury mean pressure difference between the alveolar O_2 pressure and the pulmonary capillary O_2 pressure, the $D_{L_{O_2}}$ increases toward 75 ml in individuals with an O_2-uptake capacity of about 5 liters/min.

VENTILATION AND PERFUSION

The difference in P_{O_2} between the alveolar air and arterial blood depends on several factors: the membrane component plays a role; a certain amount of admixture of bronchial and cardiac venous blood occurs; and finally, there is the effect of the passage of some blood through poorly ventilated alveoli.

Since CO_2 diffuses about 20 times more rapidly than does O_2, one cannot speak of any diffusion obstacle for CO_2. The inhaled air is not equally distributed to all the alveoli, and the composition of the gases is therefore not uniform throughout the lungs. The pulmonary capillary bed has a common blood supply, i.e., the mixed venous blood, but different areas of the lungs have an uneven perfusion. The composition of the gas in various parts of the alveolar space depends on the ventilation as well as on the blood flow, or the ratio $\dot{V}_A/\dot{Q}$.

Under extreme conditions the $\dot{V}_A/\dot{Q}$ ratio may vary from zero (when there is perfusion but no ventilation) to infinity (when there is ventilation but no perfusion). When the ratio is zero, the tensions of O_2 and CO_2 of the arterial blood are equal to those in mixed venous blood, since there is no net gas exchange in the capillaries. In the latter case no modification of the inspired air takes place. Although these extreme situations hardly occur under normal conditions, the various parts of the lung have a wide range of ventilation-perfusion ratio. The "alveolar air" actually represents various contributions from several hundred million alveoli, each possibly having slightly different exchange ratios and gas composition.

In other words, the alveolar gas tensions vary from moment to moment and from place to place within the lungs due to regional inhomogeneity of ventilation and blood perfusion; the supply of air and blood is not perfectly matched in the lungs, even in a healthy individual in any posture (Rahn and Farhi, 1964; Bates, 1965).

There are mechanisms which to some extent compensate for an uneven ventilation in relation to the blood flow in the capillary bed of the alveoli: (1) inadequately ventilated alveoles have a low P_{O_2}, which in turn causes an alveolar vasoconstriction and reduced blood flow; (2) reduced blood flow produces a reduction in the alveolar P_{CO_2} which causes a constriction of the bronchioles and, therefore, reduced gas flow. It appears, however, that there is no effective regional variation in the vasomotor tone, and it is apparent that the matching of ventilation against flow is not perfect. The alveolar-arterial oxygen pressure difference of a few millimeters of mercury which actually exists even in normal individuals is primarily a consequence of this unequal distribution. In the lungs the blood flow is greatly affected by body position. In the discussion of the lung volumes mention was made of the fact that the vital capacity increases in the standing position, compared to the lying position, due to the reduction in blood volume in the

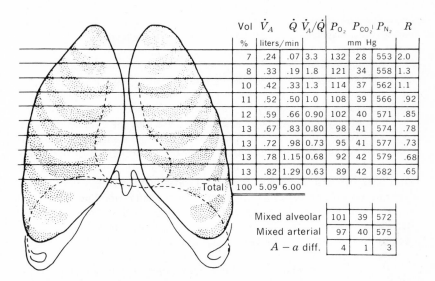

Vol	$\dot{V}_A$	$\dot{Q}$	$\dot{V}_A/\dot{Q}$	P_{O_2}	P_{CO_2}	P_{N_2}	R
%	liters/min			mm Hg			
7	.24	.07	3.3	132	28	553	2.0
8	.33	.19	1.8	121	34	558	1.3
10	.42	.33	1.3	114	37	562	1.1
11	.52	.50	1.0	108	39	566	.92
12	.59	.66	0.90	102	40	571	.85
13	.67	.83	0.80	98	41	574	.78
13	.72	.98	0.73	95	41	577	.73
13	.78	1.15	0.68	92	42	579	.68
13	.82	1.29	0.63	89	42	582	.65
Total 100	5.09	6.00					

Mixed alveolar	101	39	572
Mixed arterial	97	40	575
$A - a$ diff.	4	1	3

Fig. 7-18   Effects of observed distribution of ventilation and perfusion on regional gas tension within the lung of a normal man in sitting position. The lung is divided into nine horizontal slices, and the position of each slice is shown by its anterior rib marking. Table shows relative lung volume (Vol), ventilation ($\dot{V}_A$), perfusion ($\dot{Q}$), ventilation-perfusion ratio ($\dot{V}_A/\dot{Q}$), gas tensions (P_{O_2}, P_{CO_2}, P_{N_2}), and respiratory exchange ratio (R) of each slice. Lower table shows differences between mixed-alveolar and mixed-arterial gas tensions which would result from this degree of nonuniformity of $\dot{V}/\dot{Q}$ ratios. (From West, 1962.)

thorax in the upright position. The effect of gravity on the distribution of the blood, as well as on the perfusion, is such that in the upright position the perfusion per unit lung volume is about 5 times greater at the base than at the apex of the lung. It is true that the ventilation per unit lung volume changes in the same direction, but only slightly. As a result the $\dot{V}_A/\dot{Q}$ ratio becomes much higher in the upper lobes than in the lower lobes in the erect posture, or above 3 at the top and below 1 at the bottom of the lungs (West, 1962). This is illustrated by Fig. 7-18. As indicated, a high $\dot{V}_A/\dot{Q}$ ratio means either an overventilation or an underperfusion. The result is a highly varying composition of the alveolar air in the different parts of the lung. It has been calculated that the P_{O_2} of the uppermost alveoli may be as high as above 130 mm Hg and of the lowest alveoli below 90 mm Hg. Once again it is apparent that the lung cannot be considered as a homogeneous unit.

When the position of the body is changed from upright to supine, the perfusion of the upper lung zone increases markedly at the expense of the lower zone. In the recumbent position the calculated $\dot{V}_A/\dot{Q}$ therefore becomes quite uniform in the different pulmonary lobes.

Even during light work in the sitting or standing position the $\dot{V}_A/\dot{Q}$ ratio also becomes more uniform throughout the lungs. Certainly both upper and lower zone blood flows increase, but the former increases relatively more. The slight increase in pulmonary artery pressure that accompanies exercise is one factor that changes the balance be-tween arterial, capillary, and venous pressures on one side and the pressure outside the vessels on the other, favoring the perfusion of the upper zones of the lungs. The pressure within the pulmonary artery may vary between 7 and 20 mm Hg during cardiac cycle at rest but between 15 and 35 mm Hg during exercise with an oxygen uptake of 2.0 liters/min. The systolic pressure may exceed 50 mm Hg during maximal work. These pressure varia-tions will actually mean that the distribution of the blood flow to different parts of the lungs will not only vary with body position but also with the cardiac cycle. Even during heavy work some parts of the lungs may be unperfused during part of the diastole.

What is the effect of this difference in the topographical distribution of air and blood flow? In the final analysis this interference with the overall gas exchange is, surprisingly, rather insignificant. West (1965) states that the lung, with its uneven distribu-tion (Fig. 7-18), only wastes some 3 percent of its ventilation and 1 percent of its blood flow compared with an ideal lung. Wasted ventilation, including the effect of dead space, interferes particularly with CO_2 exchange, whereas wasted blood flow affects mainly the oxygen, and the effect may be calculated to be an impairment of the exchange of these gases by a few percent.

It is not clear to what extent the oxygen uptake may be affected by these factors during maximal work. It is evident, however, that an increased gravitational field as well as a drop in the blood pressure regardless of the cause, i.e., a vasovagal syncope, would more seriously affect the $\dot{V}_A/\dot{Q}$ ratio and therefore the gas exchange in the lungs (see Bjurstedt et al., 1968).

Summary Considerable regional inequality in ventilation of the alveoli and in blood perfusion of the capillaries exists in the lungs, causing differences in the gas exchange in different parts of the lungs. At rest in the supine or prone position, however, the ventilation distribution is rather well adjusted to follow this perfusion. In the erect posture hydrostatic forces cause a progressive decrease in perfusion from the bottom to the top of the lung without corre-sponding variations in the ventilation. Therefore, the upper lobes are relatively underperfused. During exercise the $\dot{V}_A/\dot{Q}$ ratio becomes more uniform, and the increase in pulmonary arterial pressure is at least one important contributor to this change.

OXYGEN PRESSURE AND OXYGEN-BINDING CAPACITY OF THE BLOOD

Figure 6-18 illustrates how the oxygen-binding capacity of the blood is affected by the partial pressure of oxygen in the blood. At $P_{O_2} = 100$ mm Hg, 98 percent of the hemoglobin is normally saturated with oxygen. The oxygen saturation curve is such that about half of the hemoglobin is in the form of HbO_2 and half in the form of reduced Hb at P_{O_2} in the order of 26 mm Hg (37.0°C, pH 7.40). The HbO_2 dissociation curve of the blood is affected by the CO_2 pressure, the pH, and blood temperature. Also, salts have their effects, which at least partly explains the difference in curves for various species of animals. P_{CO_2} affects pH and thereby the oxygen saturation curve, but it also has a specific effect in that CO_2 combines with Hb to form carbamino compounds ($HbNH_2 + CO_2 \rightleftharpoons HbNHCOOH$), which reduces the capacity of Hb to bind oxygen (Fig. 5-3). During muscular work the CO_2 increases, as does the temperature locally in the muscle, while at the same time the pH is lowered. This causes the liberation of O_2 to increase at a given O_2 tension. Because of this feature of hemoglobin, an effective O_2 diffusion gradient may be maintained between the capillaries and the O_2-utilizing cell. With regard to the effect of an increase in temperature, the diffusion increases about 2 percent/°C. These factors aid in the unloading of oxygen to the active muscles, but are, however, of less quantitative importance than the opening of additional capillaries in the muscles. In this manner the distance between the capillary and the muscle cell is reduced, which shortens the diffusion distance. Due to the shape of the oxygen saturation curve, a change in pH, P_{CO_2}, and blood temperature plays a relatively small role at a P_{O_2} around 100 mm Hg. During very strenuous exercise the oxygen saturation of arterial blood may, however, be reduced below 95 percent without a corresponding decrease in P_{O_2} (Fig. 6-19). At high altitude, where the alveolar P_{O_2} is lower, the arterial saturation is still more affected in a negative direction by a lowered pH and increased P_{CO_2} and temperature.

Each Hb molecule has four iron atoms. In reality the ratio between Hb and HbO_2 may vary as follows: Hb_4, Hb_4O_2, Hb_4O_4, Hb_4O_6, Hb_4O_8, in which Hb_4 is the completely reduced hemoglobin. Only in the case of Hb_4O_8 is it 100 percent saturated. The structure is such that the heme groups do not influence each other's activity, and there is no interaction between them. This means that under normal conditions there is a mixture of these combinations in proportions which are determined by such factors as P_{O_2}. As far as the volume CO_2 bound to Hb in the carbaminohemoglobin form is concerned, it varies with the degree of O_2 saturation and is greater at low O_2 saturation and less at high O_2 saturation. This may seem like a fortunate coincidence: when the O_2 uptake is great, the O_2 saturation becomes low. At the same time CO_2 is formed and the capacity to transport CO_2 from the tissues to the lungs is increased. The transport of CO_2 and O_2 was also discussed in Chap. 5. In this connection it should be recalled that reduced hemoglobin is a weaker acid than oxyhemoglobin and when reduced, it mops up H^+ ions more readily and thereby helps to prevent too large a reduction in pH through the formation of acid metabolites.

The lower part of the oxygen dissociation curve is, so to speak, a reserve which is utilized during muscular work or pathological conditions. There are very few reported measurements of oxygen content and partial pressure of oxygen in a vein draining a muscle which is working at maximum. Saltin et al. (1968) report from studies on four subjects working at maximum for about 5 min on a treadmill a P_{O_2} that averaged 12 mm Hg in blood collected from the femoral vein (O_2 content, 1.4 vol percent; pH, 7.09; and P_{CO_2}, 70 mm Hg).

Scholander (1960) points out that the myoglobin in the muscle cell may facilitate O_2 diffusion but such an aid to oxygen transport is only of significance at very low P_{O_2} near the capillary (Forster, 1967).

From an O_2 content of 20 vol percent in arterial blood, it may eventually drop during heavy work to below 1 vol percent due to the abundance of capillaries of the muscle tissue with short diffusion distances, a pH around 7.0, and a temperature exceeding 40°C. The mitochondria are apparently working efficiently even if the intracellular P_{O_2} is less than 1 mm Hg; at the end of a capillary in a working muscle the O_2 tension may probably be less than 10 mm Hg (Chance, 1964; Forster, 1964, 1967; Stainsby and Otis, 1964).

REGULATION OF BREATHING

The object of breathing is to ensure the exchange of oxygen and carbon dioxide between the blood and atmospheric air. It would therefore appear logical if these gases were to partake in the regulation of the breathing. Such is actually the case. A change in the P_{O_2}, P_{CO_2}, and H^+ concentration in the arterial blood results in a change in ventilation in such a manner as to moderate the primary change (negative feedback). Stretch receptors in the lungs and muscles affect respiration as well as a number of other factors. As in the case of changes in the circulation under different conditions, the question is often: What is the regulating and what is an eventual disturbing effect? In the case of P_{O_2}, P_{CO_2}, and H^+ concentration in the blood, a change in pulmonary ventilation may influence these factors. The temperature of the blood affects respiration, but a change in pulmonary ventilation does not cause a substantial change in the body temperature in man. In some animals, such as the dog, respiration also plays a part in temperature regulation. The effect of an increased temperature of the blood in their case also has a much more pronounced effect on respiration than in the case of man. Adrenalin affects respiration, but a change in ventilation certainly does not affect the content of adrenalin in the blood. Emotion affects respiration (e.g., the sighing of a mourning person, the rapid breathing of an excited person). A number of different theories have been advanced concerning the regulation of breathing, but none of them have as yet fully explained how the respiratory volume is adjusted to meet the demand at rest and during physical work. The interested student will find a series of review articles in Fenn and Rahn (1964, 1965).

Rest

In the central nervous system a large number of physical, chemical, and nervous variables are integrated. P_{O_2}, P_{CO_2}, and H^+ concentration, or chemical changes related to them, appear to be the most prominently controlled chemical variables. A lowering of the arterial P_{O_2} stimulates the breathing via *peripheral chemoreceptors* in the carotid and aortic bodies from where impulses go to respiratory centers in the medulla. An increase in P_{CO_2} and H^+ concentration also represents a stimulation leading to an increased ventilation, but this effect is primarily elicited from *medullary chemosensitive receptors,* the exact location and characteristics of which are not as yet clarified.

In the medulla oblongata there is a relatively well-defined *center for inspiration* as well as a *center for expiration* (Fig. 7-19). A respiratory rhythm is built in within these medullary networks. They are linked by a system of reciprocal innervation and have self-limiting mechanisms. An increased activity in the inspiratory center stimulates the inspiratory muscles and causes simultaneously an inhibition of the expiratory muscles. The activity in the inspiratory center fades away and the relaxing inspiratory muscles permit an expiration (passive at rest). With stronger stimuli of the respiratory centers, e.g., during exercise, there is a more frequent and forcible alternate activation of the inspiratory and expiratory muscles.

The inspiratory center inhibits itself by discharging impulses to a *pneumotaxic center* in the pons. This, then, dampens the activity in an *apneustic center,* also located in the pons, which otherwise stimulates the inspiratory center by spontaneous activity. There are other sources of inhibition of this apneustic center, with a reduced or absent stimulus of the inspiratory center as a consequence. When the lung tissues are stretched in the course of inspiration, stretch receptors in the inspiratory bronchioles are stimulated and, via afferent vagus fibers, inhibit the apneustic center, and the inspiration is stopped (the Hering-Breuer reflex). If the vagus is cut in an animal, the respiratory frequency becomes slower and the respiratory depth is increased. The same occurs if the connection between the pneumotaxic center and the lower areas is severed. If both these operations are combined, the respiration stops at a deep inspiratory position. Either the pneumotaxic center or the vagal impulses is capable of maintaining eupnea in the absence of the other.

There is evidence to suggest that normally the effect of the inhibitory feedback on inspiration, generated from the lung, is of no, or minor, importance. The pneumotaxic center has no intrinsic rhythmicity and depends on impulses from the outside for its operation. Such impulses originate from the inspiratory center, which through this mechanism inhibits itself and at the same time triggers the expiration. Stimulation of the pneumotaxic center therefore accelerates respiration. There are also numerous other possibilities for negative feedback mechanisms through synaptic connections with the complex and amorphous structure of the brain stem—reticular formation.

The respiratory muscles, especially the intercostal muscles, are amply supplied with muscle spindles. Their motoneurons are contacted by fibers originating in the cortex and the reticular formation, and the sensory fibers from the muscle spindles, as well as from the joint receptors, may end in the same areas. An increase in the γ activity will by a reflex mechanism activate the extrafusal fibers of the same muscle, whether it be inspiratory or expiratory. When, during

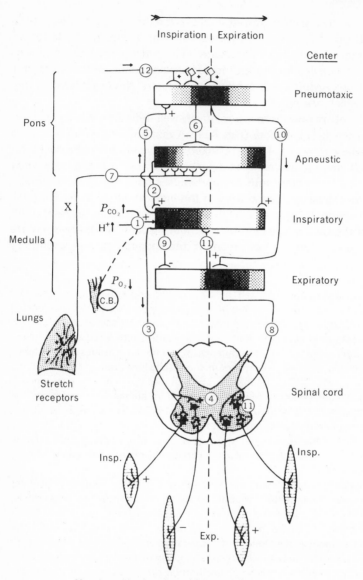

Muscles with phrenic and intercostal nerves

Fig. 7-19   Schematic presentation of part of the organization of the respiratory centers and their function at rest. The shadowed areas within blocks denoting centers show degree of nervous activity. To the left of the vertical dotted line: inspiration; to the right: expiration. The medullary centers, as well as the apneustic center, are inherently rhythmic. Under the influence of (1) stimulatory effects of the P_{CO_2} and H^+ concentration in arterial blood and cerebrospinal fluid (in some conditions also hypoxia) and (2) activity of the apneustic center, the inspiratory center discharges

a forced inspiration, the expiratory muscle fibers are stretched, this may initiate sensory impulses from their muscle spindles. This produces a facilitation of α-motoneurons of the expiratory muscles and simultaneously an inhibition of the inspiratory muscles. The same effect may also be elicited or amplified in a reflex manner by sensory impulses reaching the brain stem (Euler, 1966).

Central Chemoreceptors

The pulmonary ventilation at rest is chiefly regulated by the chemical state of the blood, particularly its CO_2 tension. An increase in the H^+ and CO_2 concentration (hypercapnia) in the arterial blood causes an increased ventilation. The exact location of the chemosensitive receptors is still unclear, whether they are located within the region of the inspiratory or expiratory centers, or in other parts of the brain stem separated from the integrative synapses of the chemoreflex area. It is possible that H^+-sensitive receptors on or near the surface of the medulla play a role which is affected by the pH of the cerebrospinal fluid (CSF) (Brooks et al., 1965). (CO_2 diffuses rapidly between blood and CSF, and an increase of P_{CO_2} in the arterial blood therefore causes a lowering of pH in the CSF as well. The barrier between blood and CSF is, however, much less permeable to ions. An administration of fixed acids to the blood will elicit an increased ventilation and a secondary decrease in P_{CO_2} of the blood and CSF. Since the acids do not easily penetrate to the CSF, the effect will actually be a transient alkalosis of the CSF and a simultaneous acidosis of the blood.) There is evidence to suggest that it is the intracellular H^+ or CO_2 which affects respiration. CO_2 diffuses more readily through the cell membrane than does H^+ but causes a secondary change in the intracellular pH($CO_2 + H_2O \rightleftharpoons H_2CO_3 \rightleftharpoons H^+ + HCO_3^-$). This may explain the fact that a certain amount of lowering of the arterial pH caused by CO_2 produces a greater increase in ventilation than would be the case had the same pH changes

---

down the pathways to the motoneurons of the inspiratory muscles (3), and inspiration occurs. Simultaneously impulses ascend to the pneumotaxic center (5), which in turn sends inhibitory impulses (6) to the apneustic center. The latter, exposed to these inhibitory impulses and eventually also to ascending inhibitory discharges (7) from pulmonary stretch receptors [via the vagal nerve (X)], ceases to activate the medullary inspiratory center. The inspiratory muscles relax, and at rest, the expiration is passive. During work the expiratory muscles become activated (8), there are no inhibitory impulses from the inspiratory center (9), and the expiratory center is exposed to facilitating impulses from the pneumotaxic center (10). With the expiratory center activated, there is a reciprocal inhibition of the inspiratory center and the inspiratory muscles (11). The two populations of neurons are linked together by a system of reciprocal innervation (4) and (11).

This seems complicated, but it is in reality even more complicated and far from clarified. All the structures are bilaterally represented. From various levels impulses reach the different centers with inhibitory or facilatory impulses (12). A reflex may exert a phasic action on each breath; a slow discharge from pulmonary stretch receptors may augment inspiration in its early phase, and later on when the discharge from the receptors becomes rapid, it may inhibit the inspiratory center. The role of the muscle spindles is not discussed in this figure, as it deals with resting conditions, but is considered in the text.

been brought about by other acids. It is difficult to ascertain definitely whether or not CO_2 has a specific effect beyond an effect via intracellular pH.

Normally the inspired air is, practically speaking, free of CO_2 (0.03 percent). The inhalation of small amounts of CO_2 causes an increase in ventilation because the alveolar P_{CO_2}, and thereby the arterial P_{CO_2}, is increased. As long as CO_2 is being inhaled, the ventilation remains elevated, since a reduced ventilation would once more increase P_{ACO_2} above the normal level of about 40 mm Hg. At a high concentration of inspired CO_2 mental confusion occurs, and even narcotic effects with concomitant reduced ventilation may be the result. A denervation of the peripheral chemoreceptors in laboratory animals under normoxic conditions hardly alters the response of the breathing to CO_2 excess, which indicates that the effect is elicited centrally.

Peripheral Chemoreceptors

The direct effect of lack of oxygen on the central nervous system is a reduction in its function. For the stimulation of respiration by hypoxia a stimulation of the peripheral chemoreceptors is required. (If the medullary center is deprived of its afferent denervation, breathing of a hypoxic mixture induces hypopnea or respiratory standstill.) The chemoreceptors in the carotid and aortic bodies consist of epithelioid cells with a rich innervation and blood supply; their oxygen usage is exceptionally high (Daly et al., 1954; Heymans and Neil, 1958).

The function of the carotid (and aortic) body can be studied by recording the impulse activity electrically in the afferent nerve connecting the chemoreceptors with the brain stem. Figure 7-20 presents an example of such a recording from a few-fiber preparation of the afferent carotid sinus nerve in a cat. The same nerve also contains afferent fibers from mechanoreceptors (see Fig. 6-12), but these fibers have in this case been removed. It is evident that the discharge of scattered impulses increases with increasing degree of hypoxia. In the upper (a) and lower (g) tracing, the animal is exposed to an hypoxia corresponding to an altitude of 4,000 m. The discharge is relatively strong, about 90 "spikes" of a certain minimal amplitude. When the oxygen tension of the air is made equivalent to sea-level conditions, the activity in the chemoreceptors is quickly reduced to about 25 spikes, and the pulmonary ventilation decreases from 0.85 liter/min to 0.66 liter/min (c). At an altitude of 6,000 m the discharge is increased, as is the pulmonary ventilation (d). After a change to oxygen breathing, the chemoreceptors almost completely cease to fire (a single spike is seen in the tracing) (f).

It has been shown conclusively that the decisive factor in the hypoxic stimulation of the chemoreceptors is a lowering of P_{O_2}, not a reduced oxygen content as such. Thus, in laboratory animals, up to 80 percent of the hemoglobin may be bound to CO without the respiration being affected, providing the P_{O_2} is kept at a normal level (Heymans and Neil, 1958).

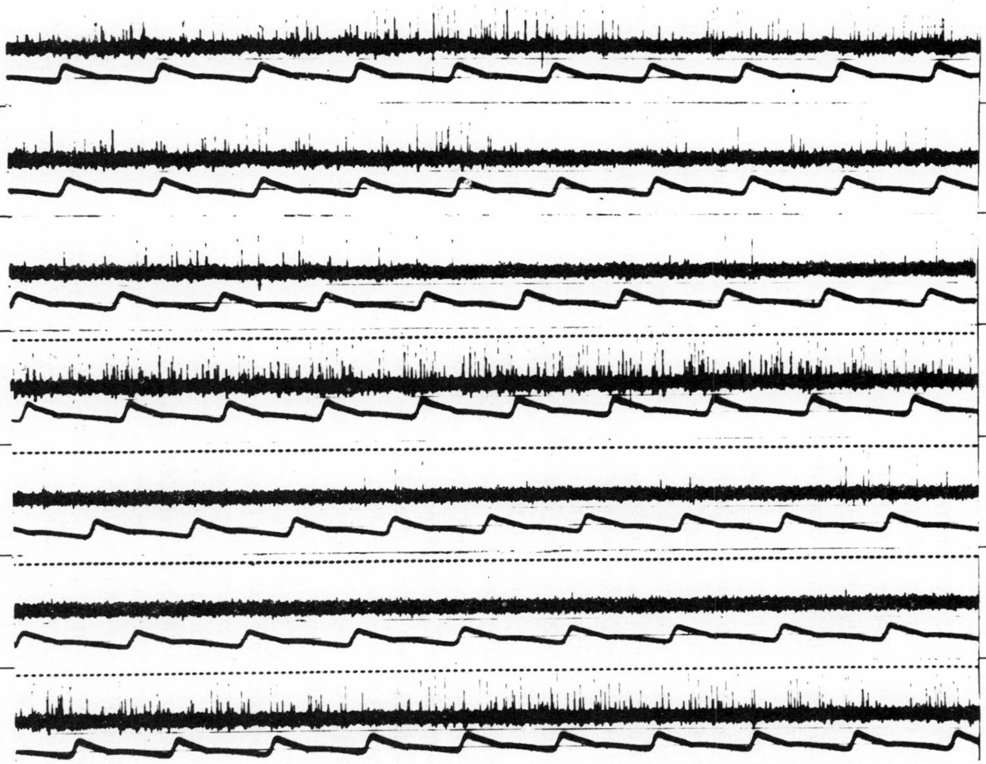

	liters/min	Number of spikes
(a) 4,000 m P_{O_2} 87 mm	0.847	93
(b) Sea level P_{O_2} 149 mm		50
(c) Sea level P_{O_2} 149 mm 4 min later	0.664	24
(d) 6,000 m P_{O_2} 68 mm	0.974	139
(e) Oxygen P_{O_2} 415 mm		11
(f) Oxygen P_{O_2} 415 mm 4 min later	0.705	1
(g) 4,000 m P_{O_2} 87 mm	0.861	90

Fig. 7-20 Action potentials recorded from a few fibers in the carotid sinus nerve in a cat subjected to different altitudes in a low pressure chamber and oxygen breathing at 4,000-m simulated altitude. Tracheal oxygen tension to the left. Records on each film strip from top downwards: time (50 c/sec), electroneurogram and arterial blood pressure (calibration lines at 150 and 100 mm Hg respectively). Right column gives pulmonary ventilation and number of action-potential spikes of an arbitrarily chosen minimal height. Note the relation between chemoreceptor activity and ventilation. (From P.-O. Åstrand, 1954.)

The discharge of the chemoreceptors increases if the blood flow through them is reduced. Stimulation of sympathetic fibers to the carotid bodies as well as blood-borne epinephrine and norepinephrine may reduce the blood flow to the chemoreceptor tissue (Lee et al., 1964; Neil and Joels, 1963). Therefore chemoreceptors may send an increasing number of nerve impulses centrally, despite an unaltered oxygen tension of the arterial blood, if there is an increased sympathetic activity, as after hemorrhage or during exercise (see below).

There are considerable individual variations in the reaction to hypoxia. Some persons may be brought to unconsciousness without pulmonary ventilation being significantly affected. In others the pulmonary ventilation may be increased to the double or more during hypoxia.

What it is that actually stimulates these peripheral chemoreceptors is not clear. If the oxygen tension in the arterial blood is high, as is the case when breathing 95 percent O_2, the arterial P_{CO_2} (by breathing CO_2) may increase and its pH drop markedly without the chemoreceptors being stimulated. So if an hypoxic drive does not exist but the oxygen saturation is adequate, these chemoreceptors then seem rather insensitive to changes in P_{CO_2} and H^+ concentration. At low oxygen tension the discharge from the chemoreceptors is increased further, however, if P_{CO_2} at the same time is high and the pH low (Hornbein and Roos, 1963; Neil and Joels, 1963). Thus, hypercapnia potentiates the effect of hypoxia, possibly by a change in the intracellular H^+ concentration or reduction in the blood flow to the area where the chemoreceptors are located (Riedstra, 1963).

Normally the P_{O_2} in persons breathing ordinary air at sea level is low enough to contribute a small but significant tonic ventilatory stimulus from the peripheral chemoreceptors (Dejours, 1964). Most subjects exhibit a sustained increase in pulmonary ventilation only when breathing 16 percent O_2 and a marked increase in ventilation only when the inspired O_2 has dropped to about 10 percent or less. Thus, the chemoreceptors have different thresholds, some of which are not stimulated markedly until the P_{O_2} has fallen to 40 or 50 mm Hg (8–10 percent O_2 inhaled). The chemoreceptors are very resistant to anoxia and can therefore maintain breathing reflexively for long periods of time.

Chemoreceptors exposed to mixed venous blood in the pulmonary artery have been suggested to be activated during exercise when the venous oxygen tension becomes reduced, and they should reflexively induce hyperpnea. Such receptors have not yet been proved to contribute to a respiratory drive (Cunningham, 1967; Neil, 1967). (With some of the hemoglobin combined with carbon monoxide the mixed venous oxygen tension becomes lower than normal, but the pulmonary ventilation is not influenced.)

There are no experimental evidences of existence of the chemoreceptors in the limbs (Dejours, 1967).

Simultaneous Effect of Hypoxia and Variations in P_{CO_2}

An hypoxic drive via the chemoreceptors increases ventilation. The result is that more CO_2 is expired than is produced. The result of this hyperventilation is that the arterial P_{CO_2} sinks (hypocapnia) and the pH rises. This means a reduced respiratory drive via Pa_{CO_2} and H^+ concentration. If this hypoxic stimulus is now eliminated by the breathing of oxygen-rich air, the ventilation is immediately

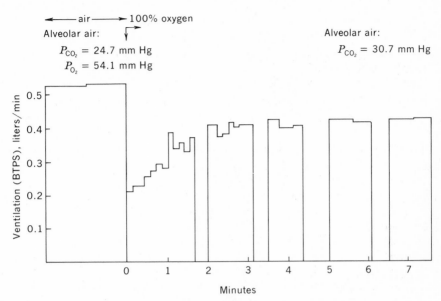

Fig. 7-21   Pulmonary ventilation of a cat acclimatized to 4,000-m altitude for 61 hours when substituting 100 percent oxygen for air. After an immediate decrease in ventilation there is a gradual increase as the alveolar CO_2 tension goes up. (From P.-O. Åstrand, 1954.)

reduced. The CO_2 is thereby accumulated, which tends to offset the effect of interrupting the P_{O_2}-dependent stimulation.

Figure 7-21 gives an example of how a certain ventilation is maintained with an alveolar P_{CO_2} of about 25 mm Hg and a P_{O_2} of 54 mm Hg. (In this case the laboratory animal is kept at an altitude of 4,000 m.) A change to breathing pure oxygen reduces the ventilation to about 40 percent of the original value. After about 2 min, the $P_{A_{CO_2}}$ has increased to about 31 mm Hg, and this increase has caused a gradual increase in ventilation, but the new level is only about 80 percent of the level which was attained during the hypoxia.

Dejours (1964) has based his evaluation of the hypoxic drive of the respiration on similar records, i.e., the initial response in pulmonary ventilation when the subject is exposed to an oxygen-rich gas mixture.

If CO_2 is added to the inspired gas in appropriate amounts to prevent the development of hypocapnia, the effect of various degrees of hypoxia can be studied. Similarly, the changes in the alveolar (and the arterial) P_{O_2} may be prevented by varying the inspired P_{O_2} synchronously with the changes in ventilation of different CO_2 mixtures. Figure 7-22 gives an example of such a series of experiments. In this case an increase in the alveolar P_{CO_2} produces a rectilinear increase in the ventilation when P_{O_2} is kept constant at 110 and 169 mm Hg

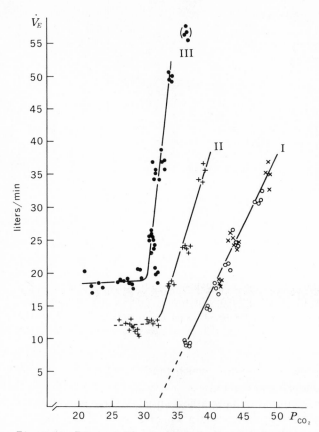

Fig. 7-22 Effect of CO_2 in inspired air on pulmonary ventilation in one subject. (I) Experiments with normal oxygen tension ($\bigcirc$ = alveolar P_{O_2}, 110 mm Hg) or high oxygen tension ($\times$ = alveolar P_{O_2}, 169 mm Hg). (II) Experiments with reduced oxygen supply ($+$ = alveolar P_{O_2}, 47 mm Hg). (III) Experiments with markedly reduced oxygen tension ($\bullet$ = alveolar P_{O_2}, 37 mm Hg). Explanations in text. (From Nielsen and Smith, 1952.)

respectively. During hypoxia corresponding to a $P_{A_{O_2}}$ of 47 mm, the ventilation at rest is higher. As a consequence of this, the P_{CO_2} falls to about 27 mm Hg compared to 36 mm during normoxia in this subject. An increase in $P_{A_{CO_2}}$ gives initially no further increase in the ventilation, but above a certain value of P_{CO_2}, a given increase in P_{CO_2} produces a greater increase in ventilation than normal, and the regression line is considerably steeper. This is even more pronounced during the marked hypoxia: $P_{A_{O_2}}$ 37 mm Hg. During severe hypoxia a moderate change in the P_{CO_2} of the blood is without effect. At higher P_{CO_2} the hypoxia potentiates the CO_2 effect. This interaction between hypoxia and hypercapnia is supposed to take place in the carotid bodies (Hornbein and Roos, 1963; Riedstra, 1963). (Kao et al., 1967, on the other hand, have results which are in support of a central interaction.) (In the discussion of the peripheral chemoreceptors it was mentioned that an elevation of P_{CO_2} or the H^+ concentration increased the discharge under hypoxic conditions. This is not quite in accordance with the constant ventilation shown in Fig. 7-22 at P_{AO_2} 37 mm Hg, in spite of an increase in the P_{ACO_2} from about 20 to 30 mm Hg.) At normal O_2 tensions the ventilation is affected by minute changes in P_{CO_2}.

Summary   The central and peripheral inputs are integrated in the respiratory centers so that P_{O_2}, P_{CO_2}, the acid-base characteristics of the blood, and the internal environment of the body are maintained at a constant level. The inspiratory and expiratory centers in the medulla oblongata are inherently rhythmic and are reciprocally coordinating the inspiratory and expiratory muscles that vary the rate and depth of breathing. Normal respiration is maintained by an interaction between these centers and centers in the pons, including an apneustic center and a pneumotaxic center. (Inspiration is interrupted by a dampening or cessation of the stimulating effect from the apneustic center on the inspiratory center.) Stretch receptors in the lungs and muscle spindles in the respiratory muscles may participate in this switch from inspiration to expiration.

The H^+ ion concentration and, especially, P_{CO_2} of the blood and CSF are continuously affecting the respiratory activity in that cells in the proximity of the respiratory center are affected. Deviation from a normal P_{aCO_2} of 40 mm Hg results in a ventilatory response minimizing the deviation. During hypoxia an increase in ventilation occurs because of discharge from chemoreceptors in the carotic and aortic bodies. A number of structures in the medulla oblongata may influence respiration. Activity of the reticular formation also increases ventilation. This effect forms a link between the proprioceptors in the muscles and joints, the cortex of the brain, and different centers of importance for the motor function, as well as the respiratory centers. Emotion, voluntary actions, and reflexes such as swallowing can easily alter the pattern of respiration.

Exercise

This topic has attracted respiratory physiologists for almost a century, and one has been searching for a "work factor" that could explain exercise hyperpnea. Efforts have been made to quantitate various factors of chemical and nervous nature with regard to their participation in the regulation of breathing during work. So far these efforts have not been too promising. Here we shall present a somewhat different approach compared to the classical concepts and we shall discuss a neurogenic factor as the primary activator of the respiratory muscles during exercise with a secondary feedback mechanism of chemical nature which will regulate and adjust the respiratory volume mainly according to the composition of the arterial blood.

As is evident from Figs. 7-9 and 7-13 the ventilation increases during muscular work almost rectilinearly with the increase in O_2 uptake up to a certain level, after which the increase in ventilation becomes steeper. As we have already stated, there is a great deal of controversy concerning the mechanism underlying this increase in ventilation during muscular work. During submaximal work the arterial P_{CO_2}, P_{O_2}, and H^+ concentrations are roughly at the same level as during rest (Fig. 7-13). During very heavy work the H^+ concentration increases, largely as the result of lactic acid production (HLa) in the working muscles. (HLa + $NaHCO_3 \rightleftharpoons NaLa + H^+ + HCO_3^-$.) Thus pH decreases and may be as low as 7.0. The relative hyperventilation which follows gives an elevated alveolar P_{O_2}, but the arterial P_{O_2} actually drops toward 90 to 85 mm Hg as compared to the normal value of 95 to 100 mm Hg. P_{ACO_2} drops toward 35 mm Hg or even lower values (Fig. 7-13). A similar drop in Pa_{O_2} and pH at rest should cause a rather moderate increase in ventilation, and a lowering of Pa_{CO_2} would in itself cause a reduced respiratory activity. *The chemical changes in the composition of the arterial blood cannot, then, explain per se the ventilation of* 100 *to* 200 *liters/min observed during heavy muscular work.*

The threshold at which the ventilation increases proportionally more than the O_2 uptake does actually vary from person to person. The individual's capacity to supply the working muscles with oxygen is of decisive importance. A person who has a low maximal aerobic power reaches this threshold value at a lower O_2 uptake than does a person who has a high maximal $\dot{V}_{O_2}$ (Fig. 7-9). During work with small muscle groups (the arms for instance), the ventilation at a given O_2 uptake is greater than is the case when larger muscle groups are engaged, for instance, with work involving the leg muscles (Stenberg et al., 1967). (During bicycle ergometer work with the arms, the ventilation was 89 liters/min at an oxygen uptake of 2.8 liters/min, but only 67 liters/min when the work was performed with the legs. The lactic acid concentration of the blood was 11 and 4 mEq/liter respectively. The heart rate was 176 and 154 respectively.)

The possibility of impulses from motor centers in the cortex or from the working muscles and joints involved in the work to the respiratory centers has been discussed (Krogh and Lindhard, 1913; Kao, 1963; Flandrois et al., 1966; Asmussen, 1967; Dejours, 1967; Kao et al., 1967). Such impulses might conceivably bring about an alteration in the threshold or set point for the sensitivity of the respiratory centers for CO_2, pH, and O_2.

When running up a flight of stairs at maximal speed or running a distance of 50 to 100 m, one does not experience an increased ventilation requirement; the task may be fulfilled during breath holding. It is only after the work has been completed that one experiences shortness of breath. Figure 7-23 illustrates how the ventilation increases with increasing work level from 300 up to 2400 kpm/min, i.e., with a requirement for O_2 uptake varying from 0.9 to 5.7 liters/min.

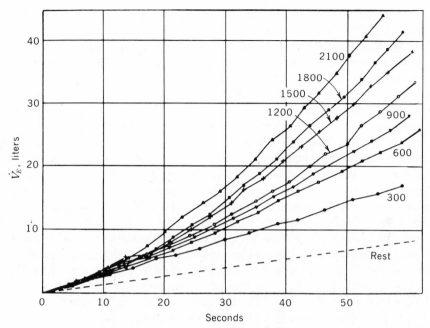

Fig. 7-23   *Increase in pulmonary ventilation during transition from resting conditions (with subject sitting on a bicycle ergometer) to a 60-sec period of work at various intensities. Figures beside the curves give the work loads in kilopond meters per minute. Each line represents the average of three experiments. Expired air was collected in a recording spirometer of 90 liters capacity. (From P.-O. Åstrand and Christensen, 1963.)*

(Maximal $\dot{V}_{O_2}$ of this subject was only about 4.5 liters/min, however.) As can be observed, there is a certain increase in ventilation at the commencement of the work, but at the end of some 10 sec it is still not differentiated in relation to the severity of the work. Any afferent impulse flow from the working muscles or motor centers in the brain would be expected to be considerably greater at 2400 kpm/min than at 300 kpm/min. Only after about 20 sec, when blood from the working muscles can be expected to have reached the lungs and the arterial side of the circulation, does the ventilation increase along with the work load. This speaks in favor of a predominantly chemical regulation of the breathing during work. On the other hand, as emphasized above, the values for the arterial P_{O_2}, P_{CO_2}, and pH are on the whole normal or not far from it. It appears that additional factors must be at play. Similar observations have been reported by Craig et al. (1963).

Campbell (1964), Eklund et al. (1964), von Euler (1966), and other investigators have emphasized the importance of the muscle spindles for the activity of the respiratory muscles. Campbell points out that the respiratory muscles are voluntary muscles subject to all the spinal and supraspinal mechanisms that affect tone, posture, and movement. In the anterior horn cells, there is an integration of respiratory and nonrespiratory drives. Muscle spindles are numerous in the intercostal muscles but much less numerous and less evenly distributed in the diaphragm. It should be recalled (Chap. 4) that an increased activity in the fusimotor γ fibers produces a contraction of the muscle spindle's (intrafusal) muscle, and if the parent muscle (the extrafusal muscle) at the same time is not activated via its α fibers, the receptors in the muscle spindles become deformed and produce an afferent impulse which, via the dorsal root, reaches the spinal cord. This afferent impulse stimulates the α motoneuron, the entire muscle is activated, and the difference in length between the receptor muscle and the muscle as a whole is reduced sufficiently so that the afferent signals cease. This γ motor spindle system represents a "follow-up-length-servo" system. Thus, any difference in length change between intra- and extrafusal muscles is perceived by this receptor of the α motoneuron through excitatory synapsis so that the misalignment will be reduced and the demanded length, and therefore the demanded volume, will be assumed. The importance of this mechanism for ventilation is shown by the fact that adding a resistance on the inspiratory side increases the force of contraction of the inspiratory muscles, thanks to an augmentation of their motoneuron discharge by the stretch reflex. In other words, the afferent nerves are a feedback design leading from the muscle spindles which provides information as to how successfully the respiratory muscles have accomplished their task.

It has been shown that the α and γ motoneurons of the inspiratory and expiratory muscles are driven from the respiratory integrating mechanisms in the medulla in a reciprocal fashion (Eklund et al., 1964). Changes in chemical respiratory drive exert reciprocal effects on the expiratory and the inspiratory α and γ excitability. In reality the γ activity in the respiratory muscles is more dominating than the α activity, so that the intrafusal muscles shorten relatively more than the extrafusal muscles and thereby the muscle spindles exert a reflex drive on the α motoneurons to achieve the demanded change of volume and pulmonary ventilation. During breathing at rest, the muscle spindles in the intercostal expiratory muscles are passive during expiration but discharge moderately when they are stretched during inspiration. During forced respiration, as during work, the muscle spindles are "called to life" even during expiration and stimulate by reflex the motoneurons of the expiratory muscles at the same time inhibiting the antagonists, that is, the intercostal inspiratory muscles and actually also the diaphragm.

This system of γ and α motoneurons, linked together via the afferent nerves of the muscle spindles and affected by supraspinal mechanisms, especially the reticular formation, may form the basis for the regulation of the ventilation of the lungs. In connection with work, the activity in the reticular formation is increased, partly through impulses from cortex cerebri and other higher centers, and partly through afferent impulses from the muscle spindles in the working muscles, etc. Since the reticular formation activity is unaffected by the conditions in the respiratory centers, it is conceivable that this may increase the activity also in the γ and α motoneurons to the respiratory muscles, with an increased volume change in the thorax as a consequence. It would be natural if the respiratory movements and respiratory frequency were adjusted according to the work rhythm, which usually is the case: the respiratory frequency is adjusted in accordance with stride frequency, pedal frequency in cycling, rowing, etc. A change in posture and work rhythm may result in a new pattern of breathing with reference to respiratory frequency and depth. The momentary increased ventilation noticed in connection with the onset of work (see Fig. 7-23) must then be adjusted according to the requirement for the elimination of CO_2 and uptake of O_2. Here chemical regulation enters into the picture. The α-γ system has a possible tendency to produce a hyperventilation, especially during the beginning of the work until the produced CO_2 has reached the lungs. Through a negative feedback elicited from the respiratory centers, if Pa_{CO_2} tends to drop, the γ and α activity driving the respiratory muscles may then be inhibited from the respiratory centers. When the CO_2 reaches the lungs without being eliminated in sufficient quantity, the P_{CO_2} of the arterial blood will rise and the inhibition becomes diminished.

This is to some extent a hypothetical discussion which postulates that in connection with muscular work, there is also an increased γ activity to the respiratory muscles. The system has, however, every possibility to produce a rhythmic coordinated switching between inspiration and expiration, partly determined by the work rhythm. It should be recalled that the respiratory muscles are involved in most activities not only to provide adequate pulmonary ventilation, but also to stabilize the trunk in posture or movements. A synchronization between the respiratory movements and work rhythm is therefore important. The γ system provides the possibility for such a coordination. [Actually there are two types of intercostal fusimotor neurons driven from different central nervous areas: rhythmic ones, with a functional link with the respiratory drive, and in addition, tonically firing ones, much more readily influenced by various reflexes from postural influence and affected by the cerebellum and other structures (Euler, 1966). This combination of systems makes it possible to integrate respiratory and postural movements at the spinal level and to correct "actual" length of the intercostal muscles to "wanted" length in accordance with the demands of breathing and also with a view to postural engagement.]

During exercise the γ-α system may create a tendency to hyperventilation; but a negative feedback from the respiratory centers may then regulate the ventilation according to the chemical composition of the arterial blood. This feedback may directly affect the motoneurons or operate via the reticular formation.

Figure 7-24 is a simplified diagram presenting components involved in exercise hyperpnea. *The neurogenic factors cannot be considered as a regulating stimulus, but act as an activator. The chemical stimuli are of decisive importance for the finer adjustment of the ventilation.* If one works very hard for 5 sec, rests for 5 sec, and so forth, the ventilation may eventually rise to 100 liters/min, but no difference in ventilation can be detected between the periods of rest and work (Christensen et al., 1960). A variation in afferent impulses from the working muscles is in this case of less importance in that the input of CO_2 to the lungs is continuously high. The greater the CO_2 production the greater the "independence" of the respiratory muscles.

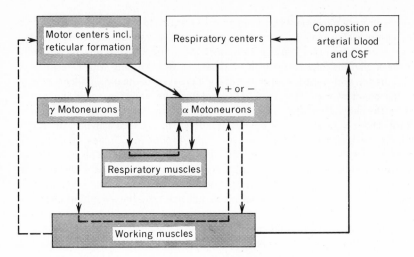

Fig. 7-24 Schematic summary of the discussion on regulation of breathing during exercise. The respiratory muscles are activated via their gamma (γ) and alpha (α) motoneurons (filled-line arrows). In a similar way the other working muscles are activated (dotted-line arrows). The different respiratory centers are influenced directly or indirectly by the chemical composition of the arterial blood and cerebrospinal fluid (CSF), mainly their P_{CO_2}, P_{O_2}, and pH. These centers can then facilitate or inhibit the motoneurons of the respiratory muscles, depending on the effectiveness of the gas exchange in the lungs. Particularly critical is the CO_2 refill from the muscles and CO_2 output from the lungs. (The afferent nerve impulses to motor centers include impulses from receptors located in tendons and joints.)

There are many intriguing experimental findings to consider when analyzing the neurogenic respiratory drive during exercise (Asmussen, 1967). In negative exercise the mechanical tension developed by the contracting muscles may be 5 to 7 times greater than in positive work, but the pulmonary ventilation, per liter oxygen uptake, is similar in the two types of work. Prolonged static effort with maintained high muscular tension does not represent an especially strong respiratory drive. The very moderate ventilatory response to the first seconds of maximal effort, e.g., a sprint, has been mentioned. However, the afferent impulses from exercising limbs and the stimulation from various parts of the brain due to the increased motor activity must be considered as coordinated activators of the respiratory muscles, but their motoneurons are subjected to various degrees of inhibition from the respiratory centers depending on the chemical composition of the blood.

It is noteworthy that any change in respiratory frequency which adapts to the rhythm of movements is, within limits, automatically followed by a change in tidal air to provide an adequate alveolar ventilation. (We have a similar situation in the regulation of cardiac output: at rest and during not too severe exercise, a change in heart rate is matched by a variation in stroke volume so that the cardiac output is maintained constant.)

Hypoxic Drive during Exercise

If during exercise the normal air is substituted by oxygen, the ventilation drops within seconds, and this effect is particularly marked during heavy work (Asmussen and Nielsen, 1946; Bannister and Cunningham, 1954; Dejours, 1964; Cunningham, 1967). The reaction is so rapid that an effect via the chemoreceptors in the carotid and aortic bodies has to be assumed. This ventilation-reducing effect of O_2 breathing is marked, however, even if the arterial P_{O_2} is at a normal level, and this fact has confused respiratory physiologists.

It has already been mentioned that the chemoreceptors have a relatively large blood flow and high oxygen uptake. They are innervated by sympathetic nerve fibers which supply the arterioles of the carotid and aortic bodies with vasoconstrictor fibers. When these fibers are activated, there is a reduction in the blood flow through the chemoreceptor areas, possibly by the blood being diverted through adjacent arteriovenous anastomoses. It is possible that a change in the P_{CO_2} and H^+ concentration of the arterial blood may contribute to modify the blood flow to the epithelioid cells of the chemosensitive areas.

During work, the sympaticus activity is increased and the concentration of epinephrine and norepinephrine in the blood increases. In fact Biscoe and Purves (1965) report immediate increase in nerve activity of the cervical sympathetic and the postganglionic nerve to the carotid body by exercise (in the cat), and these changes were abolished by femoral and sciatic nerve section. It is conceivable that the blood flow to the chemoreceptor cells is reduced to such an extent that the O_2 supply to the metabolically very active cells becomes unsatisfactory or that the removal of products of tissue metabolism becomes insufficient. This might then produce an excitation of the chemoreceptors and a stimulation of respiration despite a normal or only slightly different from normal systemic arterial P_{O_2} (Daly et al., 1954; Hornbein and Roos, 1962; Neil and Joels, 1963; Lee et al., 1964).

The heavier the work in relation to the performance capacity of the individual, the greater the sympaticus activity. Thereby the chemoreceptor drive may also increase, despite the elevated perfusion pressure. This mechanism may also contribute to the increase in ventilation during pronounced emotion as well as during work involving small muscle groups, when the ventilation is relatively high at a given oxygen uptake.

This hypothesis is not supported by Severinghaus et al. (Eisele et al., 1967). They found no difference in end-tidal P_{CO_2} and arterial pH in subjects working after bilateral blockade of the stellate ganglia with Xylocaine compared with normalcy. They conclude that "sympathetic innervation of the carotid body does not appear to contribute to the hyperpnea of exercise." The authors do not, however, state how heavy the work was in relation to the subject's maximal aerobic power. It may have been too mild. Furthermore, all sympathetic fibers to the chemoreceptor areas may not have been blocked; an eventual effect of circulating epinephrine and norepinephrine is difficult to exclude in these experiments.

Summary   During work the ventilation increases relatively rectilinearly, but this increase is relatively steeper during very heavy work. How this increase in ventilation is elicited is unknown. The comparatively small changes which are observed in the P_{O_2}, P_{CO_2}, and H^+ concentration in the arterial blood cannot explain the increase in ventilation. It is suggested as a hypothesis that muscular work as such, through afferent impulses from the engaged muscle spindles and/or from the central nervous system, increases the activity in the γ and α motoneurons of the respiratory muscles and through spinal and supraspinal reflex centers, and thus in a closely coordinated manner produces an increase in the frequency and depth of respiration, often in pace with the muscular movements. The actual regulation of the respiratory volume then takes place through a negative feedback mechanism, primarily determined by the CO_2 production in relation to CO_2 elimination during expiration. In this manner the P_{CO_2} of the arterial blood will, via the respiratory centers, determine the magnitude of the ventilation. During anaerobic work the H^+ concentration of the blood will increase, which represents a further stimulation of respiration. During heavier work the peripheral chemoreceptors in the carotid and aortic bodies will also stimulate respiration, possibly because an increased sympaticus activity will reduce the blood flow to the chemoreceptor areas so that the local P_{O_2} drops in spite of an almost normal value for P_{O_2} in the arterial blood. During maximal work there is a reduction in Pa_{O_2} which should further increase ventilation. The hyperventilation which follows may even produce a drop in Pa_{CO_2}, but the arterial pH inevitably drops (Fig. 7-13).

BREATHLESSNESS (DYSPNEA)

Comroe (1966a) presents the following definitions: *Tachypnea* is rapid breathing; *hyperpnea* is increased ventilation in proportion to increased metabolism; *hyperventilation* is ventilation in excess of metabolic requirement; *dyspnea* is difficult, labored, uncomfortable breathing. The reason why respiration may become consciously troublesome is not clear. The magnitude of the ventilation is not the determining factor: a patient may experience a pulmonary ventilation of 10 liters/min as extremely disturbing, while the athlete is not consciously troubled by a ventilation as high as 200 liters/min. In certain cases afferent vagus impulses may be the cause of the dyspnea (for instance through the collapse of some alveoli), but more commonly impulses from muscle spindles and thoracic joint receptors appear to give rise to conscious awareness and distress. It may be a question of length/tension, i.e., tension appropriateness (Campbell, 1966). Altered afferent signals of the muscle spindles and receptors in the chest wall reaching subcortical and cortical levels may cause unusual sensations. From experience we learn what to expect and how it should be felt or experienced in a certain situation, and the respiration itself, which normally proceeds unconsciously, may give rise to distress when it requires a conscious modification. Light activity at high altitude requires a ventilation which at sea level is associated with heavy work, and this difference may be experienced as distress. A person unaccustomed to muscular work may experience a ventilation of 75 liters/min as unpleasant, but after suitable training this may even appear as a pleasant sensation!

SECOND WIND

During the first minutes of work, the load may appear very strenuous. One may experience dyspnea, but this distress eventually subsides; one experiences a "second wind." The factors eliciting the distress may be an accumulation of metabolites in the working muscles and in the blood because the O_2 transport is inadequate to satisfy the requirement.

By what mechanism this changed environment is brought to consciousness is not known. During heavy work there is actually at the commencement of work a hypoventilation due to the fact that there is a time lag in the chemical regulation of the respiration. It is then actually a matter of a length/tension inappropriateness in the intercostal muscles. When the second wind occurs, the respiration is increased and adjusted according to requirement.

It appears that the respiratory muscles are forced to work anaerobically during the initial phases of the work if there is a time lag in the redistribution of blood. A stitch in the side may then develop. This is probably the result of hypoxia in the diaphragm. This is most common in untrained persons and is particularly apt to occur if heavy work is performed shortly after a large meal, when the circulatory adjustment at the commencement of work is slower. As the blood supply to the respiratory muscles is improved, the pain disappears. (This theory is not entirely satisfactory. During maximal work the oxygen tension is often somewhat reduced. It is then possible at will to increase the ventilation. This causes no further increase in oxygen uptake, so that the additional work is probably covered by anaerobic processes. In spite of this, chest pain seldom occurs. It was previously believed that the pain was caused by an emptying of the blood depots in the spleen and was caused by the contractions taking place in the spleen. In man the spleen serves no such depot function, however. Furthermore, persons who have had their spleen removed may still experience such pain.)

Well-trained athletes who have warmed up adequately prior to a muscular effort seldom experience such pain.

HIGH AIR PRESSURES, BREATH HOLDING, DIVING

High Air Pressures

(For details, see Lambertsen, 1965). While man can become acclimatized to low air pressures, there is no way to become acclimatized to high air pressures such as are encountered in deep sea diving and during escape from submarines when the survivor attempts to get from the inside of the craft where the pressure is normal to the surface through the sea where the air pressure is higher. For every 10 m (33 ft) of sea water the diver descends, an additional pressure of 1 atm is acting upon his body. As the pressure increases, more gases can be taken up by his body and dissolved in the various tissues. At a depth of about 10 m, twice as much gas will be dissolved in his blood and tissues as at sea surface. This is apt to give him trouble, mainly because of the nitrogen. The trouble with nitrogen is that it diffuses into various tissues of his body very slowly, and once dissolved, it also leaves the body very slowly when the pressure once more is reduced to the normal atmospheric pressure. This is especially bad when the pressure is suddenly reduced from several atmospheres, as may be the case during submarine escape or deep sea diving. Then the nitrogen is released from the tissues in the form of insoluble gas bubbles. These bubbles congregate in the small blood vessels where they obstruct the flow of blood. This, then, gives rise to symptoms such as pains in the muscles and joints, and even paralysis may develop if the bubbles become trapped in the central nervous system. These symptoms are known as "the bends." Obviously, the severity of the symptoms depends on the magnitude of the pressure, which means the depth to which the person has descended under water, the length of time the person has spent at that depth, and the speed with which he ascends to the surface.

The bends can be avoided to a large extent by a slow return to normal pressure so as to allow time for the tissues to get rid of their excess nitrogen without the formation of bubbles. Another way to avoid the bends is to prevent the formation of these bubbles by replacing atmospheric nitrogen with helium, which is less easily dissolved in the body. This is done by having the diver breathe a helium-oxygen gas mixture. Another advantage with this is that it is more apt to prevent the so-called nitrogen narcosis which occurs when air is breathed at 3 atm or more and results in an onset of euphoria and impaired mental activity with lack of ability to concentrate. With increasing pressures, the individual is progressively handicapped and may be rendered helpless at 10 atm. Pilots of high-flying aircraft may also suffer from bends if there is a sudden loss of pressure in the pressurized cabin, but the symptoms in these cases are usually not so severe as in the divers. In any case they usually do not occur at altitudes lower than 30,000 ft.

Prolonged breathing of 100 percent oxygen may be quite harmful, for irritation of the respiratory tract may occur after 12 hr, and frank bronchopneumonia after 24 hr. In most individuals, no harmful effects result from breathing mixtures with less than 60 percent oxygen, while newborn infants are particularly susceptible to oxygen poisoning and may suffer harmful effects with oxygen concentrations over 40 percent. The remarkable thing is that oxygen poisoning apparently is no problem when breathing 100 percent oxygen at altitudes over 18,000 feet, no matter for how long. Oxygen poisoning, therefore, is not much of a problem in aviation medicine, but it is indeed an important problem in deep sea diving where it may even affect the brain function when pure oxygen is used at depths greater than about 10 m (33 ft), but there are great individual variations in sensitivity to 100 percent oxygen. The onset of symptoms may be hastened by vigorous physical activity at great depths; it starts with muscular twitchings and a jerking type of breathing, and ends in unconsciousness and convulsions. The exact cause of this is unknown, but it is assumed that it is a matter of interference with certain enzyme systems in the tissues.

Furthermore, when breathing pure oxygen at a pressure of 3 atm or more, the oxygen dissolved in the blood covers the oxygen need of the body at rest. Therefore, at rest, the hemoglobin of the venous blood is still saturated with oxygen. This interferes with the CO_2 transportation (Chap. 5). The result will be gradual CO_2 retention in the tissue and decrease in pH.

Breath Holding—Diving

A well-nourished individual may survive without difficulty for weeks without food and for days without water. But he can only live a few minutes without oxygen. The body's ability to store oxygen is extremely limited. The blood may contain up to about 1.0 liter. In the lungs after a normal inspiration there may be about 0.5 liter oxygen; after a maximal inspiration it may be about 1.0 liter. (A total capacity of 7.0 liters and a concentration of O_2 of 15 percent $= 15/100 \times 7.0 = 1.05$ liters.)

The O_2 bound to myoglobin may amount to as much as 0.5 liter, but it can only be considered as a local store and cannot be utilized by other tissues. The hyperbolic slope of the O_2-myoglobin dissociation curve binds the oxygen effectively until its partial pressure drops to extremely low values.

When a person holds his breath at rest, a total of about 600 ml O_2 is available and can be utilized. This is enough to last about 2 min. "Normal" maximal breath-holding time is 30 to 60 sec. The arterial P_{O_2} then drops to about 75 to 50 mm Hg and the P_{CO_2} rises to 45 to 50 mm Hg. This elevated CO_2 pressure plays a greater role in forcing the individual to discontinue the breath holding than does the reduced O_2 pressure. The hypoxic drive is also an important factor, as demonstrated by the fact that the breath holding may be extended if it is preceded by O_2 breathing. Under these conditions a further increase of Pa_{CO_2} of 5 to 10 mm Hg may be tolerated. Other factors also affect the capacity for breath holding: (1) If one holds his breath with a lung volume near the total lung capacity (maximally inflated), the respiratory standstill may be extended until the P_{CO_2} has reached a value about 10 mm Hg higher than with a lung volume near the residual level (this is even the case with O_2 breathing; for this reason a different hypoxic drive may be excluded). The reason is probably that afferent impulses from the receptors in the lungs and thoracic cage produce a greater stimulus to inspiration in the expiratory position. (2) Breath holding may be prolonged by swallowing; swallowing constitutes a momentary inhibition in inspiration. (3) Single or repeated breaths after the breaking point, without change in alveolar air composition, make a new breath holding possible, and higher Pa_{CO_2} and lower Pa_{O_2} are obtained compared with the first trial (P.-O. Åstrand, 1960, Mithoefer, 1965).

If a subject voluntarily hyperventilates forcefully for about a minute, the pulmonary air is exchanged more often and the composition of its gases more closely approaches that of the inspired air. In this manner the alveolar P_{O_2} may increase to about 135 mm Hg, the P_{CO_2} may drop below 20 mm Hg, and the arterial blood will assume similar gas pressure. Because of the shape of the O_2 saturation curve (Fig. 6-18), the O_2 content of the arterial blood is hardly affected, however, and the pulmonary air will only receive an addition of 40 ml/liter pulmonary air (19 vol percent O_2 instead of 15).

A more important effect of hyperventilation in this connection is the reduction in the CO_2 content of the body. Thus the breath holding may be extended to several minutes by providing more room for CO_2. The breaking point occurs probably at a P_{CO_2} of about 40 mm Hg and a P_{O_2} of 45 mm Hg. This is still within the safe range as far as the critical O_2 pressure for the function of the nervous system is concerned, which in the case of the O_2 pressure of the arterial blood is in the order of 25 to 30 mm Hg.

If a maximal breath holding is performed during muscular work (P.-O. Åstrand, 1960), the breath-holding time is shorter than at rest, but due to the fact that O_2 uptake and CO_2 production occur at a higher rate, the composition of the arterial blood and the pulmonary air will change more rapidly. At the breaking point the alveolar P_{CO_2} may be as high as 75 mm Hg and the P_{O_2} may drop toward 40 mm Hg (the arterial gas pressures are probably at the same level). By breathing pure oxygen prior to the breath holding the $P_{A_{CO_2}}$ may exceed 90 mm Hg.

If a forceful hyperventilation precedes a breath holding during heavy work, breath holding may be prolonged until convulsions or even fainting occur (P.-O. Åstrand, 1960). The alveolar P_{O_2} may then drop toward 20 mm Hg, which explains the symptoms. (In experiments of this type in the laboratory the subject was secured in a harness in order not to fall off the bicycle ergometer!)

The various possible explanations will not be discussed here. However, on the basis of these experiments the risk involved in prolonged diving and deep diving without equipment will be emphasized. (See Lanphier and Rahn, 1963.)

If such a dive is performed following a marked hyperventilation, one may apparently hold the breath until unconsciousness occurs. It should be emphasized that the total pressure of the alveolar air at sea level is about 760 mm Hg, but at a depth of 10 m it is increased to double (i.e., twice the atmospheric pressure) by the pressure of the water on the thorax. An O_2 percentage which at sea level corresponds to an alveolar O_2 tension of about 25 mm Hg (3.5 vol percent) produces at a depth of 2 m (total pressure = 912 mm Hg) a pressure of 30 mm Hg. [$3.5/100 \times (912 - 47) = 30$.] As long as the diver remains at a depth of 2 m, the O_2 tension may thus be adequate to meet the requirement of the nerve cells. But when the diver approaches the surface, the oxygen tension may drop below the critical level. The danger is greater if the ascent is slow, in that the O_2 utilization then causes the O_2 pressure to drop further. Several cases have been described when divers attempting to beat a record, or who for various other reasons have taken their time during prolonged diving, have lost their lives or have been rescued in the nick of time (Craig, 1961; Davis, 1961). Often hyperventilation preceded the diving; the diver had then exhibited a "strange" behavior or he had simply ceased to swim and had sunk. Cases have been described when the swimmer had had no major difficulty holding his breath but then suddenly had lost consciousness.

It is thus justifiable to warn against extreme deep diving and prolonged diving without effective supervision. The diver should avoid hyperventilation prior to the dive (at the most 5 deep respirations). During diving without equipment, oxygen-breathing prior to the dive may definitely improve the performance. It represents an improvement if he has 5 liters O_2 in his lungs instead of barely 1 liter during the dive. In other types of work, the effect is negligible, however, unless the oxygen is inspired during the actual performance. In a few respirations the extra volume of oxygen is washed out, since no extra stores of oxygen can be deposited in the body. In this connection another effect of hyperventilation should be stressed. The washing out of CO_2 in connection with the hyperventilation increases the pH. The effect of this alkalosis and reduced Pa_{CO_2} is a vasoconstriction, including the vessels in the brain. Dizziness and cramps may be the result. If one holds one's breath after a hyperventilation against a closed glottis and at the same time contracts the abdominal muscles (Valsalva maneuver), the cardiac output is reduced. This in combination with the vasoconstriction in the cerebral blood vessels may produce an oxygen deprivation in the CNS sufficient to cause a transitory loss of consciousness.

Summary   During breath holding while working, a greater CO_2 content of the blood is tolerated. Since oxygen is rapidly utilized during work, there is a risk of the breath holding ending in unconsciousness. This is particularly the case if the breath holding is preceded by marked hyperventilation. The possibility of unconsciousness represents a risk during prolonged and deep diving, especially if the ascent from considerable depths takes place slowly. The intrathoracic pressure increases with the depth of the water. A given oxygen pressure in the alveoli and in the blood may therefore be adequate for the normal function of the central nervous system at a depth of a few meters but may become critically low as the diver approaches the surface.

REFERENCES

Asmussen, E.: Exercise and Regulation of Ventilation, *Circulation Res.*, **20**:1–132, 1967.

Asmussen, E., and M. Nielsen: Studies on the Regulation of Respiration in Heavy Work, *Acta Physiol. Scand.*, **12**:171, 1946.

Asmussen, E., and M. Nielsen: Physiological Dead Space and Alveolar Gas Pressures at Rest and during Muscular Exercise, *Acta Physiol. Scand.*, **38**:1, 1956.

Åstrand, I.: Aerobic Work Capacity in Men and Women with Special Reference to Age, *Acta Physiol. Scand.*, **49**(Suppl. 169), 1960.

Åstrand, P.-O.: "Experimental Studies of Physical Working Capacity in Relation to Sex and Age," Munksgaard, Copenhagen, 1952.

Åstrand, P.-O.: A Study of Chemoceptor Activity in Animals Exposed to Prolonged Hypoxia, *Acta Physiol. Scand.*, **30**:335, 1954.

Åstrand, P.-O.: Breath Holding during and after Muscular Exercise, *J. Appl. Physiol.*, **15**:220, 1960.

Åstrand, P.-O., and E. H. Christensen: The Hyperpnoea of Exercise, in D. J. C. Cunningham and B. B. Lloyd (eds.), "The Regulation of Human Respiration," p. 515, Blackwell Scientific Publications, Ltd., Oxford, 1963.

Åstrand, P.-O., L. Engström, B. Eriksson, P. Karlberg, I. Nylander, B. Saltin, and C. Thorén: Girl Swimmers, *Acta Paediat.* (Suppl. 147), 1963.

Åstrand, P.-O., and B. Saltin: Oxygen Uptake during the First Minutes of Heavy Muscular Exercise, *J. Appl. Physiol.*, **16**:971, 1961.

Bannister, R. G., and C. J. C. Cunningham: The Effects on the Respiration and Performance during Exercise of Adding Oxygen to the Inspired Air, *J. Physiol. (London)*, **125**:118, 1954.

Bargeton, D.: Analysis of Capnigram and Oxygram in Man, *Bull. de Physio-Pathologie Respiratoire*, **3**:503, 1967.

Bates, D. V.: Measurement of Regional Ventilation and Blood Flow Distribution, in W. O. Fenn and H. Rahn (eds.), "Handbook of Physiology," sec. 3, Respiration, vol. II, p. 1425, American Physiological Society, Washington, D.C., 1965.

Bates, D. V., and R. V. Christie: "Respiratory Function in Disease," W. B. Saunders Company, Philadelphia, 1964.

Bates, D. V., and J. F. Pearce: Pulmonary Diffusing Capacity, *J. Physiol. (London)*, **132**:232, 1956.

Biscoe, T. J., and M. J. Purves: Carotid Chemoreceptor and Cervical Sympathetic Activity during Passive Third Limb Exercise in the Anaesthetized Cat, *J. Physiol. (London)*, **178**:43P, 1965.

Bjurstedt, H., G. Rosenhamer, and O. Wigertz: High-G Environment and Responses to Graded Exercise, *J. Appl. Physiol.*, **25**:713, 1968.

Braus, H.: "Anatomie des Menschen," Springer-Verlag OHG, Berlin, Bd. 2, Aufl. 3, p. 134, 1956.

Brooks, C. McC., F. F. Kao, and B. B. Lloyd (eds.): "Cerebrospinal Fluid and the Regulation of Ventilation," Blackwell Scientific Publications, Ltd., Oxford, 1965.

Campbell, E. J. M.: Motor Pathways, in W. O. Fenn and H. Rahn (eds.), "Handbook of Physiology," sec. 3, Respiration, vol. I, p. 535, American Physiological Society, Washington, D.C., 1964.

Campbell, E. J. M.: The Relationship of the Sensation of Breathlessness to the Act of Breathing, in J. B. L. Howell and E. J. M. Campbell (eds.), "Breathlessness," p. 55, Blackwell Scientific Publications, Ltd., Oxford, 1966.

Chance, B., B. Schoener, and F. Schindler: The Intracellular Oxidation-Reduction State, in F. Dickens and E. Neil (eds.), "Oxygen in the Animal Organism," p. 367, Pergamon Press, New York, 1964.

Cherniack, R. M., and L. Cherniack: "Respiration in Health and Disease," W. B. Saunders Company, Philadelphia, 1961.

Christensen, E. H., R. Hedman, and B. Saltin: Intermittent and Continuous Running, *Acta Physiol. Scand.*, **50**:269, 1960.

Cole, P.: Respiratory Mucosal Vascular Responses, Air Conditioning and Thermo Regulation, *J. Laryngol. Otol.*, **68**:613, 1954.

Comroe, J. H., Jr.: "Physiology of Respiration," Year Book Medical Publishers, Inc., Chicago, Ill., 1965.

Comroe, J. H., Jr.: Some Theories of the Mechanism of Dyspnoea, in J. B. L. Howell and E. J. M. Campbell (eds.), "Breathlessness," p. 1, Blackwell Scientific Publications, Ltd., Oxford, 1966a.

Comroe, J. H., Jr.: The Lung, *Sci. Am.*, **214**(2):56, 1966b.

Cotes, J. E.: "Lung Function," Blackwell Scientific Publications, Ltd., Oxford, 1965.

Craig, A. B., Jr.: Causes of Loss of Consciousness during Underwater Swimming, *J. Appl. Physiol.*, **16**:583, 1961.

Craig, F. N., E. G. Cummings, and W. V. Blevins: Regulation of Breathing at Beginning of Exercise, *J. Appl. Physiol.*, **18**:1183, 1963.

Cunningham, D. J. C.: Regulation of Breathing in Exercise, *Circulation Res.*, **20**:1–122, 1967.

Daly, M. De B., C. S. Lambertsen, and A. Schweitzer: Observations on the Volume of Blood Flow and Oxygen Utilization of the Carotid Body in the Cat, *J. Physiol. (London)*, **125**:67, 1954.

Davis, J. H.: Fatal Underwater Breath Holding in Trained Swimmers, *J. Forensic Sci.*, **6**:301, 1961.

Dejours, P.: Control of Respiration in Muscular Exercise, in W. O. Fenn and H. Rahn (eds.), "Handbook of Physiology," sec. 3, Respiration, vol. I, p. 631, American Physiological Society, Washington, D.C., 1964.

Dejours, P.: Neurogenic Factors in the Control of Ventilation during Exercise, *Circulation Res.*, **20**:1–146, 1967.

Donevan, R. E., W. H. Palmer, C. J. Varvis, and D. V. Bates: Influence of Age on Pulmonary Diffusing Capacity, *J. Appl. Physiol.*, **14**:483, 1959.

DuBois, A. B.: Resistance to Breathing, in W. O. Fenn and H. Rahn (eds.), "Handbook of Physiology," sec. 3, Respiration, vol. I, p. 451, American Physiological Society, Washington, D.C., 1964.

Eisele, J. H., B. C. Ritchie, and J. W. Severinghaus: Effect of Stellate Ganglion Blockade upon the Hyperpnea of Exercise, *J. Appl. Physiol.*, **22**:966, 1967.

Eklund, G., C. von Euler, and S. Rutkowski: Spontaneous and Reflex Activity of Intercostal Gamma Motoneurons, *J. Physiol.*, **171**:139, 1964.

Euler, C. v.: Proprioceptive Control on Respiration, in R. Granit (ed.), "Muscular Afferents and Motor Control," p. 197, Nobel Symposium I, John Wiley & Sons Inc., New York, 1966.

Euler, C. v.: The Control of Respiratory Movement, in J. B. L. Howell and E. J. M. Campbell (eds.), "Breathlessness," p. 19, Blackwell Scientific Publications, Ltd., Oxford, 1966.

Fenn, W. O., and H. Rahn (eds.): "Handbook of Physiology," sec. 3, Respiration, vols. I and II, American Physiological Society, Washington, D.C., 1964–1965.

Flandrois, R., J. R. Lacour, J. I. Maroquin, and J. Charlot: Essai de Mise en Évidence d'un Stimulus Neurogénique Articulaire de la Ventilation lors de l'Exercise Musculaire Chez la Chien, *J. Physiol. (Paris)*, **58**:222, 1966.

Flandrois, R., R. LeFrançois, and A. Teillac: Comparaison de Plusieurs Grandeurs Ventilatoires dans Deux Types d'Exercise Musculaire, *Biotypologie*, **22**:66, 1961.

Forster, R. E.: Diffusion of Gases, in W. O. Fenn and H. Rahn (eds.), "Handbook of Physiology," sec. 3, Respiration, vol. I, p. 839, American Physiological Society, Washington, D.C., 1964.

Forster, R. E.: Oxygenation of the Muscle Cell, *Circulation Res.*, **20**:1–115, 1967.

Heymans, C., and E. Neil: "Reflexogenic Areas of the Cardiovascular System," Churchill, London, 1958.

Holmgren, A., and P.-O. Åstrand: D_L and the Dimensions and Functional Capacities of the O_2 Transport System on Humans, *J. Appl. Physiol.*, **21**:1463, 1966.

Hornbein, T. F., and A. Roos: Effect of Mild Hypoxia on Ventilation during Exercise, *J. Appl. Physiol.*, **17**:239, 1962.

Hornbein, T. F., and A. Roos: Specificity of H Ion Concentration as a Carotid Chemoreceptor Stimulus, *J. Appl. Physiol.*, **18**:580, 1963.

Howell, J. B. L., and E. J. M. Campbell (eds.): "Breathlessness," Blackwell Scientific Publications, Ltd., Oxford, 1966.

Kao, R. F.: Experimental Study of the Pathways Involved in Exercise Hyperpnoea Employing Cross-circulation Techniques, in D. J. C. Cunningham and B. B. Lloyd (eds.), "The Regulation of Human Respiration," p. 461, Blackwell Scientific Publications, Ltd., Oxford, 1963.

Kao, F. F., S. Lahiri, C. Wang, and S. S. Mei: Ventilation and Cardiac Output in Exercise, *Circulation Res.*, **20**:1–179, 1967.

Krahl, V. E.: Anatomy of the Mammalian Lung, in W. O. Fenn and H. Rahn (eds.), "Handbook of Physiology," sec. 3, Respiration, vol. I, p. 213, American Physiological Society, Washington, D.C., 1964.

Krogh, A.: "The Comparative Physiology of Respiratory Mechanisms," University of Pennsylvania Press, Philadelphia, 1941.

Krogh, A., and J. Lindhard: Regulation of Respiration and Circulation during the Initial Stages of Muscular Work, *J. Physiol. (London)*, **47**:112, 1913.

Lambertsen, C. J.: Effects of Oxygen at High Partial Pressure, in W. O. Fenn and H. Rahn (eds.), "Handbook of Physiology," sec. 3, Respiration, vol. II, p. 1027, American Physiological Society, Washington, D.C., 1965.

Lanphier, E. H., and H. Rahn: Alveolar Gas Exchange during Breath-hold Diving, *J. Appl. Physiol.*, **81**:471, 1963.

Lee, K. D., R. A. Mayou, and R. W. Torrance: The Effect of Blood Pressure upon Chemo-receptor Discharge to Hypoxia and the Modifications of This Effect by the Sympathetic-adrenal System, *J. Quart. Exp. Physiol.*, **49:**171, 1964.

Liljestrand, G.: Untersuchungen Über die Atmungsarbeit, *Skand. Arch. Physiol.*, **35:**199, 1918.

Mead, J., and G. Milic-Emili: Theory and Methodology in Respiratory Mechanics with Glossary Symbols, in W. O. Fenn and H. Rahn (eds.), "Handbook of Physiology," sec. 3, Respiration, vol. I, p. 363, American Physiological Society, Washington, D.C., 1964.

Milic-Emili, G., and J. M. Petit: Mechanical Efficiency of Breathing, *J. Appl. Physiol.*, **15:**359, 1960.

Milic-Emili, G., J. M. Petit, and R. Deroanne: The Effects of Respiratory Rate on the Mechanical Work of Breathing during Muscular Exercise, *Intern. Z. Angew. Physiol.*, **18:**330, 1960.

Miller, W. S.: "The Lung," Charles C Thomas, Publisher, Springfield, Ill., 1947.

Mithoefer, J. C.: Breath Holding, in W. O. Fenn and H. Rahn (eds.), "Handbook of Physiology," sec. 3, Respiration, vol. II, p. 1011, American Physiological Society, Washington, D.C., 1965.

Moritz, A. R., F. C. Henriques, Jr., and R. McLean: The Effect of Inhaled Heat on the Air Passages and Lungs: An Experimental Investigation, *Am. J. Pathol.*, **21:**311, 1945.

Moritz, A. R., and J. R. Weisiger: Effects of Cold Air on the Air Passages and Lungs, *Arch. Intern. Med.*, **75:**233, 1945.

Neil, E.: Baroreceptors and Chemoreceptors in the Regulation of the Blood Pressure, in "Les Concepts de Claude Bernard sur le Milieu Intérieur," p. 105, Masson et Cie, Libraires de l'Académie de Médecine, Paris, 1967.

Neil, E., and N. Joels: The Carotid Glomus Sensory Mechanism, in D. J. C. Cunningham and B. B. Lloyd (eds.), "Regulation of Human Respiration," p. 163, Blackwell Scientific Publications, Ltd., Oxford, 1963.

Nielsen, M.: Die Respirationsarbeit bei Körperruhe und bei Muskelarbeit, *Skand. Arch. Physiol.*, **74:**299, 1936.

Nielsen, M., and H. Smith: Studies on the Regulation of Respiration in Acute Hypoxia, *Acta Physiol. Scand.*, **24:**293, 1952.

Ogilvie, C. M., R. E. Forster, W. S. Blakemore, and J. W. Morton: A Standardized Breath Holding Technique for the Clinical Measurement of the Diffusing Capacity of the Lung for Carbon Monoxide, *J. Clin. Invest.*, **36:**1, 1957.

Otis, A. B.: The Work of Breathing, in W. O. Fenn and H. Rahn (eds.), "Handbook of Physiology," sec. 3, Respiration, vol. I, p. 463, American Physiological Society, Washington, D.C., 1964.

Pappenheimer, J. R.: Standardization of Definitions and Symbols in Respiratory Physiology, *Fed. Proc.*, **9:**602, 1950.

Pattle, R. E.: Surface Lining of Lung Alveoli, *Physiol. Rev.*, **45:**48, 1965.

Rahn, H.: The Sampling of Alveolar Gas, in "Handbook of Respiratory Physiology," pp. 29–37, USAF School of Aviation Medicine, Randolph Field, Texas, 1954.

Rahn, H., and L. E. Farhi: Ventilation, Perfusion and Gas Exchange: The $\dot{V}_A/\dot{Q}$ Concept, in W. O. Fenn and H. Rahn (eds.), "Handbook of Physiology," sec. 3, Respiration, vol. I, p. 735, American Physiological Society, Washington, D.C., 1964.

Riedstra, J. W.: Influence of Central and Peripheral P_{CO_2} (pH) on the Ventilatory Response to Hypoxic Chemoceptor Stimulation, *Acta Physiol. Pharmacol. Neerl.*, **12:**407, 1963.

Roughton, F. J. W.: Transport of Oxygen and Carbon Dioxide, in W. O. Fenn and H. Rahn (eds.), "Handbook of Physiology," sec. 3, Respiration, vol. I, p. 767, American Physiological Society, Washington, D.C., 1964.

Saltin, B., G. Blomqvist, J. H. Mitchell, R. L. Johnson, Jr., K. Wildenthal, and C. B. Chapman: Response to Submaximal and Maximal Exercise after Bedrest and Training, *Circulation*, **38**(Suppl. 7), 1968.

Saltin, B., and P.-O. Åstrand: Maximal Oxygen Uptake in Athletes, *J. Appl. Physiol.*, **23**:353, 1967.

Scholander, P. F.: Oxygen Transport through Hemoglobin Solution, *Science*, **131**:585, 1960.

Stainsby, W. N., and A. B. Otis: Blood Flow, Blood Oxygen Tension, Oxygen Uptake and Oxygen Transport in Skeletal Muscle, *Am. J. Physiol.*, **206**:858, 1964.

Stenberg, J., P.-O. Åstrand, B. Ekblom, J. Royce, and B. Saltin: Hemodynamic Response to Work with Different Muscle Groups, Sitting and Supine, *J. Appl. Physiol.*, **22**:61, 1967.

Turner, J. M., J. Mead, and M. E. Wohl: Elasticity of Human Lungs in Relation to Age, *J. Appl. Physiol.*, **25**:664, 1968.

Weibel, E. R.: "Morphometry of the Human Lung," Springer-Verlag OHG, Berlin, 1963.

Weibel, E. R.: Morphometrics of the Lung, in W. O. Fenn and H. Rahn (eds.), "Handbook of Physiology," sec. 3, Respiration, vol. I, p. 285, American Physiological Society, Washington, D.C., 1964.

West, J.: Regional Differences in Gas Exchange in the Lung of Erect Man, *J. Appl. Physiol.*, **17**:893, 1962.

West, J.: Topographical Distribution of Blood Flow in the Lung, in W. O. Fenn and H. Rahn (eds.), "Handbook of Physiology," sec. 3, Respiration, vol. II, p. 1437, American Physiological Society, Washington, D.C., 1965.

8

Skeletal System

contents

chapter eight
Skeletal System

The skeletal system provides the mechanical levers for the muscles. It is the supporting framework which prevents the entire body from collapsing into a heap of soft tissue; it is the protecting shell or casing for such vital, viable organs as the brain, lungs, heart, and pelvic organs; and it contains within its structures the factories for formed elements of the blood. In addition, it is the great calcium and phosphorus reserve of the body, constantly being drawn upon or added to. Bone is a living tissue, continuously undergoing changes of building up and tearing down.

For a comprehensive review of the osseous system, the reader is referred to such volumes as: "Bone, An Introduction to the Physiology of Skeletal Tissue," by McLean and Urist (1955); "The Biochemistry and Physiology of Bone," edited by G. H. Bourne (1956); and "Bone as a Tissue," edited by Rodahl, Nicholson, and Brown (1960). The purpose of this chapter is to present a summary of the structure and functions of bone as it pertains to physical performance, work, and exercise.

EVOLUTION OF THE SKELETAL SYSTEM

In the past history of vertebrates, bones and muscles are intimately related in the embryonic development and in physical activity. Primitive multicelled animals probably had only two body layers: one inside layer and one outside layer. With growth in size, a third layer developed in between; muscles and bones are products of this middle layer, from which the circulatory system also originates.

The need for a rigid supporting framework developed early in many of the multicelled animals, especially as they moved from sea to land (Romer, 1957). In most animals there is some kind of connective tissue filling the spaces between different organs, helping to bind them together. Generally such tissues contain numerous fibers felted together into a supporting structure. With the increase in size and mobility characteristics for most chordates came the need for more rigid support, such as the skeletal structures. Many of the nonchordates solved the problem with the development of an outside supporting structure, such as the shell of mollusks or the hardened superficial armor of the lobster, while in vertebrates, the major skeletal structures are internal, covered by muscles and skin.

The first type of skeletal material which developed in the vertebrates was the notochord, already typically developed in such primitive forms as the Amphioxus, in the form of a tough but flexible rod running the length of the body, along the back and below the nerve cord (Romer, 1957). This notochord not only helps to stiffen the body and support the various organs but also serves as a point of attachment for the muscles of the trunk. Although the notochord long persisted as a functional element in the vertebrates, there appeared new skeletal materials which have assumed and expanded the role of support originally provided by this notochord, in the form of cartilage and bone. Although both these substances are formed from the connective tissue, they differ considerably in their nature and appearance.

STRUCTURE

The cartilage is a translucent material formed by round cells (cartilage cells) embedded in a substance which binds them together. It is firm yet elastic and flexible and capable of rapid growth. It is therefore a useful supporting structure as long as the stress is moderate and is a common supporting structure in lower vertebrates. In the shark, for instance, it is the only skeletal material in addition to the notochord. On the other hand, in land vertebrates, such as the adult human being, it plays only a limited part in the skeletal make-up and is found only where elasticity is required, such as at the ends of the ribs, as disks between the joints of the backbone, and at the joint surfaces of the limbs.

Bone, like cartilage, is a derivative of the original connective tissue, but it is a highly specialized form and is different from connective tissue proper in that it is very hard. Like the original connective tissue, it is essentially a feltlike material containing an abundance of tiny interlacing fibers (Fig. 8-1).

Osteoblasts (one type of cell) appear on the surface of growing bone and seem to be in a continuous layer, frequently connected with one another. In the bone there are also numerous irregularly branching cells called *osteocytes*. The osteocyte is actually an osteoblast that has been surrounded by calcified interstitial substance; it may undergo transformations and assume the form of an *osteoclast*. This cell is a giant cell with a variable number of nuclei, often as many as 15 or 20. These cells are often found in close relationship to the resorption of bone and frequently lie in grooves, known as *Howship's lacunas*. This suggests that the lacunas were formed by an erosive action of the overlying osteoclasts.

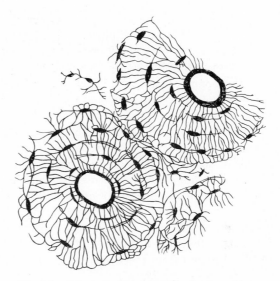

*Fig. 8-1   Schematic drawing of the structure of bone,
showing osteocytes with long branching processes that
occupy cavities in the dense bone matrix.*

The intercellular substance holding the fibers together is called *bone matrix*.
This fibrous, soft, organic matrix is impregnated with a complex mineral substance composed chiefly of calcium phosphates and calcium carbonates. Two-
thirds of the bone substance consists of salts, and this accounts for the hardness
and firmness of bone. Roughly speaking, the composition of bone may be compared to that of reinforced concrete. The fibrous reinforcements provide the
required elasticity for bone to withstand the weight we place upon the skeleton.

The ratio between inorganic and organic substances in the bone varies
during the life-span. The ratio is about 1:1 in the child. For young adults there
is about 4 times more inorganic material than organic, and in the old individual
the ratio is about 7:1. Therefore the bone is more fragile in the old than in the
young person.

Bone tissue, then, consists of osteocytes, which are cells with long branching processes that occupy cavities called *lacunas* in a dense matrix made up of
collagenous fibers laid down in an amorphous ground substance called *cement*,
which is impregnated with calcium phosphate complexes.

FUNCTION

The cellular components of bone are associated with specific functions. The osteoblasts are involved in the formation of bone, the osteocytes with the maintenance

of bone as a living tissue, and the osteoclasts with the destruction and resorption of bone. These cells are closely interrelated, and, as already mentioned, transformation may occur from one form to the other.

Although most tissues including the blood contain only 6 mg calcium/100 ml, the bone contains 10,000 mg/100 ml. Under normal conditions, a state of equilibrium exists between the calcium of the blood and that of the bones, but in disease this may be seriously altered. The calcium is absorbed from the food in the small intestine and is excreted by the large intestine and, to a lesser extent, by the kidney. The recommended daily allowance of calcium is in the order of about 1 g. The regulators of calcium metabolism are the parathyroid glands and vitamin D; the former regulate the interchange of calcium between the blood and the bones. Excess parathyroid hormone results in a rise in the blood calcium with a corresponding fall in the calcium of the bones, and a loss of calcium from the body by increased excretion. Vitamin D governs the absorption of calcium from the intestine, but excess intakes of vitamin D cause a mobilization of calcium from the bones and a deposition of calcium in soft tissues.

The maintenance of the normal mineral metabolism of the bones also depends upon the longitudinal pressure on the long bones brought about by the stress of gravity during the upright ambulatory existence (Jansen, 1920; Rodahl et al., 1966). In a series of studies of the effect of prolonged bed rest on calcium metabolism in normal young men carried out in the Division of Research, Lankenau Hospital, Philadelphia, Pa. (Fig. 8-2), it was shown that 6 weeks continuous bed rest caused a twofold increase in urinary calcium excretion. This increased urinary calcium excretion could not be prevented by bicycle ergometer work in the supine position at a load of 600 kpm/min for 1 to 4 hr a day. Quiet sitting for 8 hr daily combined with 16 hr daily supine bed rest caused no reduction in the increased urinary calcium excretion brought about by prolonged bed rest. However, quiet standing for 3 hr daily in addition to supine bed rest caused a slow drop in the elevated urinary calcium excretion produced by prolonged bed rest toward normal values. The increased urinary calcium excretion during bed rest, therefore, is not per se due to inactivity, but must be due to the absence of the normal longitudinal pressure in the long bones, which is maintained by the stress of gravity during the normal upright ambulatory existence.

Many bones, or parts of bones, are compact, solid structures, nourished through small canals carrying blood vessels and through tiny tubules connecting cell spaces with one another and with the canals. However, if all bones were solid and compact, they would be unnecessarily heavy in proportion to the strength requirements. The large bones are therefore hollow. They are solid only at the surface, and sufficient bone bars and braces extend from the solid bone exterior into the hollow interior to provide reinforcement, more or less similar to the manner in which many bridges are engineered (Fig. 8-3). So wisely is the order of nature arranged that this space in the hollow bone shafts is not wasted, but filled with bone marrow used as a factory for red blood cells and capable of a production capacity of about 2 to 3 million erythrocytes per second.

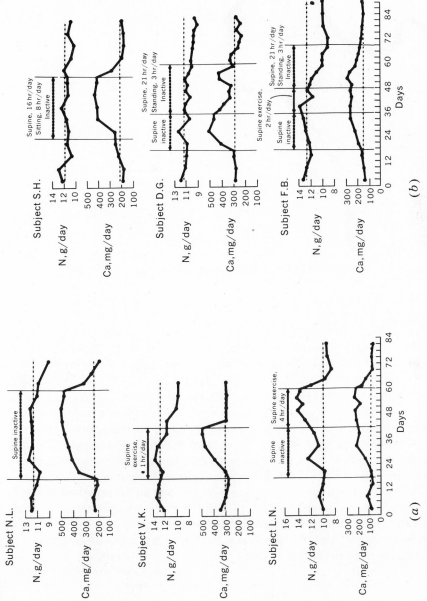

Fig. 8-2 Urinary output of nitrogen and calcium during prolonged inactive bed rest; bed rest combined with 1 hr/day supine exercise; bed rest combined with 4 hr/day supine exercise; bed rest combined with 8 hr/day quiet sitting; bed rest combined with 3 hr/day standing, and bed rest with and without supine exercise followed by 3 hr/day standing.

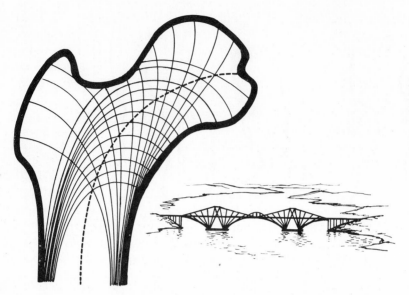

Fig. 8-3 Schematic drawing showing reinforcing bars and braces, similar to those of a bridge, of the large bones.

GROWTH

In providing rigidity, the bone has had to sacrifice the flexibility and the expand-able traits enjoyed by the cartilage. This naturally presents a problem of growth. Thus, for instance, the thigh bone of a two-year-old child has to double its length by adulthood, and although the bone may become stouter simply by adding more bone to its surface layers, it cannot stretch, nor can new bone be added to its ends, for this would interfere with its articulate surfaces forming part of the hip joint and the knee joint. This problem is solved by the insertion of a special growth section, known as the epiphysis, close to both ends of the shaft. This is an area of cartilage in which new bone may form, resulting in longitudinal growth of the bone without interference with the joint surfaces. When the indi-vidual is fully grown, the epiphyses fuse with the shaft, and growth terminates.

All the elements of the internal skeleton are first formed in the human embryo in cartilage. Later, the cartilage is destroyed and replaced by bone. It is in this replacement process that a thin film of cartilage is left, separating the ends from the shaft of the bone in the form of the epiphyses, described above.

REPAIR—ADAPTATION

Following a fracture of a bone which was initially preformed in cartilage, the first attempt at repair, starting within the first few weeks following the fracture, results in the formation of a mass of fibrocartilage known as *callus*. As pointed out by McLean and Urist (1955), the bridging of a fracture by new bone, as seen in a sagittal section, occurs in a fashion resembling that of a fixed-arch bridge, according to the principle of cantilevering frequently seen in contemporary architecture (Fig. 8-4). The new bone grows out upon the surface of the cortex on each side of the fracture line and envelops the fibrocartilaginous callus to form an arch of new bone over the fracture gap. The new bone comes down like ribs let down from the arch of a bridge to suspend the deck, and this bone gradually replaces the cartilage toward the fracture gap. Then, finally, the "deck" is laid down between the fracture ends and provides permanent union, and the superstructure disappears, leaving only the bone required for union of the fracture ends. In the case of a fracture of a long bone in a young individual this process is usually completed within a period of about 3 to 6 months.

Owing to the readiness with which bone formation and bone absorption can occur, a remodeling of bone is continually taking place. Isotope studies using P^{32} or Ca^{45} indicate that over 20 percent of the calcium and phosphate ions of bone are involved in a fairly rapid exchange in adults. (LeBlond and Greulich, 1956.)

Bone tissue is surprisingly sensitive to demands made upon it and responds readily to these demands, so that every change in the function of a bone is followed more or less by a definite change in the internal architecture. Bone bars not stressed will disappear and new ones are created where altered mechanical forces increase the demand for sturdiness. Thus as a result of training, or a fracture, the lines of stress in a bone may change with the altered direction of mechanical forces, as if it were the most plastic of structures.

Absorption of bone may occur from many causes, leading to bone rarefaction, or osteoporosis. It occurs commonly in old age, from disuse, and as a local phenomenon caused by such processes as inflammation, tumors, and aneurysms.

CONNECTIVE TISSUE

The walls of the bone-marrow cavities and of the Haversian canals of compact bone are lined with a thin layer of connective tissue known as the *endosteum*,

Fig. 8-4   Schematic drawing showing similarity between the process of bone repair and the construction of a fixed-arch bridge. (Modified from McLean and Urist, 1955.)

which also covers trabeculae of cancellous bone. It is a condensed peripheral layer of the stroma of the bone marrow. In some respects it resembles the *periosteum*, which is the connective tissue surrounding bone.

In the young rapidly growing individual, periosteum consists of a dense outer layer of collagenous fibers and fibroblasts and an inner layer of proliferating osteoblasts. In the adult, the periosteum primarily serves as a supporting function. When strong tendons or ligaments are attached to a bone, the periosteum is incorporated with them. In this manner the ends of the skeletal muscles are fused to the bones.

JOINTS

Joints are formed where two or more bones of the skeleton meet one another (Fig. 8-5). The function of the joint is the most important factor in the determination of its character and structure (Johnson, 1938). In areas like the skull, it is important that no movement should be permitted between contiguous bones; in the vertebral column, on the other hand, a slight degree of mobility is desirable, provided that it can be obtained without loss of strength or sturdiness. In other situations, the provision of a more or less wide range of movement is essential. In the first case, *fibrous joints* are provided, i.e., articulations in which the surfaces of the bones are fastened together by intervening fibrous tissue and in which there is no appreciable motion. In the second case, the connection medium between the bones concerned is white fibrocartilage; such joints are capable of a limited range of movement and are termed *cartilaginous joints* (Fig. 8-5a). In the third case, the opposed bones are separated from one another by a space lined by a special membrane which is termed *synovial membrane;* such a joint possesses a more or less wide range of movement and is known as a *synovial joint* (Fig. 8-5b,c).

In the case of the cartilaginous joints of the spinal column the opposed bony surfaces are covered with hyaline cartilage and are connected to each other by a flattened disk of fibrocartilage of a more or less complex structure. The bones are also connected by bands of white fibrous tissue known as *ligaments*, which do not form a complete capsule around the joint. A limited degree of movement is rendered possible by the compressibility of the cartilaginous disk and the degree of leverage which is available (Fig. 8-5d).

Most of the joints of the body, including all joints of the limbs, belong to the synovial group. In these joints (Fig. 8-5) the contiguous bony surfaces are covered with articular cartilage separated by a joint cavity which in a healthy individual is nothing but a tiny space. The joint is completely surrounded by an articular capsule consisting of a capsular ligament lined with a synovial membrane. The synovial membrane lines the whole of the interior of the joint, with the exception of the cartilage-covered ends of the articulating bones. The bones

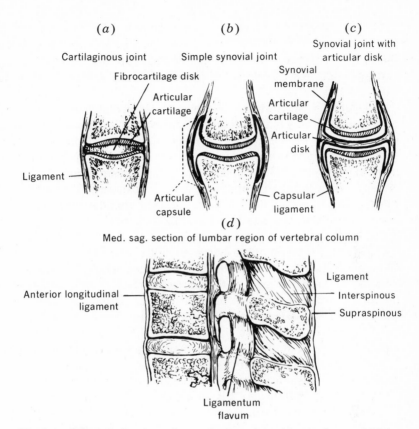

(a)   Cartilaginous joint

(b)   Simple synovial joint

(c)   Synovial joint with articular disk

Fibrocartilage disk

Articular cartilage

Synovial membrane

Articular cartilage

Articular disk

Ligament

Articular capsule

Capsular ligament

(d)

Med. sag. section of lumbar region of vertebral column

Anterior longitudinal ligament

Ligament

Interspinous

Supraspinous

Ligamentum flavum

Fig. 8-5   Schematic drawings of a cartilaginous joint (a), simple synovial joint (b), synovial joint with articular disc (c), and a sagittal section at the lumbar region of the vertebral column (d).

are usually connected by ligaments which are added to the capsular ligaments and situated superficially to the latter. Movements in such a joint may vary from a simple limited gliding movement to a wide range of movements, such as in the case of the shoulder joint. The joint cavity may be divided by an articular disk of fibrocartilage, as in the knee joint. These structures act as shock-reducing agents and serve to ensure perfect contact between the moving surfaces in any position of the joint. The synovial membrane secretes a small quantity of viscid fluid termed *synovia*. This fluid acts as a lubricant.

Holmdahl and Ingelmark (1948) have shown that in trained animals the thickness of the articular cartilage is greater than that of untrained ones. The active animals had an increase in both the cellular and intercellular components of the cartilage. Animals living in small cages restricting their possibilities to move around had in general thinner cartilages of the knee joints than animals provided with spacious cages.

The articular cartilage has no blood vessels. There is, however, direct contact between the medullary cavities of the epiphyses and the basal portions of the articular cartilage. An exchange of fluid can also take place between the cartilage and the synovial fluid of the articular space (Holmdahl and Ingelmark, 1951; Ekholm, 1951).

Figure 8-6 illustrates how the articular cartilage in the knee joint can, within minutes, vary in thickness when the animal is restricted in the movements of the joint or is exercised. After 10 min of running the increase in thickness was 12 to 13 percent compared to what it was after 60 min of immobilization (Ingelmark and Ekholm, 1948). Similar results have been obtained on humans. The explanation given for the rapid increase in thickness of the cartilage of an activated joint is that fluid flows into the cartilage from the underlying bone marrow cavity, when the cartilage is repeatedly and intermittently compressed with intermediate periods without pressure.

One consequence of the increased fluid content of the cartilage is a change in its compressibility, diminishing the incongruence between condyle and socket. This will increase the area of the contact surface of the joint in question and produce a reduced pressure per unit of area of the articular surface during compression with a given force (Ingelmark and Ekholm, 1948). Warm-up activities before vigorous exercise should include those joints which may be stressed during the activity in order to make them less susceptible to trauma.

Another advantage with regular dynamic motion of the joints is that the associated increase in supply of fluid to the articular cartilage will also provide nutriments to the cartilage. It is an old clinical observation that a joint suffers from nutritional disturbances when kept inactive.

It is not known how frequently such activities should be performed to provide an optimal nutritional situation for articular cartilages.

The movements possible in joints are usually classified as gliding angular movements (flexion-extension, abduction-adduction), circumduction (shoulder and hip joints), and rotation (as in the rotation of the humerus at the shoulder joint, or as in the movement of the radius on the ulna during pronation and supination of the hand).

Limitation of movements is affected by a number of different factors, such as the tension of ligaments or the tension of the muscles which are antagonistic to the movement. In fact, it appears that the tension of the antagonist muscles may never permit a ligament of a joint to be fully stretched. Finally, the movements of some joints are limited by the soft tissues, as in the case of flexion of the elbow, hip, and knee (Fig. 8-7). One example: A flexion in the hip joint with extended knee is limited by the length of the muscles on the back of the thigh. With a simultaneous flexion in the knee joint the flexion of the hip can be greatly extended. If in addition the flexion of the hip joint is aided by external forces, the fusion may be further increased until it is stopped by the thigh resting against the abdomen. In other words, the muscles moving a joint cannot even

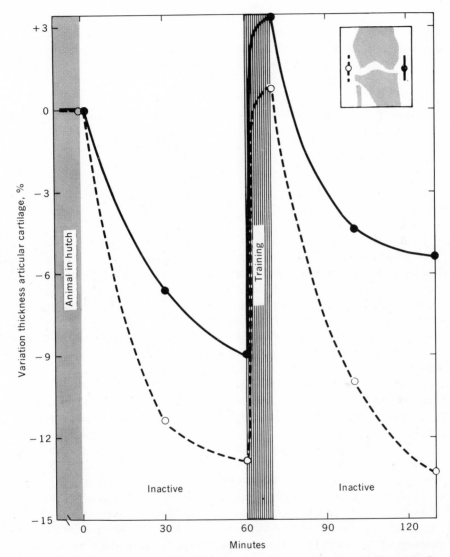

Fig. 8-6   Rabbits were taken from their spacious hutches where they could run freely. At "0" time x-ray photographs were taken for measurements of the thickness of the articular cartilage of the fibular and tibial ends of the knee joints (for symbols, see inserted figure). Measurements were repeated after 30 and 60 min of rest in a position which did not stress the knees with the animal's weight. Then the rabbits were trained on a motor-driven treadmill for 10 min, speed 40 m/min. Measurements were repeated after exercise and further periods of rest. Note a significant decrease in cartilage thickness after inactivity and a rapid increase during exercise. (Figures represent the mean of about 50 animals.) (Modified from Ingelmark and Ekholm, 1948.)

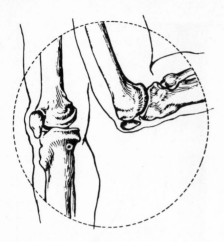

Fig. 8-7 Schematic drawing of the knee joint. The flexion of this joint is limited by the soft tissue.

with maximal force produce a movement over the full range which the joint actually permits. However, a movement in which external forces are involved may be so extreme, especially when a great force is applied suddenly, that the adjacent articular cartilages may be separated (luxation). At the same time bone, ligaments, joint capsules, soft tissues, and blood vessels may be damaged.

Since the limiting factor for flexibility is often the length of the muscles, a training that produces a lengthening of these muscles will increase the joint flexibility.

In synovial joints where the bones are connected by ligaments and muscles only, the articular surfaces are in constant apposition in all positions of the joint. The maintenance of this apposition is facilitated by atmospheric pressure and cohesion, but the muscles play a far more important role. The balance in tone between the different muscle groups which act on the joint is responsible for maintaining the articular surfaces in constant apposition, so that the stability of any joint depends on the tonus of the muscles which act on it.

Movable joints are innervated by the nerves which supply the muscles that act on them, and it is reasonable to assume that this arrangement establishes local reflex arcs which ensure stability (Johnson, 1938). The part of the articular capsule which is rendered taut on the contraction of a given muscle or group of muscles is innervated by the nerve or nerves supplying their antagonists. This may serve to prevent overstretching or tearing of the ligaments.

PATHOPHYSIOLOGY OF THE BACK

It is not within the scope of this book to discuss pathological conditions. However, there are diseases that may be influenced by activity or by inactivity. One example is backache. Muscle training, some caution when lifting and carrying loads, and application of physiological principles when planning work places may prevent or alleviate symptoms from an insufficient back. The problem of backache will therefore be discussed here.

Of all the bony structures of the human body, the spinal column plays a unique role in that it serves as a sustaining rod for maintenance of the upright position. As such, it is subjected to a complex system of forces and stresses of a wide variety of types. Frequently these forces are amazingly large.

Because of the unique nature of the structure and function of the spine, it is frequently the site of aches and pains as the result of numerous processes associated with wear and tear. It has often been said that lower backache is the price man has to pay for his upright two-legged existence. This, however, is not necessarily the case, for similar afflictions may occur in many four-legged animals.

There are few ailments which more dramatically can hinder muscular activity than backaches. Apart from being painful, it is indeed a costly condition for the individual and society in terms of lost work output, sick days, and reduced earning power. From a health standpoint, lower backaches rank as one of the most common medical complaints (Cobey, 1956). As a matter of fact, back pain, in one form or another, is experienced by almost all individuals at one time or another.

Hult (1954) has presented results of examinations of about 1,200 individuals representing different professions. They were divided into two main groups, those engaged in physically light work (e.g., white-collar workers, workers in light industry, retail trade) and in physically heavy work (e.g., longshoremen, construction workers, employees in heavy industry). The age of the subjects ranged from twenty-five to sixty years. Subjective symptoms revealed by careful notations of case history could, on the whole, be classified as belonging to one of three syndromes: two common, the stiff neck—brachialgia syndrome (in altogether 51 percent of the individuals) and the lumbar insufficiency—lumbago—sciatica syndrome (in 60 percent), and one less common, the dorsal spine syndrome (in 5 percent of the cases).

Figure 8-8 (upper part) shows (1) the marked increase in symptoms with age; and (2) a slightly higher percentage of individuals with symptoms of lower back troubles among heavy workers (64 percent) compared with those involved in light work (53 percent). The incidence of symptoms from the upper back was the same in those doing light and heavy work (51 percent).

The results of clinical and roentgenologic examinations are also summarized in Fig. 8-8 (lower part). The trend is the same as for the subjective symptoms, with the age factor being of decisive importance for the occurrence of roentgen signs of disk degeneration, which was as high as about 90 percent in the age group fifty-five to fifty-nine years. There is a consistently higher frequency of such signs in those subjects with heavy occupations, even if the difference between the two groups is moderate. Disk degeneration should be interpreted

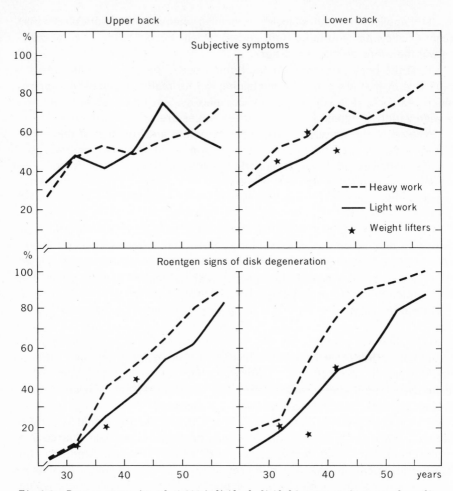

Fig. 8-8   Data on approximately 1,200 individuals divided into two main groups depending on occupation. Included are data from 56 weight lifters. The upper part of the figure shows results of case histories dealing with symptoms emanating from the upper or lower extremities and the spine. The lower part of the figure shows frequencies of objective incidence of disk degeneration and roentgen-anatomic anomalies. The graphs to the left present cases of stiff neck–brachialgia syndrome ("upper back"); the graphs to the right, lumbar insufficiency–lumbago–sciatica syndrome ("lower back"). (Data from Hult, 1954.)

as a more or less physiological process which begins early in some individuals but eventually develops in all regardless of occupation.

It was noted that in a maximum of 20 percent of those with a history of lumbago or sciatica the symptoms were provoked by an accident; in an additional 15 to 20 percent they appeared in connection with heavy lifting or a similar strain. This means that in 60 to 65 percent of those who had had lumbago or sciatica attacks the symptoms appeared without such external causative factors.

Static deformities such as differences in leg length, kyphosis, lordosis, and scoliosis had no demonstrable significance in the origin of low back trouble in these groups. Nor was there any correlation between low back trouble and body type (evaluated according to Rohrer's index), flatfoot, or foot fatigue.

There was some but not a striking relation between subjective symptoms and clinical-roentgenologic findings. Thus disk degeneration could develop and even attain advanced stages without ever having given rise to pain.

Weight lifters did not in any respect differ significantly from the other subjects.

It was estimated that in Sweden approximately 2 million working days annually were lost because of back trouble.

The occurrence of such a high frequency of symptoms from the back is not at all surprising in view of the complexity of the spinal column, its complicated joint systems, and the very heavy stress to which it may be exposed. An interesting analysis of the tolerance of the spinal column has been published by Morris et al. (1961). They point out that the spinal column may be considered to have both an intrinsic and extrinsic stability, the former being provided by the alternating rigid and elastic components of the spine bound together by a system of ligaments, while extrinsic stability is provided by the paraspinal and other trunk muscles. The stability of the ligamentous spine, which may be considered as a modified elastic rod, depends largely on the action of the extrinsic support provided by the trunk muscles. It has been shown that the critical load value at which buckling of the isolated ligamentous spine occurs, fixed at the base, is only about $4\frac{1}{2}$ lb. This is much less than the weight of the body above the pelvis (Lucas and Bresler, 1960).

In order to explain the discrepancy between the force which the spine as such can tolerate and the much larger forces to which the back is subjected in actual life situations, Morris and coworkers investigated the role of the compartments of the trunk, i.e., thorax and abdomen, in helping to provide stability of the spine and then showed that the discrepancy apparently can be explained by the role played by the trunk.

In providing extrinsic stability of the spine, Morris et al. (1961) considered that if the nucleus pulposus of the fifth lumbar disk is considered as the fulcrum of movement and a heavy weight is lifted with the hands, the arms and trunk form a long anterior lever (Fig. 8-9). The weight being lifted and the weight of the head, arms, and upper part of the trunk are counterbalanced by the contraction of the deep muscles of the back acting through a much shorter lever arm, i.e., the distance from the center of the disk to the center of the spinous process. With these factors in mind, they computed the force that results when a 170-lb man lifts a 200-lb weight and concluded that the force on the lumbosacral disk was about 2,000 lb, if the role of the trunk is omitted. This is considerably more than segments of the isolated ligamentous spine can withstand without structural failure; compression tests of two vertebral bodies with their intervening disk from subjects under forty years of age resulted in failure of the seg-

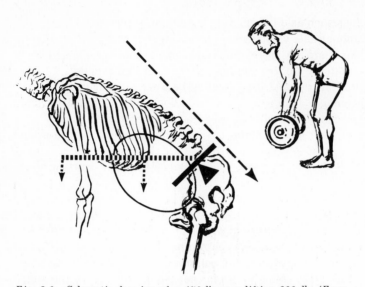

Fig. 8-9 Schematic drawing of a 170-lb man lifting 200 lb. (From Morris et al., 1961.) The nucleus pulposus of the fifth lumbar disk is considered as the fulcrum of movement. The arms and trunk form a long anterior lever. The weight being lifted is counterbalanced by the contraction of the deep muscles of the back acting on a much shorter lever (the distance from the center of the disk to the center of the spinous process). If the role of the trunk is omitted, the force on the lumbosacral disk would be about 2,000 lb, which is considerably more than segments of the isolated ligamentous spine can withstand without structural failure. When this does not happen, it is because the contracted trunk muscles convert the abdominal and thoracic cavities into semirigid cylinders which relieve the load on the spine itself.

ments of the spine under compressive loads ranging from 1,000 to 1,700 lb. In older subjects, this figure was sometimes as low as 300 lb.

In a series of experiments, Morris et al. were able to show that since the spinal column is attached to the sides of, and within, two chambers, i.e., the abdominal and thoracic cavities, the action of the trunk muscles converts these chambers into nearly rigid cylinders containing air, liquid, and semisolid material. Thus, these cylinders are capable of transmitting part of the forces generated in loading the trunk, thereby relieving the load on the spine itself. When large forces are applied to the spine, such as when lifting weight of 200 lb, there is generalized contraction of the trunk muscles, including the intercostals, the muscles of the abdominal wall, and the diaphragm. The action of the intercostals and the muscles of the shoulder girdle renders the thoracic cage a rigid structure firmly bound to the thoracic part of the spine. When inspiration increases intrathoracic pressure, the thoracic cage and spine become a solid, sturdy unit capable of transmitting large forces. By the contraction of the dia-phragm, attached at the lower margin of the thorax and overlying the abdominal wall, especially the transversus abdominis, the abdominal contents are also compressed into a semirigid cylinder. The force of weights lifted by the arms is thus transmitted to the spinal column by the shoulder girdle muscles, princi-pally the trapezius, and then to the abdominal cylinder and to the pelvis, partly through the spinal column but also through the rib cage. When larger forces are involved, increased rigidity of the rib cage and increased compression of the abdominal contents are achieved by increased activity of the trunk muscles, resulting in increased intracavitary pressures.

This is all brought about by a reflex mechanism; when a load is placed on the spine, the trunk muscles are involuntarily called into action to fix the rib cage and to compress the abdominal contents. The intracavitary pressures are thereby increased, aiding in the support of the spine.

Morris et al. (1961) concluded from their calculations of the trunk com-partments to support the spine that the actual force on the spine is much less than that considered to be present when this support by the trunk is omitted. The calculated force on the lumbosacral disk is about 30 percent less, and that of the lower thoracic portion of the spine is about 50 percent less than would be the case without support by the trunk.

Thus, the study by Morris and coworkers emphasized the important role of the trunk muscles in the support of the spine. From this it follows that well-developed trunk muscles, including the abdominal muscles, play an important role in sparing the spine and thus avoiding strain and damage. While flabby abdominal muscles may expose the spine to injurious stress, well-developed abdominal muscles, on the other hand, are an important protective device.

Man has had to carry loads from the beginning of his existence and has had time to learn how to solve this problem in keeping with the most efficient bioengineering principles. Examples of the methods used by primitive native tribes in carrying loads are well known (Fig. 8-10).

Fig. 8-10 A physiological method of carrying a load, practiced by many peoples.

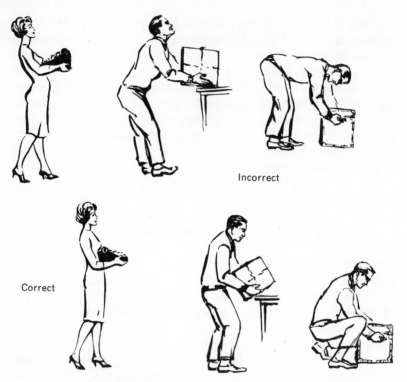

Incorrect

Correct

Fig. 8-11   Lifting and carrying a load.

It has been generally advocated that in lifting and carrying loads it is important to bring the center of gravity of the load as close as possible to the axis of the body to minimize the force movement (Fig. 8-11). Although this principle should be followed more often than is the actual case, it should be borne in mind that in many actual life situations this is not always practicable.

Often a series of motions are involved in the execution of a given task, including bending, lifting, turning, and walking with the load. However, experience has shown that symptoms of lower backache are most apt to occur when lifting is combined with a twisting or turning motion. It is therefore practical to arrange it so that the individual is facing the direction of movement before picking up the load. It is equally important to have a secure foundation for the feet during such motions, since it is a common experience for lower backache to occur during sudden bodily movements associated with slipping of the foothold. In view of the findings of Morris and coworkers, it would appear reasonable to assume that under such sudden motions there may not be sufficient time for a reflex contraction of the trunk muscles to occur, thus leaving the spinal column vulnerable.

The above considerations underline the significance of well-trained trunk muscles in the protection of the spine. On this basis, one of the most common treatments of backaches consists of training of the trunk muscles, notably the back stretchers. It now appears equally important to include the abdominal muscles in any such training program.

It cannot be overemphasized that in the case of backaches or damage of the back, prevention is the most important aspect, since the back is an inherent weakness in the human body. This brings out the need for intelligent indoctrination, starting with the children during school age, endeavoring to achieve a complete understanding of the mechanical aspects of the spinal column and methodical instruction in the proper techniques involving the use of the back in all kinds of daily activities as well as in special industrial tasks. It is true that younger individuals may abuse their bodies without causing any damage since the structures are still vital. In older individuals, however, the various tissues will become more vulnerable. It is then important that he has learned a technique which causes less strain, especially on the back.

Similar principles are also involved in other joints of the body, such as the neck and the knee. In the case of the knee joint, for instance, the importance of the quadriceps muscle in stabilizing the joint when nearly extended is well recognized. In the treatment of hydrops of the knee joint, it is often overlooked that although bed rest and immobilization cause a resorption of the fluid, the concomitant atrophy of the quadriceps muscle caused by immobilization may result in recurrence of the hydrops, at times even more severe than before, as the result of damage caused by inadequate stabilization by the atrophied quadriceps muscle. The correct treatment would be to put the patient to bed in order to relieve the joint of the burden of the body weight, but at the same time to train the leg muscles. In this connection it should again be noted that the nutrition of the cartilage may be enhanced by allowing some activity of the joint in question.

REFERENCES

Bourne, G. H. (ed.): "The Biochemistry and Physiology of Bone," Academic Press Inc., New York, 1956.

Cobey, M. C.: "Postural Back Pain," Charles C Thomas, Springfield, Ill., 1956.

Ekholm, R.: Articular Cartilage Nutrition, *Acta Anat.*, **11**(Suppl. 15-2), 1951.

Holmdahl, D. E., and B. E. Ingelmark: Der Bau des Gelenkknorpels unter verschiedenen funktionellen Verhältnissen, *Acta Anat.*, **6**:309, 1948.

Holmdahl, D. E., and B. E. Ingelmark: The Contact between the Articular Cartilage and the Medullary Cavities of the Bones, *Acta Anat.*, **12**:341, 1951.

Hult, L.: Cervical, Dorsal and Lumbar Spinal Syndromes, *Acta Orthop. Scand.* (Suppl. 17), 1954.

Ingelmark, B. E., and R. Ekholm: A Study on Variations in the Thickness of Articular Cartilage in Association with Rest and Periodical Load, *Uppsala Läkareförenings Förhandlingar*, **53**:61, 1948.

Jansen, M.: "On Bone Formation: Its Relation to Tension and Pressure," Longmans, Green & Co., Inc., New York, 1920.

Johnson, T. B. (ed.): "Gray's Anatomy," 27th ed., p. 425, Longmans, Green & Co., Ltd., London, 1938.

LeBlond, C. P., and R. C. Greulich: Autoradiographic Studies of Bone Formation and Growth, in G. H. Bourne (ed.), "The Biochemistry and Physiology of Bone," Academic Press Inc., New York, 1956.

Lucas, D. B., and B. Bresler: "Stability of the Ligamentous Spine," *Technical Report Series II*, no. 40, University oᵗ California, Biomechanics Laboratory, Berkeley and San Francisco, Dec., 1960.

McLean, F. C., and M. R. Urist: "Bone: An Introduction to the Physiology of Skeletal Tissue," The University of Chicago Press, Chicago, 1955.

Morris, J. M., D. R. Lucas, and B. Bresler: Role of the Trunk in Stability of the Spine, *J. Bone Joint Surg.*, **43A**:327, 1961.

Rodahl, K., J. T. Nicholson, and E. M. Brown (eds.): "Bone as a Tissue," McGraw-Hill Book Company, New York, 1960.

Rodahl, K., N. C. Birkhead, J. J. Blizzard, B. Issekutz, Jr., and E. D. R. Pruett: Fysiologiske forandringer under langvarig sengeleie, *Nord. Med.*, **75**:182, 1966.

Romer, A. S.: "Man and the Vertebrates," The University of Chicago Press, Chicago, 1957.

9

Physical Work Capacity

contents

chapter nine
Physical Work Capacity

DEMAND—CAPABILITY

Competitive sports events represent the classical test of physical fitness or performance capacity. Under such conditions the performance may be measured objectively in centimeters or seconds, or it may be judged subjectively, as in gymnastics, figure skating, or diving. The individual's performance is the combined result of the coordinated exertion and integration of a variety of functions. The demands of the actual event must be perfectly matched by the individual's capabilities in order to achieve top performance and championship. It is impossible to present one formula that takes into account all aspects of a man's maximal work power and capacity, since the demands set by different types of activities vary greatly. However, the following factors may serve as a frame of reference for our discussion.

Physical Performance

Energy output

Aerobic processes

Anaerobic processes

Neuromuscular function

Strength

Technique

Psychological factors

Motivation

Tactics

Natural endowment (genetic factors) probably plays a major role in a person's performance capacity, at least for those people aspiring to the levels required for the attainment of Olympic medals. Since the possible genetic combinations are astronomical in number, it is an interesting question whether a country must have a population of 100,000, 1 million, 10 million, or more, to "breed" an individual with proper endowment for top results. The more popular an event, the greater is the chance that an individual with the suitable constitution will participate and thus discover his ability. Obviously, the environment and geographic location are also important. If an individual with the perfect endowment for skiing grows up in a place where skiing is impossible, his endowment may be wasted from an athletic standpoint. The fact that an increasingly larger number of naturally endowed individuals enter the ranks of competitive athletes may in part explain the gradual improvement of athletic records.

Granted the endowment, however, definite improvement in performance may be achieved by *training*, and all the factors listed above as contributing to physical performance capacity may be modified. The very intense training programs currently employed in many fields of athletic performance contribute greatly to the improved results. Another factor explaining the gradual improvement in work output and athletic achievement over the years is the better techniques applied and the superior equipment which is becoming available through technical progress.

The athlete himself is mainly concerned with improving his ability to cut off seconds or add centimeters to his record. The scientist is interested in analyzing why the results improve or vary from time to time. Therefore, the scientific objective is (1) to evaluate quantitatively the influence of the various factors upon the performance capacity in different tasks (performance requirements); (2) to examine how these factors vary with sex, age, and body size (capacity profile); and (3) to study the effect of such factors as training and environment. It is realistic to conclude that scientists have merely begun a systematic research on the performance capacity and the many factors involved. The most advanced information concerns the energy output by aerobic processes. This may be explained by the fact that methods for quantitative measurements of energy output by the human combustion engine have long been available, actually ever since Lavoisier's discovery of the oxygen utilization by living animals. We shall therefore begin the more detailed analysis of physical work capacity with a discussion of the oxygen uptake during submaximal and maximal exercise and the maximal aerobic power (the individual's maximal oxygen uptake).

(It should be emphasized that *capacity* denotes total energy available and *power*, energy per unit of time.)

AEROBIC PROCESSES

The complexity of the capacity for aerobic muscular exercise is illustrated by Fig. 9-1. For each liter of oxygen consumed, about 5 kcal (4.7 to 5.05) will be

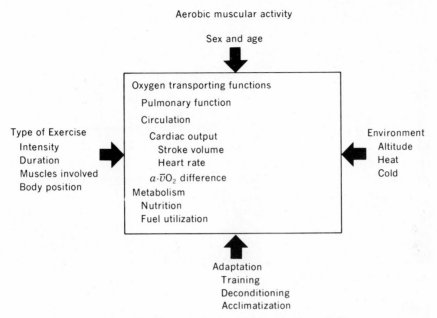

Fig. 9-1   *Factors influencing the capacity for aerobic muscular activity.*

delivered; hence, the higher the oxygen uptake, the higher the energy output. The oxygen uptake during exercise may be measured with an accuracy of ± 0.04 liter/min ($\dot{V}_{O_2} > 1$ liter/min). Figure 9-2 gives examples of how the classical Douglas bag method can be applied when studying the aerobic energy output during work or exercise.

Work Load and Duration of Work

Figure 9-3a shows how the oxygen uptake increases during the first minutes of exercise to a "steady state" where the oxygen uptake corresponds to the demands of the tissues. When the exercise stops, there is a gradual decrease in the oxygen uptake to the resting level; the oxygen debt is paid off.

The slow increase in oxygen uptake at the beginning of exercise is explained by the sluggish adjustment of respiration and circulation, i.e., the sluggish adjustment of the oxygen-transporting systems to work. The attainment of this steady state coincides roughly with the adaptation of cardiac output, heart rate, and pulmonary ventilation. A *steady-state* condition denotes a work situation where oxygen uptake equals the oxygen requirement of the tissues; consequently there is no accumulation of lactic acid in the body. Heart rate, cardiac output, and pulmonary ventilation have attained fairly constant levels. In light exercise the energy output during the first minutes of exercise can be delivered aerobically, since oxygen is stored in the muscles bound to myoglobin and in the blood per-

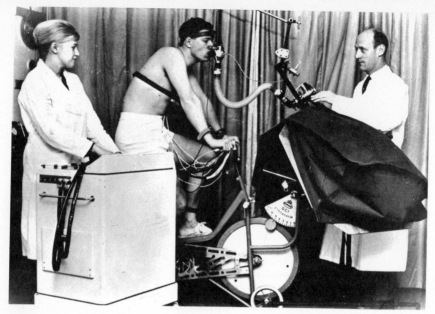

(a)

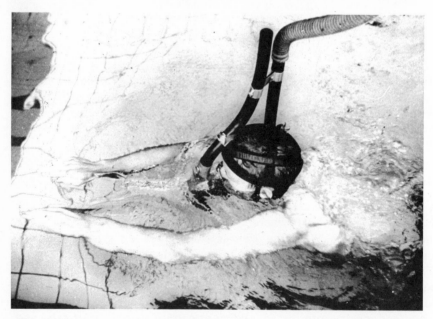

(b)

*Fig. 9-2   Application of the Douglas bag method for measuring aerobic energy output
during different types of exercise. The skier shown in (c) carries a three-way stopcock and a
stopwatch on his chest for the recording of time during which the expired air is collected
in the Douglas bag. The stopwatch automatically starts and stops when the stopcock is
turned.*

(c)

Fig. 9.2   (Continued)

fusing the muscles. During more severe exercise, anaerobic processes must supply part of the energy during the early phase of exercise, and lactic acid may be produced. [Anaerobic energy is provided by not only the glycogenolysis or glycolysis but also the breakdown of ATP and creatine phosphate (Chap. 2).] With a work load (leg work) that demands an oxygen uptake higher than 50 percent of the individual's maximal capacity and which is performed for some minutes, lactic acid (lactate) appears in the blood in a concentration that can be measured even in the arterial blood. The heavier the work load, the more important is the anaerobic energy contribution. The blood lactate concentration increases, the work becomes subjectively more strenuous, and a decrease in the body's pH affects muscular tissue, respiration, and other functions.

Figure 9-3b illustrates the linear increase in oxygen uptake, measured after about 5 min of exercise with different work loads up to a point where the maximum for oxygen transportation appears to be reached. In this case the maximal oxygen uptake is 3.5 liters/min, and this is the *maximal aerobic power* of this subject. There are two main criteria showing that this maximum has been measured: (1) there is no further increase in oxygen uptake despite further increase in work load; and (2) the blood lactate concentration is above 70 to 80 mg/100 ml of blood, or 8 to 9 mmoles/liter (P.-O. Åstrand, 1952; I. Åstrand, 1960; Rodahl and Issekutz, 1962). The presumption is that large muscle groups are involved in the exercise and that the work time exceeds 3 min.

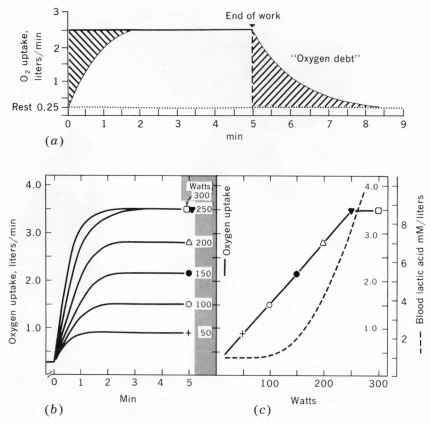

Fig. 9-3 (a) *During the first minutes of exercise the oxygen uptake increases, then levels off as the oxygen uptake has reached a level adequate to meet the demand of the tissues. At the cessation of exercise there is a gradual decrease in the oxygen uptake, as the "oxygen debt" is being paid off.*

(b) *Schematic demonstration of increase in oxygen uptake during exercise on bicycle ergometer with different work loads (noted within shadowed area) performed during 5 to 6 min.*

(c) *Oxygen uptake in the above-mentioned experiments, measured after 5 min and plotted in relation to work load. Note that 250 watts (1500 kpm/min) brought the oxygen uptake up to this subject's maximum and that 300 watts did not further increase the oxygen uptake; the increased work load was possible thanks to anaerobic processes. Maximal aerobic power = 3.5 liters/min. (For simplicity the work load which is sufficient to bring the oxygen uptake to the subject's maximum, in this case 250 watts, may be written $WL_{max\ O_2}$.) Peak lactic acid concentrations in the blood at each experiment have been included.*

From a methodological viewpoint it is important to emphasize that maximal oxygen uptake is attained at a work load that is not necessarily maximal. From Fig. 9-3c it is obvious that 250 watts were enough to reveal the subject's maximum, 3.5 liters/min, but the subject was not exhausted by this work load. He could work at a rate of 300 watts for the same period of time. *An all-out test is not necessary for the assessment of an individual's maximal aerobic power.*

There are several reasons for the delayed return of oxygen uptake to resting level after the cessation of exercise (repayment of oxygen debt): (1) refilling of the oxygen content of the body, (2) aerobic removal of anaerobic metabolites, (3) elevated metabolism due to an increase in tissue temperature and a possible increased output of adrenalin (about 13 percent elevation in metabolic rate per degree centigrade), and (4) increased oxygen demand of the activated respiratory muscles and heart. Evidently only part of the excessive oxygen uptake during recovery is used to pay off the energy debt incurred by anaerobic processes (see below). However, the heavier the work load, the more dominating is the anaerobic fraction of this excessive oxygen uptake.

The heavier the work load, the steeper is the increase in the oxygen uptake (and heart rate). This is illustrated by Fig. 9-4. After a 10-min period of work at 50 percent of maximal oxygen uptake, work loads of 1800 to 2700 kpm/min were applied until exhaustion. The tolerated work time varied from 6 min (1800 kpm/

Fig. 9-4   Curves showing increase in oxygen uptake during heavy exercise following a 10-min warm-up period. Arrows indicate time when the subject had to stop due to exhaustion. Figures indicate work load on the bicycle ergometer in kpm/min. The subject could continue the load of 1650 kpm/min for more than 8 min. (From P.-O. Åstrand and Saltin, 1961a.)

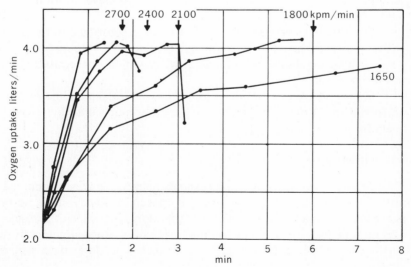

min) to less than 2 min (heaviest load). The oxygen uptake at the end was the same in all experiments, or about 4.1 liters/min. However, after 1 min of extremely heavy exercise, the oxygen uptake was 4.0 liters/min at the "super-maximal" load but only 3.0 liters/min during the less extreme but still heavy work load of 1800 kpm/min, which could not be tolerated for more than 6 min (P.-O. Åstrand and Saltin, 1961a).

There are several implications from these experiments: (1) In studies where maximal oxygen uptake is to be measured, the collection of expired air or other measurements may start after about 1 min of exercise, provided that the work load is extremely heavy (supermaximal) and is preceded by a warming-up period. For many reasons, however, it is wise to aim at a work period of about 5 min. (2) A work time of 1 min or even less may maximally load the oxygen-transporting system. (3) It is an interesting question why the ability to increase the oxygen uptake maximally within 1 min is not utilized in exercise which can be prolonged to 5 to 10 min and has a marked oxygen deficit during the first minutes. This would be an advantage, for the sooner the aerobic processes can come into full swing, the less would be the demand on the anaerobic processes, and less lactic acid would accumulate.

In repeated determinations of maximal oxygen uptake on the same subject the standard deviation is 3 percent, which includes biological and methodological variables (P.-O. Åstrand, 1952; Taylor et al., 1955; Mitchell et al., 1958; P.-O. Åstrand and Saltin, 1961a).

Summary In many types of muscular exercise the oxygen uptake increases roughly linearly with an increase in work load. The maximal oxygen uptake or *maximal aerobic power* is defined as the highest oxygen uptake the individual can attain during physical work breathing air at sea level (work time was 2 to 6 min, depending on the work load).

During heavy exercise, anaerobic processes contribute to the energy yield, not only at the beginning of work but continuously throughout the exercise period. An accumulation of metabolites will eventually necessitate the termination of the work.

In very heavy exercise, the maximal oxygen uptake and heart rate may be attained within 1 min, provided a sufficient warm-up period precedes the maximal effort.

Intermittent Work

Muscular work in industrial or recreational activities is very seldom maintained for very long at a steady rate. For this reason, a steady state, as discussed above, is rarely attained. The classical laboratory studies, with subjects working continuously for 5 min or longer on the treadmill or bicycle ergometer, represent in many ways artificial situations. However, such procedures have distinct advantages when studying the physiology of exercise or when studying patients, for they provide standardized conditions and permit comparisons to be made on repeated occasions. They may also simulate the demands placed on the body in many sport events. However, both from a practical and from a theoretical point of view, it is equally important to study the effect of intermittent work, which better mirrors the type of muscular activities encountered in industry or at home and in most types of ordinary exercise or recreational activities.

Some of the more important principles will be discussed by presenting some experiments concerning intermittent work (I. Åstrand et al., 1960*a,b*; Christensen et al., 1960).

1　A subject whose maximal oxygen uptake was 4.6 liters/min could work at 2160 kpm/min (about 350 watts) for about 8 min. Since the oxygen need was approximately 5.2 liters/min, the anaerobic processes had to provide part of the energy. When the work load was reduced to 1080 kpm/min, the work could easily be prolonged to 60 min, and the final heart rate was 135, oxygen uptake 2.45 liters/min, and the blood lactate concentration did not increase above resting level. The total oxygen uptake during the hour was 145 liters.

2　In another experiment with the same subject, the work load was again 2160 kpm/min, but now work periods of 3 min were alternated with 3-min rest periods. The subject could proceed with great difficulty for 1 hr, and the same total amount of work was performed as in Experiment 1. The oxygen uptake and heart rate were now maximal, as was the peak blood lactate concentration (120 mg/100 ml)(10 mg/100 ml = 1.1 mmoles/liter). The total energy output during the second experiment was about 10 percent higher than in the first one.

3　When the heavy work periods were shortened by introducing the rest periods more frequently, the *total* oxygen uptake over the hour was not markedly reduced. The subjective feeling of strain was less severe, however, and peak oxygen uptake, heart rate, and blood lactate concentration were now lower. Hence, with intermittent work and rest for 30 sec, respectively, the heart rate did not exceed 150, the blood lactate was only 20 mg/100 ml, and the total oxygen uptake was 154 liters during the hour. (The subject's maximal heart rate was 190.)

Figure 9-5*a* illustrates another set of experiments with the same subject. He worked on the bicycle ergometer with an extremely heavy work load of 2520 kpm/min (about 400 watts). When working continuously at this work load, he became exhausted within about 3 min. When working intermittently for 1 min and resting for 2 min, he could continue for 24 min before being totally exhausted, and the blood lactate concentration rose to 150 mg/100 ml. In another experiment, the periods of work were reduced to 10 sec and the rest periods to 20 sec. Now he could complete the intended production of 25,200 kpm within 30 min with no severe feeling of strain, and his blood lactate concentration did not exceed 20 mg/100 ml, indicating an almost balanced oxygen supply to his heavily stressed muscles. With periods of work and rest of 30 and 60 sec, respectively, intermediate results were obtained.

A prolongation of the rest periods so that the ratio between work and rest was changed to 1:4 gave, of course, a decreased total work output but had hardly any beneficial effect on the subject's fatigue. The critical factor was the length of the work periods, and the duration of the rest pauses and the total time spent resting during the 30-min period were only of secondary importance.

Figure 9-5*b* is an attempt to explain the findings presented above. When a person works for short periods at an extremely high energy output, the aerobic supply is apparently adequate despite an insufficient transport of oxygen during the burst of activity. There is, at least, no continuous increase in blood lactate concentration, so if anaerobic processes do supply energy to any marked extent, they must have been blocked before the step involving the conversion of pyruvic acid into lactic acid. It is not likely that a marked production of lactic acid could be balanced by a simultaneous disappearance rate by combustion and reconversion to glycogen of the same amount of lactic acid. A possible explanation for a predominantly aerobic oxidation might be that at the beginning of every work period the muscles have a certain volume of oxygen at their disposal. We may assume that oxygen bound to myoglobin constitutes such an oxygen store which is consumed during the initial phase of exercise, before circulation and respiration can provide an additional supply which may or may not be adequate. During the rest period the depots are refilled with oxygen. Consequently, during severe exercise it is essential that the work periods do not become so long that the oxygen supply is exhausted and the anaerobic lactic acid production is brought into play. If that is the case, the rate of work must gradually be reduced. By spacing the work so that running periods lasted for 10 sec, resting for 5 sec, a subject could actually prolong the total work and rest period to 30 min without undue fatigue at a speed that normally exhausted him after about 4 min continuous running.

Dynamic exercise is certainly an intermittent type of work, and its superiority over static work for endurance exercise may largely be explained on the basis of the muscle pump and the alternating emptying and filling of the oxygen store during alternating muscle contraction and relaxation.

Summary   The buffering effect of a hypothetical oxygen store may mean, practically speaking, that a great amount of work can be performed at an extremely heavy work load, with a relatively low peak load on the circulation and the respiration, by the introduction of properly spaced short work and rest periods ("micropauses"). The heavier the work load, the shorter should be the work periods. This physiological concept has at least three important applications:

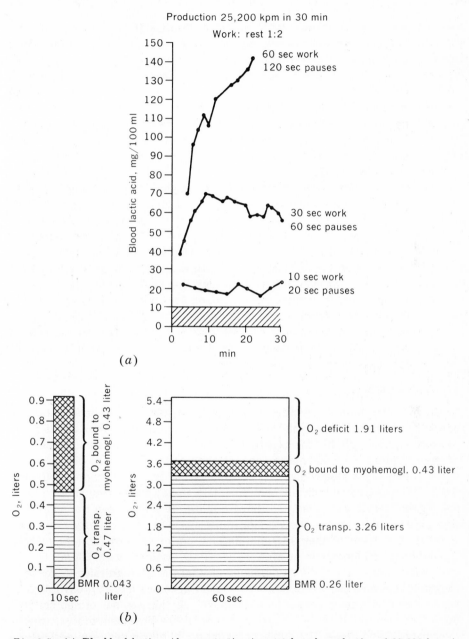

Fig. 9-5 (*a*) *The blood lactic acid concentration in a total work production of 25,200 kpm in 30 min. The work is accomplished with a load of 2520 kpm/min, the work periods being 10, 30, and 60 sec, and the corresponding rest periods 20, 60, and 120 sec respectively. (From I. Åstrand et al., 1960b.)*

(*b*) *The oxygen requirement for 10- and 60-sec work at a load of 2520 kpm/min. The schematic drawing indicates the basal metabolic rate (BMR), the calculated fraction of O_2 bound to myoglobin, transported by the blood, and the O_2 deficit. (From I. Åstrand et al., 1960b.)*

1   It may explain why older or physically disabled individuals, in spite of a reduced maximal aerobic power, can remain in jobs involving heavy work, such as forestry, farming, and construction work, or enjoy physically demanding hobbies. As long as they are free to choose the optimal length of the work and rest periods, the acute loads on respiration and circulation may not exceed the limits of their reduced capacity. It should be emphasized, however, that if the work pace is determined by a machine, even a less heavy peak load, but with relatively long work periods, may overtax the capacity of the workers whose physical work capacity is limited.

2   If the aim of a training program is to increase muscle strength, a given period of time would permit the highest load on the muscle fibers if periods of rest were frequently introduced between activity periods of 5 to 10 sec. On the other hand, a training of the oxygen-transporting system will be more effective if the exercise periods are prolonged to at least 2 to 3 min. This type of work would also adapt the tissues to high lactate concentrations, providing the exercise is severe.

3   Within limits, the difference in caloric cost between walking and running a given distance is not very great. Nor is the caloric cost of performing a given amount of work markedly dependent on the rate of work. This should be kept in mind when "prescribing" regular exercise for obese persons, or as a prevention against overweight.

Prolonged Work

When the work time is extended to 1 hr, the oxygen uptake, heart rate, and cardiac output are maintained at the same level as attained after about 5 min of exercise, provided that the oxygen uptake is not higher than about 50 percent of the maximum. The lactic acid concentration of the arterial blood is not elevated, indicating a steady state (I. Åstrand et al., 1959; I. Åstrand, 1960) (Fig. 9-6a). The well-trained individual can maintain steady state or equilibrium at a still higher relative work load, indicating a more efficient oxygen transportation and oxygen utilization in the working muscles. Elite cross-country skiers can work at 85 percent of their maximal aerobic power at least for 1 hr, oxygen uptake being 4.5 liters/min or even higher (P.-O. Åstrand et al., 1963b).

When work time is further prolonged, there is a progressive increase in oxygen intake and heart rate, and the subject becomes more or less fatigued. Figure 9-6b illustrates an experiment in which exercise was performed continuously for seven 50-min periods at an oxygen uptake of 50 percent of the subject's maximal oxygen uptake. The subjects rested for 10 min in between, and after 4 hr of work they had a break of 1 hr for lunch. The most fit subject, with a maximal oxygen uptake of 5.60 liters/min, worked with an average oxygen uptake of 2.75 liters/min; one subject with a maximal aerobic power of 2.25 liters/min worked with a work load requiring an oxygen uptake of 1.15 liters/min. The four

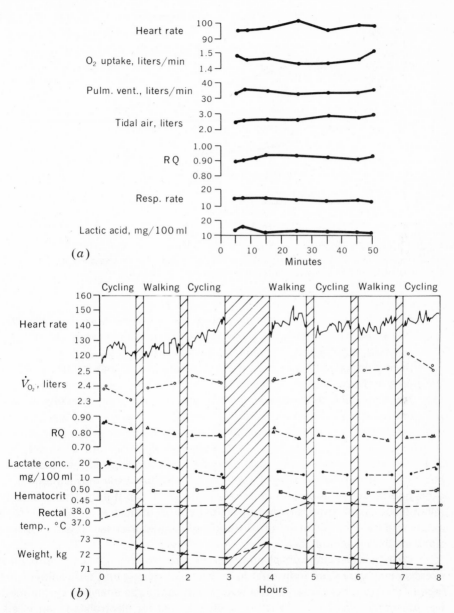

Fig. 9-6 (a) *Metabolic parameters during 1-hr work in a subject working at a load (1.5 liters O₂/min) close to 50 percent of his maximal aerobic power (2.94 liters O₂/min). (Data from I. Åstrand et al., 1959.)*

(b) *Metabolic parameters in one subject during an experiment consisting of seven work periods of 50 min each. The shaded columns represent rest periods. The subject's maximal aerobic power was 4.6 liters O₂/min. (From I. Åstrand, 1960.)*

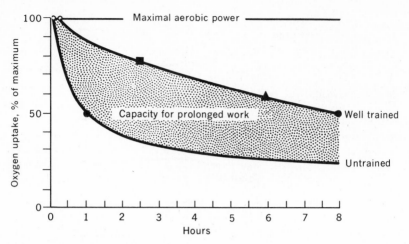

Fig. 9-7   A graphic illustration based on a few observations showing approximately the percentage of a subject's maximal aerobic power he can tax during work of different duration, and how this is affected by his state of training.

subjects participating in the experiments could fulfill the task, but they were fatigued. It appears that a 50 percent load is too high for a steady state if the physical activity is continuous for a whole working day (Fig. 9-7).

During prolonged heavy exercise, the water balance may be disturbed and the stores of available energy, particularly glycogen, may be critically low. Therefore the individual's ability to transport oxygen from air to the working muscles will not always per se be the limiting factor. It has been found that the subjective feeling of fatigue during heavy work usually coincides with a drop in blood glucose (Christensen and Hansen, 1939; Rodahl et al., 1964), and/or an emptying of the glycogen depots in the working muscles (Hultman, 1967). An increase in heart rate with reduction in stroke volume as work proceeds may sometimes be explained by dehydration due to sweating (Saltin, 1964). If dehydration and the fall in blood sugar are prevented by proper supply of fluid and sugar, performance capacity is better maintained during prolonged exercise. (Further discussions in Chaps. 14 and 15.)

There are still many unsolved problems and the limiting factor in prolonged exercise may vary from individual to individual. Training and environment may modify the work performance at a level which cannot be analyzed by the methods presently available. Thus, it is conceivable that the electrolyte balance, e.g., the K^+/Na^+ ratios, across the muscular cell membrane may be disturbed during prolonged exercise, and that the enzyme systems may be altered. As a matter of fact, in some experiments involving prolonged severe exercise, none of the physiological parameters studied correlate well with the subject's feeling of fatigue or reduction in performance capacity, e.g., blood sugar concentration, maximal oxygen uptake and cardiac output, blood lactic acid level (Hedman, 1957; Saltin, 1964).

Motivation is undoubtedly an important factor determining the endurance during heavy exercise. Well-trained, highly motivated subjects may maintain the oxygen uptake at a maximal level for at least 15 min, although most individuals feel forced to stop after 4 to 5 min at a work load which taxes the oxygen-transporting systems to a maximum.

Figure 9-7 presents the tolerance time during continuous exercise at different relative work loads, related to the individual's maximal oxygen uptake. The figure may be subject to criticism since the experimental data are scanty. The objective criteria for fatigue are often lacking or at least vaguely defined.

Summary It is obvious that the individual's maximal aerobic power plays a decisive role in his work capacity. If a given work task demands an oxygen uptake of 2.0 liters/min, the man with a maximal O_2 uptake of 4.0 liters/min has a satisfactory safety margin, but the 2.5-liter man must work close to his maximum, and consequently his internal equilibrium becomes much more disturbed. In prolonged exercise, motivation, state of training, water balance, and depots of available energy are important for the performance capacity (climatic conditions are discussed elsewhere).

Muscular Mass Involved in Exercise

The demand on the oxygen-transporting functions varies with the size of the active muscles. Since isometric work hinders the local blood flow and dynamic exercise facilitates the circulation, it follows that a greater oxygen uptake can be obtained during dynamic exercise. Usually exercise involves both static and dynamic muscle contractions. Static work produces a relatively high heart rate and arterial blood pressure; this may complicate a work evaluation based on the measurements of heart rate and blood pressure (Chap. 6).

The oxygen uptake, heart rate, and cardiac output measured during maximal leg work on a bicycle are not further increased, if at the same time the arm muscles work simultaneously on another ergometer (P.-O. Åstrand and Saltin, 1961b; Stenberg et al., 1967). The maximal oxygen uptake is approximately the same whether it is measured while running on a treadmill or during cross-country skiing or during bicycling (P.-O. Åstrand and Saltin, 1961b). In maximal work on a bicycle ergometer in the supine position the oxygen uptake is, however, only about 85 percent of the value obtained in the sitting position. But, if the subject works with both legs and arms simultaneously in the supine position, the oxygen uptake, cardiac output, and heart rate go up to the values typical for maximal exercise in the upright position (Stenberg et al., 1967). One plausible explanation for the lower work capacity for cycling in the supine position, despite an optimal venous return to the heart, is the less favorable work position, since the body weight cannot be utilized during the critical stages of the pedaling. (Consult Fig. 6-23.)

In a group of about 70 fairly well-trained women and men no difference was noted in maximal oxygen uptake in the two types of exercise: work on the bicycle ergometer and running on a motor-driven treadmill (grade 1.75 percent uphill) (P.-O. Åstrand, 1952). Rowell et al. (1964) report a significantly higher maximal oxygen uptake during running (grade 5 to 15 percent), and P.-O. Åstrand and Saltin (1961b) noticed a 5 percent difference, maximal $\dot{V}_{O_2}$ being higher when running than when cycling. It is possible that running at high speed on the treadmill with a low grade is technically so difficult that the subject fails before he has reached his maximal oxygen uptake. Another explanation for a higher oxygen uptake when running at high grades and low to moderate speeds may be that more muscles are active in the uphill running. However, the combined arm-plus-leg work should necessarily include more muscles than leg work only; yet oxygen uptake is not further increased compared with leg work.

In our experience there is not a consistently higher maximal oxygen uptake attained in running compared with cycling, and the small difference sometimes noticed has hardly any practical consequence. To obtain maximal values when using the bicycle ergometer, motivation and "stimulation" may be particularly important, due to more pronounced local fatigue in the legs (knee region) when cycling. Furthermore the work position is critical when using the bicycle ergometer as a tool. The bicyle seat should be high enough and the subject should be positioned almost vertically above the pedals. Otherwise the working position will be more or less similar to work in the supine position. (See Table 11-1.)

During swimming the highest oxygen uptake attained is about 90 percent of the maximum measured during maximal work on the bicycle ergometer (P.-O. Åstrand and Saltin, 1961b; P.-O. Åstrand et al., 1963a).

It may be concluded that the assessment of the individual's maximal oxygen uptake, i.e., his maximal aerobic power, should be made with the subject working in the upright position (running or bicycling). If the subject is tested when working in the supine position, both arms and legs should be involved in the work.

In arm exercise, the maximal oxygen uptake is about 70 percent of what is attained in leg exercise. The intra-arterial blood pressure during arm work is higher than in leg work at a given oxygen uptake or cardiac output (Fig. 6-22), and the heart rate is also higher. The consequence of this is a heavier load on the heart. For patients with heart disease or for completely untrained older individuals, heavy work with the arms (e.g., digging, shoveling snow) may therefore be hazardous. This may in part be due to the Valsalva effect during such maneuvers. The subject is apt to hold his breath while lifting the load, increasing the intrathoracic pressure which, in turn, may hinder the normal venous return to the heart.

The fact that work with arms plus legs does not further increase the oxygen uptake and cardiac output as compared with leg work alone might be interpreted

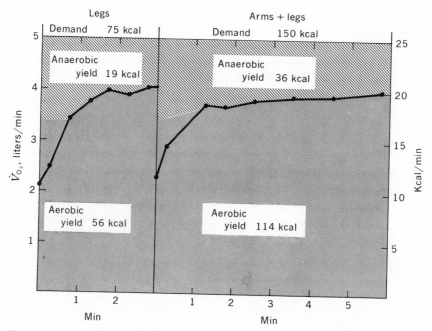

Fig. 9-8   Increase in oxygen uptake at the start of exhausting work, following a 10-min warm-up period. Left: illustrates leg work only; right: exercise with the same work load but with both arms and legs involved. Load could be tolerated twice as long with arms and legs working. Calculations of energy demand and yield are explained in text. (From data presented by P.-O. Åstrand and Saltin, 1961b; subject 1-POÅ.)

as an indication that the pumping capacity of the heart is the limiting factor during heavy exercise. However, other observations speak against this possibility. A typical example is presented in Fig. 9-8. A work load of 2100 kpm/min (350 watts) could be tolerated for about 3 min if only the leg muscles were involved. However, with 600 kpm/min for the arms and 1500 kpm/min for the legs (= 2100) the work time could be prolonged to 6 min, even if the oxygen uptake (and cardiac output) did not increase further (P.-O. Åstrand and Saltin, 1961b; Stenberg et al., 1967). Evidently the organism (inclusive heart) could tolerate a prolongation of the work period when a larger mass of skeletal muscles were activated. The subjective feeling of strain is more related to the metabolic rate per square area of muscle than to the total metabolism. Therefore, a training of the oxygen-transporting system is more efficient and is psychologically less strenuous, the larger the muscular mass involved in dynamic work. On the other hand, an untrained individual or a cardiac patient should be cautious during the beginning of a training program which includes vigorous activities exercising a large muscular mass.

ANAEROBIC PROCESSES

During light work, the required energy may be almost exclusively produced by aerobic processes, as mentioned, but during more severe work anaerobic processes are brought into play as well. Anaerobic energy-yielding metabolic processes play an increasingly greater role as the severity of the work load increases. The anaerobic energy-yielding processes have been briefly discussed in Chap. 2 (page 16). Here we shall merely summarize the more important points and present some additional comments. (We take an increase in lactic acid concentration in blood as the main indication of the involvement of anaerobic processes, since the lactic acid concentration is easy to analyze. Furthermore it should be recalled that the energy yield from the breakdown of ATP and creatine phosphate is indispensable but quantitatively the available stores of these high-energy phosphates alone can only cover the energy requirement for less than 1 min during maximal exercise, Table 2-1.)

Oxygen Deficit—Lactic Acid Production

1   During light exercise, the oxygen store in the muscle plus the oxygen supplied as the respiration and the circulation adapt to the work will completely cover the oxygen need. Most of the ordinary daily occupations belong to this category of work.

2   During exercise of moderate intensity, anaerobic processes contribute to the energy output at the beginning of the exercise until the aerobic oxidation can take over and completely cover the energy demand. Any produced lactic acid diffuses into the blood and can be traced in the venous blood draining the muscle and, eventually, in the arterial blood if the quantity of lactic acid produced is high enough. As the work proceeds, the blood lactate concentration falls again to the resting level and the work can be continued for hours (Bang, 1936).

3   During heavier exercise, the lactic acid production and, therefore, the rise in blood lactate concentration is higher and remains high throughout the work period, but it can be extended for 30 min or even longer by well-motivated individuals.

4   During very severe exercise, there is a continuously growing oxygen deficit and an increase in the lactate content of the blood because of the predominantly anaerobic metabolism. The work cannot be continued for more than a few minutes, as a rule, because the subject's muscles can no longer function.

Figure 9-9 illustrates how the arterial lactic acid concentration increases during and after severe exercise, followed by a slow decline back to the resting level. The lactic acid is produced in the muscles during the actual work, but there is a time lag for the diffusion from the working muscles and redistribution within

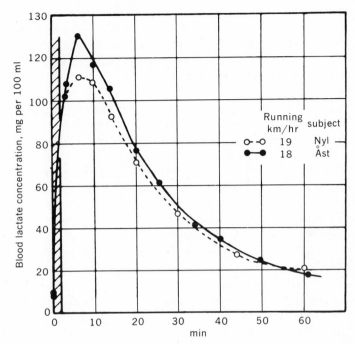

Fig. 9-9 Blood lactate concentration after severe work of 2 min duration (shaded column) in two subjects. Peak values occur several minutes following the cessation of work. (From I. Åstrand, 1960.)

the body. For a determination of peak lactic acid in the blood, samples must be taken at intervals during the first 5 to 10 min of the recovery period. It should also be noted that it takes up to 60 min or even longer before the resting level is again reached, so that if the effect of a stepwise increasing work load is studied, the samples secured at the end of the last work period do not only reflect the anaerobic component of this load, but they are also affected by the preceding work loads. Furthermore, in competitive events where the lactic acid production is high, the time between heats should be at least 1 hr to allow time for the blood lactate to return to resting values. Obviously, this decision should be based on well-established physiological principles and it should not be left to the arbitrary judging of the organizing committee whether or not the competitor should have to start a competition with a high tissue lactic acid concentration.

Lactic acid, partly buffered by the bicarbonates in the blood, lowers the pH of the blood, and the respiration is stimulated. An arterial pH as low as 7.0 has been observed after severe exercise (Chap. 7). Therefore heavy exercise causes a hyperpnea and eventually a dyspnea. In blood samples taken at rest and during and after exercise (steady state as well as maximal efforts) Keul et al. (1967) found a very high correlation between lactate concentration and pH values, and also between the sum of lactate and pyruvate concentrations and standard bicarbonate. Of the decrease in standard bicarbonate, about 95 percent was ascribed to the rise in lactate and pyruvate concentration. The remaining 5 percent was due to an increase in free fatty acids in the blood. Thanks to the buffer systems of the blood a tenfold increase in lactate concentration caused only a 1.42-fold increase in the H^+ concentration. The blood lactate concentration in men and women is on the average the same after maximal exercise and within the range of 11 to 14 mmoles/liter for twenty- to forty-year-old trained individuals. Children and older individuals usually do not attain such high values (Robinson, 1938; P.-O. Åstrand, 1952; I. Åstrand, 1960). During training, the blood lactate concentration for a given work load is lowered, but the values attained during maximal physical effort are usually higher. The blood lactate concentration may exceed 20 mmoles/liter. The highest values reported so far are in samples drawn from well-trained athletes at the end of competitive events of 1- to 2-min duration.

Figure 9-8 summarizes experiments performed on bicycle ergometers. From work load and a mechanical efficiency of 22 percent for aerobic work, the energy demand can be calculated. During 3 min exercise, at a rate representing the maximum in leg work for this subject, the energy demand was about 75 kcal. The oxygen uptake was measured continuously and was 10.7 liters. It is calculated that an additional 0.5 liter was utilized from stores bound to myoglobin and hemoglobin, refilled after the exercise. Therefore, the aerobic energy yield can be estimated to be 56 kcal (11.2 × 5). The deficit was then 75 − 56 = 19 kcal and this energy must have been derived anaerobically. A breakdown of ATP and creatine phosphate may yield 5 to 7.5 kcal, i.e., it may substitute 1 to 1.5 liters of oxygen. The remaining deficit of a minimum of about 12 kcal must have been yielded by glycogenolysis and glycolysis with a formation of lactic acid.

Since about 55 kcal are released for each six-carbon unit of glycogen which is converted into lactic acid (Chap. 2) a production of 2 moles or 180 g of lactic acid should yield 55 kcal or about 0.3 kcal/g lactic acid ($^{55}\!/_{180} = 0.3$). For a release of 12 kcal the lactic acid production must then be 39 g.

As mentioned before the subject also performed the same work load with both arms and legs, and under these conditions the work could be prolonged to 6 min before exhaustion (Fig. 9-8). At submaximal exercise the mechanical efficiency is not significantly different from ordinary cycling. (If anything the oxygen uptake tends to be higher in work performed with both arms and legs compared with leg work alone; it is also conceivable that the mechanical efficiency becomes lower at very heavy exercise since muscles which are at a mechanical disadvantage have to contribute. Therefore the calculated energy demand is probably a minimal figure.) Therefore the energy requirement is still assumed to be 25 kcal/min or 150 kcal altogether during the 6 min. The measured oxygen uptake of 22.3 liters complemented with 0.5 liter from oxygen stores within the body covers 114 kcal, leaving 36 kcal for the anaerobic processes.

A subtraction of 7 kcal as a contribution from high energy phosphate compounds leaves 29 kcal from glycogenolysis, i.e., a formation of 95 g lactic acid. (Since more muscles were working the oxygen from myoglobin and the energy yield from ATP and creatine phosphate may have been larger. Quantitatively, however, this does not alter these calculations significantly.)

The lactic acid formation necessary to explain that the subject actually could go on for 6 min on a work load of 350 watts demanding some 25 kcal/min, despite an aerobic power not exceeding about 20 kcal/min, is much higher than that reported by Margaria (1967). He concludes that the maximal production of lactic acid is 1 g/kg body weight which would be 74 g for the present subject. Margaria also gives a lower energy production/g lactic acid formed, or 0.22 kcal/g. If his figures were correct the 6-min experiment should require a release of 132 g lactic acid. A caloric value of 0.3 kcal/g seems more probable than 0.22.

Figure 9-10 presents data from a subject who worked for 2.63 min with a work load close to 400 watts with a calculated energy demand of 70 kcal. The total oxygen uptake during activity was 7.1 liters. Adding 0.5 liter from oxygen stores, the aerobic energy yield will be 38 kcal. Therefore, the anaerobic con-

Fig. 9-10   Calculated energy requirement for a 2.63-min exercise on a bicycle ergometer (column represents 70 kcal) and measured oxygen uptake during exercise and during 60 min recovery (dotted area). Horizontal lines denote the level of oxygen uptake measured at rest before exercise. Calculated aerobic energy yield during exercise: 38 kcal; anaerobic energy yield: 32 kcal. Lactic acid concentration was analyzed in blood samples and pieces of skeletal muscle obtained by needle biopsy. (Data by courtesy of B. Diamant, K. Karlsson, and B. Saltin.)

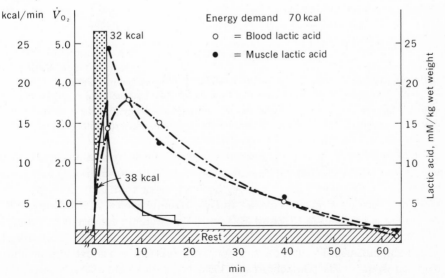

tribution was 32 kcal. Assuming that a breakdown of glycogen yielded 25 kcal, the production of lactic acid would have been 82 g.

Theoretically these productions of lactic acid are possible since the glycogen content in the muscles is normally about 15 g/kg wet weight; thus in 20 kg of muscles there is 300 g of glycogen.

Lactic Acid Distribution and Disappearance

We have seen that close to 100 g of lactic acid may be produced within a few minutes; in a well-trained top athlete it may be still higher. Lactic acid is assumed to diffuse freely into all water compartments of the body. With a water content of 40 liters, 4 g of lactic acid would give a concentration of 10 mg/100 ml of water or roughly 10 mg/100 ml or 1.1 mmoles/liter of blood. This concentration is actually the resting level of blood lactate. An additional 4 g of lactic acid would double the concentration.

A production of 39 and 95 g of lactic acid (Fig. 9-8) would increase the blood lactate concentration to 12 mmoles and 27 mmoles/liter respectively. The noticed peak concentration was, however, in both experiments 17.5 mmoles/liter blood. The discrepancy can be explained as follows: A delay in diffusion into the various water compartments of the body would give a higher blood concentration than expected; a continuous removal by resynthesis and oxidation will cause a lower blood concentration than the expected one. Muscle biopsies on man after maximal exercise have revealed a content of more than 30 mmoles lactic acid per kilogram wet muscle, but the blood concentration was just above 20 mmoles/liter (J. Karlsson, personal communication). During short-term work the lactic acid concentration in the working muscle group may be 3 to 5 times higher than that in blood (Hultman, 1967). For these and other reasons it is impossible to calculate the total production of lactic acid, i.e., the anaerobic energy yield from glycolysis from the concentration of lactic acid in the blood. A rise merely indicates that an increased formation has occurred.

In Fig. 9-10 data on lactic acid concentration is included, determined in pieces of muscles obtained by biopsies and the lactic acid concentration in blood. During early recovery it is significantly higher in the exercising muscles than in blood.

A calculation of "excess lactate" (Huckabee, 1958), by which the changes in lactate due to changes in pyruvate have been eliminated, does not provide a more accurate method to evaluate the anaerobic energy yield during exercise (Knuttgen, 1962).

At rest, it is generally assumed that most of the lactic acid produced during work, or about 85 percent, is resynthesized back to glycogen, mainly in the liver, but also in the kidneys (Levy, 1962). A gluconeogenesis from lactic acid apparently does not take place in mammalian muscles (Koeppe et al., 1964; Krebs, 1964). A fraction is oxidized to CO_2 and H_2O and this oxidation can occur in the heart muscle (cf. Newman et al., 1937; Carlsten et al., 1961; Keul et al., 1966) and also in skeletal muscles. To what degree lactic acid replaces the substrates ordinarily oxidized is uncertain (Omachi et al., 1956; Issekutz et al., 1965).

Newman et al. (1937) noticed that the removal of lactic acid, accumulated in the body after exhausting exercise, was enhanced if during recovery the subject continued to exercise, but at a lower intensity which normally did not produce any lactic acid.

The rate of removal of lactic acid increased proportionally with the metabolic rate up to some critical level of activity, different for each subject. A use of lactic acid as a fuel for the work could be one explanation for this finding, or an increased blood flow proportional to metabolic rate with more rapid transfer of lactic acid to reactive centers could be another explanation.

The loss of lactic acid with sweat and urine is negligible (cf. Newman et al., 1937).

During steady-state work of moderate intensity, with blood lactate concentrations up to about 4 mmoles/liter, the uptake of lactates in the liver has been determined (Hultman, 1967; Rowell et al., 1966). This uptake was 100 to 150 mg/min, i.e., it would take 6 to 10 min to remove 1 g of lactic acid. However, the rate of lactate uptake by splanchnic tissues appears to be proportional to its blood concentration. With a production of some 100 g of lactic acid it is necessary to presume a high disappearance rate, since the peak lactic acid concentration is down to resting level after about 1 hr.

The fate of the lactic acid produced during heavy exercise is far from revealed.

Oxygen Debt

To some extent lactic acid is eliminated during exercise (Bang, 1936; Newman et al., 1937; Rowell et al., 1966) and therefore part of the aerobic energy deficit during the early stage of exercise can soon be repaid. On the other hand there is always an oxygen debt after exercise (Fig. 9-3a), which is the oxygen uptake of the resting subject in excess of the oxygen uptake calculated for the same period of rest but not preceded by activity (oxygen debt = total oxygen uptake − resting oxygen uptake). If measured after a steady-state exercise it is of the same magnitude whether the work time is 10 or 60 min.

As discussed at the beginning of this chapter, various factors are involved in the delayed return of oxygen uptake during a recovery to the basal level. After maximal exercise of a few minutes duration this excess oxygen uptake measured during 60 min may reach values as high as about 20 liters. A refill of the oxygen stores (blood, myoglobin) will demand less than 1 liter of oxygen. At the same elevated tissue temperature and adrenalin concentration at rest which is attained during heavy exercise, up to 1 liter extra oxygen may be consumed. Increased cardiac and respiratory functions may require some 0.5 liter extra oxygen, giving a total of about 2.0 to 2.5 liters of oxygen uptake during recovery, which has nothing to do with the energy transfer in handling anaerobic end products. A breakdown of ATP and creatine phosphate may yield 5 to 7.5

kcal, i.e., it may substitute 1 to 1.5 liters of oxygen. Therefore up to 4 liters of the oxygen debt may be *alactacid*, i.e., it is not involved in the handling of lactic acid, or the *lactacid* oxygen debt (cf. Margaria et al., 1933; Margaria, 1967).

According to Krebs (1964), 1 molecule of oxygen can remove maximally about 2 molecules of lactate, of which 1.7 are resynthesized and 0.3 oxidized to CO_2 and H_2O. Therefore 22.4 liters of oxygen can remove about 180 g of lactic acid. (It takes 7 molecules of ATP to resynthesize 1 molecule of glycogen from lactate, and 1 molecule of oxygen yields, on an average, 6 of ATP.) As mentioned a production of 180 g of lactic acid from glycogen yields 55 kcal. However, an aerobic oxidation of glycogen would provide 55 kcal with only about 11 liters of oxygen used. Therefore the body has to pay about 100 percent interest in the currency of oxygen for the energy borrowed from this anaerobic bank. It is apparently important to keep the glycogen store as high as possible and since we live in an "ocean of oxygen" the payment is not a problem. Evidently we should expect the oxygen debt to be at least twice as great as the oxygen deficit (Fig. 9-3a), which is actually the case.

The calculated production of 95 g of lactic acid (Fig. 9-8) would for its elimination demand about 12 liters of oxygen; the total oxygen debt would be at the most 16 liters.

For the experiment presented in Fig. 9-10 the lactacid oxygen debt can be calculated to 10.2 liters ($82/180 \times 22.4$); with an alactacid oxygen debt of 4.0 liters the total oxygen debt would be about 14 liters. The measured oxygen uptake during recovery above the resting demand was 13 liters, a figure which is in good agreement with the one calculated from theoretical considerations.

The oxygen uptake at rest before the experiment was 0.35 liter/min, which is quite high. Another time 0.32 liter/min was measured and with this metabolic rate for 60 min the oxygen debt would have been calculated to 14.2 liters. With an oxygen uptake of 0.28 liter/min, estimated basal metabolic rate, the oxygen debt for a 60-min period would have been 17.1 liters. It should therefore be emphasized that it is very difficult to separate accurately the oxygen debt from resting oxygen uptake according to the definition.

Anyway, considering the potential energy yield for each volume of oxygen consumed, the efficiency of the anaerobic processes is apparently about 50 percent of the aerobic ones. A similar conclusion was drawn by Asmussen (1946) and Christensen and Högberg (1950). The more of the formed lactic acid which is combusted to CO_2 and H_2O, the higher would be the net mechanical efficiency of the anaerobic processes if estimated from the total oxygen uptake during work *and* recovery.

Summary   At the beginning of work and during heavy work there is a discrepancy between energy demand and the energy available by aerobic processes. The anaerobic energy yield must therefore contribute, and a breakdown of glycogen to pyruvic acid and lactic acid is quantitatively the most important step in this anaerobic process.

The heavier the work is in relation to the individual's maximal aerobic power, the larger is the oxygen deficit and the more important is the anaerobic energy yield. Under hypoxic conditions, e.g., at high altitude, it is noticed that the oxygen debt, and blood lactate concentration, is higher at a given work load compared with sea-level values (Lundin and Ström, 1947; P.-O. Åstrand, 1954; Hermansen and Saltin, 1967).

It is calculated that about 40 kcal can be provided by the glycogen-lactic acid mechanism in an athlete successful in events of 1 to 2 min duration. This covers an oxygen deficit of about 8 liters. ATP and creatine phosphate may cover an additional oxygen deficit of 1 to 1.5 liters.

Mainly during the recovery period after the exercise, about 85 percent of the lactic acid formed is resynthesized to glycogen and 15 percent oxidized to CO_2 and H_2O. For these processes extra oxygen is taken up, and this *lactacid* oxygen debt is about twice as high as the oxygen deficit. In the example given the extra oxygen uptake due to the previous lactic acid formation will be about 16 liters. In addition a few liters extra oxygen is taken up from the inspired air for the resynthesis of high energy phosphate compounds, to refill oxygen stores in the body which have been depleted during heavy exercise, and to cover aerobically the increased metabolism due to elevated tissue temperature, the effect of adrenalin, and cardiac-respiratory activity above the basal level. Altogether, the oxygen uptake above the resting, basal level after an all-out effort of a few minutes duration can exceed 20 liters.

INTERRELATION BETWEEN AEROBIC AND ANAEROBIC ENERGY YIELD

Table 9-1 presents the contribution to energy output from aerobic and anaerobic processes respectively in maximal efforts in events with large muscle groups involved. The individual's maximal aerobic power is set to 5 liters/min = 25 kcal/

TABLE 9-1

Process	Work time, maximal effort							
	10 sec	1 min	2 min	4 min	10 min	30 min	60 min	120 min
Anaerobic								
kcal	25	40	45	45	35	30	20	15
percent	85	65–70	50	30	10–15	5	2	1
Aerobic								
kcal	4	20	45	100	250	700	1300	2400
percent	15	30–35	50	70	85–90	95	98	99
Total	29	60	90	145	285	730	1300	2400

min and maximal anaerobic capacity to 45 kcal, equivalent to 9 liters of oxygen uptake in aerobic work. It is assumed that 100 percent of maximal oxygen uptake can be maintained during 10 min, 95 percent during 30 min, 85 percent during 60 min, and 80 percent during 120 min (P.-O. Åstrand et al., 1963b; Hedman, 1957).

With a work time of up to 2 min the anaerobic power is more important than the aerobic one; at about 2 min there is a 50 : 50 ratio, and with longer work time the aerobic power becomes gradually more dominating. This is graphically illustrated in Fig. 9-11.

It is very rare that an individual possesses top power for both aerobic and anaerobic processes. Therefore, the analysis in Table 9-1 should not be interpreted and applied too literally. A maximal aerobic power of 25 kcal may be coupled with a maximal anaerobic yield of 25 kcal. For this individual the proportional participation of anaerobic and aerobic processes will be different compared with the tabulated data. (He should be recommended to compete over longer distances since his relatively low maximal anaerobic power is then less of a handicap, and he should try to get rid of his rivals before the finish starts!)

Fig. 9-11   Relative contribution in percent of total energy yield from aerobic and anaerobic processes respectively during maximal efforts of up to 60 min duration for an individual with high maximal power for both processes. Note that a 2 min maximal effort hits the 50 percent mark, meaning that both processes are equally important for success.

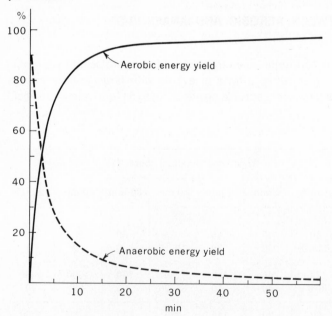

An analysis of the energetic demands of different sport events and the athlete's capabilities to fulfill these requirements may help him in his training and in his selection of suitable events.

A trained individual can work at a relatively high oxygen uptake in relation to his maximum (up to 60 to 65 percent) without any elevation in blood lactate concentration. When untrained, a rise is noted at about 50 percent of maximal aerobic power (P.-O. Åstrand, 1952; Wyndham et al., 1962; Hermansen and Saltin, 1967; Williams et al., 1967).

It is unknown which factor or factors decide the pathway of the energy yielding processes, i.e., whether the anaerobic or aerobic processes will be preferred. From studies on glycolysis in working frog muscles Karpatkin et al. (1964) conclude that it is difficult to ascribe activation of some key enzymes during stimulation to changes in the concentration of substrates, activators, and inhibitors.

Summary In maximal efforts of about 2-min duration the aerobic energy yield equals approximately the anaerobic yield. With shorter work times the anaerobic processes dominate; with longer work times the anaerobic energy yield is more important from a quantitative viewpoint.

MAXIMAL AEROBIC POWER—AGE AND SEX

It should be recalled that the maximal aerobic power is defined as the highest oxygen uptake the individual can attain during physical work breathing air at sea level. To evaluate whether or not the subject's maximal oxygen uptake has been attained, objective criteria should be used, such as measured oxygen uptake lower than expected from the work load, and blood lactic acid concentration higher than about 8 mmoles/liter.

The information provided by the assessment of maximal oxygen uptake is a measure of (1) the maximal energy output by aerobic processes, and (2) the functional capacity of the circulation, since there is a high correlation between the maximal cardiac output and the maximal aerobic power (see Fig. 6-26).

Direct measurements of the maximal oxygen uptake on 350 individuals ranging in age from four to sixty-five are presented in Fig. 9-12a. All subjects were healthy and moderately well trained; none of them was an athlete. It should be emphasized that it is almost impossible to present "normal material" since it is very difficult to define what is normal. This material is selected, but the age and sex factors which modify maximal aerobic power should be fairly evident in this homogeneous group of subjects.

Before puberty there is no significant difference in maximal aerobic power between girls and boys. Thereafter, the women's power is on an average 70 to 75 percent of that of the men. In both sexes there is a peak at eighteen to

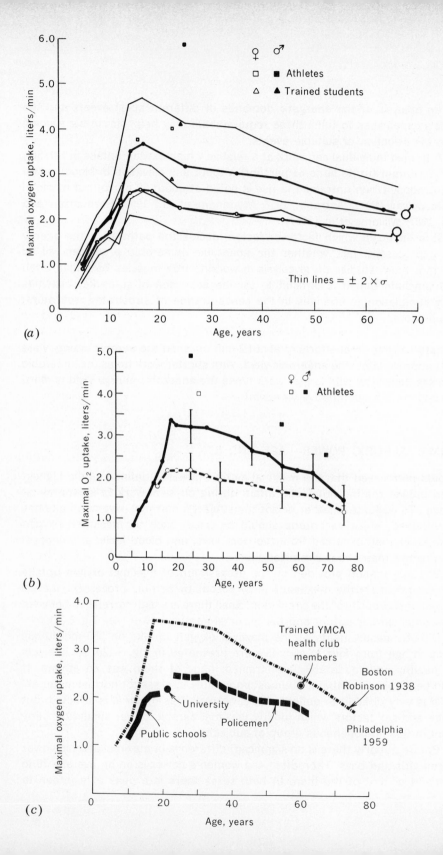

(a)

Maximal oxygen uptake, liters/min

♀ ♂
□ ■ Athletes
△ ▲ Trained students

Thin lines = ± 2 × σ

Age, years

(b)

Maximal O₂ uptake, liters/min

♀ ♂
□ ■ Athletes

Age, years

(c)

Maximal oxygen uptake, liters/min

Trained YMCA
health club
members

Boston
Robinson 1938

University

Policemen

Public schools

Philadelphia
1959

Age, years

twenty years of age, followed by a gradual decline in the maximal oxygen uptake. At the age of sixty-five, the mean value is about 70 percent of what it is for a twenty-five-year-old individual. The maximal oxygen uptake for the sixty-five-year-old man (average) is the same as that typical for a twenty-five-year-old woman.

The individual variation should be noticed. Many old subjects have a maximal power that is higher than that found in many much younger individuals. In Fig. 9-12a, the "-2 standard deviation line" for male subjects coincides closely with the average values for women, and the 95 percent range is actually ± 20 to 30 percent of the mean value at a given age.

Since regular training can in most cases increase the maximal oxygen uptake not more than 10 to 20 percent, it is evident that the natural endowment is the most important factor determining the individual's maximum. In the 1940s Gunder Hägg held many world records in middle- and long-distance running. No physiological data pertaining to him are available from this period. His body weight at that time was 69 to 70 kg. In 1963, at the age of forty-five his heart rate and oxygen uptake were recorded while he was working on a bicycle ergometer. His maximal O_2 was 4.0 liters/min and his maximal heart rate 181 beats/min. His blood lactate concentration was 13.8 mmoles/liter. His body weight was 94.5 kg. In spite of the fact that he had not trained since 1946, he had a very high maximal aerobic power (see Fig. 9-12a). This is an excellent example of how an untrained individual may have a very high aerobic power, providing his endowment is favorable.

The maximal aerobic power does not in itself reveal whether or not an individual has been physically active in the preceding years. One reservation should be made to this statement. It may be that the dimensions and functions of importance for the maximal aerobic power can be highly influenced by training early in life, between ten and twenty years of age, during the period of growth and development. This hypothesis is supported by studies on young girls training very intensively for competitions in swimming (P.-O. Åstrand et al., 1963a). They have larger hearts and higher vital capacity and maximal oxygen uptake than more sedentary girls of the same age and body size (Fig. 6-24). The girl swimmers consist of a selected group, but it is reasonable to assume that

---

Fig. 9-12 (a) Mean values for maximal oxygen uptake (maximal aerobic power) measured during exercise on treadmill or bicycle ergometer in 350 female and male subjects four to sixty-five years of age. Included are values from 3 athletes and from a group of 86 trained students in physical education. (From P.-O. Åstrand and Christensen, 1964.)

(b) Maximal aerobic power in German men and women in relation to age. The figure also includes data for male and female athletes (long-distance runners). (From Hollmann, 1963.)

(c) Similar data for American men (Philadelphia), nonathletes, compared with an earlier study in Boston (Robinson, 1938). The figure also includes mean maximal oxygen uptake in well-trained men over sixty years of age (Rodahl and Issekutz, 1962).

the dimensions and functions could be more influenced in a positive or negative way during adolescence than later on in life. In this connection, it may be pointed out that Beznak (1960) has shown in experiments on hypophysectomized rats that the size of the heart can be influenced by growth hormone, and that the strength of the heart can be influenced by thyroid-stimulating hormone.

In Fig. 9-12a are included data obtained from a group of 86 students in physical education. The mean values are definitely higher than for the average women and men, but the difference between the female and male students in maximal aerobic power is of the same magnitude as in the other material. Also included are some of the highest figures obtained so far on athletes: about 6.0 liters/min for male cross-country skiers and runners, and about 4.0 liters/min for female cross-country skiers and swimmers (P.-O. Åstrand et al., 1963a; Saltin and P.-O. Åstrand, 1967). The aerobic power of 6.0 liters (2,000 watts) should be compared with the 3.0 liters/min attained by the normal moderately trained men.

Figure 9-12b and c presents materials from other countries. The variation in maximal aerobic power with sex and age agrees quite well with the Swedish data.

The absolute values for maximal oxygen uptake will inevitably vary for different groups and populations. Selection of subjects is critical and a random sample is difficult to study successfully. For subjects with different occupations there is a definite trend that the mean values to some degree vary with the nature of the occupation. This is illustrated by Fig. 9-13 presenting the scatter of maximal oxygen uptake (determined or predicted) for different groups. The highest values are noted for forestry workers and the lowest ones for white-collar workers.

These differences in maximal aerobic power, as well as in many other parameters, are probably partly due to a selection, since those with a strong constitution are over-represented in occupations with physically demanding tasks. Furthermore such jobs may in themselves train the oxygen transport system. The more mechanized the society, the less will probably such differences become between personnel in different occupations.

Conclusions concerning the general physical standard in a country from physiological data must be drawn with caution.

An interesting question is why the best performance in endurance events is usually obtained by athletes twenty-five to thirty years of age, when the highest maximal oxygen uptake is usually reached at the age of twenty. However, there are several factors to be considered. Generally, physical activity is more regular and vigorous for those below than above twenty years of age, at least if the physical education is compulsory in school. This may explain the results presented in Fig. 9-12. On the other hand, if training is continued, the maximal aerobic power can certainly be maintained or even further increased for another 10-year period. Finally, the performance is also dependent on technique, tactics, motivation, and other factors, and intensive training and experience over the years make gradual improvement possible.

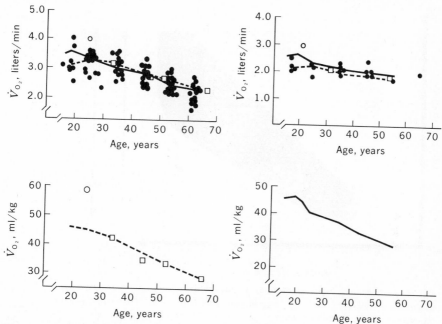

Fig. 9-13 *Group mean values of oxygen uptake in relation to age in liters/min and ml/kg ×*
min (unbroken line). Open circles denote the well-trained students presented in Fig.
9-12a. Filled dots denote predicted maximal oxygen uptake for various occupational
groups with different physical demands; their mean values are indicated by the dotted line.
Left: Males; unfilled square denotes measured oxygen uptakes in 84 building workers.
Occupational groups (1,478 individuals): the highest values were found among forestry
workers; next came construction workers, furnace workers, and gasworks employees; then
followed paper mill workers, lumberyard workers, ironworkers, and roadbuilding foremen;
next came rock-blasters, common laborers, carpenters, miners, and foundry workers. The
lowest maximal oxygen uptakes were found among drivers, repairmen, clerks, post-office
personnel, supervisors, and persons occupied in workshops.
Right: Females; unfilled square denotes measured maximal oxygen uptake in 10 housewives.
Occupational groups (1,700 individuals): the highest values were found among post-office
personnel who were physically active in their leisure time; then came housewives; followed
by physically inactive post-office personnel and other office employees; the lowest maximal oxygen
uptakes were found among shop girls and students in a nursing school at the beginning of their
studies. (From I. Åstrand, 1967b.)

Figure 9-14 shows how the performance capacity is related to the maximal
oxygen uptake in exercises with large muscle groups vigorously involved for
1 min or longer. No one can attain top results in such exercises without a high
aerobic power. On the other hand, a high power does not guarantee a good
performance, since technique and psychological factors may have a modifying
influence in a positive or negative direction.

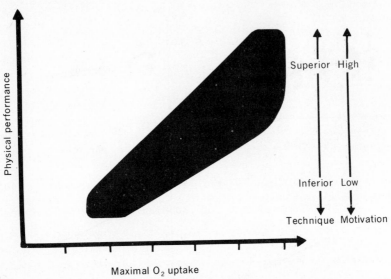

Fig. 9-14   Schematic representation of the importance of maximal aerobic power in physical performance involving large muscle groups for more than a minute, and the role played by technique and motivation in modifying top performance.

In work and exercise where the body is lifted (e.g., walking, running), the oxygen uptake should be related to the body weight. Figure 9-15 presents the same material as in Fig. 9-12a, but the maximal aerobic power is expressed in milliliters of O_2 per kilogram gross body weight. The values for young boys and girls are about the same, and the difference between average women and men is now reduced to 15 to 20 percent. The highest values so far recorded are 85 ml/kg min on a male cross-country skier (Saltin and P.-O. Åstrand, 1967), and 74 ml/kg on a female cross-country skier (Michailov and Ogoltsov, 1965).

Another way to express the individual's maximal oxygen uptake is to relate it to the dimensions of the body or to various organs. It may be of theoretical and of clinical value to examine whether or not the maximal oxygen uptake is proportional to heart size, muscular mass, lung volume, etc. Since fatty tissue is metabolically fairly inert but can constitute a large proportion of the body weight, it may be important to exclude it when evaluating the oxygen-transporting capacity.

When the weight of adipose tissue, estimated on the basis of hydrostatic weighing, is subtracted from the gross body weight of the well-trained students in Fig. 9-12a (12 kg or 20.3 percent of the body weight for the women and 8 kg or 10.6 percent for the men, Döbeln, 1956), the maximal oxygen uptake per kilogram fat-free body weight (lean body mass) can be calculated. The average figure was the same for both groups: 71.1 ml. However, since the metabolic rate at rest or during maximal muscular exercise should vary with the body weight raised to the $\frac{2}{3}$ power (Chap. 10), it is evident that the woman should actually have a higher aerobic power per kilogram lean body mass than the man. The explanation for a lower maximal oxygen uptake than expected may be the lower hemoglobin concentration in women. The lower maximal aerobic power of women may therefore be natural, since their dimensions are different from those of men, and the oxygen-binding capacity of the blood is lower. The relative increase in body-fat content in women starts at puberty.

The decrease in maximal oxygen uptake with age is not so simple to explain. Figure 9-16 shows how the maximal heart rate decreases with age from 195 for the twenty-five-year-old to about 165 for the sixty-five-year-old. The lower maximal heart rate with higher age certainly must reduce the maximal cardiac output and hence the oxygen-transporting capacity.

There is another consequence of a lower maximal heart rate. Studies on 33 building workers (bricklayers, carpenters, laborers) from thirty to seventy years of age showed that the mean heart rate during occupational activity was correlated with the individual's maximal heart rate (Fig. 9-17). For subjects with a maximal heart rate of 185 beats/min, mostly the younger workers, the mean heart rate during occupational activity was 110, and those with a maximum of 150 had a mean of 90 beats/min. The maximal aerobic power ranged from 2.2 to 3.6 liters O_2/min. In general the worker utilized the same percentage of his maximum in the work operations irrespective of his maximal oxygen uptake. In

Fig. 9-15   Mean values for maximal oxygen uptake expressed in ml O_2/kg gross body weight. Same subjects as presented in Fig. 9-12a. The standard deviation is between 2.5 and 5 ml O_2/kg body weight.

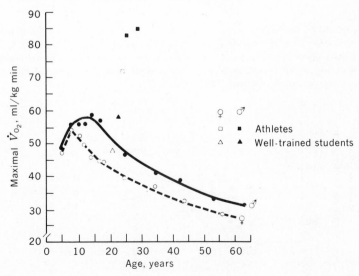

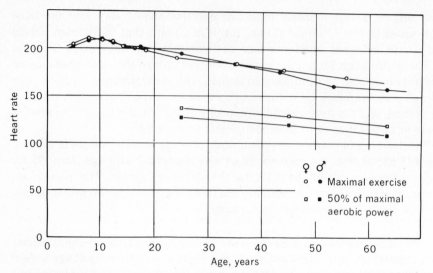

Fig. 9-16   Heart rate during maximal exercise (upper curves) and during prolonged work representing 50 percent of the individual's maximal oxygen uptake (lower curves) in the same 350 subjects represented in Fig. 9-12a. Standard deviation in maximal heart rate is ±10 beats/min. (From P.-O. Åstrand and Christensen, 1964.)

other words the older worker with a lower maximal aerobic power keeps a slower tempo than the younger one, but the relative load is the same for the two workers, or about 40 percent. The person with a high maximal heart rate can do a day's work at a higher mean heart rate than a person with a low maximal heart rate, but the relative strain may be the same on the two persons (I. Åstrand, 1967a).

Fig. 9-17   Individual values for the relationship between mean heart rate during occupational work (building) and maximal heart rate attained during work on bicycle ergometer. The heart rate was recorded by telemetry. The estimated oxygen uptakes during occupational work are presented on the right, together with their symbols. (From I. Åstrand, 1967b.)

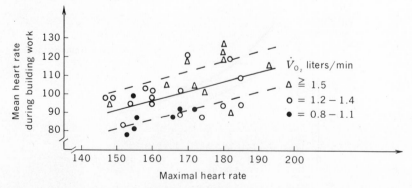

The stroke volume is lower in the older individual, but systolic and mean arterial pressures are higher; the pressures in the pulmonary artery are also higher in the aged (Strandell, 1964). The blood lactate concentration during submaximal work is higher in older individuals but lower during maximal exertion.

As an individual grows older he will usually become less physically active. Therefore part of the decrease in maximal oxygen uptake and performance is an effect of inactivity. Saltin and Grimby (1968) have studied middle-aged and old athletes and compared the data obtained with those from former athletes of the same ages who now live a sedentary life. The two groups were about equal as far as their performance in orientation racing* (including cross-country running) was concerned. It was therefore concluded that there was, at the time, no significant difference in maximal oxygen uptake. As can be seen from Table 9-2 the nonactive former athletes have now fallen behind as far as their maximal

* An orientation race is a cross-country race in which the runners must plot their own course between checkpoints. It usually covers a distance of 5 to 10 miles, often through rough terrain and may last for an hour or more.

TABLE 9-2

Data on performance of present and former athletes in orientation racing. The youngest group is currently competing. The active older athletes are still training and competing regularly. The inactive subjects had, when young, competed successfully with those of the same age who are still active; they discontinued their training more than 10 years ago due to lack of time. The figures at the top denote the number of subjects in each group.

Function	Age						
	20–30	40–49		50–59		60–69	
	active 9	active 15	non-active 10	active 14	non-active 14	active 4	non-active 5
Max O₂ uptake							
liters/min	5.4	4.0	3.3	3.4	2.9	2.7	2.6
ml/kg × min	77	57	44	53	38	43	37
Heart volume, ml		1,050	835	940	915	830	865
Max heart rate		175	182	176	175	165	170
Cholesterol, mg%		222	231	251	277	286	266
Neutral fat serum,							
mmoles/liter		0.85	1.56	0.95	1.44	1.10	1.85
Blood pressure, mm Hg		135/83	128/82	137/81	133/82	138/83	123/86
ECG IV: 1–3*		2	1	4	2	1	0

* Classified according to the Minnesota Code (see I. Åstrand et al., 1967, reference in Chap. 11).

Data from Saltin and Grimby, 1968.

aerobic power is concerned, particularly in relation to body weight. In all groups the mean values are higher than normally found for the same age groups (Figs. 9-12 and 9-15). In the sedentary group this is due to natural endowment (a highly selected group), while in those who are still active this endowment is further developed by regular physical training. There is in both groups a decline in maximal heart rate with age.

The heart volume in relation to maximal oxygen uptake was significantly larger in the older athletes and former athletes compared to healthy young males. The cholesterol level in the serum was not different from normal values, but the concentration of the serum neutral fat was lower in most athletes than the normal mean values.

In a separate study with nine well-trained athletes between the ages of forty-five to fifty-five years, cardiac output was also measured during exercise (Grimby et al., 1966). The maximal cardiac output was very high or on an average 26.8 liters/min at an oxygen uptake of 3.56 liters/min. The stroke volume was also high (average 163 ml) considering the age of the men. A low maximal arteriovenous oxygen difference seemed to be the main limiting factor for the oxygen uptake. This was partly due to a relatively low hemoglobin concentration, but probably also due to peripheral factors, possibly including an increased diffusing distance from the capillaries in the aged skeletal muscle.

During prolonged heavy physical work, the individual's performance capacity depends largely upon his ability to take up, transport, and deliver oxygen to the working muscle. Consequently, the maximal oxygen uptake is probably the best laboratory measure of a person's physical fitness, providing the definition of physical fitness is restricted to the capacity of the individual for prolonged heavy work (Herbst, 1928; Dill, 1933; P.-O. Åstrand, 1952, 1956).

For decades a discussion has been going on concerning the "limiting factors" in maximal oxygen uptake (P.-O. Åstrand, 1952, 1956), whether it is the oxygen content of the inspired air, the pulmonary ventilation, the diffusion of oxygen from alveolar space to hemoglobin, the hemoglobin content, the blood volume, the ability of the heart to pump blood, the distribution of blood flow, the ability of muscle tissues to receive the offered blood, the diffusion from capillaries to the working cells, the venous blood return, the efficiency of the mitochondria to transfer aerobic energy to the ATP-ADP machinery, access to fuel, the function of the neuromuscular system, or motivation.

We still do not know how critical the various factors are. It should be emphasized that there is a different situation in a 5-min effort as compared to a 3-hr performance.

In other sections of the book many of the listed factors are discussed and analyzed from a viewpoint of oxygen transport.

Summary The individual's maximal oxygen uptake gives a measure of the "motor effect" of his aerobic processes, i.e., his maximal aerobic power. When related to body weight, the ability to move the body can be evaluated. A calculation of the maximal oxygen uptake per kilogram fat-free body weight, or related to muscle mass, blood volume, or other such parameters, makes it possible to analyze dimensions versus function.

In prolonged exercise there is a high correlation between maximal oxygen uptake and total work output (maximal aerobic capacity). The actual oxygen uptake that can be tolerated is at a certain percentage of the maximum, this percentage being lower the longer the work time.

The maximal oxygen uptake (maximal aerobic power) increases with age up to twenty years. Beyond this age there is a gradual decline so that the sixty-year-old individual attains about 70 percent of the maximum at twenty-five years. Before the age of twelve there is no significant difference between girls and boys; thereafter the average difference in maximal oxygen uptake between women and men amounts to 25 to 30 percent. Related to the body weight the sex difference in aerobic power after puberty is 15 to 20 percent.

Top athletes in endurance events have a maximal oxygen uptake that is about twice as high as in the average man.

The gradual decline in maximal oxygen uptake with age beyond twenty is at least partly due to a decrease in maximal heart rate. Inactivity is another factor that decreases the functional range of the oxygen transporting system. Inactivity reduces the stroke volume and perhaps the efficiency of the regulation of the circulation during exercise.

It is presently impossible to point at decisive limiting factor(s) for maximal oxygen uptake.

PSYCHOLOGICAL FACTORS

Motivation or drive is the neural process which impels the individual to certain actions in pursuit of specific objectives. It plays an important role in human performance and may be the most important key to success, for abilities and physical capacities alone may be of little use unless the individual is motivated to devote all his endowment and capacity to their full limits in the attainment of specific goals. Superior performance may, on the other hand, be impossible to attain if the physical capacity of the body is limited, regardless of motivation.

It is beyond the scope of this book to analyze this exceedingly complex question, and the reader is referred to the psychological literature for a discussion of the psychological factors affecting performance.

REFERENCES

Asmussen, E.: Aerobic Recovery after Anaerobiosis in Rest and Work, *Acta Physiol. Scand.*, **11**:197, 1946.

Åstrand, I.: Aerobic Work Capacity in Men and Women with Special Reference to Age, *Acta Physiol. Scand.*, **49**(Suppl. 169), 1960.

Åstrand, I.: Degree of Strain during Building Work as Related to Individual Aerobic Work Capacity, *Ergonomics*, **10**:293, 1967a.

Åstrand, I.: Aerobic Working Capacity in Men and Women in Some Professions, *Försvarsmedicin*, **3**:163, 1967b.

Åstrand, I., P.-O. Åstrand, and K. Rodahl: Maximal Heart Rate during Work in Older Men, *J. Appl. Physiol.*, **14**:562, 1959.

Åstrand, I., P.-O. Åstrand, E. H. Christensen, and R. Hedman: Intermittent Muscular Work, *Acta Physiol. Scand.*, **48**:443, 1960a.

Åstrand, I., P.-O. Åstrand, E. H. Christensen, and R. Hedman: Myohemoglobin as an Oxygen-store in Man, *Acta Physiol. Scand.*, **48**:454, 1960b.

Åstrand, P.-O.: "Experimental Studies of Physical Working Capacity in Relation to Sex and Age," Ejnar Munksgaard, Copenhagen, 1952.

Åstrand, P.-O.: The Respiratory Activity in Man Exposed to Prolonged Hypoxia, *Acta Physiol. Scand.*, **30**:343, 1954.

Åstrand, P.-O.: Human Physical Fitness with Special Reference to Sex and Age, *Physiol. Rev.*, **36**:307, 1956.

Åstrand, P.-O., and B. Saltin: Oxygen Uptake during the First Minutes of Heavy Muscular Exercise, *J. Appl. Physiol.*, **16**:971, 1961a.

Åstrand, P.-O., and B. Saltin: Maximal Oxygen Uptake and Heart Rate in Various Types of Muscular Activity, *J. Appl. Physiol.*, **16**:977, 1961b.

Åstrand, P.-O., L. Engström, B. Erikson, P. Karlberg, I. Nylander, B. Saltin, and C. Thorén: Girl Swimmers, *Acta Paediat.* (Suppl. 147), 1963a.

Åstrand, P.-O., J. Hallbäck, R. Hedman, and B. Saltin: Blood Lactates after Prolonged Severe Exercise, *J. Appl. Physiol.*, **18**:619, 1963b.

Åstrand, P.-O., and E. H. Christensen: Aerobic Work Capacity, in F. Dickens, E. Neil, and W. F. Widdas (eds.), "Oxygen in the Animal Organism," p. 295, Pergamon Press, New York, 1964.

Bang, O.: The Lactate Content of the Blood during and after Muscular Exercise in Man, *Skand. Arch. Physiol.*, **74**(Suppl. 10):51, 1936.

Beznak, M.: The Role of Anterior Pituitary Hormones in Controlling Size, Work, and Strength of the Heart, *J. Physiol. (London)*, **150**:251, 1960.

Carlsten, A., B. Hallgren, R. Jagenburg, A. Svanborg, and L. Werkö: Myocardial Metabolism of Glucose, Lactic Acid, Amino Acids and Fatty Acids in Healthy Human Individuals at Rest and at Different Work Loads, *Scand. J. Clin. Lab. Invest.*, **13**:418, 1961.

Christensen, E. H., and O. Hansen: Arbeitsfähigkeit und Ehrnährung, *Skand. Arch. Physiol.*, **81**:160, 1939.

Christensen, E. H., and P. Högberg: Physiology of Skiing, *Arbeitsphysiol.*, **14**:292, 1950.

Christensen, E. H., R. Hedman, and B. Saltin: Intermittent and Continuous Running, *Acta Physiol. Scand.*, **50**:269, 1960.

Dill, D. B.: The Nature of Fatigue, *Personnel*, **9**:113, 1933.

Döbeln, W. v.: Human Standard and Maximal Metabolic Rate in Relation to Fat-free Body Mass, *Acta Physiol. Scand.*, **37**(Suppl. 126), 1956.

Grimby, G., N. J. Nilsson, and B. Saltin: Cardiac Output during Submaximal and Maximal Exercise in Active Middle-aged Athletes, *J. Appl. Physiol.*, **21**:1150, 1966.

Hedman, R.: The Available Glycogen in Man and the Connection between Rate of Oxygen Intake and Carbohydrate Usage, *Acta Physiol. Scand.*, **40**:305, 1957.

Herbst, R.: Der Gasstoffwechsel als Mass der körperlichen Leistungs-fähigkeit, I, Mitteilung, Die Bestimmung des Sauerstoffaufnahmevermögens beim Gesunden, *Deutsch. Arch. Klin. Med.*, **162**:33, 1928.

Hermansen, L., and B. Saltin: Blood Lactate Concentration during Exercise at Acute Exposure to Altitude, in R. Margaria (ed.), "Exercise at Altitude," Exerpta Medica Foundation, New York, 1967.

Hollmann, W.: "Höchst- und Dauerleistungsfähigkeit des Sportlers," Johann Ambrosius Barth, Munich, 1963.

Huckabee, W. E.: Relationship of Pyruvate and Lactate during Anaerobic Metabolism, II, Exercise and Formation of O_2 Debt, *J. Clin. Invest.*, **37**:255, 1958.

Hultman, E.: Studies on Muscle Metabolism of Glycogen and Active Phosphate in Man with Special Reference to Exercise and Diet, *Scand. J. Clin. Lab. Invest.*, **19**(Suppl. 94), 1967.

Issekutz, B., Jr., H. I. Miller, P. Paul, and K. Rodahl: Effect of Lactic Acids and Glucose Oxidation in Dogs, *Am. J. Physiol.*, **209**:1137, 1965.

Karpatkin, S., E. Helmreich, and C. F. Cori: Regulation of Glycolysis in Muscle, *J. Biol. Chem.*, **239**:3139, 1964.

Keul, I., D. Keppler, and E. Doll: Standard Bicarbonate, pH, Lactate and Pyruvate Concentrations during and after Muscular Exercise, *German Medical Monthly*, **12**:156, 1967.

Keul, J., E. Doll, H. Steim, U. Fleer, and H. Reindell: Über den Stoffwechsel des Herzens bei Hochleistungssportlern III. Der oxydative Stoffwechsel des trainierten menschlichen Herzens unter verschiedenen Arbeitsbedingungen, *Ztschr. Kreislaufforsch.*, **55**:477, 1966.

Knuttgen, H. G.: Oxygen Debt, Lactate, Pyruvate, and Excess Lactate after Muscular Work, *J. Appl. Physiol.*, **17**:639, 1962.

Koeppe, R. E., N. F. Inciardi, L. G. Warnock, and W. E. Wilson: Some Aspects of Metabolism of D- and L-Lactic Acid-2^{14}C by Rat Skeletal Muscle in Vivo, *J. Biol. Chem.*, **239**:3609, 1964.

Krebs, H.: "Oxygen in the Animal Organism," F. Dickens, E. Neil, and W. F. Widdas (eds.), p. 304, Pergamon Press, New York, 1964.

Levy, M. N.: Uptake of Lactate and Pyruvate by Intact Kidney of the Dog, *Am. J. Physiol.*, **202**:302, 1962.

Lundin, G., and G. Ström: The Concentration of Blood Lactic Acid in Man during Muscular Work in Relation to the Partial Pressure of Oxygen of the Inspired Air, *Acta Physiol. Scand.*, **13**:253, 1947.

Margaria, R.: Aerobic and Anaerobic Energy Sources in Muscular Exercise, in R. Margaria (ed.), "Exercise at Altitude," Excerpta Medica Foundation, New York, 1967.

Margaria, R., H. T. Edwards, and D. B. Dill: The Possible Mechanism of Contracting and Paying the Oxygen Debt and the Role of Lactic Acid in Muscular Contraction, *Am. J. Physiol.*, **106**:689, 1933.

Michailov, V. V., and I. G. Ogoltsov: The Maximal Oxygen Uptake in Russian Cross-country Skiers of Olympic Caliber, *Fed. Inst. Phys. Ed.*, Moscow, 1965.

Mitchell, J. H., B. J. Sproule, and C. B. Chapman: Physiological Meaning of the Maximal Oxygen Intake Test, *J. Clin. Invest.*, **37**:538, 1958.

Newman, E. V., D. B. Dill, H. T. Edwards, and F. A. Webster: The Rate of Lactic Acid Removal in Exercise, *Amer. J. Physiol.*, **118**:457, 1937.

Omachi, A., N. Lifson, S. L. Michel, and J. A. Swanson: Metabolism of Isotopic Lactate by the Isolated Perfused Dog Gastrocnemius, *Am. J. Physiol.*, **185**:35, 1956.

Robinson, S.: Experimental Studies of Physical Fitness in Relation to Age, *Arbeitsphysiol.*, **10**:251, 1938.

Rodahl, K., P.-O. Åstrand, N. C. Birkhead, T. Hettinger, B. Issekutz, Jr., D. M. Jones, and R. Weaver: Physical Work Capacity, *Arch. Environ. Health.*, **2**:499, 1961.

Rodahl, K., and B. Issekutz, Jr.: Physical Performance Capacity in the Older Individual, in "Muscle as a Tissue," chap. 15, McGraw-Hill Book Company, New York, 1962.

Rodahl, K., H. I. Miller, and B. Issekutz, Jr.: Plasma Free Fatty Acids in Exercise, *J. Appl. Physiol.*, **19**:489, 1964.

Rowell, L. B., H. L. Taylor, and Y. Wang: Limitations to Prediction of Maximal Oxygen Uptake, *J. Appl. Physiol.*, **19**:919, 1964.

Rowell, L. B., K. K. Kraning II, T. O. Evans, J. W. Kennedy, J. R. Blackmon, and F. Kusumi: Splanchnic Removal of Lactate and Pyruvate during Prolonged Exercise in Man, *J. Appl. Physiol.*, **21**:1773, 1966.

Saltin, B.: Aerobic Work Capacity and Circulation at Exercise in Man, *Acta Physiol. Scand.*, **62**(Suppl. 230), 1964.

Saltin, B., and P.-O. Åstrand: Maximal Oxygen Uptake in Athletes, *J. Appl. Physiol.*, **23**:353, 1967.

Saltin, B., and G. Grimby: Physiological Analysis of Middle-aged and Old Former Athletes: Comparison with Still Active Athletes of the Same Ages, *Circulation*, **38**:1104, 1968.

Stenberg, J., P.-O. Åstrand, B. Ekblom, J. Royce, and B. Saltin: Hemodynamic Response to Work with Different Muscle Groups, Sitting and Supine, *J. Appl. Physiol.*, **22**:61, 1967.

Strandell, T.: Circulatory Studies on Healthy Old Men, *Acta Med. Scand.* (Suppl. 414), 1964.

Taylor, H. L., E. Buskirk, and A. Henschel: Maximal Oxygen Uptake as an Objective Measure of Cardiorespiratory Performance, *J. Appl. Physiol.*, **8**:73, 1955.

Williams, C. G., C. H. Wyndham, R. Kok, and M. J. E. von Rahden: Effect of Training on Maximum Oxygen Intake and on Anaerobic Metabolism in Man, *Int. Z. angew. Physiol. einschl. Arbeitsphysiol.*, **24**:18, 1967.

Wyndham, C. H., C. G. Williams, and M. von Rahden: A Physiological Basis of the "Optium" Level of Energy Capacities, *Nature*, **195**:1210, 1962.

10

Body Dimensions and Muscular Work

contents

chapter ten

Body Dimensions and Muscular Work

The thrill of watching athletic competitions is partly caused by the fact that it is difficult as a rule to predict who is going to win.

It is impossible from appearance alone to tell who is an athletic champion. It is on the other hand often possible to exclude those who obviously *cannot* reach top results in certain sport events, such as shot putting, rowing, and American football. A tiny individual hardly has a chance in these events.

It is of interest to consider the human resources in relation to muscular work from a biological viewpoint. If we compare animals of different size, it is evident that certain dimensions and functional capacities are determined by fundamental mechanical necessities. In addition it may be a matter of biological adaptation.

STATICS

If we take two geometrically similar cubes of different size, the relationship between the surface and the volume of the two cubes can easily be calculated if only the scale factor between the sides of the cubes is known. If this length scale is $L:1$, the surface ratio is $L^2:1$, and the volume ratio $L^3:1$.

If we consider two geometrically similar and qualitatively identical individuals, we may expect all linear dimensions (L) to be proportional. The length of the arms, the legs, the trachea, and the individual muscles will have a ratio $L:1$. If we compare two boys, one 120 cm high, the other 180 cm high, the scale factor will be such that all lengths, levers, ranges of joint motions, and muscular contractions during a specific motion will be related as $120:180$ or as $1:1.5$ (Fig. 10-1).

Cross sections of, for instance, a muscle, the aorta, a bone, the trachea, the alveolar surface, or the surface of the body are then related as $120^2:180^2$, or

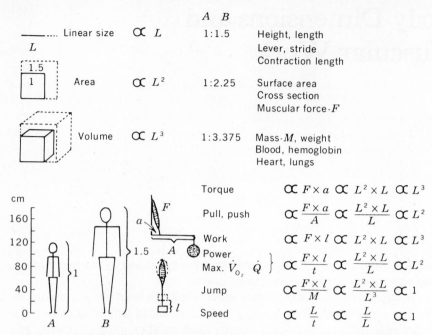

Fig. 10-1 Schematic illustration of the influence of dimensions on some static and dynamic functions in geometrically similar individuals. A and B represent two persons with body height 120 and 180 cm respectively. See text for explanation. (Partly modified from Asmussen and Christensen, 1967.)

$1^2:1.5^2$, i.e., 1:2.25. Volumes such as lung volumes, blood volumes, or heart volumes should similarly be related as $120^3:180^3$, or $1^3:1.5^3$, i.e., 1:3.375. The same applies to mass measured in units of weight, since the density of biological materials generally speaking is independent of size.

The force of gravity acts on the mass (M) of the body. If the limbs supporting this mass are proportional to the area of their cross section, there is a disproportion between the mass (which is proportional to L^3) and the support (which is proportional to L^2). In larger animals an adjustment has taken place: elephants and rhinoceroses have relatively thick, short, and straight legs. In body geometry these larger animals differ from smaller animals which have slender limbs compared to their body mass. In relation to total weight man has over double as much bone as the mouse (Thompson, 1943). Those of the larger animals which have relatively slender limbs have "compressed" bodies compared to their length (giraffe, antelope). In large animals, the design of their bodies is dominated by the necessity of supporting their weight. This was pointed out by Galileo already in 1638.

DYNAMICS

If we were to pursue the theoretical considerations above further, it is evident that the maximal force a muscle can develop (F) generally speaking is proportional to ($\propto$) its surface. It would therefore be expected that the 1.5-times taller boy should be able to produce a 2.25-times larger muscular strength, i.e., he should be able to lift a 2.25-times larger weight. The advantage of the 1.5-times longer levers (a) for the muscles to work on possessed by the taller boy is offset by the fact that the weight to be lifted also has a 1.5-times longer lever (A). (According to Fig. 10-1:

$$F \times a = M \times A$$
$$M = \frac{F \times a}{A}$$

In accordance with the discussion above $F \propto L^2$; $a \propto L$; $A \propto L$. The above equation can thus be expressed as follows: $M \propto \dfrac{L^2 \times L}{L} \propto L^2$.)

Force

The magnitude of the work to be performed is determined by the developed force ($\propto L^2$) and the distance the force is applied ($\propto L$). Consequently $W \propto L^2 \times L \propto L^3$. Thus the work which the larger boy in our example should be able to perform is accordingly 3.375 times larger than that which would be expected in the case of the smaller boy.

"Chin-ups"

If it is a matter of lifting one's own body with a mass (M), as in the case of chinning the bar, the formula can be expressed as follows: $F \times a = M \times A$, where a and A represent the levers for the muscles and the body weight respectively. In other words, the achievement is proportional to the force of the muscles and their levers, but inversely proportional to the body mass and the levers upon which it works. According to the reasoning which we have applied above, it follows that the ability to lift one's own body (i.e., do a chin-up) is proportional to

$$\frac{F \times a}{M \times A} \quad \text{or} \quad \frac{L^2 L}{L^3 L} \propto \frac{1}{L} \propto L^{-1}$$

Thus, in the case of the smaller boy the ratio will be $1:1 = 1$, but in the case of the larger boy the ratio is $1:1.5 = 0.67$, i.e., the larger and stronger boy is actually handicapped by his greater body weight when he has to lift his body, as in the case of chinning the bar.

Time

In many dynamic events an expression of the time scale (t) is necessary. It may be a matter of energy transfer per unit time, the time between two steps or between two heartbeats. It might be of interest to examine this time scale for similar animals of different sizes. If the ratio is expressed as $t:1$, the relationship between time and acceleration (a) according to the definition of the acceleration is $a \propto L \times t^{-2}$. The connection between the time scale and the length scale may then be calculated by the following formula:

$$\text{Force} = \text{acceleration} \times \text{mass}$$

$$F = a \times M \qquad a = \frac{F}{M} \qquad a \propto \frac{L^2}{L^3}$$

$$\propto \frac{1}{L} \; (\propto L^{-1})$$

(Döbeln, 1966). We may now return to the formula $a \propto L/t^2$ or $t^2 \propto L/a$. If we now replace a with L^{-1} as shown above, we arrive at the following formula:

$$t^2 \propto \frac{L}{L^{-1}} \qquad \text{or} \qquad t^2 \propto L^2 \qquad t \propto L$$

In other words: *The time scale is proportional to the length scale.*

Acceleration

Since a is proportional to L^{-1}, the taller (and heavier) person is handicapped when it is a matter of accelerating his body mass.

Frequencies

The shorter the time between two steps, heartbeats, etc., the higher the frequency (f), which is another expression of the passing of time. This may be expressed as follows: $f \propto 1/t$, and accordingly, $f \propto 1/L \; (\propto L^{-1})$.

According to this reasoning we might expect that the frequency of limb motion should vary as an inverse function of limb length. As pointed out by Hill (1950) this is generally the case. A hummingbird moves its wings about 75 to 100 times/sec while flying forward, a sparrow some 15 times/sec, and a stork only 2 or 3 times. These frequencies are roughly in inverse proportion to the linear size of the birds. "If the sparrow's muscles were as slow as the stork's it would be unable to fly. If the stork's muscles were as fast as the hummingbird's it would be exhausted very quickly" (Hill, 1950). The maximal force of a contracting voluntary muscle is roughly constant in different animals, being of the order of a few kilograms per square centimeter cross-sectional area (Chap. 4). The speed of contraction varies enormously, however, between different muscles and different animals. The balance between muscle strength, length of levers, and speed of contraction is very delicate. If man had muscles as fast as those moving wings of a hummingbird, they would soon break the bones and tear the muscles and the tendons. It is possible to swing a rod made of fragile material back and forth if it is short and relatively thick. But if it is long and with a thickness just proportional to L^2, the inertia will cause it to break if moved at the same speed. For the same reason a small motor may run at higher revolutions per minute than a large one, and the strength of the material is an important factor determining the maximal speed. For similar reasons a smaller creature can tolerate a greater quickness of movement than a larger one. In fact the margin of safety is actually quite small, so that it does happen occasionally that muscles tear, tendons break, and bones splinter during unusual strain, such as during strenuous athletic events. *Without altering the general design* it would be highly hazardous for the athlete's locomotor organs if his muscles could be altered to allow him to run say 25 percent faster.

Running Speed

The speed which may be attained in moving the body is determined, among other things, by the length of stride and the number of movements per unit of time ($\propto 1/L$). Thus, for similar animals of different size the maximal speed is proportional to $L/L = 1$, which means that the speed is the same. Short limbs with short strides move more rapidly and can therefore cover as much ground as do longer ones moving more slowly.

It is well known that "athletic animals" of different size may achieve approximately the same speed. A blue whale of 100 tons and a dolphin of 80 kg attain the same "steady state" speed of about 15 knots, and a maximal speed of about 20 knots. The speed of a whippet, a greyhound, or a racehorse, very similar in general design, is nearly the same, or about 40 mph (Hill, 1950). Gazelles and antelopes with wide variation in size are all able to reach a maximal speed of about 50 mph.

Jumping

In broad jump and high jump it is also a question of the maximal muscular force which may be developed and the distance which the muscle can shorten before the body leaves the ground. We should therefore expect the performance to be proportional to $L^2 \times L/L^3 = 1$, i.e., a small and a large animal should be equally able to lift their center of gravity. In broad jumping, kangaroos, jackrabbits, horses, mule deer, and impala antelopes actually seem to be equal to the record-holding man (Hill, 1950). Borelli drew similar conclusions some 300 years ago (1685). In high jumping, in which the aim is to lift the body as high as possible, the larger animal has an advantage, however, since its center of gravity before the jump is already at a higher level, " . . . a fact to remember sympathetically in assessing the jumping performance of small boys" (Hill, 1950).

The outstanding high jumper is without exception a tall individual. He is usually not geometrically similar to the average man, but he is long-legged, with a body weight that is not proportional to L^3 but lower.

Maximal Running Speed for Children

We might expect that the 180-cm boy would perform better in high jump than the 120-cm-tall boy, which he actually does. Their ability to move their center of gravity vertically should, on the other hand, not be different, nor should their maximal speeds. In an analysis of speed in children of different size Asmussen has divided them into age groups and plotted maximal speed, calculated from the best time on 50 to 100 m, in relation to body height (Fig. 10-2). From the age of about ten the body proportions are about the same. We may therefore consider the children represented in Fig. 10-2 as geometrically similar (Fig. 10-3).

For 11- to 12-year old boys there is no significant variation in speed with body size, which would not be expected from the discussion above. The some-

Fig. 10-2 Maximal speed in relation to body height for girls and boys of different age; almost 100,000 subjects are included in the statistics. See text for explanation. (Modified from Asmussen and Christensen, 1967.)

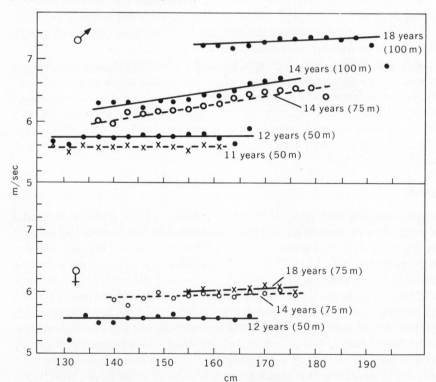

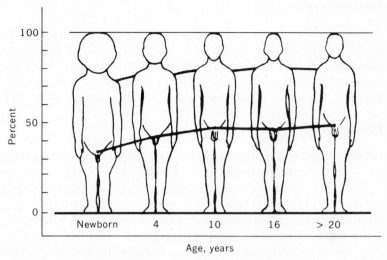

Fig. 10-3 Body proportions in different ages. Note that from the age of ten years there is no marked change in proportions. (From Asmussen and Christensen, 1967.)

what better performance of the 12-year-olds to the 11-year-old boys may be due to maturity of the neuromuscular function, improving the coordination. Even better are 14-year-old boys, but one also found that the taller boys can run faster than the shorter ones. There is a further improvement in coordination with age, but this is also probably due to sexual maturity. Their male sex hormones may have influenced their muscular strength in a positive direction. The smaller 14-year-old boys may not have reached puberty, in contrast to the taller boys. In the 18-year-old group there is again hardly any variation in results in spite of a large difference in body height. At this age all the boys have passed puberty and are sexually mature.

In the girls there is an increase in maximal speed up to the age of 14, but from then on there is no further improvement. The results are not influenced by the size of the girls in any of the age groups, which supports the assumption that the superiority of the taller 14-year-old boys is due to the effect of male sex hormones.

This independence of maximal speed with body height is in contrast to the greater muscle strength in taller children, illustrated by Fig. 4-31. However, from an anatomic viewpoint this is just what would be expected: the muscle force should increase in proportion to L^2, but the speed should be independent. As already discussed, the results obtained for boys of different size do not strictly follow the results predicted from body dimensions. Apparently biological factors may modify muscular dynamics. We have considered an age factor as well as sexual maturity, which is particularly evident for the boys' performance.

Kinetic Energy

The kinetic energy (KE) developed in a limb depends upon its mass and the square of its velocity, or KE $= M \times v^2$; but $v = L/t$ and $t \propto L$; therefore: KE $\propto M \times L^2/L^2 \propto M$ (or L^3). It follows that the work done during a single movement, calculated per unit of body weight, in producing and utilizing kinetic energy in the limbs, should be the same in large and small animals (Hill, 1950).

External Work

Hill also points out that the external work done in overcoming the resistance of air or water is proportional to the square of the linear dimensions, i.e., the surface area. Therefore the effect of this resistance should be the same in similar animals of different size. In running uphill at a given speed the effect of the slope is inversely proportional to the linear size, L. The smaller the animal the faster it should be when running uphill.

We have concluded that the maximal speed is independent of the animal's size. If one animal is 1,000 times as heavy as another, a movement will be 10 times greater than in the smaller animal. However, it only has to take $\frac{1}{10}$ of the number of steps, each step taking 10 times as long, in order to attain the same speed as the smaller animal. Since the work per movement and per unit of body weight is the same in the two animals, it follows that it will take roughly 10 times as long for the larger animal to become exhausted during a maximal run. [It may be that 10 sec to the larger animal, physiologically speaking, is the same as 1 sec to the smaller animal, so that the energy output at rest and during work per kilogram of body weight and per unit of "individual" time is the same (Hill, 1950). This individual time should be considered when comparing the life-span of animals of different size.]

Energy Supply

It is obviously important that the power output permitted by the mechanical design be matched by an equivalent supply of chemical energy.

As mentioned above, work (being the product of force and distance moved) is proportional to $L^2 \times L = L^3$, or M. The total energy output should therefore be related to the mass of the muscles and the body weight in similar animals. The *power*, i.e., work output per unit of time, must then be proportional to $L^3 \times t^{-1}$, i.e., $L^3 \times L^{-1} \propto L^2$ ($\propto M^{2/3}$).

It is well established that the *basal metabolic rate* in animals with large differences in body size, from the mouse to the elephant, follows this prediction. The resting oxygen uptake is actually proportional to $M^{0.74}$ rather than to $M^{2/3}$, but this difference is surprisingly small considering the wide variation in size, shape, and other factors (Brody, 1945). This relation tells us that smaller animals must be more active metabolically per unit of body weight than larger ones. In proportion to its weight a mouse has to eat 50 times more than a horse in order to maintain its basic activity.

Theoretically speaking it would be expected that the *maximal oxygen uptake* should be proportional to $L^2(M^{2/3})$. Since maximal *cardiac output* and *pulmonary ventilation* are also volumes per unit of time they should be proportional to $L^3 \times L^{-1} = L^2$ or $M^{2/3}$. There is a different approach to analyze this relationship giving the same result. Cardiac output is the product of the frequency of heartbeats and stroke volume. Frequency is proportional to L^{-1} and stroke volume to L^3 and therefore $\dot{Q} \propto L^{-1} \times L^3 \propto L^2$.

Similarly pulmonary ventilation is the product of respiratory frequency and tidal air: $\dot{V}_E \propto L^{-1} \times L^3 \propto L^2$.

With pulmonary ventilation proportional to $\dot{V}_{O_2}$ and with the production of CO_2 proportional to $\dot{V}_{O_2}$ we should expect the alveolar P_{CO_2} and P_{O_2} to be the same in different mammals, which is actually the case. Asmussen calculated that the vital capacity measured in subjects seven to thirty years of age (P.-O. Åstrand, 1952) was proportional to $L^{3.1}$ in males and $L^{3.0}$ in females, i.e., very close to the expected $L^{3.0}$. It may therefore be concluded that the children have lung volumes which are dimensioned to their body size.

In heavy exercise the *heat production* is very great and related to $\dot{V}_{O_2}$ or L^2, i.e., the surface of the body from which most of the excess heat is lost, at least in man. There is also heat loss via the expired air, increasing, within limits, in direct proportion to $\dot{V}_{O_2}$.

Döbeln (1956*b*) points out that a dimensional analysis should be based on body weight minus adipose tissue, since fat is metabolically inactive. For a group of 65 young female and male subjects he calculated the value of the exponent b in the equation maximal oxygen uptake $= a \times$ (body weight $-$ adipose tissue)^b and found that $b = 0.71$ (maximal oxygen uptake predicted from the Åstrand and Åstrand nomogram; adipose tissue calculated by means of hydrostatic weighing).

Maximal Aerobic Power in Children

There are very few data available on maximal values for oxygen uptake and cardiac output in animals of different size to test these hypotheses. In fully grown man the variations in body size is rather limited. For the male subjects 8 to 18 years of age, studied by P.-O. Åstrand (1952), Asmussen has calculated that the maximal attainable oxygen uptake is proportional to $L^{2.9}$ which is much higher than the expected $L^{2.0}$. In other words, children attained a relatively low maximal oxygen uptake.

The maximal oxygen uptake in 7- to 9-year old boys was on an average 1.75 liters/min or 56.9 ml/kg $\times$ min, body weight (M), 30.7 kg. For 16- to 18-year-olds the maximal oxygen uptake was 3.68 liters/min or 57.6 ml/kg $\times$ min, body weight, 64.1 kg (P.-O. Åstrand, 1952). If we calculate the maximal oxygen uptake in the children assuming that dimensions were the deciding factor according to the formula $\dot{V}_{O_2} = a \times M^{2/3}$ (where a is a constant which can be calculated from $3.68 = a \times 64.1^{2/3}$), we find that they should attain 2.26 liters/min or 70 ml/kg $\times$ min. (Calculated from the formula $\dot{V}_{O_2} = a \times L^2$, the maximal oxygen uptake of the child is 2.14 liters/min.) We may conclude that the children's maximal oxygen uptake is not as high as expected for their size and that they do not have the aerobic power to handle their weight compared to adults. It is also significant that the 8-year-old child could increase his basal metabolic rate only 9.4 times during maximal running for 5 min, but the 17-year-old boy could attain an aerobic power which was 13.5 times the basal power. Therefore the child has less in the way of a power reserve than adults. It should also be emphasized that the young subjects had a significantly higher oxygen uptake per kilogram of body weight than the older boys and adults when running at a given speed on a treadmill (P.-O. Åstrand, 1952). These two factors together may explain the fact that children have difficulty in following their parents' speed, even if maximal oxygen uptake per kilogram body weight may be the same. The children's lower efficiency can partly be explained by their high stride frequency which is an expensive utilization of energy per unit of time.

In Chap. 4 we discussed the relatively low muscular strength of children. It is conceivable that their aerobic power may be adapted to their muscular machine. In his analyses Asmussen found that muscular strength in the 8- to 16-year-old boys is proportional to $L^{2.89}$ or exactly the same as the maximal oxygen uptake ($\propto L^{2.90}$).

Since the oxygen is transported by the hemoglobin, it is of interest to compare the children's total amount of hemoglobin (Hb_T) with their body size and maximal oxygen uptake. In most of the subjects just discussed the total hemoglobin was determined. In similar animals of different size Hb_T should be proportional to M. However, per kilogram of body weight the younger boys had only 78 percent of the amount of hemoglobin of the older boys. Thus the amount of hemoglobin was definitely not proportional to body size. Assuming that the maximal oxygen uptake is proportional to Hb_T we may calculate the maximal $\dot{V}_{O_2}$ in children from the equation $\dot{V}_{O_2} = a \times Hb_T^{2/3}$ and use a value for a calculated from the data on older boys. It is found that the child's maximal oxygen uptake should be 1.92 liters/min, if the child's hemoglobin is as effective in transporting oxygen as in the adult individual. This calculated value is not far from the determined 1.75 liters/min. Using the exponent 0.74, the calculated maximal aerobic power in the 7- to 9-year-old children will be 1.78 liters/min, or very close to real maximum.

The sample of subjects selected for these analyses is limited to 21 individuals, but it is a homogeneous group. They were nonobese and in the same state of training. The results support the assumption that the maximal oxygen uptake in children and young adults is proportional to the muscular strength and to $Hb_T^{0.76}$, or roughly to $Hb_T^{2/3}$, but *not* to $M^{2/3}$.

It may be concluded that children are definitely physically handicapped compared to adults (and fully grown animals of similar size). When related to the child's dimensions, its muscular strength is low and so is its maximal oxygen uptake and other parameters of importance for the oxygen transport. Furthermore the mechanical efficiency of children is often inferior to that of the adults. The introduction of dimensions in the discussion of children's performance clearly indicates that they are not mature as working machines.

Maximal Aerobic Power in Women

For the female subjects 8 to 16 years of age, the maximal $\dot{V}_{O_2}$ is proportional to $L^{2.5}$. In light of the previous discussion the noted discrepancy from the expected L^2 is not surprising.

The reader is referred to the discussion of Fig. 6-25, the conclusion of which is that maximal stroke volume in well-trained women and men is related to heart volume as expected from the dimensions (both being proportional to L^3); the maximal cardiac output is proportional to the heart volume raised to the 0.76 power, which is close to the expected 0.7. However, the oxygen uptake in the smaller subjects, mostly women, is lower than expected from the dynamic dimensions.

Women have approximately the same maximal oxygen uptake per kilogram fat-free body mass as men. However, it should be higher in women due to their smaller size (Döbeln, 1956a and b). The lower Hb concentration in women may explain why they cannot fully utilize their cardiac output for oxygen transport.

Figure 10-4 presents data on about 227 children and young adults, stressing these points further. There is a very high correlation between maximal oxygen uptake and body weight for male, nonobese subjects (upper figure). The lower maximal oxygen uptake for female subjects above 40 kg of body weight (age about 14 years) is largely explained by their higher content of adipose tissue. The lower concentration of hemoglobin in the women's blood also contributes to the observed difference between the sexes. When one relates the total amount of hemoglobin to the maximal oxygen uptake, the difference between regression lines for the female and male subjects is insignificant (lower figure). It should be noted that the exponent b in the equation $y = a \times x^b$ (i.e., maximal $\dot{V}_{O_2} = a \times Hb_T{}^b$) is 0.76.

Maximal Cardiac Output

From this analysis it is evident that maximal $\dot{V}_{O_2}$ should be proportional to L^2 (or $M^{2/3}$). However, data on children and adults do not support this assumption, since the exponent is closer to 3. There are, nevertheless, reasons to believe that biological factors may have modified the aerobic power in the subjects studied. Since $\dot{V}_{O_2}$ must be related to cardiac output, it should be emphasized that if the total output of the heart per minute was proportional to the body weight (L^3), the blood velocity in the aorta would have to be so great in the largest mammals that the heart would be faced with an impossible task (Hoesslin, 1888; Hill, 1950). It is therefore more likely that the cardiac output is proportional, not to the body mass, but to $M^{2/3}$ (or L^2), like the basal metabolism.

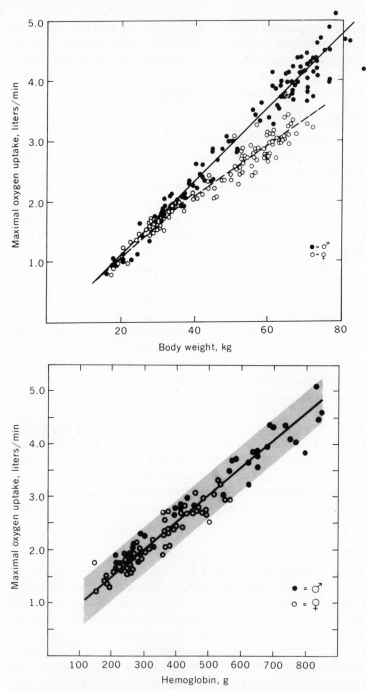

*Fig. 10-4 Upper figure: Maximal oxygen uptake measured during
bicycling or running in 227 female and male subjects 4 to 33 years*

Applying the Bernoulli theory to the work of the heart, Döbeln (1956b) came to the same conclusion.

The blood pressure is independent of body size as pointed out by Döbeln, since pressure is force per cross-sectional area, or proportional to $L^2/L^2 = 1$. In hearts working against the same blood pressure and being anatomically uniform in the sense that the coronary blood flow is a given percentage of the total blood flow, the maximal linear velocity of the blood in the aortic ostium during the period of expulsion is the same independent of the size of the heart. This means that the maximal cardiac output, and consequently the maximal aerobic power of the entire animal, is proportional to the cross-sectional area of the aortic ostium. This area is found to be proportional to L^2 (or actually $M^{0.72}$) (Clark, 1927). Therefore, in uniformly built organisms, the maximal aerobic power is proportional to body weight raised to the $\frac{2}{3}$ power (Döbeln, 1956b).

Heart Weight, Oxygen Pulse

Figure 10-5 illustrates how the heart weight is directly proportional to the body weight in mammals of the size of a mouse up to the size of a horse (heart weight = $0.0066 \times$ body weight$^{0.98}$; Adolph, 1949). In other words, over the full range of mammalian size, the heart weight is a constant fraction of the body weight. However, an animal capable of severe exercise has a heart ratio greater than 0.6 (heart ratio = heart weight $\times$ 100/body weight), while animals incapable of heavy, steady work have ratios less than 0.6 (Clark, 1927; Tenney, 1967). (Compare the data on hare and rabbit in Fig. 10-5.) For the presented parameters the best correlation and least deviation from the regression line is noticed between heart weight and hemoglobin weight. A similar picture is demonstrated for the oxygen pulse, i.e., oxygen uptake at rest/heart rate, and its relation to body weight (oxygen pulse = $0.061 \times$ body weight$^{0.99}$), blood volume, and hemoglobin respectively. Thus, oxygen pulse is a relative measure of the stroke volume.

Heart Rate

As already mentioned, the heart rate is likely to be proportional to L^{-1} (Lambert and Teissier, 1927). Therefore the larger animal should have a lower heart rate

---

of age in relation to body weight. For male subjects the exponent b = 0.76 in the equation maximal $\dot{V}_{O_2} = a \times M^b$. (For male subjects y = −0.108 + 0.060x; r = 0.980 ± 0.004; deviation from regression line = 7.5 percent.)

Lower figure: Maximal oxygen uptake for 94 of the same subjects, age 7 to 30 years, in relation to total amount of hemoglobin. In the equation $\dot{V}_{O_2} = a \times Hb_T{}^b$ the exponent b = 0.76. (For all subjects r = 0.970 ± 0.006; 2 × SD within shadowed area. For the determination of total hemoglobin a CO method was used and the absolute values may be doubtful.) The subjects are all fairly well trained, and none of them was overweight. (Modified from P.-O. Åstrand, 1952.)

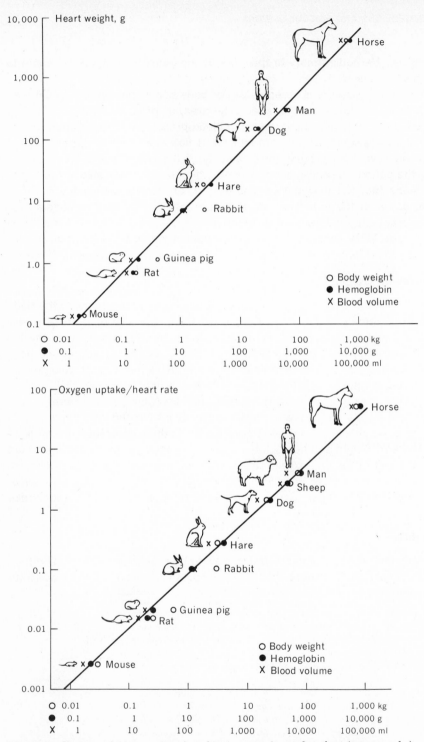

Fig. 10-5 Heart weight (upper figure) and oxygen uptake per heartbeat (oxygen pulse) in relation to body weight, total hemoglobin weight, and blood volume, respectively, in various mammals. (From various sources summarized by Sjöstrand, 1961.)

at rest, possibly also during exercise, than the smaller animal. The resting heart rate of the 25-g mouse is about 700 and of the 3,000-kg elephant, it is 25 beats/min. This difference is not a biological adaptation, but a mechanical necessity.

Summary   In the case of biological phenomena we may express the physical basic units of length, mass, and time in one single basic unit: length. This means that all units such as pressure, temperature, energy, etc. may be expressed as derivatives from this basic unit. Table 10-1 is prepared on this basis.

The conclusions which may be drawn from this table obviously do not solve any biological problems. It may, however, serve to facilitate the correct formulation of biological problems. Knowing that a fully grown man of 70 kg on an average has a maximal pulse of 195 beats/min, and that a child of 35 kg has a maximal pulse of 210 to 220 beats/min, the question is not why this latter value is greater than 195, but why it is less than 245. The latter value 245 is what might be expected from a purely dimensional consideration.

Secular Increase in Dimension

We will now consider a few additional applications of the effect of dimensions on human performance. Since in many countries man has been growing taller

TABLE 10-1

Dimensions in physics and physiology

Quantity	Dimension	
	Physical	Physiological
Length	L	L
Mass	M	L^3
Time	t	L
Surface	L^2	L^2
Volume	L^3	L^3
Density	$L^{-3}M$	L^0
Velocity	Lt^{-1}	L^0
Frequence	t^{-1}	L^{-1}
Flow	L^3t^{-1}	L^2
Acceleration	Lt^{-2}	L^{-1}
Force	LMt^{-2}	L^2
Pressure	$L^{-1}Mt^{-2}$	L^0
Temperature*	L^2t^{-2}	L^0
Energy	L^2Mt^{-2}	L^3
Power	L^2Mt^{-3}	L^2

* Physical dimension from J. C. Georgian, The Temperature Scale, *Nature,* **201**:695, 1964.
SOURCE: By courtesy of W. v. Döbeln.

in recent generations, some improvement in athletic performance is to be expected. This steady secular increase in growth is typical of countries with a satisfactory nutritional status. Besides the increase in bodily dimensions, such as height and weight, at all ages from birth to adulthood, the maturation of certain physiological functions, notably those connected with sexual maturity, is also accelerated. There has been a steady decrease in age of menarche, from about 17 years of age in 1840 to $13\frac{1}{2}$ years of age in 1960 (Tanner, 1962). A similar trend of earlier maturation of boys is also apparent from the available data; boys now reach their maximal height at an earlier age than was the case a generation ago. The influence of sexual maturity on performance is evident from Fig. 10-2. Asmussen and his collaborators (1955, 1964, 1967) point out that although the height and weight for children of a given age have shifted upward during recent decades, the weight-height curves have remained practically unaltered during the last couple of decades. It is also a fact that in spite of the general increase in height and weight of Olympic athletes during the last 30 years, their body proportions are remarkably constant. It may therefore be assumed that in spite of the increased dimensions, the present-day taller athletes are geometrically no different from those of earlier generations.

If the height of an athlete is 184 cm (the mean height of the participants in decathlon in Rome in 1960), and the height of an athlete 30 years ago was 176 cm (mean height of an athlete about 30 years ago), their heights will compare as 1.06:1, and their muscle strength as 1.13:1 (Asmussen, 1964). This means that, due to different dimensions, the average top athlete now is 6 percent taller than the top athlete of 30 years ago, but his muscular strength should be 13 percent greater than 30 years ago. This means that the maximal work that the muscles of the athlete in decathlon could perform should be 20 percent greater than 30 years ago, for the maximal work a muscle can produce is the product of its maximal force and the distance it can shorten. When the size of the oxygen-transporting organs is the limiting factor, the taller athlete should consequently be able to deliver 13 percent more oxygen to his muscles per unit of time than the smaller athlete.

In such events as throwing the javelin or putting the shot, the increase in bodily dimensions may influence the achievements in two ways: In the first place, the strength of the athlete increases in proportion to the second power of his height. This will tend to improve the results, particularly since the weight of the equipment is constant and not varied with the weight of the thrower. Secondly, the greater height from which the javelin or shot start their flight will cause them to travel further. These two factors, and particularly the first one, would result in better records, and may partly account for the improvements in records which have taken place. Anyway, there is clearly a good physiological basis for the selection of tall throwers.

This discussion is included to demonstrate that in some events part (but only part) of the improvement in results over the years may be due to the athletes' dimensional change. Khosla (1968) finds that the winners in different throwing events in the Olympic Games in Rome and Tokyo were on the average definitely taller and heavier than their competitors. Similarly, the winners in jumping and running, with the exception of 10,000-m and marathon running, were taller. This, he claims, is unfair to peoples who are less tall, and suggests the classification of the competitors according to height and weight in events in which body size may influence the results.

Old Age

Fig. 10-6 summarizes data from the literature on different static and dynamic dimensions in man from twenty up to sixty years of age. It is evident that a strict interrelation of these functions based on dimensions alone does not occur. The body height is maintained constant, but body weight, heart weight, and heart volume increase with age. Blood volume and total amount of hemoglobin are not markedly changed. Heart rate at a given submaximal work load is the same in the old and the young, i.e., the oxygen transport per heartbeat (oxygen pulse) is constant. However, maximal oxygen uptake, heart rate, stroke volume, pulmonary ventilation, and muscular strength decrease significantly with age. Apparently an older individual of the same body size as a younger one is in many ways different both in structure and in function.

SUMMARY

We have given a number of examples proving that static and dynamic functions in animals of different sizes in many cases have dimensions which are mechanically meaningful and desirable. The strength of a muscle is adjusted to the strength of bones, tendons, joints, connective tissue, and the muscle itself as a matter of safety. Furthermore, it provides a reasonable efficiency of the movements. There is, in general, a remarkable adjustment of the links involved in the chain of oxygen supply and energy output, so that none of the individual links are much stronger or weaker than necessary. On the other hand, there are examples showing that organisms of different size are not in all respects similar and uniform; there are deviations from the general trend. Deviations may sometimes be of a physical nature; in larger animals, for instance, the weight of the skeleton is relatively high. It is also known that training may markedly improve physical performance. Sometimes this improvement is accompanied by changes in organic dimensions, but this is not always the case (see Chap. 12). The performance of children is lower than expected from their dimensions, making it evident that biological factors are involved. Women and old individuals have a relatively low maximal oxygen uptake compared to a twenty-five-year-old man.

It is very fruitful, however, to consider whether differences in performance of animals of different size, including children and adults, can be explained by

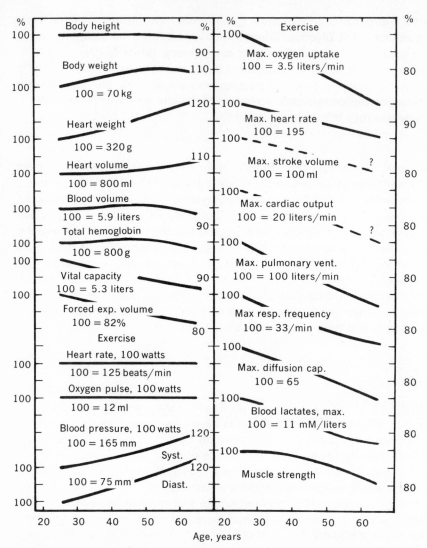

Fig. 10-6   Variation in some static and dynamic functions with age. Data have been collected from various studies, including healthy male individuals. For data on the same function only one study was consulted. The values for the 25-year-old subjects = 100 percent; for the older ages the mean values are expressed in percentage of the 25-year-old individuals' values. The mean values cannot be considered as "normal values," but their trends illustrate the effect of aging. Note that the heart rate and oxygen pulse at a given work load (100 watts or 600 kpm/min, oxygen uptake about 1.5 liters/min) are identical throughout the age range covered, but the maximal oxygen uptake, heart rate, cardiac output, etc., decline with age. The data on cardiac output and stroke volume are based on relatively few observations and are therefore less certain.

purely dimensional factors. If such a consideration fails to account for the differences, biological adaptations would appear probable.

As stated at the beginning of this chapter, it is usually not possible to tell who might be the best athlete without testing his ability. The top skier or runner in endurance events has a maximal aerobic power which is about twice that of ordinary nonathletes of similar age (6.0 liters/min compared with 3.0 liters/min). Yet the top athlete's body height is not 250 cm, which would have to be the case if the body dimensions alone were to determine his oxygen-transporting ability (maximal $\dot{V}_{O_2} \propto L^2$). In fact these top athletes are of average height, and some of their dimensions are very similar to those of an average man, while others are different (see Chap. 12).

REFERENCES

Adolph, E. F.: Quantitative Relations in the Physiological Constitution of Mammals, *Science*, **109**:579, 1949.

Asmussen, E.: Growth and Athletic Performance, *FIEP*, **34**(4):22, 1964.

Asmussen, E., and K. Heebøll-Nielsen: A Dimensional Analysis of Physical Performance and Growth in Boys, *J. Appl. Physiol.*, **7**:593, 1955.

Asmussen, E., and E. H. Christensen: "Kompendium i Legemsövelsernes Specielle Teori," Köbenhavns Universitets Fond til Tilvejebringelse af Läremidler, Köbenhavn, 1967.

Åstrand, P.-O.: "Experimental Studies of Physical Working Capacity in Relation to Sex and Age," Ejnar Munksgaard, Copenhagen, 1952.

Borelli, J. A.: De Motu Animalium, *Lugduni in Batavis, Apud Danielem à Guerbeeck . . .* , 1685.

Brody, S.: "Bioenergetics and Growth," Reinhold Book Corporation, New York, 1945.

Clark, A. J.: "Comparative Physiology of the Heart," University Press, Cambridge, 1927.

Döbeln, W. v.: Human Standard and Maximal Metabolic Rate in Relation to Fat-free Body Mass, *Acta Physiol. Scand.*, **37**(Suppl. 126), 1956a.

Döbeln, W. v.: Maximal Oxygen Intake, Body Size, and Total Hemoglobin in Normal Man, *Acta Physiol. Scand.*, **38**:193, 1956b.

Döbeln, W. v.: Kroppsstorlek, Energiomsättning och Kondition, in G. Luthman, U. Åberg, and N. Lundgren (eds.), "Handbok i Ergonomi," Almqvist & Wiksell, Stockholm, 1966.

Galilei, G.: Discorsi et Dimonstrazioni Matematiche, Interne à due Nueve Scienze, *Elzévir*, 1638.

Hill, A. V.: The Dimensions of Animals and Their Muscular Dynamics, *Proc. Royal Inst. Great Britain*, **34**:450, 1950.

Hoesslin, H. v.: Ueber die Ursache der scheinbaren Abhängigkeit des Umsatzes von der Grösse der Körperoberfläche, *Arch. f. Anat. Physiol., Physiol. Abt.*, p. 323, 1888.

Khosla, T.: Unfairness of Certain Events in the Olympic Games, *Brit. Med. J.*, 4:111, 1968.

Lambert, R., and G. Teissier: Théorie de la Similitude Biologique, *Ann. Physiol.*, **3**:212, 1927.

Sjöstrand, T.: Relationen Zwischen Bau und Funktion des Kreislaufsystems Unter Pathologischen Bedingungen, *Forum Cardiologicum,* Boehringer & Soehne, Mannheim-Waldhof, Heft 3, 1961.

Tanner, J. M.: "Growth and Adolescence," Blackwell Scientific Publications, Ltd., Oxford, 1962.

Tenney, S. M.: Some Aspects of the Comparative Physiology of Muscular Exercise in Mammals, *Circulation Res.,* **20:**1–7, 1967.

Thompson, D. W.: "On Growth and Form," Cambridge University Press, New York, 1943.

11

Evaluation of Physical Work Capacity on the Basis of Tests

contents

chapter eleven

Evaluation of Physical Work Capacity on the Basis of Tests

Generally speaking, there have been two main approaches to the assessment of physical performance: (1) physical fitness tests with scoring of actual performance in situations which represent basic performance demands, and (2) studies of cardiopulmonary function at rest and/or during exercise.

The capacity for muscular work is dependent on a variety of functions, some of which are entirely unrelated to one another (Chap. 9). Furthermore, the demands made on physical work capacity vary greatly with different work tasks.

PHYSICAL FITNESS TESTS

Since most of the so-called fitness tests, including evaluation of flexibility, skill, strength, etc. are related to special gymnastic or athletic performance; they are really not suitable for an analysis of basic physiological functions. Practice and training in the performance of the actual test may greatly influence the results. The Kraus-Weber test may serve as an example. According to the results of studies in which the Kraus-Weber test was used, American girls and boys were definitely inferior to European children (Kraus and Hirschland, 1954). However, the activities included in the test are typical for activities commonly used in European physical education classes. This placed the European children at an advantage over the American ones. Had the testing procedure included activities popular in the United States, the results might have been reversed, rating the American children superior to the Europeans.

The fact that there may be significant correlation between the results from complicated test batteries applied to a group of individuals, or that the scores are related to certain parameters characteristic of the subjects, does not necessarily mean a direct relationship. Such data may cause confusion rather than solve problems. From a physiological and medical viewpoint, any test battery for the evaluation of physical fitness is rather meaningless unless it is based on sound physiological considerations. The widespread use of such test batteries in physical education can be justified from a pedagogic and psychological viewpoint. It may help the teacher or coach to stimulate the athlete's interest in his training. Furthermore, any progress can be evaluated objectively. The selection of such activities and tests should therefore be based on pedagogic and psychological considerations with adaptation to local facilities. If they cannot be justified from these viewpoints it is better to exclude them from the curriculum altogether. Too often the tests are incorrectly claimed to serve a physiological purpose. Actually, from a physiological viewpoint, application of a test battery may sometimes be unsuitable, since the performance of the tests usually demands maximal exertion of a subject who may be completely untrained.

It may be concluded that test batteries represent applied psychology, and may have no physiological foundation.

For a review of the commonly used physical fitness tests the reader is referred to: The A.A.H.P.E.R. Fitness Test Manual (1965) and Fowler and Gardner's paper (1963).

TESTS OF MAXIMAL AEROBIC POWER

The significance of a high maximal oxygen uptake in many types of physical activity was discussed in a previous chapter.

Direct Determination

In laboratory experiments three methods of producing standard work loads have been mainly applied: running on a treadmill, working on a bicycle ergometer, and using a step test. In Chap. 9 the general methodological criteria were discussed in some detail: the work should involve large muscle groups; and the measurement of the O_2 uptake should be started when the work has lasted a few minutes (the severity of the work load has to be taken into consideration when judging the duration of the work period). The work load should be sufficiently heavy so that the O_2 uptake does not entirely correspond with the energy requirement of the work load (non-steady–state work), i.e., the anaerobic energy yield should contribute even at the end of the work period. Preferably, several submaximal and maximal work tests should be performed in experiments extended over several days (Fig. 9-3b). Ideally, a definite plateau should be reached when relating O_2 uptake to speed or work load. Specific suggestions concerning instrumentation are given in the appendix.

The critical question is whether or not the different types of work mentioned above give the same maximal O_2 uptake. A number of studies have been made to clarify this question (P.-O. Åstrand, 1952; P.-O. Åstrand and Saltin, 1961; Glassford et al., 1965; Chase et al., 1966; Kasch et al., 1966; Wyndham et al., 1966; Stenberg et al., 1967; Hermansen and Saltin, 1968).

It appears that by running on the treadmill uphill ($\geq 3°$ inclination), the O_2 uptake may be brought to a maximum, while running horizontally or at a slight inclination results in a somewhat lower maximal O_2 uptake, (Taylor et al., 1955). They conclude: "Raising the grade with the speed constant (7 mph), is the more satisfactory method of increasing work load with the motor driven treadmill to attain a maximal oxygen uptake."

Bicycling produces, on the average, a lower O_2 uptake, at least compared to running uphill. In studies in which objective criteria have been used to determine whether or not the maximal O_2 uptake had been reached for the type of work in question, the values for running are on an average 5 to 8 percent higher than for bicycling.

Table 11-1 summarizes the mean values from several studies.

TABLE 11-1

Study	Number of subjects	Running		Bicycling		
		Uphill $\geq 3°$	"Horizontal" 0 to 1°	Legs		Arms + legs
				Sitting	Supine	
Åstrand, 1952	♀33	...	2.89	2.76		
	♂34	...	4.04	4.03		
Åstrand and Saltin, 1961	5	4.69	...	4.47	3.85	
	6	...	...	4.23	...	4.24
Stenberg et al., 1967	5	...	...	3.87	3.42	3.95
Chase et al., 1966	18	3.86	...	3.28		
Wyndham et al., 1966	40	...	3.08	2.84		
Glassford et al., 1965	24	3.76	...	3.49		
Hermansen and Saltin, 1969	5	4.22	...	4.06		
	6	4.68	4.48	4.34		
	55	4.16	...	3.90		

Comments to Table 11-1: Chase et al. applied the technique described by Taylor et al. (1955) in their treadmill study, which means that a "leveling off" of the oxygen uptake with increasing load was established and several experiments were conducted. When the bicycle ergometer was used as a tool, the work load was increased intermittently every minute until the subject claimed he could no longer pedal at the required rate. Therefore the same objective criteria used in treadmill experiments were not applied in the bicycle ergometer study. The greater difference in maximal oxygen uptake between running and cycling observed in this study compared to other studies may conceivably be due to the different methods used. Wyndham et al. (1966) had their subjects sitting behind the cicycle ergometer with their legs almost horizontal. This posture is more similar to cycling in the supine position, which is less efficient for loading the oxygen transport system (P.-O. Åstrand and Saltin, 1961).

Hermansen and Saltin (1968) found a higher maximal oxygen uptake for 5 subjects who pedaled on the bicycle ergometer at 60 rpm, compared with 50 rpm (4.06 and 3.94 liters/min respectively). When running uphill, the subjects averaged a maximum of 4.22 liters/min. They also noticed a higher maximal oxygen uptake when running uphill than when running horizontally, thus confirming the observations by Taylor et al. (1955). This may explain why Åstrand (1952) found no significant difference in maximal oxygen uptake for his subjects when comparing treadmill and bicycle ergometer experiments. Hermansen and Saltin (1968) used a pedaling rate of 50 rpm for their 55 subjects, which was perhaps unfortunate in view of their finding that 60 rpm might have further increased the maximal oxygen uptake.

According to Kasch et al. (1966), a step test may bring the oxygen uptake to the same value as running on a treadmill as long as the treadmill is running horizontally. Their subjects performed the step test with a bench 12 in. high. They started with a rate of 24 steps/min. The stepping rate was then increased at intervals in increments of 3 to 6 steps/min until the maximal rate was attained.

It is not possible at present to explain the reason for the somewhat higher oxygen uptake when running uphill compared with running horizontally or bicycling. It can hardly be caused by the activation of a larger muscle mass during running uphill, since simultaneous work with both arms and legs does not increase the maximal aerobic effect compared with work with the legs only (Table 11-1). The faster work rhythm during running may enhance the venous return, but if this is the case, running uphill should be no more advantageous than running horizontally.

During bicycling a feeling of local fatigue may often be experienced or a sensation of pain in the thighs or knees, which may be disturbing. This may cause the work effort to be interrupted before the oxygen-transporting organs have been fully taxed. The work position is of critical importance. The subject should be sitting almost vertically over the pedals. The seat should be high enough so that the leg is almost completely stretched when the pedal is in its lowest position. In the Swedish studies (Hermansen and Saltin, Table 11-1) there was no difference between trained and untrained subjects when comparing bicycling and running. However, in this material even the untrained subjects

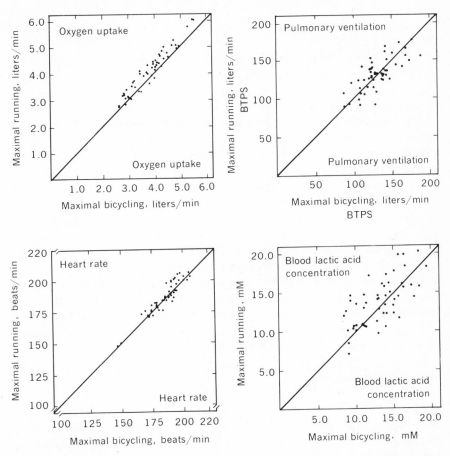

Fig. 11-1   Individual values for some functions studied in 55 subjects during maximal running uphill on a treadmill ($\geq$ 3°) and work on a bicycle ergometer (50 rpm) in a sitting position for about 5 min. Line of identity is drawn. A trend is noted for a somewhat higher maximal oxygen uptake during running compared with cycling, but pulmonary ventilation, heart rate, and blood lactic acid concentration were not different. It should be emphasized that a pedaling rate of 60 rpm may give a higher maximal oxygen uptake than is the case with a pedaling rate of 50 rpm. (From Hermansen and Saltin, 1968.)

were familiar with bicycling. In the case of persons who have never used a bicycle before, a *maximal test* on the bicycle may be undesirable as a method to assess maximal oxygen uptake. Motivation and stimulation of the subject are especially important in the case of bicycling. When a person runs on the treadmill, it is, so to speak, a matter of all or nothing: the subject is forced to follow the speed of the belt or he has to jump off. On the bicycle it is possible to continue to work at a reduced rate in most types of bicycle ergometers. Figure 11-1 summarizes results of studies by Hermansen and Saltin (1968). Despite the

7 percent difference in maximal oxygen uptake between running and bicycling, the maximal pulmonary ventilation, heart rate, and blood lactate concentration were not significantly different in the two procedures.

Normally a test of maximal aerobic power starts with a submaximal load which also serves as a warming-up activity. After this the load may be increased in one of several ways: (1) The load may be immediately increased to a level which in preliminary experiments has been found to represent the predicted maximal load for the subject. (2) The load may be increased stepwise with several submaximal, maximal, or "supermaximal" loads, the subject working 5 to 6 min at each load, with or without resting periods between each load. (3) The load may be increased stepwise every or every other minute until exhaustion (see appendix, where examples are presented). When any one of these procedures is carefully conducted, they give the same maximal oxygen uptake (own observations; Binkhorst and Leeuwen, 1963). From a physiological viewpoint the second method (2) is preferable (see Fig. 9-3b). It is often of interest to obtain steady-state conditions when measuring oxygen uptake, pulse rate, ventilation, etc., at submaximal work loads. This requires a work period of at least 5 min. The more or less continuously increasing work load (procedure 3) is a quick method which may reveal the subject's maximal oxygen uptake. However, since steady-state conditions are not attained at submaximal work, this procedure does not provide reliable information as to how the oxygen-transport problem is solved at different levels of physical effort, a type of information which may be of considerable interest.

Summary   Ideally, any test of maximal oxygen uptake should, at least, meet the following general requirements: (1) the work in question must involve large muscle groups; (2) the work load must be measurable and reproducible; (3) the test conditions must be such that the results are comparable and repeatable; (4) the test must be tolerated by all healthy individuals; and (5) the mechanical efficiency (skill) required to perform the task should be as uniform as possible in the population to be tested.

The magnitude of the external work can be expressed exactly and it may be reproduced with a high degree of accuracy in the case of the bicycle ergometer and the treadmill. For these reasons the use of the bicycle ergometer or the treadmill is preferable to that of the step test. In the case of comparative studies the same method of producing the test load should be used. Running on the treadmill should be uphill, with an inclination of 3° or more. When using the bicycle ergometer the subject should be placed in a sitting position directly above the pedals. The seat should be sufficiently high, and the pedal frequency should be about 60 rpm. By preference, the work intensity should be selected so that the subject can proceed for at least 3 min. It has repeatedly been emphasized that the correct criterion for an attained maximal oxygen uptake should be a final "leveling off" of the oxygen uptake despite an increasing work load, i.e., a failure of a higher work load to significantly increase oxygen uptake. Additional remarks concerning the different types of work will be presented in the following section on submaximal tests.

Prediction from Data Obtained at Rest or Submaximal Test

Although the aerobic power in terms of maximal oxygen uptake can be determined with a reasonable degree of accuracy, the method is rather time consuming. It requires fairly complicated laboratory procedures and demands a high degree of cooperation from the subject. Although this is the method of choice for any scientific investigation, it is by no means a method which may conveniently be applied routinely in the office of a physician.

The practicing physician (especially the industrial physician), the coach, the physical therapist, or anyone interested in physical performance are nevertheless often faced with the need to assess a person's circulatory fitness. This has created the need for a simple test for such an evaluation of an individual, based upon submaximal work stress. In the case of older individuals as well as in the case of certain patients, the physician may be reluctant to expose the patient to the risk of an exhausting maximal work load. This is true whether one is considering job placement, fitness for continued employment, or retirement.

Rest   No objective measurements made on the resting individual will reveal his capacity for physical work or his maximal aerobic power. Even a simple questionnaire may actually reveal more useful information than could be obtained from measurements made at rest. A low heart rate at rest, a large heart size, or similar parameters may indicate a high aerobic power, but they may on the other hand be a symptom of disease.

A significant correlation between any parameter and maximal oxygen uptake indicates a direct or indirect dependence. Figure 10-4 (lower part) illustrates such a relationship. The range of the observed values is of decisive importance for the numerical value of the correlation coefficient, but for an evaluation of the individual case the *standard deviation* from the regression line is the critical factor. The deviation may be large, i.e., a prediction of one parameter from the other is very uncertain, despite a high correlation between the parameters in question. From the data presented in Fig. 10-4 (lower part) we find that the correlation coefficient between maximal oxygen uptake and total amount of hemoglobin (Hb_T) is as high as 0.970, but the standard deviation of hemoglobin weight is still as high as 10.5 percent at an oxygen uptake of 2.6 liters/min. The figure shows that one girl with 350 g Hb_T can transport up to 2.7 liters O_2/min, but another girl with the same Hb_T only reaches an oxygen uptake of 1.9 liters/min. This is actually to be expected statistically, since 350 g Hb represents an oxygen uptake of 2.25 liters/min with such a standard deviation that 95 out of 100 subjects of a similar group are expected to fall within the range 1.8 to 2.7 liters/min and 5 subjects will lie outside these values.

In conclusion it may be stated that the correlation between parameters is of interest when analyzing biological interactions, but for an evaluation of an individual from indirect methods the standard deviation from the regression line indicates the accuracy of the prediction.

Heart rate response to standardized work The simplest and most extensively applied way of testing the circulatory functional capacity is to determine the heart rate during or after exercise (step test, treadmill, or bicycle ergometer test). From the heart-rate response the circulatory capacity can be evaluated. In the following material certain basic principles involved in cardiopulmonary function tests will be discussed.

During the Second World War, the Harvard step test was developed as a screening test to select individuals according to their physical fitness. The height of the bench (50 cm, or 20 inches) and the stepping frequency (30 steps/min) were selected so that only roughly one-third of the subjects should be able to perform the test for a 5-min period. The heart rate was counted during recovery from the exercise with the subject sitting on the bench (from 1 to $1\frac{1}{2}$ min). There was also a treadmill version of the Harvard fitness test (Johnson et al., 1942). The lower the number of heartbeats during the recovery and the longer the work time, the higher the score. In general it was noticed that the subjects with high score had also a better performance in many activities demanding a high aerobic power than those with lower scores. However, the results of this test, as well as all other similar tests demanding an all-out effort, depend on a number of factors besides a high maximal aerobic power, such as anaerobic power, technique, and especially motivation, which largely affect the duration of the work period. These factors do not help a person very much, however, in the performance of heavy, prolonged work as long as his maximal aerobic power is low. If objective criteria are not used to control the degree of exhaustion or the circulatory load, it is impossible to conclude whether the test was interrupted because the subject was exhausted or just unwilling to exert himself. This fact tends to render the test less useful in the testing of conscripts.

It should also be emphasized that the recording of pulse rate in connection with a work load should take place *during* work. In the same individual the correlation coefficient between the pulse rate *during* submaximal work and 1 to $1\frac{1}{2}$ min *after* the work may be high. In one study it was $r = 0.96$ with a deviation from the regression line of 5 percent (see Fig. 11-2). For a group of subjects the correlation coefficient for steady-state heart rate and recovery rate was reduced to 0.77 and the deviation increased to 10 percent (I. Ryhming, 1953). From these data it is evident that the recovery heart rate only gives a rough idea of the heart rate attained *during* work if the results are compiled from different subjects.

It has been a prominent tendency in the development of testing procedures to analyze separately the various factors of importance for the physical work capacity. In particular tests, procedures have been developed designed to assess

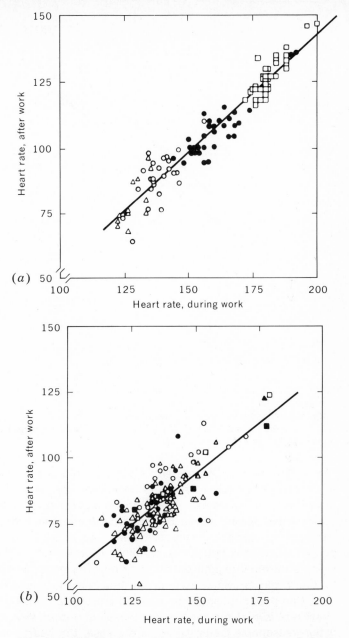

Fig. 11-2 (a) *Heart rate 1 to 1½ min after work in relation to heart rate during work for one subject (female).* △ = *step test,* ○ = *600 kpm/min,* ● = *900 kpm/min,* □ = *1200 kpm/min.* (b) *Heart rate 1 to 1½ min after work in relation to heart rate during work. (Presented are 61 individual average values from step test and 66 from bicycle test.) The figure* △ = *step test,* ● = *900 kpm/min, and* ○ = *600 kpm/min. The figures* ■ *and* □ *symbolize the average values of 15 different tests with the intensities of 600, 900, and 1200 kpm/min for two female subjects. The symbol* ▲ *shows the average from six step tests done by a male subject with 30 steps/min on a bench 40 cm high. (From Ryhming, 1953.)*

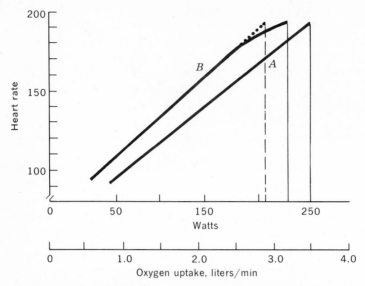

Fig. 11-3   The increase in heart rate with increasing work load (and oxygen uptake) is linear within a wide range. In some subjects (B), the oxygen uptake may increase relatively more than the heart rate as the work load becomes very heavy. The prediction of maximal oxygen uptake by an extrapolation to the subject's presumed maximal heart rate (195 in this case) suggests a maximum of 2.9 liters/min (dotted line), but the actual maximal aerobic power is 3.2 liters/min. The individual's maximal heart rate is also a critical factor in an extrapolation.

the oxygen-transporting system. Most such modern circulatory exercise tests are based on a linear increase in heart rate with increasing O_2 uptake or work load. Figure 11-3 gives examples of this relationship for two subjects with different maximal aerobic power. If the slope of the heart rate–oxygen uptake line can be determined from measurements made during submaximal exercise and an extrapolation is made to a probable value for maximal heart rate, the individual's maximal oxygen uptake may be predicted (for subject A in Fig. 11-3, 3.5 liters/min, provided that his maximal heart rate is 195).

One presumption for an assessment of the circulatory capacity based on heart rate is that the cardiac output ($\dot{Q}$) at a given oxygen uptake only varies within reasonable limits. If this is the case, the heart rate (HR) will inversely vary with the individual's stroke volume (SV); i.e., the larger the stroke volume, the lower the heart rate, since $HR \times SV = \dot{Q}$. The maximal cardiac output and oxygen uptake should then be finally determined by the individual's maximal heart rate. However, there are several factors which exert a decisive influence on the maximal circulatory capacity and aerobic power, and experimental studies show that there are considerable sources of error in any prediction of the efficiency of the oxygen transporting system from submaximal tests:

1 *The linear increase in heart rate* with increase in oxygen uptake is a typical feature. There are, however, many exceptions. In some cases the oxygen uptake increases relatively more than the heart rate as the work load becomes very heavy (subject B in Fig. 11-3). One possible explanation for this phenomenon may be that an efficient redistribution of blood, giving the working muscles an appropriate share of the cardiac output, is not brought about until the very heavy work loads are reached. The consequence of this is that in this subject the maximal oxygen uptake will be underestimated by an extrapolation from the heart rate response to submaximal loads.

2 *The maximal heart rate* declines with age (Fig. 9-16). Therefore if old and young subjects are included in the same study, the circulatory capacity of the older subjects will be consistently overestimated compared with that of the younger subjects. By introducing an age factor, a correction can be made (see appendix). However, the standard deviation for maximal heart rate within an age group is about ± 10 beats/min, i.e., 50 percent of the tested subjects will be more or less overestimated and the remainder underestimated. In Fig. 11-3 subject A was assumed to have a maximal heart rate of 195, maximal oxygen uptake = 3.5 liters/min. If the maximal heart rate was 170, the maximal oxygen uptake will be only 2.9 liters/min. An extrapolation of the heart rate to 215 for a subject with the same slope for the relation heart rate to oxygen uptake as subject A will reach an oxygen uptake of 4.0 liters/min.

3 In cases where the oxygen uptake is predicted from the work load, assuming a fixed *mechanical efficiency*, it should be kept in mind that the mechanical efficiency may vary by ± 6 percent (bicycle ergometer). In Fig. 11-3, 150 watts are indicated with a mean O_2 uptake of 2.1 liters/min. An oxygen uptake as low as 1.9 or as high as 2.3 liters/min at the same work load would not be unusual, however. The consequence of this is that in a subject with a low mechanical efficiency (whose oxygen uptake at the submaximal work load is relatively high), maximal oxygen uptake will be predicted to be lower than it actually is, since his heart rate is influenced by the extra oxygen transport.

4 The last factor to be considered is based on Fig. 6-17. The *cardiac output* is not strictly related to the oxygen uptake but shows individual variations. For the prediction of maximal oxygen uptake from heart rate at a submaximal load, this variation does not matter. The oxygen uptake (measured or predicted) per heartbeat is actually evaluated during the test. When consideration is given to the maximal heart rate, the maximal oxygen uptake is "calculated." However, if the work test is conducted for an evaluation of cardiac performance, e.g., stroke volume, the individual variation in cardiac output and arteriovenous oxygen difference must be taken into consideration.

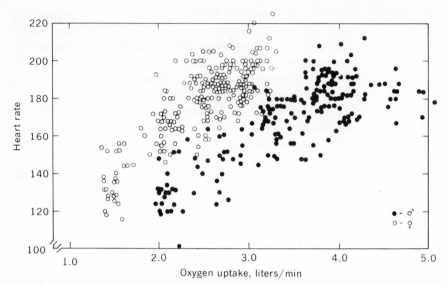

Fig. 11-4   Heart rates in relation to oxygen uptake for 86 adult female and male subjects. Maximal as well as submaximal values are represented. (From P.-O. Åstrand, 1952.)

All these factors must be considered when assessing the efficiency of a person's oxygen-transporting system.

The development of a principle for the prediction of the maximal O_2 uptake will be briefly described. Figure 11-4 gives individual values for the pulse rate at different O_2 uptakes, up to maximal O_2 uptake. The material includes 86 subjects, all of whom were physical-education students. As is evident from the figure, the scatter of the data is considerable. At an oxygen uptake of 3 liters/min there are pulse rates from 140 to 220. A pulse rate of 180 beats/min represents for some female subjects an oxygen uptake of 2.0 liters/min and as much as 5.0 liters/min for some male subjects. A closer examination of these data revealed that in the case of men the pulse rate was on the average 128 at an oxygen uptake representing 50 percent of the maximal O_2 uptake, and 154 at an O_2 uptake representing 70 percent of the maximum. The corresponding values for females were 138 at $\dot{V}_{O_2}$ = 50 percent and 168 at 70 percent (Fig. 11-4). The standard deviation (SD) was in the order of 9 beats/min. This being the case, it may be argued that if a male subject has a pulse rate of 128 at an O_2 uptake of 2.3 liters/min, his maximal O_2 uptake ought to be the double of this value, or about 4.6 liters/min. On the basis of Fig. 11-5 a nomogram was constructed for the prediction of maximal O_2 uptake from submaximal pulse rates (120 to 170) (P.-O. Åstrand and Ryhming, 1954). This nomogram has subsequently been modified. This adjusted version is shown in Fig. 11-6. I. Åstrand (1960) has further critically examined the nomogram. She found that persons over 25 years of age consistently had their maximal O_2 uptake overestimated. This could be explained on the basis of the reduction in the maximal pulse rate with age. A correction factor for age was therefore introduced (see appendix). Generally speaking, the application of the nomogram represents an extrapolation to a maximal pulse rate typical for the subject's age, as discussed in connection with Fig. 11-3. In the original group of subjects the maximal pulse rate was 195. It is therefore understandable that 50-year-old subjects with a maximal pulse rate of about 170 will have their maximal O_2 overestimated. Empirically it was found that the predicted maximal O_2 uptake came closer to the actually measured maximal O_2 uptake if a correction factor of 0.75 was introduced for pulse rates of 170 (see appendix). If possible, several tests should be performed at different test loads, and the mean figures should be calculated according to the nomogram. However, if need be, the nomogram may be adapted for the use of only one test load.

Fig. 11-5   Relationship between heart rate during work (bicycle ergometer) and oxygen uptake expressed in percent of subject's maximal aerobic power. Left of ordinate, heart rates of women; right of ordinate, those of men. Thin lines denote one standard deviation. (From P.-O. Åstrand and Ryhming, 1954.)

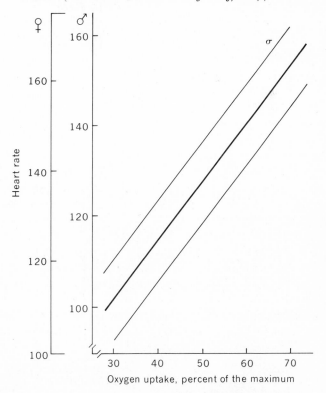

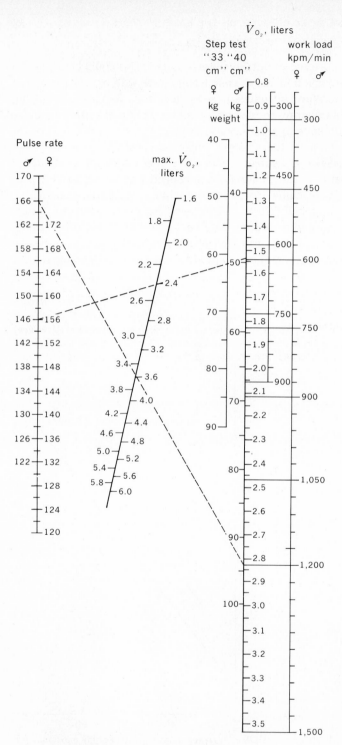

(*See legend on opposite page*)

The standard error of the method for the prediction of maximal oxygen uptake from submaximal exercise test and the nomogram (Fig. 11-6) is about 10 percent in relatively well-trained individuals of the same age, but up to 15 percent in moderately trained individuals of different ages when the age factor for the correction of maximal oxygen uptake is applied (I. Åstrand, 1960). Untrained persons are often underestimated; the extremely well-trained athletes are often overestimated. With a maximal aerobic power predicted to 3.0 liters/min the actual O_2 uptake for 5 out of 100 subjects is then less than 2.1 or higher than 3.9 liters/min (SD = ± 15 percent). It is important to keep in mind this limitation in accuracy, and this drawback holds true for any submaximal cardio-pulmonary test described so far. The validity of the nomogram has been tested in other laboratories. In some cases there has been good agreement between the actually measured and predicted maximal O_2 uptake from the nomogram (Glassford et al., 1965; Teräslinna et al., 1966). In other studies the subject's maximal O_2 uptake has been found to be underestimated when the nomogram was used (Rowell et al., 1964; Chase et al., 1966). Figure 9-13 gives an example of how the mean value for predicted maximal O_2 uptake in groups of subjects on the whole coincides with the actually measured values.

The nomogram discussed is also adapted to a step test (Ryhming, 1953) and running on a treadmill.

In a study of 84 persons, age 30 to 70 years old, von Döbeln et al. (1967) found the nomogram to underestimate the maximal O_2 uptake by 0.15 liter/min. The standard deviation was 17 percent. Utilizing electronic data-handling techniques they observed that the introduction of a new correction factor for age eliminated the systematic difference between measured and predicted maximal oxygen uptake. The best prediction was obtained if submaximal heart rate, maximal heart rate, and age were used (SD = 8.4 per-cent). The precision was only slightly reduced if maximal heart rate was omitted. They also found that the age factor could not be substituted by maximal heart rate without the loss of accuracy in the prediction. Measurement of body size did not contribute to the prediction.

It would be reasonable to assume that some variation in the accuracy may be encountered from one group of subjects to another, both with regard to the difference between predicted and measured maximal oxygen uptake and the scatter of the data. Considering the considerable error of the method and the fact that it is only applied, at

Fig. 11-6   The adjusted nomogram for calculation of maximal oxygen uptake from submaximal pulse rate and O_2-uptake values (cycling, running or walking, and step test). In tests without direct O_2-uptake measurement it can be estimated by reading horizontally from the "body weight" scale (step test) or "work load" scale (cycle test) to the "O_2 uptake" scale. The point on the O_2-uptake scale ($\dot{V}_{O_2}$, liters) shall be connected with the corresponding point on the pulse rate scale, and the predicted maximal O_2 uptake read on the middle scale. A female subject (61 kg) reaches a heart rate of 156 at step test; predicted max. $\dot{V}_{O_2}$ = 2.4 liters/min. A male subject reaches a heart rate of 166 at cycling test on a work load of 1200 kpm/min; predicted max. $\dot{V}_{O_2}$ = 3.6 liters/min (exemplified by dotted lines). (From I. Åstrand, 1960.)

best, as a screening test, a consistent difference between measured and predicted maximal O_2 uptake of a few 100 ml/min is of no importance.

Maritz et al. (1961) and Wyndham et al. (1966) have applied a similar principle in their predictions. Their subjects perform four submaximal work loads (usually step tests). From measurements of heart rate and oxygen uptake the researchers fit a straight line to the four pairs of plots and extrapolate to a mean maximal heart rate for the population in question. Margaria et al. (1965) have also introduced a nomogram based upon similar concepts.

In some test procedures the physical work capacity (PWC) is evaluated from data on work load, oxygen uptake, or oxygen pulse at a given heart rate, such as 180, 170, or 150 beats/min (Sjöstrand, 1947; Wahlund, 1948; Balke, 1954). The methodological error must necessarily be about the same in these tests as in the application of the Åstrand-Åstrand nomogram (I. Åstrand, 1960). The calculated PWC_{170}, etc., is related to the maximal stroke volume of the heart, but it is *no measure of effectiveness or maximal power or rate of work output*. In any such estimate, the maximal heart rate must be considered. If the oxygen uptake is measured during the submaximal test, it is illogical to disregard individual variation in mechanical efficiency by expressing the capacity in watts or kilopond meters per minute at heart rate 170. However, the most important error is introduced when individuals of different ages are compared or evaluated without correcting for the decline in maximal heart rate with age. The mean value for heart rate at a given submaximal oxygen uptake is the same for individuals of the same sex and state of training regardless of age (from twenty-five up to at least seventy years of age) (I. Åstrand, 1960). By definition, therefore, the calculated PWC_{170} or oxygen pulse at a given work load is the same (Fig. 10-6). The real performance capacity, however, is declining with age. Furthermore, the subjective feeling of strain is higher at a given heart rate the older the individual. Recent studies have also confirmed that there is a low correlation between the oxygen uptake or work load per minute achieved at a heart rate of 170 or 150 and the measured maximal oxygen uptake, cardiac output, heart size, or blood volume in individuals from twenty to seventy years of age (Strandell, 1964).

The conclusion is that an evaluation of the maximal effect of the oxygen-transporting system, based on studies at submaximal work loads or oxygen uptake, should be done with the utmost caution, especially when persons of different age groups are considered. Figure 10-6 presents mean values for various functional parameters in relation to age. It is evident that the decline in physical performance capacity is not related to a similar change in heart size, blood volume, or heart rate during a standard work load, etc.

There are at least two situations, however, when the submaximal work test has proved to be very useful:

1   In the clinical examination of patients or presumably healthy individuals, it is often important to include an exercise test in order to examine the

cardiovascular system under functional stress. The work test should be slightly higher than the work load encountered during the patient's regular daily activities for reasons which will be discussed later.

2   In our experience, the submaximal exercise test is a very useful tool in evaluating whether or not a training program has been effective in improving the individual's circulatory capacity. It has been widely applied in top athletes, in trained and untrained adults, and in children. In such cases the individual is his own control; it is a matter of comparing the individual with himself at repeated tests over months or years. In the simple work test on the bicycle ergometer, a counting of the heart rate is all that is needed for the evaluation. Figures 11-7 and 11-8 present examples of such applications of the test. A gradually decreasing heart rate at a standard load as the training progresses may stimulate the individual to continue his efforts to improve his circulatory capacity further. The individual tested is usually interested in knowing the result of his test, and will compare it with other data. In such cases, a prediction of the maximal oxygen uptake in absolute figures and per kilogram of body weight is definitely not essential but may be justified if the conductor of the test is aware of the limitations of the method. In these cases, the simple work test is not a research instrument but a training guide.

It should be noted that the pulse rate at a single submaximal test does not reveal anything about the subject's state of training, since constitutional factors play a greater role than merely the state of training. A person may have a low pulse at a standard load and yet be entirely untrained, and a well-trained individual may show a high pulse rate. The test procedure is discussed in detail in the appendix.

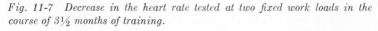

Fig. 11-7   Decrease in the heart rate tested at two fixed work loads in the course of 3½ months of training.

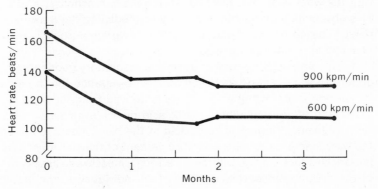

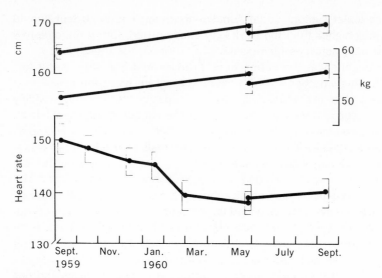

Fig. 11-8 Presents data on 163 fourteen-year-old boys collected over a period of 1 year with a work test on a bicycle ergometer (600 kpm/min, oxygen uptake about 1.5 liters/min). The decrease in heart rate, observed in steady state of work, suggests an 8 percent increase in maximal oxygen uptake per kilogram body weight from September 1959 until May 1960. After summer vacation for 2½ months, the heart rate for 100 retested boys was somewhat higher and so also was the body weight. The maximal oxygen uptake per kilogram weight was now 5 percent lower; the boys apparently did not train as hard when on holiday as they did in school.

Respiratory quotient, lactic acid   During heavy exercise of short duration (up to 5 min) the respiratory quotient determined on the expired air exceeds 1.0. Issekutz et al. (1962) have suggested that the work respiratory quotient (RQ) under standardized conditions may be used as a measure of physical work capacity. It was shown that ΔRQ (work RQ minus 0.75) increases logarithmically with the work load, and maximal O_2 uptake is reached at a ΔRQ value of 0.40. This observation offers the possibility of predicting the maximal O_2 uptake of a person, based on the measurement of RQ during a single 5-min bicycle ergometer test at a submaximal load.

This proposed method of using the respiratory quotient in the assessment of aerobic work capacity is based on the observation that a part of the expired CO_2 during short lasting work efforts is derived from the body bicarbonate pool as a result of accumulation of lactic acid during exercise.

The concentration of lactic acid in the blood starts to increase as soon as the work load exceeds about 50 to 60 percent of the individual's maximal aerobic power. This fact may be utilized as a rough measure to assess the effectiveness of the oxygen-transporting system of an individual during standardized work.

If the blood lactate is not increased, it may indicate that the individual's maximal O_2 uptake is at least twice as high as the O_2 uptake measured during the test load in question.

Limitations   Even when the tests are carried out during strictly standardized conditions, the methodological error in the prediction of the maximal aerobic power is, as previously mentioned, considerable (SD = 10 to 15 percent). By standardized conditions we mean: The subject must be free of any infection. Several hours must have elapsed between the last meal and the test. The subject should not have been engaged in any physical work heavier than the test load the last few hours prior to the test. He should not be allowed to smoke the last 2 hr before the test. The temperature of the room where the test is to be performed should be 18 to 20°C; the room should be adequately ventilated. An electric fan should be available, especially if the test involves prolonged work. The subject should be relaxed.

Under such standardized conditions the variation from day to day in heart rate at a given oxygen uptake is less than 5 beats/min providing the state of training is the same. The mean value for a group of subjects undergoing repeated tests remains almost exactly at the same level under these conditions (Ryhming, 1953).

It appears that a number of situations may cause a marked increase in the pulse rate at a submaximal work load, without the maximal oxygen uptake capacity being significantly reduced. The following examples may be mentioned:

1   Dehydration during heavy physical work or during exposure to heat (Saltin, 1964; Chap. 15).
2   Prolonged heavy exercise (Saltin, 1964; Chaps. 14 and 15).
3   Work in a hot environment (Williams et al., 1962).
4   After pyrogen-induced fever (Grimby and Nilsson, 1963).

In these situations the performance capacity may actually be reduced. Thus under these conditions the excessive heart rate response to a given submaximal load is a better criterion for a reduced work capacity than the maximal oxygen uptake. However, the effect is presently impossible to determine quantitatively from such tests. It should again be emphasized that the maximal aerobic power is only one of several factors which determines performance capacity. It should also be stressed that fear, excitement, and related emotional stress may also cause a marked elevation of the heart rate at a submaximal work load without either maximal O_2 uptake or performance capacity being affected. However, the heavier the work load the less pronounced is this nervous effect on the pulse rate. It is usually recommended that the test load should be sufficiently high so as to bring the pulse rate up to or above 150 beats per min in the case of younger subjects.

In some cases the pulse rate at a standard work load may be unchanged while the maximal O_2 uptake and the performance capacity are reduced, for instance:

1 Following acclimatization for a certain period at high altitude (Christensen, 1937; P.-O. Åstrand, 1954; P.-O. Åstrand and I. Åstrand, 1958; Saltin, 1967)
2 During semistarvation (Keys et al., 1950)

These examples illustrate further the danger of drawing conclusions concerning maximal O_2-uptake capacity and physical-performance capacity from data obtained during submaximal tests.

Summary   The submaximal exercise test can be applied as a valuable screening test for the evaluation of the functional capacity of the oxygen-transporting system. However, for research purposes it is not accurate enough to substitute the actual measurements of this capacity. It may be a useful method for selecting out the best, the worst, and the average men from a group. In the examination of patients the exercise test makes it possible to include observations and studies during physical activity which might simulate the work load of the patient's daily activities. Heart size, electrocardiogram, blood pressure, etc., should be evaluated together with results from work tests. However, any prediction of the work capacity should be avoided if the patient has not been subject to maximal exercise. Repeated submaximal tests on the bicycle ergometer are very useful in controlling the effectiveness of a physical training program. The results obtained may highly motivate the subject to continue his training.

Type of exercise   In most cases the preferable instrument for routine tests or studies of physical work capacity is, in our opinion, the *bicycle ergometer*. The technique involved is simple. The caloric output or the oxygen uptake can be predicted with greater accuracy than for any other type of exercise. Within limits, the mechanical efficiency is independent of body weight. This is a definite advantage in studies which require repeated examinations over the years. The work load can, however, simply be selected according to the subject's gross body weight, calculated lean body mass, etc. (for example, 5 or 10 kpm/min × kg). The bicycle ergometer operated with a mechanical brake is inexpensive (e.g., "Monark" bicycle ergometer). It is easy to move from place to place, and is not dependent on the availability of electrical power. Since the subject on the bicycle ergometer exercises in a sitting or lying position with arms and chest relatively immobile, it is quite simple to obtain good ECG tracings and to perform studies with indwelling catheters. During *submaximal* work a pedal frequency of from 40 to 50 revolutions/min produces the lowest O_2 uptake, i.e., the greatest mechanical efficiency, and therefore also a relatively low pulse rate (Grosse-Lordemann and Müller, 1937; Eckermann and Millahn, 1967).

The variation in O_2 uptake with different pedal frequencies at a standard work load should be kept in mind if the O_2 uptake is not measured but merely calculated from the work load used. Bicycle ergometers producing a constant load even with relatively large variations in pedal frequency have certain advantages. It should be clearly realized, however, that the O_2 uptake is not strictly determined by the load (watt) but varies with the pedal frequency. Respiration is also affected by the pedal frequency (Chap. 7). Usually the subject is asked to try to maintain a certain pedal frequency, such as 60 rpm. In our opinion, however, it is best to insist on a fixed pedal frequency such as 50 rpm, since this frequency produces an optimal mechanical efficiency. (Acoustical signals such as those produced by a metronome are easier to follow than visual signals.) It should be emphasized, however, that it is by no means essential that the chosen pedal frequency is optimal, as long as the mechanical efficiency is known. The majority of the data concerning O_2 uptake at different loads on the bicycle ergometer is obtained from experiments using a pedal frequency of 50 rpm (see appendix).

In *heavy* or *maximal* work the optimal pedal frequency is higher, e.g., the use of lower gear when traveling uphill during bicycle racing (P.-O. Åstrand, 1953). As mentioned, 60 rpm may produce a somewhat higher maximal O_2 uptake than 50 rpm (and 70 rpm). Thus one should change from 50 rpm to 60 rpm in the case of heavy work, if the object is to measure maximal oxygen uptake. The type of bicycle ergometer which produces a constant load regardless of pedal frequency has the drawback during maximal work, when the subject is tired and is unable to maintain the tempo, that the load increases for each pedal revolution. The chances are that the subject may be forced to discontinue work before it is possible to complete a critical measurement. With the use of a bicycle ergometer in which the work load varies with the pedal frequency (e.g., Krogh, Monark), the load admittedly drops when the tempo no longer can be maintained, but the subject can in any case continue. (It is important to be able to record the pedal frequency continuously if the work load is critical for the experiment.)

Figure 11-9 presents oxygen demand (kcal demand) based on average values at various loads on a bicycle ergometer. As a general guide some common equivalent activities are included for comparison. It should be pointed out that in some of these activities the individual's body weight may modify the energy expenditure.

The *treadmill* is, however, preferable in studies of yound children below the age of about ten. The work load is dependent on the body weight, which may be a disadvantage in longitudinal studies. Since the caloric output per kilogram and kilometers per hour is more variable than it is at a given work load on the bicycle ergometer, the oxygen uptake should be measured during walking and running on the treadmill (coefficient of variation is about 15 percent) (Mahadeva

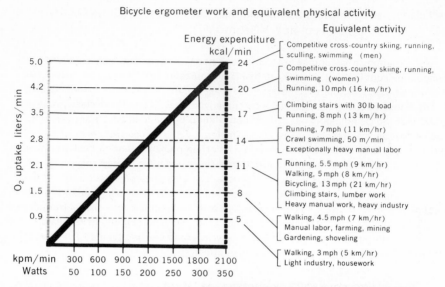

Bicycle ergometer work and equivalent physical activity

Fig. 11-9 Bicycle ergometer work and equivalent physical activity.

et al., 1953; I. Åstrand, 1960). This may limit its application. Older individuals may have some difficulty in walking on the treadmill. The provision of a handrail for support will make the work load still more unpredictable. The learning factor is more evident in the case of the treadmill than in the case of the bicycle ergometer. The treadmill is expensive and immobile. Recordings of ECG and other measurements may be more complicated to accomplish during walking or running than during bicycle-ergometer riding.

Formulas are available for calculating the energy cost of walking or running (Passmore and Durnin, 1955; Bobbert, 1960*b*; Margaria et al., 1963). The *step test* has a more limited application, since it is poorly standardized and offers limited provisions for varying the load on the oxygen-transporting system. However, methods have been developed for gradational step tests (Maritz et al., 1961; Nagle et al., 1965; Kasch et al., 1966; Shephard et al., 1966). Step tests are particularly useful in field studies and in studies of large numbers of subjects.

At submaximal work loads there is no difference in the pulse rate at a given oxygen uptake whether the subject is using a step test or a bicycle ergometer test (Ryhming, 1953). Nor does a comparison between walking, running, and bicycling reveal any significant difference in the pulse rate at a given O_2 uptake (Berggren and Christensen, 1950; Bobbert, 1960*a*; Hermansen and Saltin, 1968).

WORK TESTS ON PATIENTS

Principles

Functional tests have gained wide application in medical practice, and have proved to be of great diagnostic value. The basis for all such functional tests is the assumption that an organic abnormality or functional inadequacy is more apt to become apparent when the organ or organ system is subject to functional stress than is the case at rest when the demand is minimal (Bruce et al., 1963; Reindell et al., 1967; Sjöstrand, 1967).

In the case of the heart, which plays a vital role in the function of the human machine, functional stress tests are not as yet routinely applied. In the overriding majority of cases the heart is examined at rest. It is evident, however, that any malfunction of the heart as a pump is less likely to become apparent as long as the demand on the heart is limited to a cardiac output of 5 to 7 liters of blood/min. An inadequacy is more likely to be detected if the requirement is increased to 10 or 15 liters/min. Although the heart in many cases may be capable of maintaining a cardiac output quite adequate for resting conditions, it may be insufficient to meet the circulatory requirements encountered during different daily activities. In other cases, the limited capacity may become apparent only when the "reserve capacity" of the individual is taxed, i.e., when he is engaged in work of an intensity which exceeds that of his usual daily life. Finally, it is possible by exposing the heart to a work load to determine whether a roentgenologically detected increase in heart size or an electrocardiographic deviation from the normal is of any functional significance. In this situation a determination of stroke volume or oxygen pulse during exercise has proved to be of great value (even if it does not reveal the maximal circulatory capacity).

As we have already mentioned, the maximal aerobic power varies within wide limits even in a relatively homogeneous group of normal individuals of the same age. For this reason it may be quite difficult to ascertain if an individual's physical work capacity is low because of pathological changes or because the individual in question represents the extreme variant within the so-called normal limits. Furthermore, the state of training is of importance, since training affects physical work capacity. Thus, 6 weeks of bed rest in a normal individual may reduce his physical work capacity to about one-half that of his pre-bed-rest level (Chap. 12). If the bed rest is necessitated by disease processes in a vital organ, the deterioration of the physical work capacity may be even greater. It is thus apparent that even if the physical work capacity of an individual is known prior to the onset of a disease, it may be difficult to determine whether a deterioration in physical work capacity is due to the disease or inactivity, or both. In any case, it is of considerable practical importance to be able to follow the patient's physical work capacity during the convalescent period and during periods of retraining.

Although electrocardiographic tracings have been recorded immediately following exercise for a number of years, notably with the aid of the so-called Master step test (Master et al., 1953), it is only during recent years that electrocardiographic examination *during* the performance of physical work has become a subject of considerable interest. The Master step test does not provide for the opportunity of gradually increasing the work load so that the severity of the work can be adjusted according to the safe limits of the patient's capacity.

The degree of the ST-depression on the ECG is related to the circulatory load of the individual in such a way that the ST-depression increases gradually with the work load or the heart rate and reaches the most pronounced depression at maximal load. This has been shown by Blomqvist (1965) among others. In order to be able to compare results from different studies the work tests must be carried out in such a way that a well-defined work load is reached. The work load should be defined both with regard to absolute load, for instance in kpm/min, watt, or O_2 uptake, and if possible, in relation to the maximal oxygen uptake of the tested subject.

For a number of reasons it is usually preferable to test the subject on a submaximal load only. Two principally different procedures are used to obtain a standardization of such a submaximal test.

One way is to expose all patients to the same work load, for instance to a certain multiple of the basal metabolic rate. Another way is to stress the subject in proportion to his assumed maximal capacity.

The first method can be used if the work load is chosen so that a certain predicted oxygen uptake is reached, which for instance might be 8 times the basal metabolic rate. The best results with this method would be expected with a type of exercise that gives the least variation in mechanical efficiency. For this reason the bicycle ergometer test is preferable, since the error in the prediction of the oxygen uptake is small. However, there are several drawbacks with a test performed in this way, since many subjects will be tested on a comparatively low load while others will not be able to complete the test.

The second procedure implies the choice of a load that is related to the maximal capacity of the individual. The body weight has often been included in this prediction. If the intention is to validate the individual's ability to move himself, it is important to take the body weight into consideration. The step test and the treadmill test are then the methods of choice. However, it should be noted that the bicycle ergometer can also be used for such a test. The work load can be chosen according to the patient's body weight, for instance 10 kpm/min per kilogram of body weight. Since the intensity should be related to the circulatory capacity of the individual it is, however, important to investigate the correlation between body weight and this capacity. In children and in young, healthy, lean subjects there is a good correlation between body weight on the one hand and maximal oxygen uptake (cardiac output), total hemoglobin, etc., on the other (P.-O. Åstrand, 1952). However, in subjects 30 to 70 years old the correlation between body weight and maximal $\dot{V}_{O_2}$ is low (Fig. 10-6). The reason for this is partly the fact that the degree of obesity varies to a great extent and partly the fact that aging is accompanied by a deterioration of the oxygen-transporting function which is not reflected in the dimensions of the individual. Let us take some examples: (1) An adult who increases his body weight by 20 kg does not increase his circulatory capacity to a corresponding degree. This means that he will be tested on a load that is closer to his maximal capacity if the load is chosen in relation to the body weight. (2) If a thirty-year-old man and a seventy-year-old man with similar body weights walk on a treadmill with the same speed and grade, they will reach the same oxygen uptake and the same heart rate, but the relative load will be very different. The reason for this is the decrease in maximal cardiac output, maximal oxygen uptake, and maximal heart rate with increasing age (Fig. 10-6). On the other hand, the mechanical efficiency and the heart rate at a given submaximal load do not change in normal subjects.

The simplest way of producing a load that is related to the individual's circulatory capacity is to increase the load until the subject's heart rate has reached a certain level in relation to the maximal heart rate that is typical for his age (Fig. 9-16). If the limit for young persons 20 to 29 years of age is set at a heart rate of 170 beats/min, it ought to be

160/min for 30 to 39 years old, 150/min for 40 to 49 years old, 140/min for 50 to 59 years old, and 130/min for 60 to 69 years old (I. Åstrand et al., 1967). It appears that the best sub-maximal test would be one in which the patient is exposed to work loads on the bicycle ergometer or the treadmill for 5- to 6-min periods, starting with a low load and increasing the load in a stepwise fashion until the desired heart rate is reached.

A test performed in this fashion permits a classification of the ECG both at a fixed load independent of the individual's capacity and at a load related to the individual's circulatory capacity. Other advantages with this procedure are: (1) The test is easy to perform from a practical point of view, and (2) the physical demands of different occupational or recreational activities can be reproduced (Fig. 11-9).

In the case of patients with atrial fibrillation or other types of arrhythmias the relative load cannot be assessed as described above. From Blomqvist's studies (1965) it is evident that an ST-depression during work is not always accompanied by a change *after* work. From Furberg's studies (1967) it is clear that an ST-depression that appears only after work might be of a sympathicotonic type. Therefore the ST changes recorded during work are the most essential manifestations. The recording of the patient's ECG during work, in addition to before and after work, not only provides the physician with an opportunity to monitor the ECG for reasons of safety during the actual test, but it also may reveal changes which otherwise may be undetected.

From this discussion it may be concluded that, if at all possible, a standardized procedure should be applied and that a bicycle ergometer or a treadmill should be used for exercise tests involving work electrocardiography. Studies of the oxygen-transporting system are useful in order to obtain information (1) pertaining to the physical work capacity (aerobic power) of the individual, and (2) for diagnostic purposes. The physical work capacity has been discussed previously. This discussion will therefore be limited to work tests in combination with electrocardiography, i.e., work electrocardiography.

There is no doubt that coronary insufficiency may be detected by the use of ECG recordings in connection with work tests. The work ECG may be of particular value in cases where the history is unclear and the ECG is normal at rest. Furthermore, arrhythmias of various kinds which occur in connection with physical activity and which cause symptoms or concern may be verified with the aid of work ECG. Premature ventricular or supraventricular contractions or various degrees of atrioventricular or intraventricular conduction disturbances are examples of this category. As pointed out by Öhnell (1944), it may in many cases be easier to diagnose preexcitation during work than at rest.

Certain individuals reveal abnormally high blood pressures during exercise despite a normal pressure at rest. This kind of response may be a preliminary stage of hypertension. There are also those who react in the opposite manner. Both ECG and blood pressure should therefore be recorded during work.

For diagnostic purposes, or in order to evaluate indications for the operation of different types of cardiac abnormalities, it may be helpful to combine electrocardiography with intravascular measurements of pressure and to determine cardiac output during work. This is also true in the case of different types of disturbances in peripheral circulation, in which determination of oxygen uptake may also be combined with determination of P_{O_2}, P_{CO_2}, pH, and lactic acid in the arterial and venous blood from the extremity in question during an exactly controlled work load. (For, example, see Pernow and Zetterquist, 1968.)

From these examples it is evident that there are certain abnormalities or disease processes which may be revealed with the aid of work ECG and other measurements during work. But there are also conditions which cannot be diagnosed with the aid of ECG. A patient with advanced coronary heart disease may have a normal ECG, and an individual with so-called abnormal ECG (according to the norms) may be free from symptoms and other signs of heart disease. Well-trained athletes are often characterized by high-amplitude R- and T-waves, probably as a result of ventricular hypertrophy, and minor intraventricular conduction defects are also common; the PQ-interval may be prolonged or of varying duration (Smith et al., 1964; Vernerando and Rulli, 1964; Grimby and Saltin, 1966).

In all cases, however, where cardiac abnormalities are demonstrated or suspected and a work test is to be performed, the ECG should be monitored continuously during the work test to enable the immediate detection of the occurrence of arrhythmias and ST-segment changes.

Evaluation of Results

When the work test in combination with the recording of ECG has been subject to criticism, it has been because the criteria which have been used for the evaluation of the work ECG have not been adequately analyzed. This has resulted in a large number of false-positive and false-negative diagnoses. It is obvious that in all studies where the recording of ECG changes is one of the basic parameters to be compared, well-defined, specific, and carefully graded criteria for the classification of such factors as the ST depressions have to be established. This is particularly important when the results of different studies carried out by different investigators are compared. It has been clearly demonstrated that conventional clinical methods of ECG interpretation carry a large inter- and intra-individual variability (Acheson, 1960; Davies, 1958; Epstein et al., 1961). It has also been shown that strict adherence to a set of well-defined ECG criteria will reduce observer errors (Blackburn et al., 1960; Blackburn, 1965). The ECG classification system proposed by Blackburn et al. (1960), the *Minnesota Code*, has been extensively used in epidemiological studies of cardiovascular disease, and it has wider acceptance than any other classification method. This method has been modified by I. Åstrand et al. (1967), who also adapted the code to the CR and CH leads, which are preferable to V leads.

It has been emphasized that ECG observations during work stress are more likely to reveal abnormalities than is the case at rest. Such changes are more frequent among individuals over 50 years of age. Whether these changes are the result of aging processes or whether they are caused by disease processes is not clear. In several studies a critical evaluation and a coding of the ECG's have been made according to the original or modified Minnesota Code. The frequency of segmental ST changes, which usually mean changes that have a prognostic significance, was less than 10 percent in men below 40 years of age, about 15 percent at 40 to 50 years of age, about 20 percent at 55 years of age, and about 35 percent at 60 years of age. For women, the incidence at 40 to 45 years of age was about 20 percent, at 50 to 55 years of age about 30 percent, and over 55 years of age about 50 percent (Lepeschkin and Surawics, 1958; Lepeschkin, 1960; Rumball and Acheson, 1960; I. Åstrand, 1963, 1965).

Summary

The best procedure for the evaluation of the ECG during work stress is probably to increase the work load in a stepwise fashion with a work period of about 6 min at each load until a certain heart rate is reached which is related to the maximal heart rate typical for the individual's age. The ECG should be recorded both during and after exercise. The ECG should be classified according to an objective system which makes possible a separation of different changes and a quantitative analysis of these changes. The Minnesota Code (modified) offers the best norms in this respect. CR and CH leads are preferable to V leads.

With the procedure outlined here the frequency of ST-segment depressions in an unselected group of "healthy" Swedish men 60 years of age is about 35 percent. In Swedish women of the same age the frequency is about 50 percent.

REFERENCES

Acheson, R. M.: Observed Error and Variation in the Interpretation of Electrocardiograms in an Epidemiological Study of Coronary Heart Disease, *Brit. J. Prevent. & Social Med.*, **14**:99, 1960.

American Association for Health, Physical Education and Recreation: The AAHPER Fitness Test Manual, rev. ed., Washington, D.C., 1965.

Åstrand, I.: Aerobic Work Capacity in Men and Women with Special Reference to Age, *Acta Physiol. Scand.*, **49**(Suppl. 169), 1960.

Åstrand, I.: Exercise Electrocardiograms in a 5-year Follow-up Study, *Acta Med. Scand.*, **173**:257, 1963.

Åstrand, I.: Exercise Electrocardiograms Recorded with a 8-year Interval in a Group of 204 Women and Men 48–63 Years Old, *Acta Med. Scand.*, **178**:27, 1965.

Åstrand, I., et al.: The "Minnesota Code" for ECG Classification. Adaptation to CR Leads and Modification of the Code for ECG's Recorded during and after Exercise, *Acta Med. Scand.*(Suppl. 481), 1967.

Åstrand, P.-O.: "Experimental Studies of Physical Working Capacity in Relation to Sex and Age," Munksgaard, Copenhagen, 1952.

Åstrand, P.-O.: Study of Bicycle Modifications Using a Motor Driven Treadmill-bicycle Ergometer, *Arbeitsphysiol.*, **15**:23, 1953.

Åstrand, P.-O.: The Respiratory Activity in Man Exposed to Prolonged Hypoxia, *Acta Physiol. Scand.*, **30**:343, 1954.

Åstrand, P.-O., and I. Åstrand: Heart Rate during Muscular Work in Man Exposed to Prolonged Hypoxia, *J. Appl. Physiol.*, **13**:75, 1958.

Åstrand, P.-O., and Irma Ryhming: A Nomogram for Calculation of Aerobic Capacity (Physical Fitness) from Pulse Rate during Submaximal Work, *J. Appl. Physiol.*, **7**:218, 1954.

Åstrand, P.-O., and B. Saltin: Maximal Oxygen Uptake and Heart Rate in Various Types of Muscular Activity, *J. Appl. Physiol.*, **16**:977, 1961.

Balke, B.: Optimale Körperliche Leistungsfähigkeit, ihre Messung und Veränderung infolge Arbeitsermüdung, *Arbeitsphysiol.*, **15**:311, 1954.

Berggren, G., and E. H. Christensen: Heart Rate and Body Temperature as Indices of Metabolic Rate during Work, *Arbeitsphysiol.*, **14**:255, 1950.

Binkhorst, R. A., and P. van Leeuwen: A Rapid Method for the Determination of Aerobic Capacity, *Intern. Z. Angew. Physiol.*, **19**:459, 1963.

Blackburn, H.: The Electrocardiogram in Cardiovascular Epidemiology, Problems in Standardized Application, *Am. N.Y. Acad. Sci.*, **126**:882, 1965.

Blackburn, H., A. Keys, E. Simonsen, P. Rautaharju, and S. Punsar: The Electrocardiogram in Population Studies: A Classification System, *Circulation*, **21**:1160, 1960.

Blomqvist, G.: The Frank Lead Exercise Electrocardiogram: A Quantitative Study Based on Averaging Technic and Digital Computer Analysis, *Acta Med. Scand.*, **178**(Suppl. 440), 1965.

Bobbert, A. C.: Physiological Comparison of Three Types of Ergometry, *J. Appl. Physiol.* **15**:1007, 1960a.

Bobbert, A. C.: Energy Expenditure in Level and Grade Walking, *J. Appl. Physiol.*, **15**:1015, 1960b.

Bruce, R. A., J. R. Blackmon, J. W. Jones, and G. Strait: Exercise Testing in Adult Normal, Subjects and Cardiac Patients, *Pediatrics*, **32**(Suppl.):742, 1963.

Chase, G. A., C. Grave, and L. B. Rowell: Independence of Changes in Functional and Performance Capacities Attending Prolonged Bed Rest, *Aerospace Med.*, **37**:1232, 1966.

Christensen, E. H.: Sauerstoffaufnahme und respiratorische Funktionen in grossen Höhen, *Skand. Arch. Physiol.*, **76**:88, 1937.

Christensen, E. H., and W. H. Forbes: Der Kreislauf in grossen Höhen, *Skand. Arch. Physiol.*, **76**:75, 1937.

Davies, L. G.: Observer Variation in Reports on Electrocardiograms, *Brit. Heart J.*, **20**:153, 1958.

Döbeln, W. v., I. Åstrand, and A. Bergström: An Analysis of Age and Other Factors Related to Maximal Oxygen Uptake, *J. Appl. Physiol.*, **22**:934, 1967.

Eckermann, P., and H. P. Millahn: Der Einfluss der Drehzahl auf die Herzfrequenz und die Sauerstoffaufnahme bei konstanter Leistung am Fahrradergometer, *Int. Z. angew. Physiol. einschl. Arbeitsphysiol.*, **23**:340, 1967.

Epstein, F. H., J. T. Doyle, A. A. Pollack, H. Pollack, G. P. Robb, and E. Simonson: Observer Variation in Interpretation of Electrocardiograms: Suggestions for Objective Interpretations, *J. Am. Med. Ass.*, **175**:847, 1961.

Fowler, W. M., Jr., and G. W. Gardner: The Relation of Cardiovascular Tests to Measurements of Motor Performance and Skills, *Pediatrics*, **32**(Suppl.):778, 1963.

Furberg, C.: Adrenergic Beta-blockade and Electrocardiographical ST-T Changes, *Acta Med. Scand.*, **181**:21, 1967.

Glassford, R. G., G. H. I. Baycroft, A. W. Sedgwick, and R. B. J. Macnab: Comparison of Maximal Oxygen Uptake Values Determined by Predicted and Actual Methods, *J. Appl. Physiol.*, **20**:509, 1965.

Grimby, G., and N. J. Nilsson: Cardiac Output during Exercise in Pyrogen-induced Fever, *Scand. J. Clin. Lab. Invest.*, **15**(Suppl. 69):44, 1963.

Grimby, G., and B. Saltin: A Physiological Analysis of Physically Well-trained Middle-aged and Old Athletes, *Acta Med. Scand.*, **179**:513, 1966.

Grosse-Lordemann, H., and E. A. Müller: Der Einfluss der Tretkurbellänge auf das Arbeitsmaximum und den Wirkungsgrad beim Radfahren, *Arbeitsphysiologie*, **9**:619, 1937.

Hermansen, L., and B. Saltin: Oxygen Uptake during Maximal Treadmill and Bicycle Exercise, *J. Appl. Physiol.*, **26**:31, 1969.

Issekutz, B., Jr., N. C. Birkhead, and K. Rodahl: Use of Respiratory Quotients in Assessment of Aerobic Work Capacity, *J. Appl. Physiol.*, **17**:47, 1962.

Johnson, R. E., L. Brouha, and R. C. Darling: A Test of Physical Fitness for Strenuous Exertion, *Rev. Canad. Biol.*, **1**:491, 1942.

Kasch, F. W., W. H. Phillips, W. D. Ross, J. E. L. Carter, and J. L. Boyer: A Comparison of Maximal Oyxgen Uptake by Treadmill and Step-Test Procedures, *J. Appl. Physiol.*, **21**:1387, 1966.

Keys, A., J. Brozek, A. Henschel, O. Michelsen, and H. L. Taylor: "The Biology of Human Starvation," University of Minnesota Press, Minneapolis, pp. 675, 735, 1950.

Kraus, H., and R. P. Hirschland: Minimum Muscular Fitness Tests in School Children, *Res. Quart.*, **25**:178, 1954.

Lepeschkin, E.: Exercise Tests in the Diagnosis of Coronary Heart Disease, *Circulation*, **22**:986, 1960.

Lepeschkin, E., and B. Surawicz: Characteristics of True-positive and False-positive Results of Electrocardiographic Master Two-step Exercise Tests, *New England J. Med.*, **258**:511, 1958.

Mahadeva, K., R. Passmore, and B. Woolf: Individual Variations in the Metabolic Cost of Standardized Exercises: The Effect of Food, Age, Sex, and Race, *J. Physiol.*, **121**:225, 1953.

Margaria, R., P. Aghemo, and E. Rovell: Indirect Determination of Maximal O_2 Consumption in Man, *J. Appl. Physiol.*, **20**:1070, 1965.

Margaria, R., P. Cerretelli, P. Aghemo, and G. Sassi: Energy Cost of Running, *J. Appl. Physiol.*, **18**:367, 1963.

Maritz, J. S., J. F. Morrison, J. Peter, N. B. Strydom, and C. H. Wyndham: A Practical Method of Estimating an Individual's Maximal Oxygen Intake, *Ergonomics*, **4**:97, 1961.

Master, A. M., L. Pordy, and K. Chesky: Two-step Exercise Electrocardiogram, *J. Am. Med. Ass.*, **151**:458, 1953.

Nagle, F. J., B. Balke, and J. P. Naughton: Gradational Step Tests for Assessing Work Capacity, *J. Appl. Physiol.*, **20**:745, 1965.

Öhnell, R. F.: Pre-excitation: A Cardiac Abnormality, *Acta Med. Scand.*(Suppl. 152), 1944.

Passmore, R., and J. V. G. A. Durnin: Human Energy Expenditure, *Physiol. Rev.*, **35**:801, 1955.

Pernow, B., and S. Zetterquist: Metabolic Evaluation of the Leg Blood Flow in Claudicating Patients with Arterial Obstructions at Different Levels, *Scand. J. Clin. Lab. Invest.*, **21**:277, 1968.

Pirnay, F., J. M. Petin, R. Bottin, R. Deroanne, J. Juchmes, and G. Belge: Comparaison de deux méthodes de mesure de la consommation maximum d'oxygène, *Intern. Z. Angew. Physiol.* **23**:203, 1966.

Reindell, H., K. König, and H. Roskamm: "Funktionsdiagnostik Des Gesunden und Kranken Herzens," H. Thieme Verlag, Stuttgart, 1967.

Rowell, L. B., H. L. Taylor, and Y. Wang: Limitations to Prediction of Maximal Oxygen Uptake, *J. Appl. Physiol.*, **19**:919, 1964.

Rumball, C. A., and E. D. Acheson: Electrocardiograms of Healthy Men after Strenuous Exercise, *Brit. Heart J.*, **22**:415, 1960.

Ryhming, I.: A Modified Harvard Step Test for the Evaluation of Physical Fitness, *Arbeitsphysiol.*, **15**:235, 1953.

Saltin, B.: Aerobic Work Capacity and Circulation at Exercise in Man, *Acta Physiol. Scand.*, **62**(Suppl. 230), 1964.

Saltin, B.: Aerobic and Anaerobic Work Capacity at 2300 Meters, *Med. Thorac.*, **24:**205, 1967.

Shephard, R. J.: The Relative Merits of the Step Test, Bicycle Ergometer, and Treadmill in the Assessment of Cardio-Respiratory Fitness, *Intern. Z. Angew. Physiol.*, **23:**219, 1966.

Sjöstrand, T.: Changes in the Respiratory Organs of Workmen at an Ore Melting Works, *Acta Med. Scand.*(Suppl. 196), 687, 1947.

Sjöstrand, T. (ed.): "Clinical Physiology," Svenska Bokförlaget, Bonniers, 1967.

Smith, W. G., K. J. Cullen, and I. O. Thorburn: Electrocardiograms of Marathon Runners in 1962 Commonwealth Games, *Brit. Heart J.*, **26:**469, 1964.

Stenberg, J., P.-O. Åstrand, B. Ekblom, J. Royce, and B. Saltin: Hemodynamic Response to Work with Different Muscle Groups, Sitting and Supine, *J. Appl. Physiol.*, **22:**61, 1967.

Strandell, T.: Circulatory Studies on Healthy Old Men with Special Reference to the Limitation of the Maximal Physical Working Capacity, *Acta Med. Scand.*, **175**(Suppl. 414), 1964.

Taylor, H. L., E. Buskirk, and A. Henschel: Maximal Oxygen Intake as an Objective Measure of Cardiorespiratory Performance, *J. Appl. Physiol.*, **8:**73, 1955.

Teräslinna, P., A. H. Ismail, and D. F. MacLeod: Nomogram by Åstrand and Ryhming as a Predictor of Maximum Oxygen Intake, *J. Appl. Physiol.*, **21:**513, 1966.

Vernerando, A., and V. Rulli: Frequency Morphology and Meaning of the Electrocardiographic Anomalies Found in Olympic Marathon Runners and Walkers, *J. Sport Med.*, **4:**135, 1964.

Wahlund, H.: Determination of the Physical Working Capacity, *Acta Med. Scand.*(Suppl. 215), 1948.

Williams, C. B., G. A. G. Bredell, C. H. Wyndham, N. B. Strydom, J. F. Morrison, P. W. Flemming, and J. S. Wood: Circulatory and Metabolic Reaction to Work in Heat, *J. Appl. Physiol.*, **17:**625, 1962.

Wyndham, C. H., N. B. Strydom, W. P. Leary, and C. G. Williams: Studies of the Maximum Capacity of Men for Physical Effort, *Intern. Z. Angew. Physiol.*, **22:**285, 1966.

12

Physical Training

contents

chapter twelve

Physical Training

INTRODUCTION

In several sections of this book it has been shown that different organs or organ systems may be affected by a variety of factors. The effect may be transitory or it may last for a considerable period of time, i.e., an adaptation takes place. Figure 9-1, for instance, illustrates that the aerobic muscular capacity is affected by training, deconditioning, and acclimatization (altitude, heat, cold). In this chapter we shall discuss how physical activity and physical inactivity affect the body morphologically as well as functionally.

Conclusions concerning the effect of physical training have often been drawn from studies of well-trained persons, and the data obtained have been compared with similar data from studies on sedentary individuals. The disadvantage of such *cross-sectional studies* is that it may be impossible to determine whether any difference observed depends on constitutional dissimilarities or on the training as such. It is in any case quite obvious that the great maximal aerobic power which is characteristic for the top athlete in endurance largely depends on organic advantages which are endowed. Thus a person with a maximal oxygen uptake of 45 ml/(kg) (min) cannot, under any circumstance, no matter how much he trains, attain a maximal oxygen uptake of 80 ml/(kg) (min), which is required for Olympic medals in certain sport events (Fig. 12-15). It may not be quite fair, but it is nevertheless a fact that the "choice of parents" is important for progress.

For these reasons *longitudinal studies* must be designed, in which the same individual is followed for shorter or longer periods of time. However, such studies are rare. Since they are difficult to perform only a few subjects have usually been included in each study. Even this approach with long observation periods of the same individuals is not without objections from a scientific point of view.

1   If the object is to examine such problems as the effect of physical train-
    ing, the selection of subjects is critical. If the selection is not done on
    a strictly random basis, the material may not be representative. If a
    selection is done from a group of volunteers, it must be remembered
    that those who volunteer for such studies may do so for special reasons
    which may be medical, social, psychological, or personal.

2   Any such study may well represent an intervention of the subject's
    normal pattern of life. This may in itself bring about different effects:
    the subject may perhaps change his diet, his smoking habits, or other
    habits which may elicit effects on the organism which, per se, have
    nothing to do with the training. It may also be difficult to avoid affecting
    the control group by the mere fact that the group is being studied and
    therefore subject to special attention. In the case of patients it is espe-
    cially complicated to arrive at a clear-cut experimental situation in that
    ethical considerations in such cases especially may be brought into the
    foreground.

3   Experience has shown that the drop-out frequency is rather high in the
    case of training studies.

4   It is quite difficult to establish a person's physical condition objectively
    prior to the training, and any training effect obviously depends on the
    initial level at the start of the training. As mentioned on several occa-
    sions, there is at present no method of investigation available which
    can definitely reveal and separate the influence of constitutional factors
    on the one hand, and the effects of training on the other. Thus one is
    largely forced to rely on the individual's own statements. However,
    what one person may characterize as a physically active life may to
    another represent a sedentary life. Occupation and recreational activ-
    ities may in themselves only represent indications rather than precise,
    meaningful information regarding degree of physical activity, etc. The
    duration and intensity of such activities are important factors to be
    considered.

5   The intensity (tempo) of the actual training is difficult to assess, define,
    or reproduce. In any case it is a time-consuming and complicated task
    to record this variable, especially in the case of large-scale investigations.

On the basis of these considerations it is understandable that different
investigators arrive at different results concerning the effect of physical activity
or inactivity both with regard to qualitative as well as quantitative changes.

In this chapter we will consider (1) the physiological basis for the development of a training program, (2) the biological long-term effects of different levels
of physical activity. The discussion will be limited to data which are fairly well
documented. (3) In a subsequent section quantitative aspects will be presented,
based on results from a few limited studies concerning the effect of bed rest on

the one hand and training on the other. (In Chap. 18 the relationship between physical activity and the state of certain diseases will be considered.)

Table 12-1 summarizes effects of training on organs and organ functions. Some of the data were obtained on animals, others on man. There are still many open questions as indicated in the table. At the end of this chapter the table will be discussed in some detail.

PRINCIPLES FOR TRAINING BASED ON PHYSIOLOGY

The basic finding in Christensen's studies (1931) was that regular training with a given standard work load gradually lowered the heart rate, say from 180 to 160 beats/min. Further training did not modify this heart rate response. After a period of training on a heavier load, the original standard work load could then be performed with a still lower heart rate, say from 140 to 150 beats/min. This general principle is apparent during training of a number of functions: *An adaptation takes place to a given load; in order to achieve further improvement the training intensity has to be increased.* This principle has been elucidated by several studies summarized in the other sections of this chapter, and it is in full agreement with practical experience. There is no linear relationship between amount of training and the training effect. For instance, 2 hr training per week may cause an increase in maximal O_2 uptake, say by 0.4 liter/min. If the training is twice as much, that is, 4 hr per week, the increase in O_2 uptake will not be twice as great, that is, 0.8 liter/min, but possibly 0.5 to 0.6 liter/min (see Fig. 12-1). Obviously there is a limit to the increase, and the rate and magnitude of the increase vary from one individual to the next.

One Swedish cross-country skier had a maximal oxygen uptake of 5.48 liters/min when he was studied in 1955. In 1963 it was about the same, or 5.60 liters/min, but he had trained almost daily during the intervening eight years and successfully participated (winning gold medals) in two Olympic games and two World Championships (Saltin and P.-O. Åstrand, 1967). In repeated tests another Swedish skier Sixten Jernberg never exceeded 5.88 liters/min in maximal aerobic power, measured in 1955. He also trained intensively and competed very successfully up until 1964, winning a gold medal in the 50-km race in that year's Olympic Games. There are many similar examples indicating a definite ceiling of the aerobic power (see Ekblom, 1969).

Older individuals (above fifty years of age) may be less trainable than younger ones. On the other hand, it should be kept in mind that some effect of training may be noticed even at very old age. It is important for the individual who wishes to improve his general state of fitness, to ascertain what amount of training may produce the most satisfactory result. One has to weigh the time available for training against the effects achieved by the training. If one has reached such a level of physical fitness that several additional hours of training

TABLE 12-1

Effects of training on organs and organ functions

Organ or function	Increase	Decrease	No effect	References
Locomotive organs				
Strength of bones and ligaments	x			Ingelmark, 1948; 1957; Viidik, 1966; Tipton et al., 1967
Thickness of articular cartilage	x			Holmdahl and Ingelmark, 1948
Muscle mass (hypertrophy)	x			Marpurgo, 1897; Siebert, 1929; Vannotti and Pfister, 1934; Man-i et al., 1967
Number of muscle cells	?		x	Linge, 1962; Reitsma, 1965.
Muscle strength	x			See Hettinger, 1968; Table 12-7
ATP, creatine phosphate, muscle	x			Palladin and Ferdmann, 1928; Yakovlev, 1958
Myoglobin	x			Whipple, 1926
Potassium, muscle	x			Nöcker et al., 1958
Capillary density, muscle	x			Vannotti and Pfister, 1934; Vannotti and Magiday, 1934; Petrén et al., 1936 (Fig. 12-9)
Arterial collaterals, muscle	x			Schoop, 1964; 1966
Circulation				
Heart volume	x		x	Roskamm et al., 1966; Reindell et al., 1967; Sjöstrand, 1967; Ekblom, 1969; Figs. 12-10, 12-15
Heart weight	x			Siebert, 1929; Thörner, 1949; Liere and Northup, 1957
Capillary density, heart	x			Petrén et al., 1936; Fig. 12-9
Coronary collaterals	?			Eckstein, 1957; Tepperman and Pearlman, 1961
Blood volume, total hemoglobin	x			Deitrick et al., 1948; Taylor et al., 1949; Hollman and Venrath, 1963; Miller et al., 1964; Sjöstrand, 1967; Saltin et al., 1968
Alkali reserve			x	See P.-O. Åstrand, 1956
Hemoglobin concentration			x	See P.-O. Åstrand, 1956
Plasma protein concentration			x	See P.-O. Åstrand, 1956

TABLE 12-1 (*continued*)

Organ or function	Increase	Decrease	No effect	References
Cardiac output, rest			x	See P.-O. Åstrand, 1956
Submaximal work		?	?	Freedman et al., 1955; Rowell, 1962; Frick et al., 1963; Tabakin et al., 1965; Andrew et al., 1966; Ekblom et al., 1968; Saltin et al., 1968; Fig. 12-14
Maximal work	x		?	Rowell, 1962; Ekblom et al., 1968; Saltin et al., 1968; Fig. 12-11, 12-14
Heart rate, rest		x		See Steinhaus, 1933
Submaximal work		x		Christensen, 1931; see ref. under "Cardiac output", Figs. 12-14, 12-15
Maximal work		?	x	Robinson and Harmon, 1941; Knehr et al., 1942; Ekblom, 1969
Stroke volume, rest	x			Ekblom et al., 1968; Saltin et al., 1968; Fig. 12-14
Submaximal work	x			See "Cardiac output"; Fig. 12-14
Maximal work	x			See "Cardiac output"; Fig. 12-14
a-$\bar{v}O_2$ difference, rest			x	See "Cardiac output"
Submaximal work	?		?	See "Cardiac output"; Fig. 12-14
Maximal work	x			See "Cardiac output"; Figs. 12-11, 12-14
Oxygen uptake, rest			x	See P.-O. Åstrand, 1956
Given work load		x	x	See P.-O. Åstrand, 1956
Maximal work	x			Robinson and Harmon, 1941; Knehr et al., 1942; Taylor et al., 1949; Rowell, 1962; Ekblom et al., 1968; Ekblom, 1969; Saltin et al., 1968; Table 12-4; Figs. 12-11, 12-13
Blood lactic acid, rest			x	See P.-O. Åstrand, 1956; Williams et al., 1967; Ekblom, 1969; Fig. 12-12
Given work load		x		See P.-O. Åstrand, 1956; Fig. 12-12
Maximal work	x			
Local blood flow, working muscle	?			Elsner and Carlson, 1962; Rohter et al., 1963

TABLE 12-1 (*continued*)

Organ or function	Increase	Decrease	No effect	References
Arterial blood pressure, rest			x	See P.-O. Åstrand, 1956
Submaximal work	?		?	Ekblom et al., 1968; Tabakin et al., 1965; Frick et al., 1963
Maximal work	?		?	Ekblom et al., 1968; Ekblom, 1969
Respiration				
Lung volumes, adults			x	See P.-O. Åstrand, 1956
Lung volumes, adolescents	?		?	See P.-O. Åstrand, 1956
Pulmonary ventilation, rest			?	See text
Submaximal work			x	See text
Maximal work	x*	x		Ekblom et al., 1968
Tidal air, rest	?		x	See text
Submaximal work	?		x	
Maximal work	?		?	
Respiratory rate, rest			x	See text
Submaximal work		?		
Maximal work		?	x	
Diffusing capacity, rest	x		x	Anderson and Shephard, 1968; Reuschlein et al., 1968; Saltin et al., 1968
Submaximal work			x	See "rest"
Maximal work	x*			See "rest"
Miscellaneous				
Specific gravity of body	x			Pařízková and Poupa, 1963; Skinner et al., 1964
Serum cholesterol concentration		?	x	Skinner et al., 1964; Fox and Haskell, 1967; Karvonen, 1967
Serum triglycerides		x	x	Skinner et al., 1964; Fox and Haskell, 1967

* Secondary to the increase in maximal oxygen uptake.

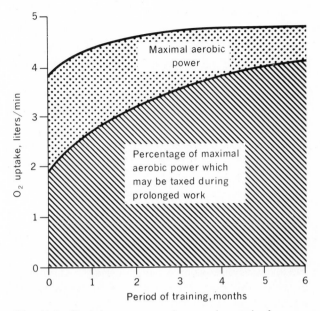

Fig. 12-1   Training causes an increase in maximal oxygen uptake. With training a subject is also able to tax a greater percentage of his maximal oxygen uptake during prolonged work.

per week would be necessary in order to attain a further improvement of a few percent or less, it would hardly be worth the effort. One has to accept the fact that the daily fluctuations in one's state of fitness may exceed this amount of difference.

It is a general experience that it takes several years for an athlete to achieve top results. It is not clear what qualitative and quantitative changes take place in different organ functions which may explain the slow, gradual improvement of the results. An analysis of the demands which a particular athletic event places on the body should form the basis for the training program, taking into consideration whatever deficiencies there may be in the athlete's resources or capabilities to meet these demands.

A certain amount of training of the oxygen-transporting organs is necessary for all categories of athletes, regardless of the nature of the athletic event. Thus he will be better able to cope with the special training required for the event. Furthermore, even the warm-up prior to the event requires a certain amount of fitness. Included in the general training which all individuals, irrespective of profession, age, and sex must undergo, should be (1) a training of the O_2-transporting function (Chap. 18); (2) a muscle training, especially of the back and abdominal muscles (Chap. 8); (3) a training aimed at maintaining joint mobility and the enhancement of the metabolism of the articular cartilage (Chap. 8). In addition, (4) low-caloric consumers should be stimulated to increase their metabolism through regular exercise so as to eventually become high-caloric consumers (Chap. 14). As far as the O_2-transporting function is concerned, distinction should be made between factors primarily involved in the heart and central circulation, and factors involved in the peripheral circulation. In the case of the *central circulation* the training is effective and less strenuous if as large a muscle mass as possible is engaged in the training. In the case of the *peripheral circulation* it is a matter of training the muscles which will be engaged in the performance of the type of event or activity in which an improvement is desired.

Continuous versus Intermittent Exercise

It has often been discussed whether physical training is most effective when the work is accomplished continuously or intermittently, i.e., with periods of more intensive muscular work followed by periods of mild exercise or even rest. In the following discussion we shall present a summary of results from a few studies in which the physiological effects of these types of work are compared (I. Åstrand et al., 1960; Christensen et al., 1960).

One subject was made to accomplish a certain amount of work (64,800 kpm) in the course of 1 hr. This could be done either by uninterrupted work with a load of 1080 kpm/min, or by intermittent work with a heavier work load, interrupted by rest periods at regular intervals. The double work load was chosen, i.e., 2160 kpm/min; thus the required amount of work (64,800 kpm) could be accomplished by 30 min of work within the span of 1 hr. Working continuously without any rest periods, the subject could only tolerate this high work load for 9 min, at the end of which he was completely exhausted. If, instead, he worked for 30 sec, rested for 30 sec, worked for 30 sec, and so on, he could complete the work with moderate exertion. The longer the work periods, the more exhausting appeared the work, even though the rest periods were correspondingly increased. Some of the results of these studies are summarized in Table 12-2. It appeared that with work periods of 3 min interrupted by 3-min rest periods, the load on the oxygen-transporting organs was maximal and the degree of exertion particularly high (note the high lactate concentration in the blood).

Three conclusions may be drawn from these experiments: (1) For the total energy metabolism the amount of work is far more decisive than is the work intensity (the difference in total oxygen utilization during 1 hr was about 10 percent when comparing continuous work with intermittent work with 3-min work periods and 3-min rest periods; the degree of exertion and the load on the oxygen transporting organs was, however, very different in the two situations. (2) In order to force the muscles to work against a large resistance without experiencing exhaustion, the work periods should be of short duration (less than 1

TABLE 12-2

Data on one subject performing 64,800 kpm on a bicycle ergometer within 1 hr with different procedures

Type of exercise		Oxygen uptake		Pulmonary ventilation, liters/min	Heart rate, beats/min	Blood lactic acid, mg/100 ml
		liters/1 hr	liters/min			
Continuous						
1080 kpm/min		146	2.44	49	134	12
2160 kpm/min*			4.60	124	190	150
Intermittent						
2160 kpm/min						
Work	Rest					
½ min	½ min	154	2.90†	63†	150	20
1 min	1 min	152	2.93†	65†	167	45
2 min	2 min	160	4.40	95	178	95
3 min	3 min	163	4.60	107	188	120

* Could only be performed for 9 min.
† Measured during ½ min.
SOURCE: I. Åstrand et al., 1960.

min). (3) For the purpose of taxing the oxygen-transporting organs maximally, work periods of a few minutes' duration represent an effective type of work.

This experiment, as well as a similar experiment with heavier work loads (Fig. 9-5), was briefly reviewed in Chap. 9. A further study will be summarized in this connection (Christensen et al., 1960). Table 12-3 presents the main data on one of the subjects of this study. It is striking that when the subject ran for 10 sec followed by a 5-sec pause, he could run for 20 min during the 30-min period at a high speed without undue fatigue and with a low blood lactic acid concentration. At the end of each running period the load on the oxygen transporting system was maximal or 5.6 liters/min. On the average, the oxygen uptake when the subject was running was 5.1 liters, but the oxygen requirement can be calculated to be 7.3 liters per work minute. It is still an open question how the deficit of about 11 kcal, $(7.3 - 5.1) \times 5$, is covered. The low blood lactate concentration indicates that anaerobic glycogenolysis was not the important energy supplier. The high energy-containing phosphate compounds may have served as a buffer supported by aerobic processes utilizing oxygen bound to the myoglobin in addition to the amount of oxygen transported during the running. At rest the oxygen stores can easily be refilled (Fig. 9-5). It should be pointed out that in all the experiments the oxygen uptake and pulmonary ventilation were also high during the interspersed resting periods. However, the results indicate that the duration and spacing of exercise and resting periods are rather critical with respect to the peak load on the oxygen-transport system. If the

TABLE 12-3

Data on one subject during intermittent running for 30 min at 20 km/hr on a treadmill. Between the running periods, which were varied, the subject was standing beside the treadmill. During continuous running he could proceed for 4.0 min covering a distance of about 1,300 m. Oxygen uptake: 5.6 liters/min; pulmonary ventilation: 158 liters/min; blood lactic acid concentration: 150 mg/100 ml.

Periods work–rest, sec	Distance, m	Oxygen uptake, liters/min			Pulmonary ventilation, liters/min			Blood lactate, mg/100 ml
		Work		Rest	Work		Rest	
		Highest	Average		Highest	Average		
5–5	5,000	. . .	4.3	4.5	. . .	101	101	23
5–10	3,330	. . .	3.4	3.0	. . .	81	77	16
10–5	6,670	5.6	5.1	4.9	157	142	140	44
10–10	5,000	4.7	4.4	3.8	109	104	95	20
15–10	6,000	5.3	5.0	4.5	140	139	144	51
15–15	5,000	5.3	4.6	3.8	110	90	95	21
15–30	3,330	3.9	3.6	2.8	96	79	64	16

SOURCE: Data from Christensen et al., 1960.

resting period of 5 sec is prolonged to 10 sec (running for 10 sec), the peak oxygen uptake observed will be reduced from 5.6 to 4.7 liters/min. Running at the same speed for 15 sec, then resting for 15 sec, was not enough to bring the oxygen uptake (5.3 liters/min) to a maximum.

Figure 12-2 presents another example of how critical the work intensity is for the load on the oxygen-transporting organs in work of short duration. During running at a speed of 22.75 km/hr for 20 sec, 10-sec rest, followed by running, and so on, the oxygen uptake became maximal. If the speed is reduced to 22.0 km/hr, the oxygen uptake is reduced to about 90 percent of the maximum. On the other hand the subject can continue for about 60 min at the lower speed as against only 25 min at the higher speed of 22.75 km/hr. It is an important but unsolved question which type of training is most effective: to maintain a level representing 90 percent of the maximal oxygen uptake for 40 min, or to tax 100 percent of the oxygen uptake capacity for about 16 min.

Figure 12-3 illustrates an experiment with the same subject. Running intermittently 400 m in 70 sec with only a 20-sec rest between each running period brought the oxygen uptake almost to a maximum. There was a gradual increase in the blood lactic acid concentration to 12 mmoles/liter. Another day the subject performed repeated 400-m runs, still in 70 sec, but the resting periods were prolonged from 20 to 60 sec. In this case the oxygen uptake did not exceed 80 percent of his maximum. The lactic acid concentration was also lower, or on a 4 to 5 mmole level.

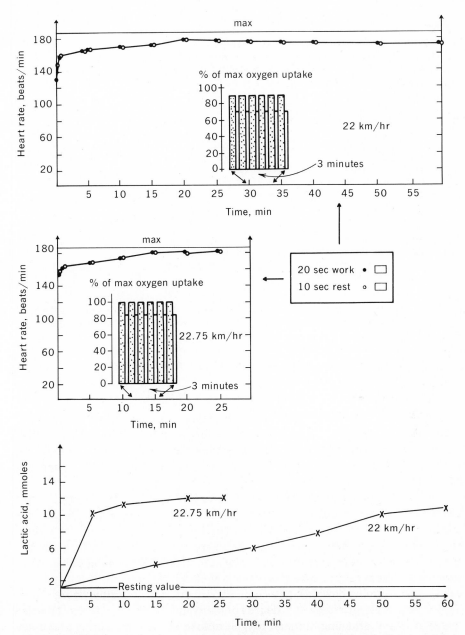

Fig. 12-2 Oxygen uptake, heart rate, and blood lactate concentration during a training program, running on the treadmill, with short work and rest periods (20 and 10 sec respectively) at two different speeds: 22.0 and 22.75 km/hr. Note that the heart rate and oxygen uptake are not maximal in the first case (22.0 km/hr) but that they are in the second case (22.75 km/hr). It should also be noted that the total work time is reduced to half in the latter case. The lactic acid concentration in the blood is about the same in both cases when the work is discontinued. (From Karlsson et al., 1967b.)

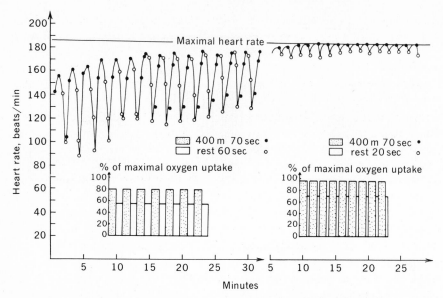

Fig. 12-3   An analysis of the effect of varying lengths of rest periods on the oxygen uptake during running on a treadmill with constant speed and work time. The longer rest period produced a significantly lower oxygen uptake and heart rate during work as well as during rest compared with the experiments in which shorter rest periods were interspersed between each running period. In the latter case the figures approached maximal values. (From Karlsson et al., 1967b.)

The type of training typical for cross-country skiers during the summer and fall is cross-country running through a terrain where there are both short steep slopes and longer and more gentle inclines. Figure 12-4 presents examples of the oxygen uptake during such running, which may extend over a distance of 20 to 40 km (12 to 25 miles).

One day the speed was recorded at different sections of the course as the subjects were running at their usual speeds. On another day the oxygen uptake was measured while the subjects were coached so as to maintain the same speed as before. In the case of these two subjects the oxygen uptake reached maximal values (checked against laboratory studies) when running uphill. When running on the level the oxygen uptake was about 80 percent of the maximal value; when running downhill it was only 50 percent of the maximal figure. There is no doubt, however, that a high degree of motivation is required to exert oneself to the extent of attaining maximal oxygen uptake at intervals during prolonged work. (It may also be a question of the availability of glycogen—see Chap. 14.)

Summary A series of studies have shown that maximal oxygen uptake (and cardiac output) may be attained in connection with repeated periods of work of very high intensity of as short duration as 10 to 15 sec, providing the rest periods between each burst of activity are very short (of equal or shorter duration than the work periods). In more prolonged work, of several minutes duration, the duration of the rest periods is less critical. In this case it is the intensity of the work that determines the load. If the work periods exceed about 10 min or so, a high level of motivation is required in order to attain maximal oxygen uptakes. In the case of continuous work the high tempo required for the severe taxation of the oxygen-transporting system must alternate with periods of reduced tempo,

In the case of very short work periods, about 30 sec or shorter, a very severe load may be imposed upon both muscles and oxygen-transporting organs without the engagement of anaerobic processes leading to any significant elevation of the blood lactate. It is thus possible to select the proper work load and work and rest periods in such a manner that the main demand is centered on (1) muscle strength without a major increase in the total oxygen uptake; (2) aerobic processes without significantly mobilizing anaerobic processes; (3) anaerobic processes without maximal taxation of the oxygen-transporting organs; (4) both aerobic and anaerobic processes simultaneously. The alternatives 2 and 4 do not entail maximal taxation of muscle strength; alternative 3 does not neces-

Fig. 12-4 Oxygen uptake in percent of the maximum during training by cross-country running. Measurements were carried out only during a limited section of the approximately 20-km track. Mean values on two cross-country skiers, both Olympic gold medal winners. (From Karlsson et al., 1967b.)

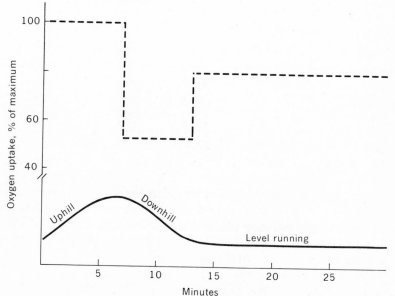

sarily require maximal strength. In the following we will consider how these principles may be applied.

Training of Muscle Strength

The main point of training is primarily to develop strength and endurance in the type of work in which an improvement is sought, whether it be static or dynamic. For a general muscular development, work against high resistance is desirable, with muscle contractions of no more than a few seconds' duration, followed by relaxation, repeated 5 to 10 times. The untrained individual may well refrain from maximal loads. The training should take place at least several times a week.

Patients confined to bed rest or prolonged inactivity of parts of the body may avoid muscular atrophy by subjecting the muscles involved to contractions of a few seconds' duration corresponding to more than about one-third of the maximal strength once a day. The same applies to future astronauts during prolonged space flights. It may be difficult to determine when a muscle contraction is sufficient to correspond to a third of the maximal strength of the muscle. The point is, however, that the contraction does not have to be maximal. This means that most patients may engage in sufficient training to prevent muscular atrophy. It is beyond the scope of this book to consider how patients with neurological lesions may train the neuromuscular function by utilizing certain reflex mechanisms and combinations of movements (see Chap. 4).

As long as the diet is adequate and varied, the protein requirement of the body during training may be met. Additional amounts of protein over and above the normal requirement in the form of protein tablets or an extreme meat diet do not stimulate an increase in muscle mass (see Chap. 14).

Training of Anaerobic Power

Theoretically it is conceivable that an improvement of the processes which depend on the high-energy phosphate compounds may be achieved through maximal work of very short duration, up to 10 to 15 sec, since the energy for this type of work primarily is delivered through these processes. The rest periods should not be too short in order to prevent a major mobilization of glycogenolysis. It is impossible to present exact general guidelines, but the rest periods between each maximal effort should probably be at least a few minutes. It is essential that the training involves the same muscle groups which are engaged in the event in which an improvement is sought.

The training of the anaerobic processes which also involves the splitting of glycogen to lactic acid may no doubt be accomplished rather effectively by adhering to the following guidelines: maximal effort for about 1 min, followed by 4 to 5 min rest, then a further period of 1 min maximal effort, followed by 4 to 5 min rest, and so on. At the end of 4 to 5 such work periods a highly motivated athlete (runner) may gradually attain lactic acid concentrations in the blood in excess of 20 mmoles/liter and an arterial pH approaching 7.0 or lower. No doubt

TRAINING

this form of training is psychologically very strenuous. If a training of the anaerobic motor power also requires the muscles to be exposed to a high lactic acid concentration, this training is rational providing the proper muscle groups are engaged. In order to produce major changes in the cellular milieu, affecting the respiratory function leading to dyspnea, large muscle groups have to be involved in activities such as running.

A training of the anaerobic motor power is important for many groups of athletes. Since this form of training is psychologically very exhausting, it should preferably not be introduced until a month or two prior to the competitive season. Such strenuous training is not recommended for ordinary people.

Training of Aerobic Power

It has already been pointed out that physical activity ranging from repeated work periods of a few seconds' duration up to hours of continuous work may involve a major load on the oxygen-transporting organs and thereby induce a training effect. The following method appears to be more "foolproof" than most other methods: One should work with large muscle groups for 3 to 5 min, rest or engage in light physical activity for an equal length of time, then proceed with a further work period, etc., as required depending on ambitions and the objective of the training. The tempo does not have to be maximal during the work periods in question. It is not necessary to be exhausted when the work is discontinued. Figure 9-3 offers an explanation for this. Results are presented from experiments on the bicycle ergometer with work periods of 5 min. A load of 250 watts represented a maximal oxygen uptake for this subject without being exhausting. An increased load did not produce a further increase in oxygen uptake, but represented a further demand on the anaerobic energy yield and a more pronounced sensation of fatigue. When the work was repeated several times at a load of 250 watts a gradual accumulation of blood lactate took place. The longer the rest periods, the lower the blood lactate. Mild exercise, e.g., jogging, between the heavier bursts of activity may be of an advantage, since the elimination of lactic acid seems then to be faster than at complete rest (Newman et al., 1937; Gisolfi et al., 1966). It has been shown experimentally that the cardiac output and the stroke volume attain their highest values at a load which produces the maximal oxygen uptake, in this case 250 watts. During "supermaximal" work, the oxygen uptake as well as the cardiac output and stroke volume may even attain lower values than is the case at a slightly lower work load. The heart rate may on the other hand be somewhat higher. In this case the training effect may conceivably be slightly reduced. There is no evidence to support the assumption that it is of importance to engage the anaerobic processes to any extreme degree in order to train the aerobic motor power.

The justification for a submaximal tempo in the optimal training of the oxygen-transporting system may further be elucidated by Fig. 12-5. For six subjects an individual speed was determined which brought them to complete

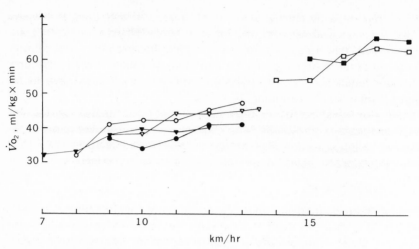

Fig. 12-5 Individual data on oxygen uptake in relation to speed for two female (triangles) and four male subjects. Treadmill was set at an angle of 3°. The highest speed for each subject could be maintained for just 4.0 min. (From Karlsson et al., 1967a.)

exhaustion at the end of the fourth minute of running. Other days the speed of the treadmill was decreased stepwise by 0.5 to 1 km/hr without changing the total distance of the run. A reduction in speed by as much as 3 km/hr for some of the subjects did not, as demonstrated in Fig. 12-5, decrease the oxygen uptake. Therefore, since maximal oxygen uptake can be attained at a submaximal speed, this lower speed may be sufficient and probably optimal as a training stimulus. A highly motivated individual and someone with a high anaerobic capacity will show a wide plateau; others may just be able, or willing, to push themselves to the point where the maximal oxygen uptake is reached (for example, 250 watts in Fig. 9-3b), or not even that far. It is thus impossible to delineate a border value where the greatest load on the oxygen-transporting organs is attained. In the case of healthy young persons the speed of running may be reduced to about 80 percent of the maximum, which may be maintained for a period of 3 to 5 min. If, in other words, the distance which may be covered by running, swimming, or bicycling in a matter of say 3.0 min is covered instead in about 3.5 min, the demand on the aerobic processes remains the same. Thus the stopwatch in such cases should be used to maintain a reduced tempo, not to stimulate the trainee to attain a better achievement in terms of better timing.

As an aid to determine whether or not the load has been maximal or nearly maximal, the heart rate during the work, i.e., the work pulse, may be used. It should not differ more than 10 beats/min from the individual's maximal heart rate, assessed during controlled laboratory experiments. A person accustomed to heavy work can also sense when the pulmonary ventilation has reached the steep part of the slope on the curve relating pulmonary ventilation to oxygen uptake (see Fig. 7-9). This type of training is certainly far more pleasant than is the case when the tempo is higher. Thus, one should definitely distinguish between training of the aerobic and the anaerobic processes. In competitive events in which a superior effect is required in both these processes, the training obviously should also include this combination. This aspect, however, should be postponed until a few months prior to the start of the season, otherwise the athlete may be unable to keep up his training program for psychological reasons.

Figure 12-6 shows how this type of training may be arranged, and its effect on heart rate and oxygen uptake. In the case of the athlete it may be important to adhere strictly to an established schedule. In the case of the ordinary individual who trains for the sake of his own pleasure, it is certainly not necessary to adhere so strictly to a fixed program.

The type of training described for the oxygen-transporting system, with a submaximal tempo for periods of 3 to 5 min, may increase the maximal oxygen uptake and is probably also effective in eliciting many of the side effects which have been described earlier in this book. The ability to work for prolonged periods of time utilizing the largest possible percentage of the maximal oxygen uptake (see Fig. 12-1) may probably primarily be developed just by working con-

Fig. 12-6   Heart rate and oxygen uptake recorded in two subjects during training with alternating 3-min running and 3-min rest. The efforts were not maximal, but the oxygen uptake reached maximal values, as did the heart rate. (From Saltin et al., 1968.)

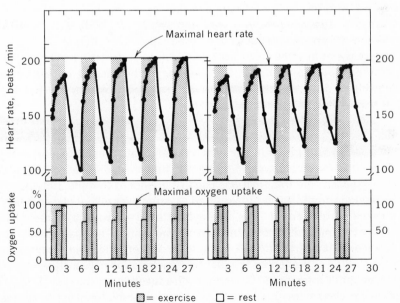

tinuously during long periods of time (endurance training). The capacity to store glycogen in the muscles and the ability to mobilize and to utilize free fatty acids play a major role in prolonged work (see Chap. 14). To what extent these may be developed through endurance training is still obscure.

Whatever "special effects" these two types of training may elicit in order to improve the maximal aerobic power and the maximal aerobic capacity, a certain overlapping in the elicited effect is highly probable. (It is an interesting but unsolved question as to what physiological training effects may be obtained with continuous work representing 80 percent of the maximal cardiac output (say 30 liters/min) in the course of 1 hr, compared to 20 min of maximal cardiac output, and 40 min with only 40 percent of the maximal output. The amounts of blood pumped by the heart are 1,440 liters and 1,080 liters respectively.)

The athlete often needs a certain amount of variation in his training. According to the above there are considerable possibilities for variation, even though certain programs are more critical than others concerning work time, rest periods, and intensity.

In the case of patients and completely untrained individuals, it is out of the question to prescribe an accelerated tempo in connection with the training of circulation. A previously bedridden patient should be satisfied with a training load which commences by elevating the heart rate by about 30 beats/min above the resting value (to about 100 beats/min). In the case of the habitually sedentary individual an elevation of the heart rate by about 60 beats/min may be a suitable initial intensity. The principle of intermittent work is also valid for these categories of individuals. With daily training periods of up to 30-min duration, or even with only a few training periods per week, the tempo may gradually be increased. The individual's health, age, and interest may determine how strenuous the training should be. It should be pointed out that in many cases it is not necessary, or in some cases not even desirable, to attain maximal oxygen uptake and cardiac output during the training. In the case of patients and ordinary individuals the "endurance training" may conveniently be accomplished by such activities as walking or bicycling.

It has repeatedly been emphasized that large muscle groups should be engaged when training the central circulation. Figure 12-7 may be of interest in this connection. In a number of top Swedish athletes belonging to the National Team, the maximal oxygen uptake was determined by laboratory experiments. In most cases the treadmill was used. It is evident that the athletes who had the highest maximal oxygen uptakes had selected events that placed heavy demands on their aerobic power. This means that these events also represent an excellent form of training of the oxygen-transporting system. Ball games fall rather low on the scale (data for handball, basketball, soccer, and lawn tennis support this). This may be explained by the fact that each period of activity with a high tempo is frequently interrupted by periods of reduced tempo or rest (Table 12-2). Competitive calisthenics entail work periods up to 1 min. Needless to say, calisthenics may be performed in such a manner that they entail an excellent training effect of the aerobic power (straddle jump, sequences of movements engaging large muscle groups). Intensive activities involving small muscle groups (chinning the bar, push-ups, weight lifting) may be harmful for untrained individuals and cardiac patients since these activities produce a high heart rate and high blood pressure at a given cardiac output and oxygen uptake (Chap. 6).

"Circuit training" (Morgan and Adamson, 1962) entails a series of activities, for instance eight different activities performed one after the other. At the end of the eighth activity one starts from the beginning again and carries on until the entire series has been repeated altogether three times. By a preliminary test in which many of the activities may be performed at maximal exertion, the number of repetitions of each activity is determined. The time required for the three repetitions is about 12 min. The advantage with this circuit training is that every individual undergoes a program adjusted to his level of fitness. He may follow his own improvement in that the time for the three repetitions is recorded and he endeavors to shorten the time required. At the end of a few weeks' training a new test is performed of the number of repetitions of each activity he can manage (for instance push-ups). This training produces a high degree of motivation. The program may be accomplished in a limited space. The disadvantage is that an untrained individual is exposed to tests requiring maximal exertion. It has been found that a correctly devised program varying the involvement of large and small muscle groups, mixing static and dynamic work, does not produce maximal oxygen uptake measured on the bicycle ergometer, but only about 80 percent of the maximal O_2 uptake (Hedman, 1960). In spite of this, the heart rate is almost maximal, the lactic acid concentration in the blood is very high, and the degree of exertion is considerable. Circuit training may be included in a training program, not only for athletes (especially those who fall within the lower part of Fig. 12-7) but also for school children for the sake of variation and for experience.

The Canadian Air Force program (1962) is essentially devised as a form of circuit training and has, on the whole, the same advantages and disadvantages as does circuit training in general.

Year-round Training

In all cases regularity in the performance of training is important. It is possible within a month to develop a reasonable level of fitness, strength, and so on, but this disappears when the training is discontinued. As already pointed out, less effort is required to maintain a certain level of fitness than to develop this level in the first place. In the case of ordinary people it may therefore be worthwhile, when time permits, to devote more time to training, and then later on to endeavor to maintain the acquired level of fitness by only a few training periods per week.

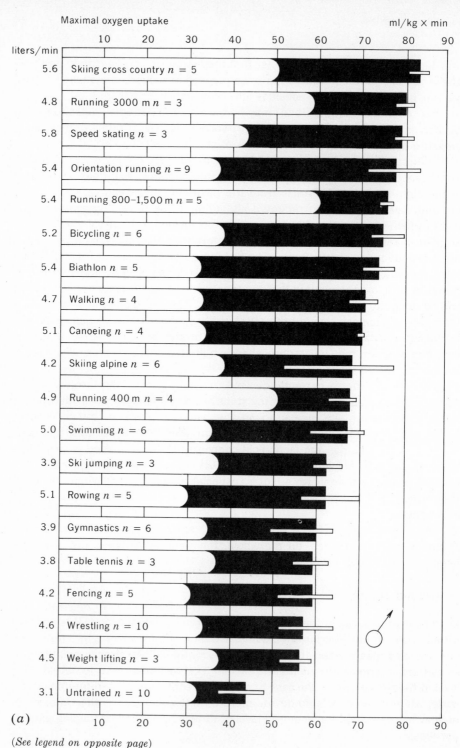

Maximal oxygen uptake   ml/kg × min

liters/min		
5.6	Skiing cross country n = 5	
4.8	Running 3000 m n = 3	
5.8	Speed skating n = 3	
5.4	Orientation running n = 9	
5.4	Running 800–1,500 m n = 5	
5.2	Bicycling n = 6	
5.4	Biathlon n = 5	
4.7	Walking n = 4	
5.1	Canoeing n = 4	
4.2	Skiing alpine n = 6	
4.9	Running 400 m n = 4	
5.0	Swimming n = 6	
3.9	Ski jumping n = 3	
5.1	Rowing n = 5	
3.9	Gymnastics n = 6	
3.8	Table tennis n = 3	
4.2	Fencing n = 5	
4.6	Wrestling n = 10	
4.5	Weight lifting n = 3	
3.1	Untrained n = 10	

(a)

(See legend on opposite page)

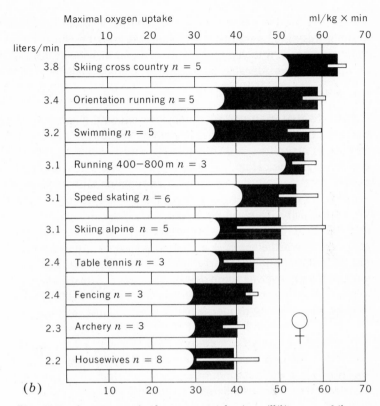

Fig. 12-7 Average maximal oxygen uptake in milliliters per kilogram body weight times minutes for male (a) and female (b) Swedish national teams in different sports. The horizontal thin white bar at the top of each black bar denotes range; n = number of subjects. To the left average maximal oxygen uptake in liters per minute is also given. (From Saltin and P.-O. Åstrand, 1967.)

The athlete often has to train many different functions (aerobic and anaerobic power, strength, endurance, technique). It may be practical and sometimes necessary because of time limitations at certain periods to concentrate on some of these functions. This particular type of training has to be continued, however, even though it may be at a reduced intensity, when he devotes himself to the next function. Otherwise he will be losing the improvement he has attained. The problem is to decide how intensively the different functions have to be trained in order to maintain a satisfactory level.

The keen competition of today necessitates year-round training. For a variety of reasons we believe that this is important especially with regard to the oxygen-transporting system. This is particularly important for the aging athlete. The reason why many thirty- to thirty-five-year-old athletes may continue to rank among the world elite in endurance events is often due to a relatively hard training during all seasons of the year. He may be what is known as "hard to train" compared with younger athletes, and if he allows his fitness to deteriorate, he may have considerable difficulty in regaining his level of training.

Psychological Aspects

It appears that man has a tendency to become inactive after he has reached puberty. Unfortunately, physical activity without a definite purpose becomes from then on rather rare. In the past, and in some countries and occupations even today, physical activity was a part of one's daily life and work. Inasmuch as the need for physical activity in connection with most daily occupations is about to be abolished, it will be necessary to devote some of the leisure time to physical activity. This will be further discussed in Chap. 18. In this connection it should be emphasized that effective physical training does not have to be very stressful. Nevertheless it is a common misconception that physical training is unpleasant and difficult to arrange. In many cases the ideal solution is to acquire a hobby which involves some degree of physical activity, even though the training in such cases may not be very intensive. As a rule most people can be motivated to exercise providing proper facilities are available. Examples of simple types of training will be given in Chap. 18. Psychologically speaking it is far more attractive to walk or run a 5-km course than to cover a 1-km course 5 times, to say nothing of 100 laps on a 50-m course. It is more acceptable to cover a distance which offers a certain amount of variation than to spend an equal number of calories indoors on a stationary bicycle.

Tests

Objective tests to determine the effect of a training on different functions are both important and desirable. Such tests may be an aid in the development of the program and they encourage the individual to continue his training. Apart from the bicycle ergometer test, there are no simple reproducible tests available for field use (Chap. 11). Figures 11-7 and 11-8 presented examples of the heart rate during a standardized work load during a period of training. Figure 12-8 presents further examples of the application of the submaximal work test, in these cases with elite athletes. The cross-country skier Sixten Jernberg (maximal oxygen uptake, 5.88 liters/min) was followed for several years. During his career he won several Olympic medals. Considerable seasonal variations in his heart rate at a standard work load were observed, the highest heart rate being observed during the summer and the lowest values during the winter season. It was noted that the heart rate dropped with the advancing years. He trained more intensively, even during the summer, from 1955. The second example in Fig. 12-8 was a champion canoeist. His physical condition improved during winter and early spring; training included cross-country running and skiing. When the lakes were cleared of ice, the athlete took up canoeing almost

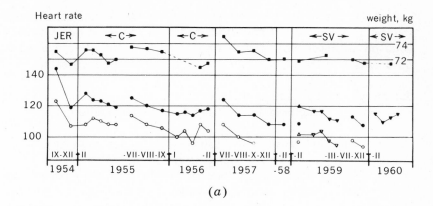

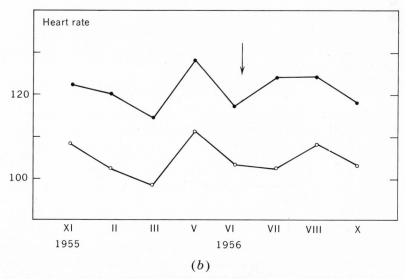

Fig. 12-8 (a) Body weight (squares) and heart rate during work on the bicycle ergometer on various occasions from 1954 to 1960 for Sixten Jernberg, successful cross-country skier. Symbols: C = Cortina, SV = Squaw Valley, open symbols = 900, and closed symbols = 1200 kpm/min (200 watts). Left triangle = test in low pressure chamber simulating 1,800-m altitude = Squaw Valley.

(b) Heart rate at 900 (○) and 1200 (●) kpm/min during repeated tests on a canoeist training for the November 1956 Olympic Games in Melbourne. Cross-country running was interrupted from April to May. Arrow indicates broken rib which hindered training.

exclusively. Due to the engagement of smaller muscle groups, the intensity of training could not be maintained, as shown by the heart rate response (May). Later on, a broken rib hindered training for more than $1\frac{1}{2}$ months. (He was unsuccessful at the Olympic Games due to a broken paddle at the start!)

Summary   From a practical standpoint it may be logical to list four components of a rational training program aimed at developing the different types of power:

1   Bursts of intense activity lasting only a few seconds may develop muscle strength and stronger tendons and ligaments.
2   Intense activity lasting for about 1 min, repeated after about 4 min of rest or mild exercise, may develop the anaerobic power.
3   Activity with large muscles involved, less than maximal intensity, for about 3 to 5 min, repeated after rest or mild exercise of similar duration may develop the aerobic power.
4   Activity at submaximal intensity lasting as long as 30 min or more may develop endurance, i.e., the ability to tax a larger percentage of the individual's maximal aerobic power.

BIOLOGICAL LONG-TERM EFFECTS

It should be emphasized that by long-term effects we mean changes which may require a certain amount of time to be developed (weeks, months, or years).

In this section we will discuss in some detail the various training effects indicated in Table 12-1.

Locomotive Organs

In Chap. 8 it was pointed out that *bones, ligaments,* and *joint cartilages* are affected by use as well as by disuse. Bone structures which are not stressed may disappear and new bone trabeculae may be created where altered mechanical forces increase the demand for sturdiness. The hard interstitial substance consists of carbonates and phosphates of calcium, and may constitute up to about 60 percent of the dry, fat-free weight in adult life. After a long period of inactivity it may be as low as about 40 percent (Ingelmark, 1957). The increased urinary calcium and phosphorus during prolonged bed rest depends on a demineralization of the bones (Deitrick et al., 1948; Rodahl et al., 1967). The thickness of the articular cartilage is greater in trained animals than in untrained ones. The compressibility of the cartilage will therefore increase with training, providing greater possibilities of compensating for any incongruence of the articular surfaces of the cartilages. The contact surface will increase, and when stressed, the force per unit of surface will thus decrease (Holmdahl and Ingelmark, 1948; Ingelmark, 1957). Training causes a hypertrophy of the intercellular substance of connective tissue, increasing the volume of tendons and ligaments, and enhances their tensile strength. The hypertrophy of the tendons is accompanied by a hypertrophy of the muscles attached to them (Ingelmark,

1948). In the bone-ligament-muscle system the attachment of the ligament to the bone is the weakest point, although they tend to become stronger in trained animals (Viidik, 1966).

During training there is an increase in *muscle mass* by an enlargement of the already existing fibers. Most investigators find no increase in the number of muscle cells with training through a division of already existing cells. Results from animal experiments have recently been published, however, which indicate that prolonged training may create new muscle fibers, even though the most important factor in the volume change of the muscle is a true hypertrophy of the existing muscle fibers (Linge, 1962; Reitsma, 1965). The total amount of protein in the muscle increases with training and decreases with inactivity. Disuse-atrophy is associated with a decline in myofibrils, and the proportion of sarcoplasm proteins rises (Helander, 1958). There is in other words a selective reduction in the contractile myofibril proteins, and this is more pronounced than the reduction which takes place in the total amount of protein. With inactivity there is an increase in the fat content of the muscle, while it decreases with training.

Both training and disuse of a muscle are associated with *biochemical changes* in the muscle (Yakovlev, 1958). According to Nöcker et al. (1958), training causes an increase in the potassium content of the skeletal muscles. During exhausting work, however, the potassium content drops to lower levels in trained than in untrained individuals. Enzyme systems in the muscles may be affected by training, which may partly account for the increased power that occurs with training (Holloszy, 1967; Kraus and Kirsten, 1969). It has been claimed that the serum level of various enzymes is raised after exercise, but this increment is more pronounced in untrained than in trained individuals (Fowler et al., 1962; Garbus et al., 1964). This finding suggests a difference in cellular permeability.

Several investigators have found an increase in the number of capillaries in the muscle as the result of training (Fig. 12-9). An increase in the number of capillaries reduces the tissue cylinder around a capillary, increasing the capillary surface area for an exchange of materials. It would also tend to raise P_{O_2} and lower P_{CO_2} and the concentration of metabolites in the interstitial fluid around the muscle fiber.

Training may also develop the arterial tree probably by opening potential collateral vessels (Schoop, 1964, 1966). The increased blood flow with a vasodilation distal to the arterial tree in question is believed to be an important factor in the development of these collaterals.

In Chap. 4 it was pointed out that the measured *muscle strength* may vary greatly from test to test in the same subject. The explanation is probably that some of the training takes place in the motoneurons associated with the muscles being trained. The less the inhibition of these motoneurons the more motor units may contract at tetanus frequency. In certain situations (danger, competition, etc.) and as the result of training, this inhibition may itself be inhibited, resulting in the development of a larger muscular strength (Ikai and Steinhaus, 1961). These two factors—a variation in the composition of the muscle tissue and a variation in the number of muscle cells which may be engaged through maximal exertion of the subject's willpower—may explain the fact that the effective area of the cross section of a muscle is not a dependable measure of its maximal strength (McMorris and Elkins, 1954; Rasch and Morehouse, 1957).

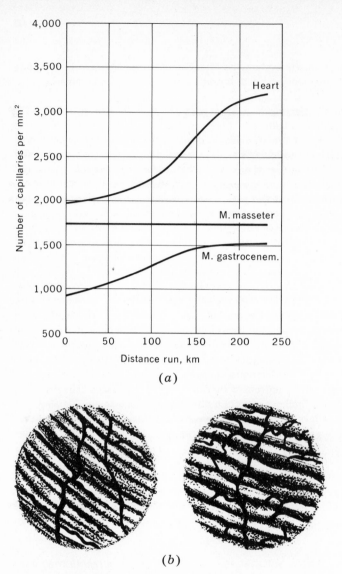

(a)

(b)

Fig. 12-9   Increased vascularization of a muscle as the result of training.
(a) Capillary density in three muscles of guinea pigs in the course of daily training on a treadmill at a speed that was gradually increased to 60 m/min, running distance about 1,800 m each time. Animals were analyzed after different training times. Note that in the masseter muscle there was no change in capillary density, in contrast to the systematically exercised muscles. (Modified from Petrén, 1935.)
(b) Left: Untrained muscle with relatively few capillaries and few anastomoses. Right: Trained muscle with thickened muscle fibers, increased number of capillaries, and numerous anastomoses. (Redrawn from Vannotti and Pfister, 1934.)

As discussed in Chap. 4, one may distinguish between maximal static (isometric) and dynamic strength (sometimes incorrectly called isotonic strength). In addition, the endurance is also of interest, both with reference to static and dynamic exercise. Figure 4-28 shows that only when the exerted force is reduced to a level below 15 percent of the maximal strength is it possible to maintain the contraction "indefinitely." During short efforts lasting only for seconds, the anaerobic processes play a dominating role, with high phosphate compounds involved in the energy yield. In the case of more prolonged efforts the aerobic processes assume a more dominating role, in which case an effective circulation is of major importance. For this reason, and in view of the important role played by the nervous system in the development of muscular strength (with a special nervous circuit for every single movement of the body), it is obvious that the effect of the training primarily is related to those parts of the body and the movements which are involved in the training. The effect is considerably less pronounced in unfamiliar activities, even though the same muscle groups may be engaged (Rasch and Morehouse, 1957). A considerable confusion has resulted from the fact that some investigators have used the same activity both for testing and for training, while others have used a test situation which has not corresponded to the actual training.

It has been shown that maximal isometric training to a great extent increases the isometric strength without the endurance being significantly affected. Compared to the increase in strength, the endurance may actually be reduced. This may be explained by the occurrence of a muscular hypertrophy without a corresponding increase in vascular-ization. The result may be an increase in the tissue cylinder surrounding the capillary.

A training involving submaximal strength and repeated dynamic contractions increases primarily the dynamic endurance (Petersen et al., 1961; Hansen, 1967), while a training with repeated isometric contractions develops endurance merely in this par-ticular type of work (Hansen, 1961).

According to Rasch et al. (1961) and Hettinger (1968), a general improvement of isometric strength is achieved which is unrelated to the length of the muscle during the training (for instance training of the biceps muscle with an angle of 45, 90, or 135° flexion of the elbow joint).

The mechanism eliciting the changes which produce an increased strength as the result of a training program is unknown. It is of interest to note that hypoxia per se does not appear to be of decisive importance. Training carried out while the circulation to the muscle in question is occluded does not produce any better effect than is the case when the circulation is free (Hettinger, 1968).

In Chap. 4 it was pointed out that the structural and metabolic changes in a denervated muscle are not identical with those of an inactivated muscle with an intact nerve supply. In a muscle subject to disuse atrophy caused by a cast, the total nitrogen content is reduced to a lesser extent than is the case after neurotomy (Helander, 1958). A stimulation of a denervated muscle will not lead to a hypertrophy, as is the case in a normal animal subjected to training (Gutman et al., 1961). Apparently the nerve cell has a trophic function essential for the muscle metabolism and protein synthesis.

There is only a slight correlation between static strength and speed of movements. A training of the progressive resistance type improving the strength may, however, also increase the speed of movement, but the change is not large (Clarke and Henry, 1961). A strength training has no influence on reaction time ability (Beers, 1935; Clarke and Henry, 1961).

Summary   The degree of activity may to a great extent affect the locomotor organs, morphologically as well as biochemically. Training is associated with an increased strength of the muscle attachment of the bone, and an increase in the contractile force of the muscle. The processes which are the basis for the energy yield are affected in a favorable manner, and the peripheral circulation is improved. In all probability the central nervous system is also affected, leading to improved coordination and making it possible for more motor units to be engaged simultaneously at tetanus frequency.

A person's potential for development of muscle strength is established at birth, since a given individual is born with a certain fixed number of muscle fibers, which on the whole, remains unaltered throughout life. While training increases the maximal muscular strength, inactivity reduces this strength. This is related to changes in the thickness of each muscle fiber and therefore also to the cross section of the muscle as a whole.

Muscular endurance is related to the exerted force in relation to the maximal strength which the muscle can develop; when the exerted force is below 15 percent of the muscle's maximal strength, the contractions may be continued more or less indefinitely. In the training of a muscle the objective of the training should be kept in mind in that the physiological processes which are the basis for the development of muscular strength and which are affected through training are different, depending on the duration of the muscular effort. Brief efforts depend primarily on anaerobic processes. In prolonged efforts aerobic processes play a major role.

Isometric training increases the isometric strength without the endurance being significantly affected. Repeated dynamic contractions increase primarily dynamic endurance. Hypoxia per se does not appear to be of decisive importance in the development of muscular strength. Apparently the nerve cell has an essential trophic function governing the metabolism and protein synthesis associated with the increase in muscular strength through training.

Oxygen-transporting System

Like the skeletal muscle, the heart is adaptable to variations in the individual's physical activity. The changes just described also occur in essence in the heart muscle. It is typical for athletes in endurance events to have a large *heart volume* (Figs. 6-24, 12-10). In relation to body weight the rabbit has a small heart compared to the hare (Fig. 10-5). With training, the rabbit becomes more like the hare as far as its physical constitution is concerned (Arshavsky, personal communication). Animal experiments have demonstrated an enhanced vascularization of the heart muscle as a consequence of training (Fig. 12-9). An important question which is not as yet fully answered is whether an enhancement of collateral formation in the heart muscle is limited to already vascularly handicapped areas or whether it may also occur in a healthy heart.

There is a high correlation not only between heart volume and maximal O_2 uptake in persons of a certain age but also between *blood volume* or *total hemoglobin* and maximal O_2 uptake (Fig. 10-4). As we have already pointed out, however, there are considerable individual variations in these parameters. The older person may largely retain his circulatory dimensions although he has a reduced maximal aerobic power (Fig. 10-6). This is true even in the case of older, former athletes, especially as regards the heart size (Holmgren and Strandell, 1959; Grimby and Saltin, 1966; Reindell et al., 1967). Training may increase and bed rest decrease these parameters, but the interrelationship between these parameters may not always be consistent (Deitrick et al., 1948; Taylor et al., 1949; Hollman and Venrath, 1963; Miller et al., 1964; Sjöstrand, 1967; Saltin et al., 1968). Bed rest primarily causes a decrease in the plasma volume, and a reduction in the red cell volume only occurs after prolonged bed rest (Miller et al., 1964; Vogt et al., 1967).

It has long been established that individuals known to have considerable endurance usually have a slow resting *heart rate*. Hoogerwerf (1929) found a mean pulse of 50 beats/min in 260 athletes participating in the Amsterdam Olympic Games (1928), the lowest value being 30 beats/min. In one cross-country skier the resting heart rate was repeatedly as low as 28 beats/min, while a heart rate of 170 was recorded during heavy exercise (unpublished results). Habitual training enables a person to achieve a certain cardiac output at rest, as well as during work, with a slow heart rate and a large stroke volume. This improves

Fig. 12-10   The heart volume per kilogram body weight for members of different national teams of Germany and a group of untrained individuals of the same age. Each dot represents one subject. Note the large individual variations within a group. (From Roskamm, 1967.)

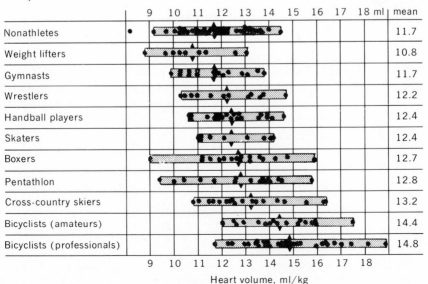

	9 10 11 12 13 14 15 16 17 18 ml	mean
Nonathletes		11.7
Weight lifters		10.8
Gymnasts		11.7
Wrestlers		12.2
Handball players		12.4
Skaters		12.4
Boxers		12.7
Pentathlon		12.8
Cross-country skiers		13.2
Bicyclists (amateurs)		14.4
Bicyclists (professionals)		14.8

9   10   11   12   13   14   15   16   17   18

Heart volume, ml/kg

the economy of the heart muscle as far as energy requirement and oxygen demand are concerned (Chap. 6; Berglund et al., 1958; Raab, 1966). The origin of this sinus-mediated bradycardia is not clear (Hall, 1963).

It is assumed that training causes an increased centrogenic vagal cholinergic drive combined with a sympathoinhibitory mechanism. Prolonged inactivity on the other hand causes preponderance of the oxygen-wasting adrenergic system. Herrlich et al. (1960) have shown in rats that prolonged training in a running cage was followed by a significant increase of the atrial acetylcholine content (Tipton et al., 1966). This may be the result of an increased vagal discharge. Trautwein et al. (1960) have presented evidence of a spontaneous release of acetylcholine in the right atrium of the dog's heart. It is conceivable that this production may also be affected by training. According to Tipton (1961), however, an intact vagus is necessary for a hypertrophy to occur of the heart muscle, and the difference in resting heart rate between trained and untrained animals does not occur if the vagal nerve is severed.

Whether or not the stretch receptors in the dilated and hypertrophic atria might elicit a bradycardia has also been discussed. There is no evidence, however, in support of such a mechanism. A cardiac patient may also have a large heart, but this is not associated with any bradycardia. The increased blood volume caused by training also causes an improved venous return and enhances the filling of the heart during diastole.

It may be concluded that the mechanism underlying the effect of training on the work of the heart is by no means clear. In addition to an effect on the contractile force of the heart muscle and an effect on the relationship between sympaticus and parasympaticus, there is also an effect on the enzyme system and the electrolyte balance in the myocardial tissue. The final result is a diminished demand on the oxygen consumption and thereby on the blood flow through the heart muscle at a given cardiac output. There are good indications to suggest that the blood supply is enhanced in the trained heart.

At rest the *pulmonary ventilation* per liter of oxygen consumed is more or less unchanged after training. Possibly the depth of respiration is somewhat increased and the respiratory rate correspondingly reduced. (It should be pointed out that any measurement of pulmonary ventilation at rest is very difficult to achieve without affecting the subject.)

It is stated in several reports that the *cardiac output* at submaximal work and at a given oxygen uptake is not significantly affected by training. Others have found that it is somewhat lowered. Bevegård et al. (1963) found well-trained athletes to have the same cardiac output at a given oxygen uptake as did untrained persons. On the whole, the maximal heart rate appears to be the same at different levels of training. The fact that the mean heart rate is somewhat lower in athletes engaged in endurance events (about 10 beats/min, Saltin and P.-O. Åstrand, 1967) may be due to constitutional factors. The possibility cannot be excluded, however, that many years of training, perhaps especially when this is achieved with a marked increase in heart volume, might also bring about a drop in the maximal heart rate (Ekblom, 1969).

An increased stroke volume combined with an unaltered maximal heart rate means an increased maximal cardiac output as a consequence of habitual physical activity. Since training in addition improves the possibility for the tissues to utilize the available oxygen volume, there is a dual basis for an *increased maximal oxygen uptake:* (1) an increased maximal cardiac output; (2) an increased a-$\bar{v}O_2$ difference. Quantitatively these two factors appear to contribute about equally toward increasing the maximal aerobic power.

Table 12-4 summarizes data pertaining to maximal O_2 uptake in connection with training on the one hand and prolonged periods of bed rest on the other. A certain amount of variation in the results is evident. This may be explained by the fact that the subjects were not all equally "sedentary" prior to the training, or that the training intensity varied, and that different individuals undoubtedly react differently on training. It should be noted that the maximal O_2 uptake in three subjects (Saltin et al., 1968) increased from 1.74 liters/min following bed rest to 3.41 liters/min, or by an amount equivalent to 100 percent after training (Fig. 12-11).

At a given submaximal O_2 uptake the content of *lactic acid* in the blood is lower in a trained subject than in an untrained one, as shown by a series of studies (Fig. 12-12). This may be interpreted as an expression of a more effective oxygen transport during the beginning of the work, leading to a diminished anaerobic energy yield. When a certain increase in blood lactate is observed at an O_2 uptake corresponding to 50 percent of the maximal O_2 uptake in an

TABLE 12-4

Variations in measured maximal oxygen uptake during prolonged bed rest and training, respectively, in some longitudinal studies (training period up to 6 months). Control values were obtained before bed rest and training. If not otherwise stated, subjects were twenty to thirty years of age

Authors	Number of subjects	Maximal oxygen uptake, liters/min		
		Control	After bed rest	After training
Robinson and Harmon, 1941 (Dill et al., 1966)	9	3.36	. . . .	3.90 +16%
Knehr et al., 1942	14	3.45	. . . .	3.69 +7%
Taylor et al., 1949	2	3.85	3.18 −17%	
Rowell, 1962	7	3.47	. . . .	3.93 +13%
Ekblom et al., 1968	8	3.15	. . . .	3.68 +16%
Saltin et al., 1968	3	2.52	1.74 −31%	3.41 +33%
Saltin et al., 1969	42*	2.90	. . . .	3.43 +18%
Saltin et al., 1969	8†	2.25	. . . .	2.67 +19%
I. Åstrand and Kilbom, to be published	11‡	1.90	. . . .	2.18 +15%

* Age 34 to 50 years.
† Age 50 to 63 years.
‡ Women, age 19 to 27 years.

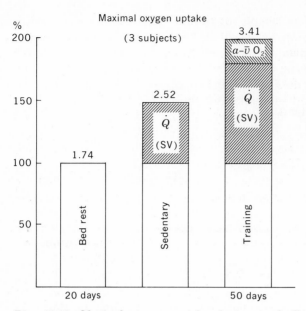

Fig. 12-11 Maximal oxygen uptake during treadmill running for three subjects after bed rest (= 100 percent), when they are habitually sedentary, and after intensive training, respectively. The higher oxygen uptake under sedentary conditions compared with bed rest is due to an increased maximal cardiac output ($\dot{Q}$). The further increase after training is possible due to a further increase in maximal cardiac output and arteriovenous O_2 difference ($a\text{-}\bar{v}O_2$). The maximal heart rate was the same throughout the experiment; therefore the increased cardiac output was due to a larger stroke volume (SV). (From data obtained by Saltin et al., 1968.)

untrained individual, this percentage may be elevated to 60 or 65 percent in a well-trained individual (see Chap. 9; Williams et al., 1967).

No consistent data are available concerning the blood pressure during work before and after training. In the case of persons without elevated blood pressure eventual changes are in any case moderate (Ekblom, 1969).

During maximal work, the well-trained individual usually achieves a higher concentration of lactic acid in the blood and a lower blood pH than an untrained individual. To what extent this may be due to an increased physiological tolerance for lactic acid with its side effects, or a greater psychic ability to exert oneself, is not clear.

The enlargement of muscle fibers induced by training and, thereby, the increased muscular strength must inevitably increase the energy demand including an increased anaerobic energy yield.

During submaximal work of relatively low intensity the *pulmonary ventilation per liter O_2 consumed* does not change materially with training. Apparently the depth of respiration is increased somewhat, associated with a corresponding reduction in the respiratory rate. During heavier work the ventilation per liter O_2 uptake is reduced, but reaches a higher level during maximal work. A study of Fig. 7-9 facilitates the explanation of these findings. During light work, the level of the pulmonary ventilation is primarily determined by the CO_2 production which is directly related to the O_2 utilization. During heavier work the pH is also altered, primarily by the increase in blood lactate. This causes a relatively steeper increase in the pulmonary ventilation. Since the blood lactate concentration is generally lower during submaximal work following training, the respiratory drive is reduced, and the result is a lower pulmonary ventilation. The higher maximal ventilation is partly due to the increased maximal aerobic power, leading to an increased CO_2 production, and is partly due to the higher maximal lactic acid level. Thus, the untrained individual and individuals with low maximal aerobic power fall within the left side of the shaded area in Fig. 7-9, while training moves the curve downward to the right. Since the vital capacity does not change with training (in adults), no major changes in the depth of respiration are to be expected during work. It should be recalled that this is up to 50 to 55 percent of the vital capacity in trained as well as in untrained individuals (Chap. 7). At any

Fig. 12-12 To the left: Progressive decrease of blood lactate for a standard amount of exercise: running on a treadmill at 7 mph for 10 min. During the first 20 days (A) training consisted of running daily on the treadmill for 20 min at 7 mph. A steady level of blood lactic acid is reached around 3 mmoles. During the following thirty days (B), training is increased to running at 8.5 mph for 15 min daily. Blood lactic acid decreases further, and a new steady level is reached around 1.5 mmoles after the standard test. (From Edwards et al., 1939.)

To the right: Peak concentration of lactic acid in blood in relation to oxygen uptake up to maximum before (dotted line) and after (solid line) 16 weeks of physical training. Mean values on 8 subjects exercising on a bicycle ergometer. Vertical bars indicate ±1 standard deviation. (From Ekblom et al., 1968.)

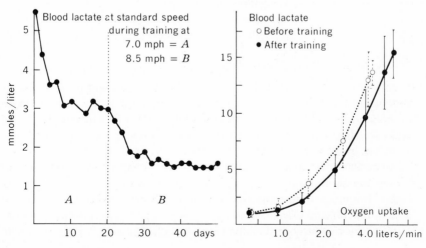

given level of pulmonary ventilation the mechanical work of breathing is the same for untrained and trained individuals (Milic-Emili et al., 1962).

Figure 12-1 illustrates schematically how the maximal aerobic power may increase with training. Another effect of training is that the individual, during prolonged work, may tax a larger percentage of his maximal O_2 uptake than is the case when he is untrained. The reason for this is difficult to assess. A more effective O_2 supply to the working muscle due to an enlargement of the vascular bed, an increased potassium content of the muscle, an increased glycogen content, and a higher psychic "fatigue threshold" may contribute in a positive direction. With training, the individual becomes more accustomed and willing to push himself closer to his limit. It is important to keep in mind, however, that achievements in endurance events may be improved beyond that which are mirrored in changes in the maximal aerobic power.

Recovery from Exercise

There are very few longitudinal studies concerning the payment of oxygen debt, return to resting level of heart rate, cardiac output, temperature, etc., before and after training. The fast recovery of athletes after muscular work is, on the other hand, well established.

It has been recently observed by Hartley and Saltin (1968) that untrained subjects who (1) work for 6 min at an O_2 uptake of about 40 percent of their maximal aerobic power, then (2) work for 6 min at a load of about 70 percent of their maximum, followed by (3) a 10-min rest, and then (4) repeat the 40 percent load, now have the same O_2 uptake and cardiac output as under (1) but they now have a significantly higher heart rate, in many cases as much as 20 beats/min higher, and consequently also have a smaller stroke volume. Following training, this increase in heart rate is much less, and well-trained athletes may accomplish the entire procedure listed above with the results from 1 and 4 being almost identical. This type of testing appears quite promising as a method of assessing the level of physical fitness or the level of habitual physical exercise.

Mechanical Efficiency, Technique

In activities which are relatively uncomplicated technically, such as walking, running, or bicycling, there is a very slight increase in efficiency with training, but this increase is less than the variability between individuals. There is no definite difference in the consumption of calories per kilogram body weight and kilometer in differently trained runners (Böje, 1944; Erickson et al., 1946; P.-O. Åstrand, 1956). On a bicycle ergometer both Olympic medal bicyclists and untrained persons have the same mechanical efficiency at submaximal work loads. The more complicated the exercise, the greater are the individual variations in mechanical efficiency and the greater is the improvement with training. It should be pointed out that the mechanical efficiency of a person performing heavy work may be overestimated if the assessment is based on measurement of O_2 uptake, since anaerobic processes may have contributed to the energy yield.

In many achievements the aim of the training is not primarily to reduce the energy expenditure during the event in question, but to attain an improvement "at all costs," for instance by increasing the developed power. Lauru (1957) gives an example with an athlete doing a broad jump from a "force platform" before and after training. (With the aid of this force platform, force phenomena during movement arising vertically, frontally, and transversely can be recorded and analyzed.) Before training, the push against the platform during the jump reached 103 kp. During the training, the push increased to 145 kp. The performance was at least partially improved by an improved coordination and a smoother sequence of motions.

It is an interesting question to what extent the unskilled individual directs his muscle movements primarily by the alpha motoneurons as the leading nerve track in contrast to the trained individual innervating his muscles via the gamma loop (Chap. 4).

It should be pointed out that motor learning is so specific in nature that one cannot speak of a general learning ability with reference to motor coordination in skilled movements.

Orthostatic Reaction

It is well established that prolonged bed rest produces a marked deterioration in the cardiovascular response to posture as measured by heart rate and blood pressure changes produced by an upright position, e.g., tested on a tilt-table; this test shows a reduced head-up tolerance with fainting or a tendency to faint as a consequence (Deitrick et al., 1948; Taylor et al., 1949; Birkhead et al., 1964; Miller et al., 1965; Stevens et al., 1966; Saltin et al., 1968). This reduced tolerance develops already after a few days of bed rest, especially in older individuals (Chebutarev, personal communication). On the other hand it disappears within a few days of ambulation following 14 days of recumbency (Vogt et al., 1967).

In a later section of this chapter we will discuss factors of importance for the impaired tilt-table tolerance. In this connection we shall only point out that work in a recumbent position does not counteract the development of tilt-table intolerance (Birkhead et al., 1964; Miller et al., 1965). A few hours' ambulation per day, or the application of negative pressure to the lower body in the recumbent position, simulating effects of orthostasis on the cardiovascular system, may nearly completely maintain the orthostatic tolerance during a period of prolonged bed rest (Birkhead et al., 1964; Stevens et al., 1966).

Body Composition

During inactivity or habitual training, the body weight may remain relatively constant although inactivity often produces a gradual weight gain. During intensive training the specific gravity of the body increases while the skin-fold thickness decreases (Pařízková and Poupa, 1963; Skinner et al., 1964). Increased specific gravity (i.e., changed from 1.068 to 1.077; from 1.058 to 1.063 in the studies mentioned above) indicates a reduction in the fat content. This is also supported by the reduced skin-fold thickness observed. At the same time there is a certain increase, especially of the muscle mass and the blood volume.

In the case of fat-free body mass, however, it is only a matter of variations of a few kilograms in the body weight with different states of training. It is therefore evident that change in body weight alone is an inadequate index of possible alterations in body composition.

Blood Lipids

It appears that individuals with normal values for serum cholesterol and phospholipid levels do not change significantly with training (Skinner et al., 1964; Fox and Haskell, 1967; Table 9-2). Thus, the serum cholesterol level of lumberjacks has been found to be practically identical with that of other men in the same geographical areas; the serum cholesterol level is practically unaffected by occupation or by the physical activity classification (Karvonen, 1967). There may be one reservation: endurance skiers who practice an extreme form of physical activity have lower serum cholesterol values than their less active counterparts (Karvonen, 1967).

Psychological Changes

"There is a general assumption that physical fitness has psychological correlates. The concept of body-mind relationships is age-old. Psychosomatic research has indicated that physical changes result from continued psychological states; it seems logical to assume the reverse: that psychological changes result from physical states, such as fitness. Although there is a general assumption that this is so and considerable claims rather vaguely documented, there are surprisingly few firmly validated data" (Hammett, 1967). There are several studies demonstrating positive correlations between athletic ability on the one side and the general level of social adjustment and many psychological factors on the other. However, as emphasized by Hammett, they cannot be regarded as valid indications of psychological change related to increasing physical activity since it is equally possible that they reflect predilections, i.e., persons with certain psychological characteristics may gravitate to physical training programs. There are to date too few longitudinal studies to permit us to draw any firm conclusions.

It is a common observation that a person engaged in physical training eventually experiences a given work load as being lighter, and even work at a given heart rate may appear lighter. Breathlessness is probably largely a question of lack of experience (Chap. 7). Thus the subjective discomfort associated with a pulmonary ventilation of 50 or 100 liters/min may be less (a form of conditioning).

It is important, especially in the treatment of patients, that the feeling of anxiety at a given work load may be reduced during regular training. To some extent a positive connection between physical training and "body image" is of importance (Hammett, 1967); the simple fact, observed by Dawson (1920), that "when the trained or practiced individual is engaged in physical exercise, he naturally accomplishes more work with less apparent exertion and less subjective distress" is more meaningful to the individual than the recording of a series of physiological changes in connection with a training period. According to Chebutarev (personal communication), older individuals are greatly affected even after a few days bed rest. A marked feeling of malaise develops with loss of appetite, pains in the heart region and the muscles, and gastrointestinal troubles. These older individuals are more difficult to rehabilitate than younger ones.

EXAMPLES OF STUDIES OF BED REST AND TRAINING

Oxygen-transport System

Saltin et al. (1968) carried out extensive studies on the effect of a 20-day period of bed rest followed by a 50-day period of physical training in five male subjects, aged nineteen to twenty-one. Three of the subjects had previously been sedentary and two of them had been physically active. The maximal oxygen uptake fell from an average of 3.3 in the control study (before bed rest) to 2.4 liters/min after bed rest, i.e., there was a 27 percent drop (see Fig. 12-13). The stroke volume during supine exercise on a bicycle ergometer at 600 kpm/min (oxygen uptake, 1.5 liters/min) decreased from 116 to 88 ml, or about 25 percent. The heart rate increased from 129 to 154 beats/min. The cardiac output at this standard load fell from 14.4 to 12.4 liters/min. Thus the arteriovenous O_2 difference became somewhat increased. Also during upright exercise at submaximal loads there was a reduction in cardiac output (15 percent) and stroke volume (30 percent) (Fig. 12-14). An oxygen uptake that could normally be attained with a heart rate of 145 required a heart rate of 180 beats/min after bed rest. During maximal treadmill exercise, the cardiac output fell from 20.0 to 14.8 liters/min (a 26 percent reduction). It should be recalled that the maximal oxygen uptake was now 27 percent less. Since the oxygen content of arterial blood did not change, the maximal arteriovenous O_2 difference was therefore not modified by the bed rest. These findings are illustrated in Fig. 12-14. The maximal heart rate was not altered, so the fall in maximal cardiac output was due to a reduction of stroke volume.

The physical training produced an increase in maximal oxygen uptake from 2.52 to 3.41 liters/min in the previously sedentary subjects (an increase of 33 percent), and from 4.48 to 4.65 liters/min in the previously active subjects (Fig. 12-13). The most dramatic improvement in maximal aerobic power was noted with the three usually sedentary subjects when comparing the "after bed rest" values with the posttraining ones. They actually increased their maximal oxygen uptake by 100 percent, or from 1.74 to an average of 3.41 liters/min as mentioned. This study illustrates how critical the level of physical activity is before a training regime for the evaluation of the effectiveness of a training program to improve the maximal aerobic power. In the previously physically active subjects the improvement was only 4 percent. Starting with the "after bed rest" level, the increase was, however, 34 percent (from 3.48 to 4.65 liters/min). For the three previously sedentary

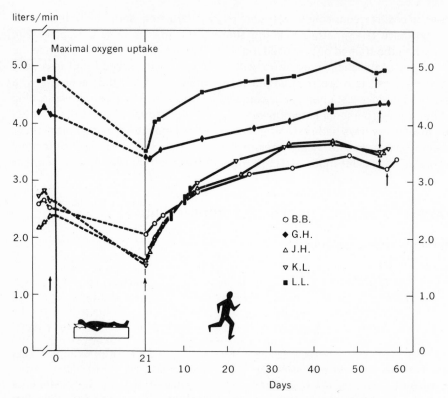

liters/min

Fig. 12-13   Changes in maximal oxygen uptake, measured during running on a motor-driven treadmill, before and after bed rest and at various intervals during training; individual data on 5 subjects. Arrows indicate circulatory studies. Heavy bars mark the time during the training period at which the maximal oxygen uptake had returned to the control value before bed rest. (From Saltin et al., 1968.)

subjects the improvement was 33 and 100 percent respectively. In other words, the training program applied may be said to have caused the maximal oxygen uptake of the participants to increase from 4 to 100 percent depending on how the initial level is defined. From Fig. 12-13 it is evident that the three sedentary subjects exceeded their control values as early as about 10 days after the commencement of the training. The two previously active subjects required 30 to 40 days to achieve the noted improvement.

Figure 12-11 illustrates the variation in maximal oxygen uptake for the three normally inactive subjects. The higher maximum noticed during their normal sedentary life, compared with the more extreme inactivity when immobilized in bed, was due to a higher cardiac output. The further improvement in maximal oxygen uptake with intensive training was partly due to a still higher cardiac output and partly to an increased arteriovenous O_2 difference by a more complete extraction of oxygen from the blood in the tissues.

Figure 12-14 summarizes the circulatory data for the five subjects. The training did increase the stroke volume and decrease the heart rate at submaximal work loads. The

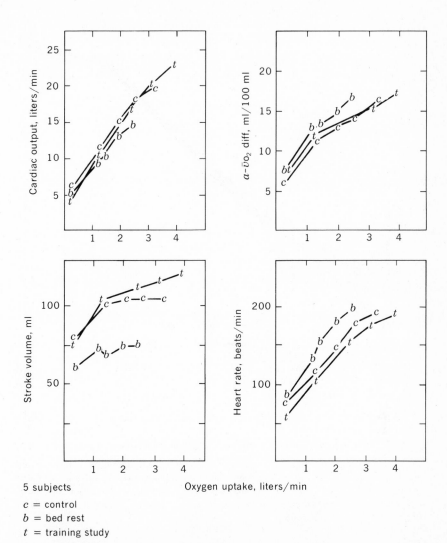

5 subjects

c = control
b = bed rest
t = training study

Fig. 12-14   Mean values of cardiac output, arteriovenous oxygen difference, stroke volume, and heart rate in relation to oxygen uptake during running at submaximal and maximal intensity before (c) and after (b) a 20-day period of bed rest and again after a 50-day training period (t). (Modified from Saltin et al., 1968.)

maximal heart rate was, however, not modified by the training. The maximal cardiac output in the sedentary subjects decreased from 17.2 to 12.3 liters/min during bed rest; after the training it rose to 20.2 liters/min. The arteriovenous oxygen difference was in the same experiments 14.7, 14.9, and 17.0 ml/100 ml blood respectively. The stroke volume fell from 90 to 62 ml during bed rest, but after training it increased to 105 ml. The changes in heart rate at various levels of oxygen uptake due to inactivity and caused by activity

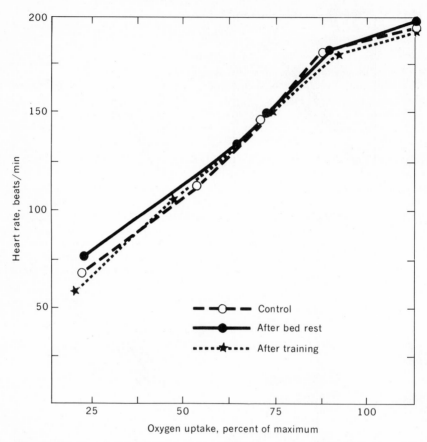

Fig. 12-15   Mean values of heart rate in relation to oxygen uptake in percent of the maximal aerobic power. Five subjects were studied during running. Note that the heart rate was about 130 at 50 percent of maximal oxygen uptake, but the actual volume of oxygen transported varied from an average of 1.2 up to 1.95 liters/min. (From data obtained by Saltin et al., 1968.)

are obvious. However, when the heart rate was related to the oxygen uptake in percent of the maximum, the heart rate response remained the same in the different conditions (Fig. 12-15).

Figure 12-16 shows the changes in heart volume. The mean value for the three sedentary subjects was 740 ml. After bed rest it was 690 ml and increased to 810 ml at the end of the training, an increase of 17 percent. The increase in stroke volume was 69 percent, however.

Blood volume decreased significantly during bed rest (from 5.06 to 4.70 liters in the 5 subjects, i.e., a 7 percent reduction). The fall in plasma volume was slightly more pronounced than in the red cell mass. During training the plasma volume and red cell mass increased again, in most subjects above the control values.

The lean body mass changed from 66.3 to 65.3 kg after bed rest and increased to 67.0 kg after training. There were no significant changes in ultrastructural morphology of voluntary muscle (quadriceps) during the experimental period.

The basal heart rate recorded during sleep was on the average 51.3 beats/min after bed rest and decreased to 39.7 at the end of the training. During bed rest the median heart rate during the day was 71 beats/min but a certain percentage of all the beats recorded were faster than 100/min.

The training program in this study was rather intensive and continuously supervised. The weekly schedule included two workouts daily, Monday through Friday; on Saturdays there was only one workout, and Sundays were free. The workouts consisted of both interval and continuous exercise, mainly outdoor running of from 2.5 up to 7 miles. In the interval exercise the speed was chosen so that the oxygen demand during 2- to 5-min periods of running was at or near the individual's maximal oxygen uptake. During the continuous running, usually for more than 20 min, the oxygen uptake varied from 65 to 90 percent of the subject's maximal value. One of the subjects who had not trained before was covering an average of 40 miles per week (64 km), and one of the previously trained subjects covered about 50 miles per week (80 km).

Fig. 12-16   Individual data on estimated heart volume before and after bed rest and during a period of physical training. (Saltin et al., 1968.)

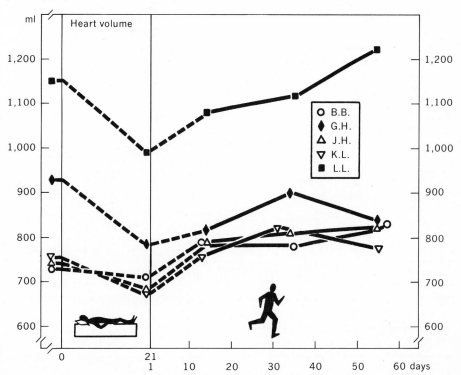

The authors concluded that the most striking effect of prolonged bed rest was the pronounced impairment of the circulatory adaptation to exercise, confirming earlier studies. The decrease in circulatory fitness also noticed in supine exercise indicates that it is not only a matter of reduced tolerance to the upright position. There may be a myocardial effect of prolonged bed rest in addition to an impairment in control of venous capacitance vessels. It cannot be excluded, however, that prolonged bed rest also causes a reduced parasympathetic activity which may facilitate an enhancement of the sympaticus activity.

The authors also emphasized that "there is in most normal subjects the potential of a very wide range of adaptation to physical stress. In a given individual, on whom prolonged immobilization is imposed, the ability to adapt may be impaired to a remarkable degree. On resumption of sedentary activity, some degree of the impairment is overcome. The same subject, after a short intensive training period shows further improvement in his capacity for physical stress."

There are no data available showing how far the maximal oxygen uptake (and cardiac output) can be increased by intensive training for longer periods of time. Ekblom et al. (1968) studied eight subjects before and after 16 weeks of rather intensive training. The maximal oxygen uptake increased by 16 percent, or from 3.16 to 3.68 liters/min, partly due to an increased arteriovenous oxygen difference (from 138 to about 145 ml/liter) and partly due to an increased cardiac output (from 22.4 to about 24.2 liters/min). Since the maximal heart rate was unchanged, the direct cause of the higher cardiac output was the greater stroke volume (from 112 to 127 ml). (There is some uncertainty in the circulatory data obtained during maximal work after the training in the above-mentioned study, since the oxygen uptake was on the average 0.24 liter/min less when the circulatory data were collected than was the case at a test a few days earlier.)

Two of the subjects continued their training up to about 30 months. The main data are presented in Table 12-5.

TABLE 12-5

Subject	Maximum values						
	$\dot{V}_{O_2}$ liters/min	$\dot{Q}$ liters/min	HR	SV ml	$\dot{V}_E$ liters/min	$a\text{-}\bar{v}O_2$ diff. ml/liter	HV ml
I.S.	3.07	23.9	205	122	128	128	920
	4.36	27.6	185	151	158	158	1060
Diff, %	42	15	−10	24	23	23	15
L.M.	3.58	25.5	206	124	116	140	960
	4.53	30.5	205	149	160	149	950
Diff, %	26	20	0	20	38	6	−1

SOURCE: Ekblom, 1969.

With reference to Table 9-2 it should be emphasized that physical training may counteract the decline in maximal aerobic power which normally accompanies aging (after age 30). Assuming that the subjects who were habitually active and those who were habitually inactive at the time of examination all had about the same maximal oxygen uptake when they were 25 years old, physical training should theoretically produce approximately a 20 percent improvement. (This improvement would be even higher if related to body weight.)

Hollmann (1965) has also presented data from a longitudinal study showing that a regular physical training can rather effectively counteract the age-induced decrease in maximal aerobic power. From 1949 to 1952 the maximal oxygen uptake was determined in 56 subjects. They were retested in 1964. Of the subjects, 39 had lived a sedentary life for the last 12 to 15 years, and 17 had been training on an average of twice a week. In the inactive group (mean age, 59 years) the maximal oxygen uptake had declined from an average of 3.05 to 2.11 liters/min, or 31 percent. In the still physically active group (56 years of age) the decline was from 3.21 to 2.90 liters/min, or only about 10 percent. The body weight averaged 75.9 kg in the active group and 81.5 kg in the inactive one.

Hollmann (1965) studied six previously sedentary subjects who trained four times a week, 20 to 30 mln each time during a period of 10 weeks. During the first weeks the heart rate during training was about 125 beats/min. During this period the maximal oxygen uptake increased from an average of 2.90 to 3.07 liters/min (6 percent) without the heart volume being affected. The resting heart rate dropped from 73 to 63 beats/min. During the subsequent 5 weeks, the training intensity increased so that the heart rate increased to 150 to 180 beats/min. The maximal oxygen uptake increased altogether 21 percent, or to 3.51 liters/min, and the heart volume increased from 730 to 820 ml (13 percent). The resting heart rate dropped to an average of 59 beats/min. Karvonen et al. (1957) in studies of six subjects (trained by running on a treadmill ½ hr daily, four to five times a week over a period of 4 weeks) noticed that "the heart rate during training has to be more than 60 percent of the available range from rest to the maximum attainable by running or above approx. 140 per minute, in order to produce a decrease of the working heart rate."

Jones et al. (1962) noticed that 5 min of rope skipping daily for a 4-week period reduced the mean heart rate at a fixed ergometer test load of 450 kpm/min from 159 to 141 beats/min in a group of seven habitually sedentary women 19 to 42 years of age. (This difference is statistically significant at less than the 0.01 level.) The mean heart rate during the rope skipping was 168 beats/min at the beginning of the training, but dropped to 145 beats/min.

Roskamm (1967) summarizes experimental studies conducted in Freiburg, Germany. Altogether, 80 soldiers took part in the study. They were divided into four groups: Three of the groups trained ½ hour daily, 5 days per week for 4 weeks. The total amount of training, measured in kpm, was the same in all these groups. All the training was done on bicycle ergometers. One group trained continuously with a work load which brought the heart rate to a value of about 70 percent of the difference between the heart rate at rest and during maximal work. (Example: if the heart rate at rest was 65 beats/min and the maximum 195, the "70 percent load" should increase the heart rate to about 155 beats/min: 195 − 65 = 130; 70/100 × 130 = 91; 65 + 91 = 156.) A second group trained at alternating work loads: for 1 min the load was 50 percent higher than this 70 percent load, followed by a 1-min period when the load was 50 percent lower. The third group performed a similar training with varying work loads but each period was extended to

$2\frac{1}{2}$ min. The fourth group served as a control group without additional exercise. As an index of the training effect the maximal load which could be achieved per pulse beat during work on the bicycle ergometer was used (maximal watt-pulse, similar to maximal oxygen pulse). It was found that all groups that trained improved their values significantly compared to the control group. Almost all the trained soldiers had an increase of more than 10 percent. The group which carried out the $2\frac{1}{2}$-min interval training showed the greatest improvement of the maximal watt-pulse. On the other hand, the decrease in heart rate at the standard load (100 watts) was most pronounced for the subjects training continuously at the 70 percent load.

In another study the Freiburg researchers noticed a 20 percent increase in the maximal watt/pulse beat in a group of 18 younger persons who were training $\frac{1}{2}$ hour daily for 4 weeks. Some of the subjects then stopped their training, and their increased level of physical performance was partially lost within 2 weeks. The other subjects continued to train but only every third day. This amount of training did not further improve their physical condition but was enough to maintain the attained level by the previous training. This finding is in agreement with the experience that it takes less training effort to maintain a level of physical fitness than it does to attain the approved level in the first place.

The final study by Roskamm et al. (1966) which will be discussed here involved 18 men, aged 16 to 18, 20 to 30, and 50 to 60 years, and 6 women, aged 20 to 30 years. The training and evaluations were done according to the procedure outlined above. The heart rate during training was in the case of the younger subjects about 150 beats/min, and in the case of the older subjects about 130 beats/min (lower maximal heart rate). After 4 weeks' training every group showed a significant increase in the maximal watt-pulse, but in the case of the 50- to 60-year-old subjects the training effect was less pronounced than in the case of the younger subjects.

Bed Rest: Orthostatic Reaction, Calcium Output

The results of studies by Dietrick et al. (1948) and by Taylor et al. (1949) have indicated a variety of changes attributable to prolonged bed rest: muscle wasting, negative nitrogen balance, demineralization of bones, increased urinary calcium and phosphorus elimination, kidney stones, urinary bladder infection, constipation, lowered basal metabolism, reduced blood volume, tilt-table intolerance, and markedly reduced physical work capacity (Vallbona et al., 1965).

In a series of studies conducted in the Division of Research at Lankenau Hospital, Philadelphia (Birkhead et al., 1963, 1964; Rodahl et al., 1967), normal young men were confined to the horizontal position in bed for up to 63 days. In the first experiment, four subjects were studied before and after a 6-week period of continuous bed rest. This caused a marked impairment of tilt-table tolerance and a marked reduction in physical work capacity, as judged by heart-rate response to a standard work load. The physical work capacity barely returned to pre-bed rest level following an 18-day retraining period. There was a twofold increase in urinary calcium output.

These initial experiments, then, confirmed the conclusions of previous studies that prolonged bed rest is indeed harmful to the individual's functional capacity. The next task was to determine whether all these changes are caused by inactivity as such or whether some of them may be attributable to the maintained horizontal position, rather than inactivity per se. In the next series of experiments, therefore, four healthy young men were confined to prolonged bed rest as in the previous experiment, except that they now worked for 1 hr per day on a bicycle ergometer set at 600 kpm/min. Two of the subjects worked in the horizontal position and two worked in the sitting position. It was observed that this amount of physical activity, whether it was performed lying down or sitting up, prevented any appreciable deterioration in physical work capacity, as measured by a comparison of the subject's maximal oxygen uptake before and after bed rest. Working in the horizontal position did not prevent the development of tilt-table intolerance, but working in the upright position partly prevented the development of orthostatic hypotension. In all four subjects the increased urinary calcium elimination persisted in spite of the exercise. In a subsequent experiment it was observed that ergometer work of 600 kpm/min for 4 hr in the horizontal position failed to counteract the development of tilt-table intolerance or increased urinary calcium excretion. From these experiments it may be concluded that although the deterioration of physical work capacity in prolonged bed rest is caused by inactivity, the observed changes in tilt-table tolerance and urinary calcium output produced by horizontal bed rest are not caused by inactivity as such, but may be attributable to the effect of absence of gravitational stresses.

It is conceivable that the maintenance of a normal calcium metabolism may in some way be related to the forces of gravity acting constantly or intermittently upon the skeletal system. To test this notion, four healthy young men were confined to bed until the increased urinary calcium output had developed. They were then allowed to sit inactively for 8 hr a day in a wheel chair; during the rest of the 24-hr period they were confined to horizontal bed rest for a period of 3 weeks. This prevented the development of tilt-table intolerance in three of the four cases. There was a slight deterioration in physical work capacity, but the elevated urinary calcium persisted during the 3 weeks of inactive sitting. It is of clinical importance to note that no benefit is achieved for the patient by lifting him out of bed and placing him in a chair, as far as the urinary loss of calcium is concerned.

When the subjects who had developed a twofold increase in urinary calcium elimination by continuous bed rest for about 3 weeks were allowed to stand upright, but without moving about, for 2 hr a day, while spending the rest of the 24-hr period in bed, the urinary calcium excretion showed a tendency to return toward normal levels within 3 to 4 weeks. It definitely returned to normal levels in subjects who were allowed to stand for 3 hr a day. Their tilt-table tolerance, however, was still partially impaired. In another two subjects a pressure equal to their body weight was exerted on their bodies longitudinally for 3 hr daily with the aid of a spring arrangement, while they remained in horizontal bed rest for 3 weeks. In one of the two subjects this abolished the increased urinary calcium produced by continuous bed rest. This suggests that the cause of the increased urinary calcium excretion is associated with absence of pressure on the skeletal structures.

It is evident from these studies that the three major effects of prolonged bed rest, decreased maximal aerobic power, tilt-table intolerance, and increased urinary calcium loss, are separable. Of these changes only the reduced aerobic power appears to be due to inactivity as such. It does show, however, that prolonged continuous bed rest is not the best form of rest. It appears that man, through evolution, has become adjusted to his upright, two-legged, gravity-stressed existence, and ambulatory activity is necessary to maintain normal functions.

Even more dramatic changes have been produced by water submersion experiments in which subjects dressed in a rubber suit and wearing a partial-pressure-type helmet are maintained floating in a water tank, oxygen being supplied by a pipe connected with the helmet (Graveline, 1961). Thus supported by water, normal weight sensation is altered, and movement is relatively effortless and without the usual resistance caused by normal gravitational situations. Tilt-table, centrifuge, and heat-chamber studies demonstrated significant cardiovascular deterioration even after 6-hr exposure, becoming progressively more severe in 12- and 24-hr experiments.

On the basis of these studies, it has been suggested (Graveline et al., 1961) that during prolonged space flight under true weightless conditions, the organism may attain a critical state of deconditioning which will seriously attenuate its tolerance for reentry stresses and the normal gravitational environment. It should be borne in mind, however, that water submersion is by no means similar to zero gravity states. Only observations made during prolonged space flight will reveal the true nature of the effect of prolonged weightlessness and whether or not an exercise program may counteract any of these adverse effects.

Confirming the previous findings of Birkhead et al. (1964), both Miller et al. (1965) and Stevens et al. (1966) observed that work in the supine position (6 × 10 min/day) during 4 weeks of bed rest did not improve the postural tolerance compared to absolute bed rest. On the other hand, application of a negative pressure (−30 mm Hg) to the lower body in the recumbent position at the end of a long period of bed rest caused a rapidly increasing salt and water retention as manifested by hemodilution, weight gain, and plasma volume expansion. These changes were associated with nearly complete maintenance of orthostatic tolerance. However, the maximal aerobic power remained decreased whether or not the plasma volume was restored by the use of the lower body negative pressure.

In *summary*, then, prolonged bed rest in the recumbent position, whether completely immobilized or not, causes deterioration in physical work capacity, orthostatic hypotension, and increased urinary calcium elimination. Different mechanisms may be involved in the development of these changes, since 1-hr ergometer exercises daily may eliminate the deterioration of physical work capacity, but does not affect calcium metabolism or tilt-table tolerance. Whatever the mechanism is, it appears that a minimum amount of stimuli is required to maintain normal cardiodynamic and metabolic functions.

Muscular Training

The question to what extent the muscle mass is affected by inactivity and by training has for obvious reasons primarily been studied in animal experiments. Linge (1962) denervated the triceps surae muscle in rats and implanted the plantar muscle into the tuberosity of the calcaneum. The plantar muscle, with a weight that was originally only about 18 percent of that of the triceps, was thus forced to do much heavier work. After some time it doubled its weight and trebled its force, and the training evidently induced a strong protein synthesis.

Siebert (1929) trained rats on a treadmill. In one experiment the animals were divided into three groups. All of them were running on a treadmill at a given speed, 9 m/min. One group trained 1 hr a day for 3 months, a second group 3 hr a day for 3 months, a third group 3 hr a day for 6 months. In all groups there was, after the training, an increase in the weight of the heart and the gastrocnemius muscle, but no difference was noted between the groups. He concludes that 1-hr-daily training was enough to adapt the muscles to this rate of work and no additional stimulus to hypertrophy was caused by the longer periods of training. In a second experiment the rats ran a given distance but at different speeds. It was found that the higher the training speed, the heavier became the heart and gastrocnemius muscles. In other words these experiments indicate that speed was more critical than the distance covered as a stimulus to muscular hypertrophy.

It has been suggested that the atrophy which takes place in a muscle is most pronounced when the muscle is denervated. When an extremity is immobilized in a cast, the loss of muscle strength is greater than is the case when the individual is immobilized by prolonged bed rest. Müller and Hettinger (1953) have reported experiments in which a muscle that was immobilized by a cast lost 20 percent of its maximal strength within a week. These authors introduced, around 1950, a training theory according to which a muscle contraction of a few seconds' duration, representing at least one-third of the maximal strength of the muscle, repeated once daily could counteract an atrophy of the muscle. Only when the force of contraction was below 20 percent of the maximal strength did the atrophy eventually develop. A muscle contraction representing two-thirds of the maximal strength of about 6-sec duration once a day produced the same training effect as many training periods daily with maximal effort and until muscular exhaustion (Hettinger and Müller, 1953). This finding is not in accordance with the experience from, for instance, the training of weight lifters, who definitely apply their maximal strength during their training sessions. Other investigations have failed to verify the results of Hettinger and Müller. Müller has together with Rohmert (1963) taken up these studies for reevaluation, and he has modified his standpoint. As general advice for the improvement of muscle strength, Hettinger (1968) recommends: Maximal strength effort for about 2- to 3-sec duration repeated 1 to 5 times daily depending on the purpose of the training. He states that one training period every 2 weeks did not produce any lasting increase in muscle strength but that one training session per week produced approximately 40 percent of the effect produced by daily training. The individual reactions to training, with regard to different muscle groups and to different individuals, is emphasized by Hettinger.

Zimkin (1960) states that in the initial stages of training of individual muscle groups in man, the best effect was obtained when the weight of the load to be raised and the rate of movements were not excessive and the intervals between training sessions were sufficiently long. However, the load had to eventually approach the maximal level in order to achieve further improvement in muscle strength.

Petersen et al. (1961) have conducted several series of investigations concerning different types of training. The results are summarized in Table 12-6. They show convincingly what has been pointed out in several connections: that the improvement in performance primarily is tied to the function which is being trained. A training of static endurance increases the ability to attain static endurance; a training of dynamic endurance produces a marked improvement in this form of work without improving endurance for static work. Some improvement, although very slight, is observed in the maximal strength through training of endurance with lower tension developed at each contraction. Of importance for the improvement of endurance is a more abundant blood supply to the muscle in question. It is therefore natural that a type of training which produces an increase in the flow rate through the muscle also represents the most effective stimulus for the development of the vascular bed, i.e., dynamic work. The maximal strength on the other hand depends on the neuromuscular function and especially on the ATP-ADP machinery and not on the aerobic energy yield.

TABLE 12-6

*Measurement before and after training of maximal strength, arm flexor, in (1) isometric and
(2) dynamic contractions. Endurance tested with load that was 60 percent of the maximal strength
and number of contractions (kpm) counted before fatigue made further successful contractions
impossible for (3) isometric and (4) dynamic exercise. Training restricted to one type of muscle
work at a time. The figures give the average performance before (e.g., 378 kp × cm) and after
(e.g., 400 kp × cm) training as well as the percentage difference*

Training	Maximal strength		Endurance 60% of maximum		Study
	(1) Isometric, kp × cm*	(2) Dynamic, kp†	(3) Isometric, No. of contractions	(4) Dynamic, kpm	
Dynamic "60%" 150 contr./day 30 days	378 to 400 +6%	8.8 to 11.4 +29%		16.4 to 843 +5040%	Petersen et al., 1961
Isometric "60%" 150 contr./day 30 days	425 to 441 +4%	10.7 to 11.3 +6%	29 to 336 +1060%	17 to 24 +41%	Hansen, 1961
Dynamic "60%" 100 contr./hour 30 days		12.1 to 13.7 +13%	100 to 98 −2%	60 to 438 +630%	Hansen, 1967

* Torque.
† Weight was lifted 25 cm.

In this connection the observations by Ikai et al. (1964) are of interest. He examined
the number of contractions of arm flexor and knee extensor performed once a second
against a load which was one-third of the maximal strength and the exercise was continued to exhaustion. Table 12-7 summarizes some of the results.

It is noteworthy that throwers were stronger than the rest of the subjects, but
relatively speaking, they had the lowest endurance, especially in the case of the leg
muscles. The male middle- and long-distance runners had no stronger leg muscles than
did the average men, but they were characterized by an impressive endurance; the performance of their arm muscles, however, did not differ from the average. If anything,
the endurance test was inferior in this case. These results probably reflect differences
in training in these groups of subjects.

As far as the specificity of a training effect is concerned, one is once again reminded
of the fact that the nervous control plays a decisive role for the developed strength. An
inhibition of the motoneuron may be more or less pronounced. It appears as if a certain
strength developed in the course of a long period of time is maintained largely by the
same motor units (Carlsöö, 1952). An increase in the developed force in the same activity
means that the same motor units became engaged with greater frequency and that new
motor units are recruited in addition. Only at maximal effort are the motor units representing the "last reserve" thrown into play, especially in emergency situations. It is

possible that these reserve units may become more easily engaged as the result of train-ing. In order to train these motor units in the motions in question, maximal exertion is required. (Many individuals usually exert themselves far from maximally. The reason why one does not find atrophied muscle cells in sedentary individuals may be that the motor units, which in the mentioned cases are only recruited during maximal effort, are already engaged at moderate exertion in a different motion pattern.)

An untrained individual should not subject himself to a maximal load at the start of the training, especially if he wants to avoid sore muscles.

In *summary:* the guidelines for the improvement of muscular strength which were advanced by de Lorme in 1945 are in our judgment still valid: "In order to obtain rapid hypertrophy in weakened, atrophied muscle, the muscle should be subjected to strenuous exercise, and, at regular intervals, to the point of maximum exertion." Low-repetition (5 to 10 times), high-resistance exercises produce power; high-repetition, low-resistance exercises produce endurance.

Nonspecific Effects

Zimkin (1960, 1964) has reviewed different studies, mainly by Russian researchers, which show that training constitutes a factor of great importance for increasing the nonspecific resistance of the body to various unfavorable factors. He mentions that in this way it is

TABLE 12-7

Muscle strength and endurance

Subjects*	Leg extensor		Arm flexor	
	Max strength, kp	No. of controls ⅓ of max strength	Max strength, kp	No. of controls ⅓ of max strength
Men				
Average men (5;10)	55	48	17	75
Sprinters (7)	71	52	19	65
Middle- and long-distance runners (6)	55	399	19	48
Hurdlers (3)	61	67	19	46
Jumpers (8)	68	49	21	45
Throwers (8)	88	38	26	51
Women				
Average women (10)			9	70
Sprinters (3)	48	71	12	71
Middle- and long-distance runners (4)	53	68	10	68
Hurdlers (5)	48	67	10	67
Jumpers (2)	56	57	12	58
Throwers (1)	68	43	17	43

* Number of subjects in parentheses.
SOURCE: Ikai, 1964.

possible to increase the resistance to extreme cooling and heating (rats), to several toxic substances (rats), to the action of x-rays (rats), to infections and other illnesses (rats and humans). He quotes a study by Shernyakov: "By analysing examination findings for several thousands of physically trained and untrained individuals doing mental work Shernyakov has shown that medical assistance was sought by 58.1 percent of the individuals who did not do physical exercises, by 38.0 percent of those who did them irregularly, and by 20.8 percent of the individuals who did regular exercises." He emphasizes that the training period should be of relatively long duration and the exercises themselves not excessive but yet sufficiently intense. Both animal experiments and observations on athletes indicate that excessive loading during training may be followed not only by absence of any increase of nonspecific resistance, but even by a reduced resistance. The increased resistance to unfavorable agents persisted for a limited time after the training ended; the shorter the duration of training, the shorter was this period. The explanation for the improved nonspecific resistance could be chemical changes on the cellular level and effects on the endocrine glands induced by the physical training. It is certainly a critical question how far extrapolations can be extended to humans from results obtained on rats. The observed difference in sick calls between trained and untrained employees is difficult to evaluate. It may be a question of reduced susceptibility to certain diseases in the trained individuals, but it may also be a matter of less pronounced subjective symptoms, or a conscious supression of symptoms. Whatever the reason may be, the net effect appears to be positive. If on the other hand suppression of symptoms was more common among trained individuals, this may not necessarily be an advantage since it may postpone medical attention and adequate treatment.

REFERENCES

Anderson, T. W., and R. J. Shephard: Physical Training and Exercise Diffusing Capacity, *Intern. Z. Angew. Physiol.*, **25**:198, 1968.

Andrew, G. M., C. A. Guzman, and M. R. Becklake: Effect of Athletic Training on Exercise Cardiac Output, *J. Appl. Physiol.*, **21**:603, 1966.

Åstrand, I., P.-O. Åstrand, E. H. Christensen, and R. Hedman: Intermittent Muscular Work, *Acta Physiol. Scand.*, **48**:448, 1960.

Åstrand, P.-O.: Human Physical Fitness with Special Reference to Sex and Age, *Physiol. Rev.*, **36**:307, 1956.

Beers, L. B.: The Acute and Chronic Effects of Exercise on the Latent Period of the Gastrocnemius Muscle in Man, *Arbeitsphysiol.*, **8**:539, 1935.

Berglund, E., H. G. Borst, F. Duff, and G. L. Schreiner: Effect of Heart Rate on Cardiac Work, Myocardial Oxygen Consumption and Coronary Blood Flow in the Dog, *Acta Physiol. Scand.*, **42**:185, 1958.

Bevegård, S., A. Holmgren, and B. Jonsson: Circulatory Studies in Well-trained Athletes at Rest and during Heavy Exercise, with Special Reference to Stroke Volume and the Influence of Body Position, *Acta Physiol. Scand.*, **57**:26, 1963.

Birkhead, N. C., J. J. Blizzard, J. W. Daly, G. J. Haupt, B. Issekutz, Jr., R. N. Myers, and K. Rodahl: Cardiodynamic and Metabolic Effects of Prolonged Bed Rest, *Technical Documentary Report AMRL-TDR-63-37*, Aerospace Medical Research Laboratories, Wright-Patterson Air Force Base, Ohio, May, 1963.

Birkhead, N. C., J. J. Blizzard, J. W. Daly, G. H. Haupt, B. Issekutz, Jr., R. N. Myers, and K. Rodahl: Cardiodynamic and Metabolic Effects of Prolonged Bed Rest with Daily Recumbent or Sitting Exercise and Sitting Inactivity, *Technical Documentary Report AMRL-TDR-64-61*, Aerospace Medical Research Laboratories, Wright-Patterson Air Force Base, Ohio, August, 1964.

Böje, O.: Energy Production, Pulmonary Ventilation and Length of Steps in Well-trained Runners Working on a Treadmill, *Acta Physiol. Scand.*, **7:**362, 1944.

Carlsöö, S.: Nervous Coordination and Mechanical Function of the Mandibular Elevators, *Acta Odont. Scand.*, **10**(Suppl. 11), 1952.

Christensen, E .H.: Beiträge zur Physiologie schwerer körperlicher Arbeit, *Arbeitsphysiol.*, **4:**1, 1931.

Christensen, E. H., R. Hedman, and B. Saltin: Intermittent and Continuous Running, *Acta Physiol. Scand.*, **50:**269, 1960.

Clarke, D. H., and F. M. Henry: Neuromotor Specificity and Increased Speed from Strength Development, *Res. Quart.*, **32:**315, 1961.

Deitrick, J. E., G. D. Whedon, and E. Shorr: Effects of Immobilization upon Various Metabolic and Physiologic Functions of Normal Men, *Am. J. Med.*, **4:**3, 1948.

De Lorme, T. L.: Restoration of Muscle Power by Heavy-resistance Exercise, *J. Bone Joint Surg.*, **27:**645, 1945.

Dill, D. B., E. E. Phillips, Jr., and D. MacGregor: Training: Youth and Age, *Am. N.Y. Acad. Sci.*, **134:**760, 1966.

Eckstein, R. W.: Effect of Exercise and Coronary Artery Narrowing on Coronary Collateral Circulation, *Circulation Res.*, **5:**230, 1957.

Edwards, H. T., L. Brouha, and R. T. Johnson: Effect de l'entraînement sur le taux de l'acide lactique au cours du travail musculaire, *Le Travail Humain*, **8:**1, 1939.

Ekblom, B.: Effect of Physical Training on Oxygen Transport System in Man, *Acta Physiol. Scand.*, Suppl. 328, 1969.

Ekblom, B., P.-O. Åstrand, B. Saltin, J. Stenberg, and B. Wallström: Effect of Training on Circulatory Response to Exercise, *J. Appl. Physiol.*, **24:**518, 1968.

Elsner, R. W., and L. D. Carlson: Postexercise Hyperemia in Trained and Untrained Subjects, *J. Appl. Physiol.*, **17:**436, 1962.

Embden, G., and H. Habs: Beitrage zur Lehre vom Muskeltraining, *Scand. Arch. Physiol.*, **49:**122, 1926.

Erickson, L., E. Simonson, H. L. Taylor, H. Alexander, and A. Keys: The Energy Cost of Horizontal and Grade Walking on the Motordriven Treadmill, *Am. J. Physiol.*, **145:**391, 1946.

Fowler, W. M., S. R. Chowdhury, C. M. Pearson, G. Gardner, and R. Bratton: Changes in Serum Enzyme Levels after Exercise in Trained and Untrained Subjects, *J. Appl. Physiol.*, **17:**943, 1962.

Fox, S. M., III, and W. L. Haskell: Population Studies, *Can. Med. Ass. J.*, **96:**806, 1967.

Freedman, M. E., G. L. Snider, P. Brostoff, S. Kimelblot, and L. N. Katz: Effects of Training on Response of Cardiac Output to Muscular Exercise in Athletes, *J. Appl. Physiol.*, **8:**37, 1955.

Frick, M. H., A. Konttinen, and H. S. S. Sarajas: Effects of Physical Training on Circulation at Rest and during Exercise, *Am. J. Cardiol.*, **12:**142, 1963.

Garbus, J., B. Highman, and P. D. Altland: Serum Enzymes and Lactic Dehydrogenase Isoenzymes after Exercise and Training in Rats, *Am. J. Physiol.*, **207:**467, 1964.

Gisolfi, D., S. Robinson, and E. S. Turrell: Effects of Aerobic Work Performed during Recovery from Exhausting Work, *J. Appl. Physiol.*, **21:**1767, 1966.

Graveline, D. E.: Physiologic Effects of a Hypodynamic Environment: Short-term Studies, *Aerospace Med.*, **32:**726, 1961.

Graveline, D. R., B. Balke, R. E. McKenzie, and B. Hartman: Psychobiologic Effects of Water-immersion-induced Hypodynamics, *Aerospace Med.*, **32:**387, 1961.

Grimby, G., and B. Saltin: Physiological Analysis of Physically Well-trained Middle-aged and Old Athletes, *Acta Med. Scand.*, **179:**513, 1966.

Gutman, E., R. Beránek, P. Hník, and J. Zelená: Physiology of Neurotrophic Relations, *Proc. 5th Nat. Congr. Czech. Physiol. Soc.*, 1961.

Hall, V. E.: The Relation of Heart Rate to Exercise Fitness: An Attempt at Physiological Interpretation of the Bradycardia of Training, *Pediatrics*, **32**(Suppl.):723, 1963.

Hammett, V. B. O.: Psychological Changes with Physical Fitness Training, *Can. Med. Ass. J.*, **96:**764, 1967.

Hansen, J. W.: The Training Effect of Repeated Isometric Muscle Contractions, *Intern. Z. Angew. Physiol.*, **18:**474, 1961.

Hansen, J. W.: Effect of Dynamic Training on the Isometric Endurance of the Elbow Flexors, *Intern. Z. Angew. Physiol.*, **23:**367, 1967.

Hanson, J. S., and B. S. Tabakin: Comparison of the Circulatory Response to Upright Exercise in 25 "Normal" Men and 9 Distance Runners, *Brit. Heart J.*, **27:**211, 1965.

Hartley, L. H., and B. Saltin: Reduction of Stroke Volume and Increase in Heart Rate after a Previous Heavier Submaximal Work Load, *Scand. J. Clin. Lab. Invest.*, **22:**217, 1968.

Hedman, R.: Fysiologiska Synpunkter på Cirkelträning. *Tidskrift i Gymnastik*, **87:**1, 1960.

Helander, E.: Adaptive Muscular "Allomorphism," *Nature*, **182:**1035, 1958.

Helander, E.: Muscular Atrophy and Lipomorphosis Induced by Immobilizing Plaster Casts, *Acta Morphol. Neerl. Scand.*, **3:**92, 1960.

Herrlich, H. C., W. Raab, and W. Gigee: Influence of Muscle Training and of Catecholamines on Cardiac Acetylcholine and Cholinesterase, *Arch. Intern. Pharmacodyn.*, **129:**201, 1960.

Hettinger, Th.: "Isometrisches Muskeltraining," Georg Thieme Verlag, Stuttgart, 1968.

Hettinger, Th., and E. A. Müller: Muskelleistung und Muskeltraining, *Arbeitsphysiol.*, **15:**111, 1953.

Hollmann, W.: "Körperliches Training als Prävention von Herz-Kreislaufkrankheiten," p. 62, Hippokrates-Verlag, Stuttgart, 1965.

Hollmann, W., and H. Venrath: Die Beeinflussung von Herzgrösse, maximaler O_2-Aufnahme und Ausdauergrenze durch ein Ausdauertraining mittlerer und hoher Intensität, *Der Sportarzt*, **14:**189, 1963.

Holloszy, J. O.: Biochemical Adaptations in Muscle, *J. Biochem. Chem.*, **242:**2278, 1967.

Holmdahl, D. E., and B. E. Ingelmark: Der Bau des Gelenkknorpels unter verschiedenen Funktionellen Verhältnissen, *Acta Anat.*, **6:**309, 1948.

Holmgren, A., and T. Strandell: The Relationship between Heart Volume, Total Hemoglobin and Physical Working Capacity in Former Athletes, *Acta Med. Scand.*, **163:**149, 1959.

Hoogerwerf, S.: Elektrokardiographische Untersuchungen der Amsterdamer Olympiakämpfer, *Arbeitsphysiol.*, **2:**61, 1929.

Ikai, M.: The Effects of Training on Muscular Endurance, *Proc. Int. Congr. Sport Sci.*, p. 109, 1964.

Ikai, M., and A. H. Steinhaus: Some Factors Modifying the Expression of Human Strength, *J. Appl. Physiol.*, **16**:157, 1961.

Ingelmark, B. E.: Der Bau der Sehnen während verschiedener Altersperioden und unter wechselnden funktionellen Bedingungen I, *Acta Anat.*, **6**:113, 1948.

Ingelmark, B. R.: Morpho-physiological Aspects of Gymnastic Exercises, *FIEP-Bull.*, **27**:37, 1957.

Jones, D. M., C. Squires, and K. Rodahl: Effect of Rope Skipping on Physical Work Capacity, *Res. Quart.*, **33**:236, 1962.

Karlsson, J., P.-O. Åstrand, and B. Ekblom: Training of the Oxygen Transport System in Man, *J. Appl. Physiol.*, **22**:1061, 1967a.

Karlsson, J., L. Hermansen, G. Agnevik, and B. Saltin: Energikraven vid Löpning, *Idrotts-fysiologi*, rapport nr. 4, Framtiden, Stockholm, 1967b.

Karvonen, M. J.: Nutrition in Heavy Manual Labour, in G. Blix (ed.), "Nutrition and Physical Activity," Almqvist & Wiksells, Uppsala, p. 59, 1967.

Karvonen, M. J., E. Kentala, and O. Mustala: The Effects of Training on Heart Rate, *Am. Med. Exp. Fenn.*, **35**:307, 1957.

Knehr, C. A., D. B. Dill, and W. Neufeld: Training and Its Effects on Man at Rest and at Work, *Am. J. Physiol.*, **136**:148, 1942.

Kraus, H., and R. Kirsten: Effects of Exercise on Structure and Metabolism of Skeletal Muscle at the Cell Level, *Pflügers Arch.*, **308**:57, 1969.

Lauru, L.: Physiological Study of Motion, *Advanced Management*, **22**:17, 1957.

Liere, E. J. van, and D. W. Northup: Cardiac Hypertrophy Produced by Exercise in Albino and in Hooded Rats, *J. Appl. Physiol.*, **11**:91, 1957.

Linge, B. van.: The Response of Muscle to Strenuous Exercise, *J. Bone Joint Surg.*, **44**:711, 1962.

Man-i, M., K. Ito, and K. Kikuchi: Histological Studies of Muscular Training, *Res. Physical Education*, **11**:153, 1967.

Marpurgo, P.: Über Aktivitäts-Hypertrophie der willkürlichen Muskeln, *Virchows Arch.*, **150**:522, 1897.

McMorris, R. O., and E. C. Elkins: A Study of Production and Evaluation of Muscular Hypertrophy, *Arch. Phys. Med.*, **35**:420, 1954.

Milic-Emili, G., J. M. Petit, and R. Deroanne: Mechanical Work of Breathing during Exercise in Trained and Untrained Subjects, *J. Appl. Physiol.*, **17**:43, 1962.

Miller, P. B., R. L. Johnson, and L. E. Lamb: Effects of Four Weeks of Absolute Bed Rest on Circulatory Functions in Man, *Aerospace Med.*, **35**:1194, 1964.

Miller, P. B., R. L. Johnson, and L. E. Lamb: Effects of Moderate Physical Exercise during Four Weeks of Bed Rest on Circulatory Functions in Man, *Aerospace Med.*, **36**:1077, 1965.

Morgan, R. E., and G. T. Adamson: "Circuit Training," G. Bell and Sons, Ltd., London, 1962.

Müller, E. A., and Th. Hettinger: Über Unterschiede der Trainingsgeschwindigkeit atrophierter und normaler Muskeln, *Arbeitsphysiol.*, **15**:223, 1953.

Müller, E. A., and W. Rohmert: Die Geschwindigkeit der Muskelkraft-Zunahme bei isometrischen Training, *Intern. Z. Angew. Physiol.*, **19**:403, 1963.

Muscatello, U., A. Margreth, and M. Aloisi: On the Differential Response of Sarcoplasm and Myoplasm to Denervation in Frog Muscle, *J. Cell. Biol.*, **27**:1, 1965.

Newman, E. V., D. B. Dill, H. T. Edwards, and F. A. Webster: The Rate of Lactic Acid Removal in Exercise, *Am. J. Physiol.*, **118:**457, 1937.

Nöcker, J., D. Lehmann, and G. Schleusing: Einfluss von Training und Belastung auf den Mineralgehalt von Herz und Skelettmuskel, *Intern. Z. Angew. Physiol.*, **17:**243, 1958.

Palladin, A., and D. Ferdmann: Über den Einfluss der Trainings der Muskeln auf ihren Kreatingehalt, *Hoppe-Seylers Z. Physiol. Chem.*, **174:**284, 1928.

Par̆ízková, J., and O. Poupa: Some Metabolic Consequences of Adaptation to Muscular Work, *Brit. J. Nutr.*, **17:**341, 1963.

Petersen, F. B., H. Graudal, J. W. Hansen, and N. Hvid: The Effect of Varying the Number of Muscle Contractions on Dynamic Muscle Training, *Intern. Z. Angew. Physiol.*, **18:**468, 1961.

Petrén, T.: Die totale Anzahl der Blutkapillaren im Herzen und Skelettmuskulatur bei Ruhe und nach langer Muskelübung, *Verhandl. Anatom. Gesellsch.* (Suppl. *Anat. Anz., 81*), p. 165, 1936.

Petrén, T., T. Sjöstrand, and B. Sylvén: Der Einfluss der Trainings auf die Heufigkeit der Capillaren in Herz- und Skelettmuskulatur, *Arbeitsphysiol.*, **9:**376, 1936.

Raab, W.: Training, Physical Inactivity and the Cardiac Dynamic Cycle, *J. Sports Med. Phys. Fitness*, **6:**38, 1966.

Raab, W., and H. J. Krzywanek: Cardiac Sympathetic Tone and Stress Response Related to Personality Patterns and Exercise Habit, in W. Raab, "Prevention of Ischemic Heart Disease," Charles C Thomas, Publisher, Springfield, Ill., 1966.

Rasch, P. J., and L. E. Morehouse: Effect of Static and Dynamic Exercise on Muscular Strength and Hypertrophy, *J. Appl. Physiol.*, **11:**29, 1957.

Rasch, P. J., W. R. Pierson, and G. A. Logan: The Effect of Isometric Exercise upon the Strength of Antagonistic Muscles, *Intern. Z. Angew. Physiol.*, **19:**18, 1961.

Reindell, H., K. König, and H. Roskamm: "Funktionsdiagnostik des gesunden und kranken Herzens," Georg Thieme Verlag, Stuttgart, 1967.

Reitsma, W.: "Regeneratie, Volumetrische en Numerieke Hypertrophie van skeletspieren bij Kikker en Rat," Acad. Proefschrift, Vrije Universiteit Te Amsterdam, 1965.

Reuschlein, P. S., W. G. Reddan, J. Burpee, J. B. L. Gee, and J. Rankin: Effect of Physical Training on the Pulmonary Diffusing Capacity during Submaximal Work, *J. Appl. Physiol.*, **24:**152, 1968.

Robinson, S., and P. M. Harmon: The Lactic Acid Mechanism and Certain Properties of the Blood in Relation to Training, *Am. J. Physiol.*, **132:**757, 1941a.

Robinson, S., and P. M. Harmon: The Effects of Training and of Gelatin upon Certain Factors Which Limit Muscular Work, *Am. J. Physiol.*, **133:**161, 1941b.

Rodahl, K., N. C. Birkhead, J. J. Blizzard, B. Issekutz, Jr., and E. D. R. Pruett: Physiological Changes during Prolonged Bed Rest, in G. Blix (ed.), "Nutrition and Physical Activity," p. 107, Almqvist & Wiksell, Stockholm, 1967.

Rohter, F. D., R. H. Rochelle, and C. Hyman: Exercise Blood Flow Changes in the Human Forearm during Physical Training, *J. Appl. Physiol.*, **18:**789, 1963.

Roskamm, H.: Optimum Patterns of Exercise for Healthy Adult, *Can. Med. Ass. J.*, **22:**895, 1967.

Roskamm, H., H. Reindell, and K. König: "Körperliche Aktivität und Herz und Kreislaufer-krankungen," Johann Ambrosius Barth, Munich, 1966.

Rowell, L. B.: "Factors Affecting the Prediction of Maximal Oxygen Intake from Measure- ments Made during Submaximal Work with Observations Related to Factors Which

May Limit Maximal Oxygen Intake" (thesis), University of Minnesota, Minneapolis, 1962.

Royal Canadian Air Force: "Exercise Plans for Physical Fitness," *This Week Magazine*, Mount Vernon, N.Y., 1962.

Saltin, B., and P.-O. Åstrand: Maximal Oxygen Uptake in Athletes, *J. Appl. Physiol.*, **23:**353, 1967.

Saltin, B., G. Blomqvist, J. H. Mitchell, R. L. Johnson, Jr., K. Wildenthal, and C. B. Chapman: Response to Submaximal and Maximal Exercise after Bed Rest and Training, *Circulation*, **38**(Suppl. 7), 1968.

Saltin, B., L. H. Hartley, Å. Kilbom, and I. Åstrand: Effect of Physical Conditioning on Oxygen Uptake, Heart Rate, and Blood Lactate Concentration by Submaximal and Maximal Exercise in Middle-aged and Old Men, *J. Scand. Clin. Lab. Invest.*, to be published.

Schoop, W.: Bewegungstherapie bei peripheren Durchblutungsstörungen, *Med. Welt*, **10:**502, 1964.

Schoop, W.: Auswirkungen gesteigerter körperlicher Aktivität auf gesunde und krankhaft veränderte Extremitätsarterien, in Roskamm et al. (eds.), "Körperliche Aktivität und Herz und Kreislauferkrankungen," p. 33, Johann Ambrosius Barth, Munich, 1966.

Siebert, W. W.: Untersuchungen über Hypertrophie des Skelettmuskels, *Z. Klin. Med.*, **109:**350, 1929.

Sjöstrand, T. (ed.): "Clinical Physiology," Svenska Bokförlaget, Stockholm, 1967.

Skinner, J. S., K. O. Holloszy, and T. K. Cureton: Effects of a Program of Endurance Exercises on Physical Work, *Amer. J. Cardiol.*, **14:**747, 1964.

Steinhaus, A. H.: Chronic Effects of Exercise, *Physiol. Rev.*, **13:**103, 1933.

Stevens, P. M., P. B. Miller, T. N. Lynch, C. A. Gilbert, R. L. Johnson, and L. E. Lamb: Effects of Lower Body Negative Pressure on Physiologic Changes Due to Four Weeks of Hypoxic Bed Rest, *Aerospace Med.*, **37:**466, 1966.

Tabakin, B. S., J. S. Hanson, and A. M. Levy: Effect of Physical Training on the Cardiovascular and Respiratory Response to Graded Upright Exercise in Distance Runners, *Brit. Heart J.*, **27:**205, 1965.

Taylor, H. L., A. Henschel, J. Brozek, and A. Keys: Effects of Bed Rest on Cardiovascular Function and Work Performance, *J. Appl. Physiol.*, **2:**223, 1949.

Tepperman, J., and D. Pearlman: Effects of Exercise and Anemia on Coronary Arteries of Small Animals as Revealed by the Corrosion-Cast Technique, *Circulation Res.*, **9:**576, 1961.

Thörner, W.: Neue Beiträge zur Physiologie des Trainings, *Arbeitsphysiol.*, **14:**95, 1949.

Tipton, C. M.: The Effect of Muscular Activity on the Heart Rate of the Rat, *The Physiologist*, **4:**123, 1961.

Tipton, C. M., R. J. Barnard, and G. D. Tharp: Cholinesterase Activity in Trained and Untrained Rats, *Intern. Z. Angew. Physiol.*, **23:**34, 1966.

Tipton, C. M., R. J. Schild, and R. J. Tomanek: Influence of Physical Activity on the Strength of Knee Ligaments in Rats, *J. Appl. Physiol.*, **212:**783, 1967.

Trautwein, W., W. J. Whalen, and E. Grosse-Schulte: Elektrophysiologischer Nachweis spontaner Freisetzung von Acetylcholin in Vorhof des Herzens, *Pflügers Arch.*, **270:**560, 1960.

Vallbona, C., F. B. Vogt, D. Cardus, W. A. Spencer, and M. Walters: The Effect of Bedrest on Various Parameters of Physiologic Function, part I, Review of the Literature on the Physiological Effects of Immobilization, *NASA Contractor Report NASA CR-171*, March, 1965.

Vannotti, A., and M. Magiday: Untersuchungen zum Studium des Trainiertseins, *Arbeitsphysiol.*, **7:**615, 1934.

Vannotti, A., and H. Pfister: Untersuchungen zum Studium des Trainiertseins, *Arbeitsphysiol.*, **7:**127, 1934.

Viidik, A.: Biomechanics and Functional Adaptation of Tendons and Joint Ligaments, in F. G. Evans (ed.), "Studies on the Anatomy and Function of Bone and Joints," Springer-Verlag OHG, Heidelberg, p. 17, 1966.

Vogt, F. B., P. B. Mack, P. C. Johnson, and L. Wade, Jr.: Tilt Table Response and Blood Volume Changes Associated with Fourteen Days of Recumbency, *Aerospace Med.*, **38:**43, 1967.

Whipple, G. H.: The Hemoglobin of Striated Muscle, 1, Variations Due to Age and Exercise, *Am. J. Physiol.*, **76:**693, 1926.

Williams, C. G., C. H. Wyndham, R. Kok, and M. J. E. von Rahden: Effect of Training on Maximum Oxygen Intake and on Anaerobic Metabolism in Man, *Intern. Z. Angew. Physiol.*, **24:**18, 1967.

Yakovlev, N. N.: Problem of Biochemical Adaptation of Muscles in Dependence on the Character of Their Activity, *J. Gen. Biol. USSR* (Eng. Transl.), **19:**417, 1958.

Zimkin, N. V.: The Importance of Size of Load in Rate of Performance and Duration of Exercises and of the Intervals between Sessions in Relation to Effective Muscular Training, *Sechenov Physiol. J. USSR* (Eng. Transl.), **46:**1000, 1960.

Zimkin, N. V.: Stress during Muscular Exercises and the State of Non-specifically Increased Resistance, in E. Jokl and E. Simon (eds.), "International Research in Sport and Physical Education," p. 448, Charles C Thomas, Publisher, Springfield, Ill., 1964.

13

Energy Cost of Various Activities

contents

chapter thirteen

Energy Cost of
Various Activities

INTRODUCTION

Precise information pertaining to the energy cost of different kinds of physical work is not only of theoretical interest. Such information is essential as a basis for an intelligent and fair administration of any rationing program in case of war or other emergencies. It is also of interest in connection with the discussion of the cost of living, especially in underdeveloped areas where economy and food availability are greatly restricted. These questions are less critical in countries where the living standard is high and where food is abundantly available. But in areas where the sources of food available to the ordinary individual are limited, a worker such as a lumberjack who has to spend 5000 kcal per day obviously needs to spend more money for food than a clerk requiring only 3000 kcal or less to maintain nutritional balance.

A basic requirement for the maintenance of physical work output under conditions of dietary limitations is to cover caloric need. It has been shown in studies in underdeveloped countries and among prisoners of war that in the long run, output is related to supply of food calories. It is true that motivation also plays a role, but under conditions of severe dietary restrictions food availability is the most important limiting factor (Kraut and Müller, 1946). As food availability increases, motivation assumes an increasingly important role.

The human body is an expensive machine to operate compared to a tractor or a bulldozer because of the high cost of the fuel required in the form of food to provide its energy. If the human body could utilize fuel oil as a source of energy, as in the case of a tractor, a man could work for a whole year, expending 1 million kcal for less than $3, but because of the high cost of food, it takes an average worker 150 times that amount to operate his human machine for 1 year.

However, this figure could be materially reduced if man could adjust himself to a simpler diet, for it is possible to provide a balanced and adequate diet yielding 3000 kcal, 70 gm protein, all essential amino acids, and the required amounts of minerals and vitamins for 60 cents/day in the form of a formula diet consisting of cereal, dried milk powder, corn oil, margarine, and sugar, with vitamin and mineral supplements. Such a diet has been used routinely and exclusively in the metabolic ward of Lankenau Hospital, and, in some cases, has been consumed by the same individual for as long as 2 years without any adverse effect of any kind.

There are a number of ways to arrive at an estimate of average 24-hr caloric cost for different kinds of physical work. For scientific purposes there is a choice between (1) the assessment of daily caloric intake required to maintain body weight during normal activity and (2) the estimation of 24-hr energy expenditure on the basis of time-activity data (usually a record of activity at 15-min intervals throughout the day). Obviously the best approach is to employ both methods simultaneously. Ideally, therefore, estimates of energy costs should include both assessment of energy expenditure and measurement of caloric consumption. Passmore and Durnin (1955) have presented data showing a simultaneous assessment of energy intake and energy expenditure. The results arrived at independently agreed within 10 percent. Even closer agreements were obtained in simultaneous individual food weighings and energy expenditure estimates among soldiers and Eskimos in Alaska (Rodahl, 1960).

ESTIMATES OF CALORIC INTAKE

As we have just shown, the energy cost of different occupations can be estimated with a reasonable degree of accuracy by individual food weighings during a weekly period, repeated at intervals, and correlated with records of body weight. In the past, such estimates have often been based on large-scale food surveys, food inventories, or even by questionnaires dealing with dietary habits and rough estimates of intakes provided by the subjects themselves. Such procedures, although covering larger population samples, are subject to considerable error, however. Individual food weighings, on the other hand, will yield more precise data regarding food actually consumed by the subject. This slow procedure limits the number of subjects who can be studied at any one time. Nevertheless, it is as a rule preferable to have a few accurate figures than a large number of inaccurate ones, and if care is taken to select subjects who are fairly representative for the group in question, this procedure is quite satisfactory. Ideally, if the problem is to assess requirements in healthy individuals, a complete medical examination of each subject is desirable in order to exclude any pathological states which may interfere with the study.

The subjects are allowed to select the food as usual in unlimited quantities. Preferably, each food item is collected on separate paper plates and in paper cups and accurately weighed. At the end of the meal, each plate or cup, together with the remaining unconsumed food, is again weighed and the actual food con-sumption recorded. In-between-meal consumption must also be recorded. For the consumption of the results, the values may be taken from standard tables (U.S. War Department Technical Manual, 1945; McCance and Widdowson, 1947, 1960; Wooster and Blanck, 1949; Mattice, 1950; Watt and Merrill, 1963).

Numerous investigations have been carried out by different workers in which the values obtained by computation from food composition tables have been compared with those given by direct chemical analysis. These have, in general, shown that the results are sufficiently close to warrant the use of food composition tables in dietary surveys of families or groups of individuals (Food and Agriculture Organization of the United Nations, 1949, 1957). In the case of calories there is certainly no doubt about the general validity of estimating the caloric value of a diet from tables of food composition, provided complete and accurate data pertaining to the quantities of all foods consumed are obtained (Hunter et al., 1948). To facilitate the slow process of computing the nutritional composition of the meal, the food consumed at each meal may be entered directly onto specially prepared data sheets from which the figures may be punched on data cards for machine handling. This will eliminate lengthy transcription and reorganization of the data, eliminate much error, and save considerable time.

Summary   With the aid of available food tables a suitable diet may be composed and the caloric content estimated with an error no greater than 15 percent (Durnin and Passmore, 1967). A similar error is of no importance in developing ration scales for the Armed Services, prisons, schools, and other institutions or as guides for doctors and dietitions in prescribing diets for patients. However, for accurate scientific studies, such an error might be critical. In this type of work it is desirable to analyze duplicate samples of the diets actually eaten.

Recording of body weight is essential, since the food consumed can only be considered as representing actual requirements if it keeps the body weight constant (within ± 0.5 kg).

ESTIMATES OF ENERGY EXPENDITURE

Since the validity of using oxygen uptake as a basis for measuring energy expen-diture has been established, this indirect calorimetry has been used to determine the energy cost of a great variety of human activities. With the development of highly portable devices for collecting expired air under field conditions and rapid methods of analyzing the oxygen and carbon dioxide content of the air samples, our knowledge of the energy cost of physical work has greatly increased. The magnitude of the available data is indeed apparent from Passmore and Durnin's review of the subject (1955) (Durnin and Passmore, 1967).

The standard methods most commonly used for the determination of O_2 uptake during physical work, especially bicycle ergometer or treadmill work, are described in the appendix. In field studies, the classical method is to collect the expired air in Douglas bags carried on the subject's back. At present there are also other methods available by which the volume of the expired air is measured with flow meters, and aliquot samples of the expired air are collected in a small rubber bladder (the Max Planck respirometer, Müller and Franz, 1952; the integrating motor pneumotachygraph, or IMP, Wolff, 1958). The expired air is then analyzed for O_2 and CO_2 using conventional gas-analysis technique (Haldane or Micro-Scholander) or electronic O_2 and CO_2 analyzers. Various methods are described and discussed by Consolazio et al. (1963) and Durnin and Passmore (1967).

It should be borne in mind that the O_2 uptake measured during physical activity of high intensity does not represent the total energy cost of that activity if anaerobic processes contribute materially to the total energy metabolism. Following the cessation of the work in such cases there is a "repayment" of the "oxygen debt" incurred during the work period, and this amount of oxygen should really be added to the oxygen uptake measured during the activity to arrive at the total cost of that activity (Chap. 9). The expression of aerobic energy expenditure in terms of liters O_2/min may be preferable to kcal/min because it is the former which is measured. Furthermore, the caloric coefficient of O_2 varies and is difficult to measure precisely during work, so that the transformation of O_2/min to kcal/min is subject to error. However, in view of the fact that most data in the literature are given in terms of calories, the values in this discussion are quoted in this way. Simply by dividing the figure for caloric expenditure by 5, an approximate value for O_2 is obtained.

The method of indirect calorimetry enables the energy expenditure of an individual to be measured during a definite activity for a limited period of time. If determinations are made of the metabolic cost of each activity throughout the entire day, a fairly good idea can be obtained of the mean energy cost as well as of its variation. On the basis of such data for the different kinds of activities undertaken by an individual, together with time-activity records, it is possible to arrive at a fairly good approximation of the total 24-hr energy expenditure by using the formula:

Daily energy expenditure = time spent in each activity (min) × metabolic
cost of each activity (kcal/min)

It is essential that the different activities are accurately recorded, especially the actual duration of each activity. Under ideal circumstances, such estimates may be as accurate as most nutritional surveys, probably with an error of not more than about 10 percent (Harries et al., 1962). Durnin and Passmore (1967) point out that to organize and execute measurements of daily energy expenditure presents difficulties comparable to those encountered in surveys of daily caloric consumption by the individual dietary survey technique.

It is customary to base the estimate on recorded activities for the entire week. In any such estimates it is important to include careful records not only of activities during the working day, but also off-duty energy expenditures and recreational activities. With the shortening of the work week, off-duty energy expenditures will assume an even more important role.

In such estimates it is important to concentrate on obtaining accurate figures for those activities which constitute a major part of the 24-hr energy expenditure. A considerable error in the estimated energy cost of an activity which occupies only a very short period of time, say once in a week or once in a day, will represent a very small part of the total energy expenditure, and errors will be insignificant. In such cases the use of available data from tables of energy expenditure would indeed be justifiable. In the case of major activities, on the other hand, which constitute a large proportion of the occupation in question, it may be advisable to make several measurements of oxygen uptake under the different phases of such activities rather than relying entirely on tables of caloric expenditures, since minor inaccuracies here may result in considerable errors. Under ideal conditions, such as in metabolic wards where the subject is under constant surveillance and the activity recorded by nurses and technicians, the estimate of energy expenditure is quite accurate, probably with an error less than 5 percent.

Since in a given person there is generally a linear relationship between O_2 uptake and heart rate, the heart rate or pulse rate under certain standardized conditions may be used as a rough index of O_2 uptake in a given task (Lundgren, 1946; I. Åstrand, 1967). By comparing the pulse rate obtained during work with the pulse rate during different bicycle ergometer work loads, a rough estimate of the severity of the work load can be made. It should be borne in mind, however, that the pulse rate may be significantly affected by factors other than metabolic rate, such as environmental temperature, muscle groups involved in the exercise, work position, and emotional stress (see below and Chap. 6, "Heart Rate").

CALORIC OUTPUT AT A GIVEN TASK

Passmore and Durnin (1955) point out that there are no important individual variations in the mechanical efficiency with which a standardized muscular movement is made. They suggest that the obvious difference between energy expended by old and young is brought about mainly by variations in the economy of muscular movement and in the amount of surplus activity. It is generally true that children are unnecessarily vigorous in their movements, while needless physical activity tends to be avoided with advancing age. These factors must be taken into account when considering energy expenditure of persons of widely different ages engaged in the same type of industrial work.

I. Åstrand (1960) has shown that there is no age difference in mechanical efficiency in subjects working at submaximal work loads on the bicycle ergometer, but women attain a somewhat lower O_2 uptake at a given work load compared to men. On the treadmill, however, younger children were at a mechanical disadvantage, and the net O_2 uptake per kilogram of body weight during treadmill work at a given speed fell progressively as the age increased from five to fifteen years (P.-O. Åstrand, 1952).

Available evidence also shows that the effect of environmental temperature, as such, on the metabolic cost of work is very small (Consolazio, 1963). Gray et al. (1951) reported variations in metabolism up to 4 percent in men working at a fixed bicycle ergometer work load at ambient temperatures −15°C and 32°C. This variation, however, is probably within the experimental error. It appears that any increase in the caloric cost of work in a cold environment is primarily due to the hobbling effect of the heavy clothing and the energy cost of bodily movement in bulky clothing through difficult terrain, snow, etc. Nelson et al. (1948) found that metabolic heat production for a given amount of work remained unchanged in three men walking in seven hot environments between 32°C and 49°C.

ENERGY EXPENDITURE DURING VARIOUS ACTIVITIES

Data for O_2 uptake, expressed as kilocalories per minute, of a variety of activities taken from various sources, but mostly from Passmore and Durnin (1955) and Spitzer and Hettinger (1958), are summarized in Fig. 13-1. The figures represent gross values and include the values for the specific dynamic action (SDA) of food, as was the case in Passmore and Durnin's paper. Attempts to allow for the SDA only invite a false impression of accuracy, for the SDA of food is indeed variable and will change with the time of day as related to meal times. Furthermore, the SDA will in any case only represent a small fraction of the total 24-hr metabolism in a physically active individual.

Sleeping

In the past it has been difficult to obtain accurate values for energy expenditure in a sleeping subject due to the discomfort and restriction imposed by noseclip and mouthpiece, or the technical problem of leaks when using a face mask. The use of the open-circuit method with a plastic hood covering the entire head of the patient and the Noyons basal metabolism apparatus (Issekutz et al., 1963) offers many distinct advantages when making metabolic measurements in a resting or sleeping individual. With this method, repeated samples can be taken throughout the period of observation without disturbing the subject in any way.

Generally, one-third of the 24-hr period is spent in bed sleeping or resting. The energy spent in this way accounts for about one-tenth to one-quarter of the daily expenditure, if the basal metabolic rate (BMR) is accepted as a measure of the metabolic rate when in bed, whether the subject is asleep or awake.

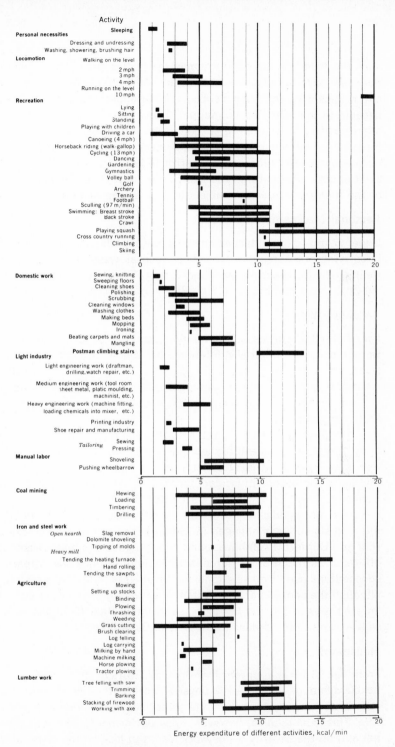

Fig. 13-1 Energy expenditure of different activities (kcal/min). Bars denote range of data presented in literature (see text). A prerequisite for a high energy output is certainly a high maximal aerobic power.

It is true that the metabolism of the fasting subject as a rule may fall below the BMR when he is asleep. At the start of the night's sleep, however, as pointed out by Passmore and Durnin (1955), the metabolism may be above the basal level because of the SDA effect of supper. These two factors thus tend to cancel each other out so that the energy expenditure throughout the night is not far from that of the BMR value. In any case, a deviation of 10 percent above or below the normal basal level represents an error of less than about 3 percent of the total 24-hr energy expenditure. The BMR value may therefore be taken as a measure of the metabolic rate of a subject in bed, asleep or awake.

Personal Necessities

Under normal circumstances an individual spends not more than 1 hr of the day in carrying out personal necessities, such as washing, shaving, brushing hair, cleaning teeth, dressing, and undressing. Energy expenditure values for such activities are summarized in Fig. 13-1.

Sedentary Work

For all practical purposes, the energy expenditure of mental work, including office work, etc., is not materially different from that of sitting or standing unless such occupations involve a great deal of physical activity, such as walking from one place to another, bending, opening drawers, etc. Passmore and Durnin (1955) list a mean value of 1.6 kcal/min for miscellaneous office work sitting, and 1.8 kcal/min standing.

Although the brain utilizes a substantial part of the total O_2 uptake of the body at rest, it is well established that mental work requires only an insignificant rise in O_2 uptake, at least as long as the mental effort is not associated with markedly increased muscular tension or emotional stress (Benedict and Benedict, 1933; Benedict and Carpenter, 1909). Benedict and Benedict (1933) observed no significant difference in metabolism at rest and during mental effort in six subjects engaged in 15-min periods of arithmetic exercises. Others (Eiff and Göpfert, 1952) have found an average rise in metabolism during mental efforts of as much as 11 percent, but this difference was apparently due to a concomitant rise in muscle tone. However, it is indeed interesting to note that although the central nervous system is extremely sensitive to lack of oxygen and lowered oxygen tension, an increase in intellectual functions is not associated with a significant increase in the overall O_2 uptake.

Housework

From Fig. 13-1 it is apparent that domestic work involves many tasks which may be classified as fairly heavy physical work, although modern equipment has contributed greatly to making life somewhat easier for the present-day house- wife. This is all relative, however, since performance of physical work depends not only on the severity of the work load, but also on the physical work capacity of the individual. For this reason, the lighter work load of present-day domestic occupations due to modernizations may represent, relatively speaking, as heavy a load on the present-day housewife as that of her grandmother, if the latter were more physically fit.

Figure 13-2 presents an example of a continuous recording of heart rate during a day's work. The data on heart rate alone do not necessarily tell anything about energy expenditure, but the figure shows at least that the housewife was physically active!

Light Industry

A fair amount of data is available regarding the energy cost of different kinds of manual labor and industrial tasks. With the development of automation, the physical work load of the industrial worker on the whole has been greatly reduced. This is evident from an early study by Kagan et al. (1928) who compared energy expenditure by men assembling machinery entirely by hand with those using a conveyer system. In the former case, the energy expenditure varied from 5.2 to 6.4 kcal/min; in the latter case it varied between 1.8 and 4.7 kcal/min. According to Passmore and Durnin (1955; Durnin and Passmore, 1967) a wide variety of industrial activities classified by them as light industry demanded energy expenditure rates between 2 to 5 kcal/min for men and 1.5 to 4 kcal/min for women, pointing out that further subdivision or classification is quite impracticable. Examples of such light industry are given in Fig. 13-1. During an 8-hr working day the caloric output is about 1200 kcal for men and 900 kcal for women.

Manual Labor

A number of studies have shown that the energy expended during the performance of similar types of work may vary greatly, depending on the technique used in accomplishing the work. This is even true for relatively simple activities such as carrying a load, depending on how it is carried, the size of the shovel, etc. Bedale (1924) showed that when carrying a load, the energy expenditure was minimal when a yoke across the shoulder was used; it was maximal when the load was carried on the hip under the arm. When carrying a load, the energy expenditure rises markedly with the increased speed of walking (Brezina and Kolmer, 1912; Cathcart et al., 1923). The energy cost of climbing up and down stairs with a load is, according to Crowden (1941), $11\frac{1}{2}$ times that of walking on the level carrying the same load. In a study of the energy cost of transporting a load with the aid of wheelbarrows, Hansson (1968) showed that the energy expenditure is higher the smaller and softer the wheel, and that a two-wheel cart is more efficient than a wheelbarrow with a single wheel (Fig. 13-3).

There is little data available to assess the overall energy expenditure of building work. Accurate assessment in this case is particularly difficult because of the variety of individual operations involved. Figure 13-4 gives an example of a

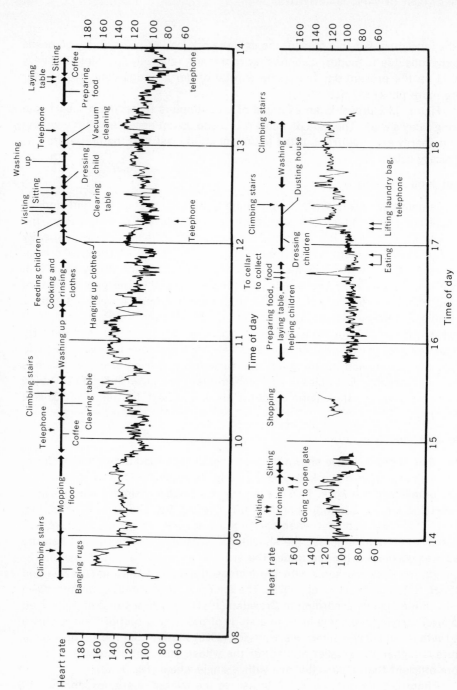

Fig. 13-2 Heart rate in a housewife during an ordinary day of housework.

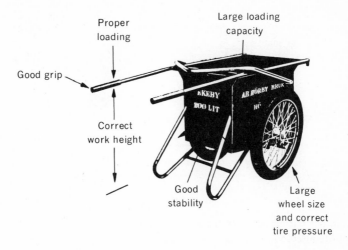

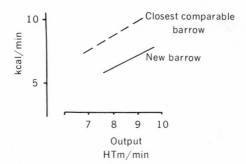

HTm/min = Horizontal movement of one ton a distance of one meter

Fig. 13-3   Study of separate details of the construction and capacity of the wheelbarrow, indicating how a "correctly designed" barrow should look. The result was an experimental wheelbarrow having a 40 percent greater capacity at a similar physiological load. (From Hansson, 1968.)

markedly increased production in a given task in spite of reduced energy expenditure and effort. A common procedure in grouting precast concrete units for the building of houses was to work in a forward-inclined position as illustrated to the left in Fig. 13-4. Construction of a new tool (to the right) made it possible to work in a more upright and therefore more comfortable position. This resulted in an increased production at lower caloric cost.

Another example from the building industry may be given. The influence of the choice of tools on the rate of work and work effort was studied in a group of men engaged

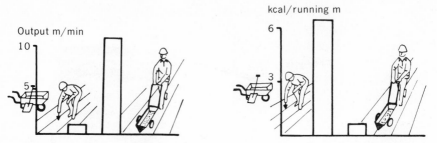

Fig. 13-4   Pouring concrete between precast concrete units (grouting) by two different methods. It is apparent that the task has been made easier and production increased by the new method. (From Hansson, 1968.)

in the nailing of impregnated paper under standardized conditions (Hansson, 1968). The rate attained in nailing when a stapler was used was 3 to 4 times higher than when using an ordinary hammer. There were no significant differences, however, in oxygen uptake per minute, quality of the work, or estimated degree of fatigue when using the different tools. I. Åstrand et al. (1968) have further studied the energy cost of hammer nailing at three different heights: at bench level, into a wall at head level, and into a ceiling above the head. The number of hammer strokes did not differ significantly in the three situations. However, the number of nails driven in per minute was lower when nailing into the wall (10.6 nails/min) than when nailing into the bench (14.6), and still lower for nailing into the ceiling (4.5 nails/min), indicating that the strokes became less powerful or were less well aimed when nailing into the wall and ceiling compared to bench nailing. Nailing in the three positions resulted in an oxygen uptake of about 1.0 liter/min in each case. The 11 subjects, all of whom were skilled carpenters, also performed leg work on a bicycle ergometer with the same oxygen uptake. Nailing into wall or ceiling resulted in a greater elevation of the heart rate than nailing at bench level; the heart rate during bicycle exercise at a comparable oxygen uptake was lower (or 102 beats/min) than for all types of nailing (130 beats/min when nailing into ceiling). It is interesting to note that leg exercise with an oxygen uptake of 1.4 liters/min gave approximately the same rise in heart rate as nailing with an oxygen uptake of only 1.0 liter/min. The intra-arterially measured blood pressures during nailing were higher than during leg exercise at a given oxygen uptake (and probably cardiac output), the difference being most pronounced between nailing into ceiling and bicycle work ($P < 0.01$).

These examples are presented to illustrate that (1) the work output may vary with the tools used and with the working positions, even though the caloric expenditure is the same; and (2) that the physiological effects of a given caloric demand may vary considerably depending on tools and techniques.

Studies of coal miners in different countries have shown a good general agreement for the work with pick and shovel. It appears that the energy expenditure of shoveling ranges, as a rule, from 6 to 7 kcal/min. In a German study (Lehmann et al., 1950), the mean gross energy expenditure during actual coal mining was 5 kcal/min, and the mean expenditure of calories per minute for the total time spent underground was 3.5. Walking to and from the coal face in a stooping position may, according to Passmore and Durnin (1955), require as much as 10 kcal/min. It appears that in spite of the increased mechanization, mining is still hard physical work. Studies in the iron and steel industry (Lehmann et al., 1950; Christensen, 1953) show wide variations in energy expenditure, so that generalizations are not justified. Although some of the workers periodically may work very hard, many are only doing light work in most modern plants.

Several studies have verified the general impression that farming is hard work, at least in the busy season, especially where the advantage of mechanization is not available (Durnin and Passmore, 1967). Milking by hand requires about 4.5 kcal/min, as against 3.5 kcal/min for machine milking. Horse ploughing entails 6 kcal/min, while tractor ploughing requires less than 5 kcal/min (Hettinger et al., 1953a and b).

It is generally recognized that lumber work involves heavy expenditure of energy. In fact, lumbering is probably the hardest form of physical work, requiring sometimes as much as 6000 kcal/day (Lundgren, 1946).

Hansson et al. (1966) report from studies in India that the mean value of energy expenditure per square meter in cross-cutting trees was 620 kcal/m² when the workers used their local saws, but only 280 kcal/m² when sawing with correctly maintained saws. The work output was more than 100 percent higher than the mean output with the local saws. Particularly in areas with food shortage the use of better equipment and instructions in its proper maintenance are very important. When spending about 7 kcal/min, which was the spontaneously chosen rate of work, the lumberjack managed to cut 1,000 cm²/min when using a power chain saw when felling trees, but only 200 cm²/min with a two-man crosscut saw.

In another study Hansson (1965) noted that lumberjacks with very high earnings were characterized by a particularly high maximal aerobic power compared to the normal earner. With regard to muscle strength and precision in a variety of standardized work operations there was no significant difference between the two categories. Thanks to his higher aerobic power the top worker could attain a higher work output, he did not take as long breaks, and he became less tired at a given caloric output compared with his less-productive colleague. It should again be emphasized that a worker involved in manual labor, who is more or less free to set his own pace, normally accepts working with a caloric output which is approximately 40 percent of his maximal aerobic power (I. Åstrand, 1967; Chap. 9).

Recreational Activities

A general summary of measurements of energy expenditure of a variety of common recreational activities is given in Fig. 13-1, ranging from light indoor recreation and games to strenuous outdoor sports (see also Chap. 16). It should be pointed out that the listed figures can be taken as approximate values only, in view of the many factors which may have a profound influence upon the energy expenditure of such activities. Individuals pursue their recreations with very different degrees of vigor. Swimming, ball games, dancing, gardening, etc., can be enjoyed with an energy expenditure which ranges from low to very high. Playing volley ball or taking a swim does not mean the same degree of physical activity for all persons (Bullen et al., 1964). Anyone, even the partially incapacitated person, can find a suitable recreation.

Data for energy expenditure of playing children have been published by Taylor et al. (1948, 1949) and Cullumbine (1950). The reported values range from 1 to 3 kcal/min. (Walking and running are discussed in Chap. 16.)

CALORIC OUTPUT AND WORK STRESS

It should be kept in mind that the values discussed here represent the energy cost of work as such and do not reflect the actual strain imposed upon the individual performing the work.

Various systems have been devised for the grading of the physical effort required in industrial work. The terms *light, moderate, heavy, very heavy,* and *unduly heavy* are used, and they are based on the caloric expenditure, heart rate, body temperature, and/or sweating rate measured during work (Christensen, 1953). Such classifications are very helpful in many connections. However, when considering the strain of work in relation to a particular individual, his work capacity and aerobic power have to be considered. A work load demanding 10 kcal/min represents a severe work stress for an individual whose maximal aerobic power is only 15 kcal/min, but it is only a slight work load for one whose maximal aerobic power is 25 kcal/min.

We have previously pointed out that a heart rate of, say, 120 does not mean the same for an individual with a maximal heart rate of 195 as for an individual with a maximal heart rate of 160 (Figs. 9-16, 9-17). Body temperature and sweating rate may be a good index of the relative work load but do not indicate the magnitude of caloric expenditure (Chap. 15). As far as the requirement for physical strength and working capacity of the worker is concerned in the degree of strain imposed upon him by the work, the peak load of the tasks is more important than the mean caloric expenditure. A steel worker may expend 10 kcal/min during 1 hr of shoveling gravel or dolomite, but during the rest of his 8-hr shift, his caloric output may be only 2 to 2.5 kcal/min. His 8-hr caloric expenditure is then $600 + 900 = 1500$ kcal. A worker with a job which demands a consistent rate of work without any peak loads may attain a higher 8-hr caloric expenditure (e.g., $480 \times 4 = 1920$ kcal), without requiring as strong a physique as does the steelworker. Heavy work or awkward working position may often hamper the recruitment for certain types of work, even though these factors may only be operating for short periods of time.

DAILY RATES OF ENERGY EXPENDITURE

Table 13-1 summarizes data presented by Durnin and Passmore (1967) on daily rates of caloric output by individuals with various occupations. They point out

TABLE 13-1

Daily rates of energy expenditure by individuals with various occupations

Occupation	Energy expenditure, kcal/day		
	Mean	Minimum	Maximum
Men			
Elderly retired	2330	1750	2810
Office workers	2520	1820	3270
Colliery clerks	2800	2330	3290
Laboratory technicians	2840	2240	3820
Elderly industrial workers	2840	2180	3710
University students	2930	2270	4410
Building workers	3000	2440	3730
Steel workers	3280	2600	3960
Army cadets	3490	2990	4100
Elderly peasants (Swiss)	3530	2210	5000
Farmers	3550	2450	4670
Coal miners	3660	2970	4560
Forestry workers	3670	2860	4600
Women			
Elderly housewives	1990	1490	2410
Middle-aged housewives	2090	1760	2320
Laboratory technicians	2130	1340	2540
Assistants in department stores	2250	1820	2850
University students	2290	2090	2500
Factory workers	2320	1970	2980
Bakery workers	2510	1980	3390
Elderly peasants (Swiss)	2890	2200	3860

that the mean values reflect the common opinion about the relative amounts of physical activity required by the different occupations. There is, however, a wide variation in each group, largely due to differences in physical activity involved in the subject's chosen recreations, but differences in the nature of the work done are also important.

Due to the great increase in the world population and the difficulties in many areas in feeding the people, it is very important to develop a worldwide program of food supply that matches the requirements. An estimation of food requirements is naturally an essential basis for such a program of agricultural planning. The data presented in Table 13-1 are examples of such measures. The United Nations have a Food and Agricultural Organization (FAO) which endeavors to advise and assist governments in these matters (1949, 1957). A committee in the United States (Food and Nutrition Board, 1964) has presented their "reference" man, 25 years old, with a body weight of 70 kg, and a reference woman, 25 years of age, and 58 kg. Both the man and the woman are presumed to live in an environment with a mean temperature of 20°C. They are considered to be "moderately" active physically, with occupations that could be characterized neither as sedentary nor as hard physical labor. A revision was made of the FAO recommendations (1957) so that the daily allowances were adjusted downwards from 3200 to 2900 kcal/day for the reference man, and from 2300 to 2100 kcal for the reference woman. (The FAO references were a man who weighed 65 kg and a woman with a body weight of 55 kg.) The revision was made since it was thought that the original allowance was unrealistic for the average American man, living a more sedentary life. (However, as pointed out by Durnin and Passmore, 1967, the original FAO recommendations may be physiologically more sound and intended for an energy requirement of men and women who are sufficiently active for full health; what Americans may need is not less food, but more physical activity.) Table 13-2 presents arbitrary examples of energy expenditure by reference man and woman (Food and Nutrition Board, 1964).

Adjustments must be made when individuals or population averages differ from the "reference" in characteristics of age, body size, or activity.

TABLE 13-2

Activity	Time, hr	Man		Woman	
		Rate, kcal/min	Total	Rate, kcal/min	Total
Sleeping and lying (1)	8	1.1	540	1.0	480
Sitting (2)	6	1.5	540	1.1	420
Standing (3)	6	2.5	900	1.5	540
Walking (4)	2	3.0	360	2.5	300
Other (5)	2	4.5	540	3.0	360
Total	24		2880		2100

1 Essentially basal metabolic rate plus some allowance for turning over or getting up or down.
2 Includes normal activity carried on while sitting, e.g., reading, driving an automobile, eating, playing cards, and desk or bench work.
3 Includes normal indoor activities while standing and walking spasmodically in limited areas, e.g., personal toilet, moving from one room to another.
4 Includes purposeful walking, largely outdoors, e.g., home to commuting station to work site, and other comparable activities.
5 Includes spasmodic activities in occasional sports exercises, limited stair climbing, or occupational activities involving light physical work. This category may include weekend swimming, golf, tennis, or picnic using 5 to 20 kcal/min for limited time.

There are recommendations for these adjustments, and for the special energy demands of pregnancy and lactation. It is proposed that the caloric allowance be reduced by 5 percent per decade between ages 35 to 55 and by 8 percent per decade from ages 55 up to 75. It is also proposed that daily allowances be increased by 200 kcal during pregnancy. Some women, however, may so reduce physical activity during pregnancy that the extra demands for calories may be largely compensated without additional food calories. The additional energy requirements of lactation is recommended to be about 120 kcal for each 100 ml of milk produced.

It is also noted that it is common in the United States to encounter persons with reduced requirements because of a relatively sedentary existence. While anyone who is physically active usually adjusts his caloric intake to energy expenditure, there is evidence to suggest that inactive persons tend to consume more calories than they need in order to remain in energy balance.

SUMMARY

An estimate of daily caloric expenditure is of importance for the calculation of caloric need. This is of particular interest in areas with shortage of food. Such an estimate can be made by two methods: (1) assessment (by food weighing) of daily caloric intake required to maintain body weight; (2) estimation based on time-activity data and measurements (or predictions) of the energy cost of all activities. Both methods are equally accurate or reliable, with an error of no more than about 10 percent.

As expected, there is a wide individual variation in caloric output depending on profession and leisure activity. The range of daily rates of energy expenditure presented in Table 13-1 is from 1340 up to 5000 kcal. About 3000 kcal/day is a reasonable expenditure for a man who is not engaged in manual labor but who is regularly active during leisure time; a reasonable figure for his wife would be about 2300 kcal/day.

In the majority of professional activities including office work, most tasks in household duties, light industry, laboratory work, hospital work, retail and distribution trade, the caloric output is less than 5 kcal/min.

In the building industry, agriculture, the iron and steel industry, the Armed Services, there are many jobs which occasionally demand a caloric expenditure of up to 7.5 kcal/min or even higher, particularly if mechanical aids are few and prefabricated materials are only utilized to a small extent.

Still higher energy demands are found in fishing, forestry work, mining, and dock labor, where figures up to or exceeding 10 kcal/min have been obtained.

The energy expenditure in recreational activity naturally covers the whole range from a few calories per minute up to utilization of the full power of aerobic and anaerobic processes, depending on the type of activity and degree of vigor with which it is pursued.

The physiological and psychological effects of a given caloric output (per minute; per 8 hr; per day) is determined by the individual's maximal aerobic power, size of the engaged muscle mass, posture, whether work is intermittent at high rate or continuous at a lower intensity, and by environmental conditions.

Even if her output is relatively low, a housewife may strain herself as much during domestic work as a lumberjack, farmer, or miner whose work requires a higher caloric output.

REFERENCES

Åstrand, I.: Aerobic Work Capacity in Men and Women with Special Reference to Age, *Acta Physiol. Scand.*, **49:**(Suppl. 169), 1960.

Åstrand, I.: Degree of Strain during Building Work as Related to Individual Aerobic Work Capacity, *Ergonomics*, **10:**293, 1967.

Åstrand, I., A. Guharay, and J. Wahren: Circulatory Responses to Arm Exercise with Different Arm Positions, *J. Appl. Physiol.*, **25:**528, 1968.

Åstrand, P.-O.: "Experimental Studies of Physical Working Capacity in Relation to Sex and Age," Munksgaard, Copenhagen, 1952.

Bedale, E. M.: "Comparison of the Energy Expenditure of a Woman Carrying Loads in Eight Different Positions," Medical Research Council Industrial Fatigue Research Board, no. 29, 1924.

Benedict, F. G., and C. G. Benedict: "Mental Effort in Relation to Gaseous Exchange, Heart Rate and Mechanics of Respiration," Carnegie Institute, Washington, D.C., no. 446, 1933.

Benedict, F. G., and T. M. Carpenter: "Influence of Muscular and Mental Work on Metabolism and Efficiency of the Human Body as a Machine," U.S. Department of Agriculture Office Experimental Stations Bull. 208, 1909.

Brezina, E., and W. Kolmer: Über den Energieverbrauch bei der Geharbeit unter dem Einfluss verschiedener Geschwindigkeiten und Verschiedener Belastungen, *Biochem. Z.*, **38:**129, 1912.

Bullen, B. A., R. B. Reed, and J. Mayer: Physical Activity of Obese and Nonobese Adolescent Girls: Appraisal by Motion Picture Samples, *Amer. J. Clin. Nutr.*, **14:**211, 1964.

Cathcart, E. P., D. T. Richardson, and W. Campbell: Maximum Load to Be Carried by the Soldier, *J. Roy. Army Med. Corps.*, **40:**435, **41:**12, **87:**161, 1923.

Christensen, E. H.: "Physiological Valuation of Work in the Nykoppa Iron Works," Ergonomic Society Symposium on Fatigue, W. F. Floyd and A. T. Welford (eds.), p. 93, Lewis, London, 1953.

Consolazio, C. F.: The Energy Requirements of Men Living under Extreme Environmental Conditions, in G. H. Bourne (ed.): "World Review of Nutrition and Dietetics," vol. 4, p. 53. Pitman Medical, London, 1963.

Consolazio, C. F., R. E. Johnson, and L. J. Pecora: "Physiological Measurements of Metabolic Functions in Man," McGraw-Hill Book Company, New York, 1963.

Crowden, G. P.: Stair Climbing by Postmen, *The Post* (London), p. 10, July 26, 1941.

Cullumbine, H.: Heat Production and Energy Requirements of Tropical People, *J. Appl. Physiol.*, **2**:640, 1950.

Durnin, J. V. G. A., and R. Passmore: "Energy, Work and Leisure," William Heinemann, Ltd., London, 1967.

Eiff, A. W., and H. Göpfert: Ausmass und Ursachen der Energieumsatz-Veränderungen bei geistiger Arbeit, *Z. Ges. Exp. Med.*, **120**:72, 1952.

Food and Agriculture Organization of the United Nations: "Dietary Surveys, Their Technique and Interpretation," Washington, D.C., 1949.

Food and Agriculture Organization of the United Nations: "Caloric Requirements," Nutritional Studies no. 15, Rome, 1957.

Food and Nutrition Board: "Recommended Dietary Allowances," Nat. Acad. Sci., Nat. Res. Coun. Washington, D.C., publication 1146, 1964.

Gray, E. L., C. F. Consolazio, and R. M. Karl: Nutritional Requirements for Men at Work in Cold, Temperate and Hot Environments, *J. Appl. Physiol.*, **4**:270, 1951.

Hansson, J.-E.: The Relationship between Individual Characteristics of the Worker and Output of Work in Logging Operations, *Studia Forestalia Suecia*, no. 29, Skogshögskolan, Stockholm, 1965.

Hansson, J.-E.: Work Physiology as a Tool in Ergonomics and Production Engineering, AI-Rapport 2, *Ergonomi och Produktionsteknik*, National Institute of Occupational Health, Stockholm, 1968.

Hansson, J.-E., A. Lindholm, and H. Birath: Men and Tools in Indian Logging Operations: A Pilot Study in Ergonomics, *Rapporter och Uppsatser*, Inst. för Skogteknik, Skogshögskolan, Stockholm, no. 29, 1966.

Harries, J. M., E. A. Hobson, and D. F. Hollingsworth: Individual Variations in Energy Expenditure and Intake, *Proc. Nutr. Soc.*, **21**:157, 1962.

Hettinger, Th., and W. Wirths: Der Energieverbrauch beim Hand-und Motorplügen, *Arbeitsphysiol.*, **15**:41, 1953a.

Hettinger, Th., and W. Wirths: Über die Körperliche Beanspurchung beim Hand und Maschinemelken, *Arbeitsphysiol.*, **15**:103, 1953b.

Hunter, G., J. Kastelig, and M. Ball: Assessment of Diets: Analysis versus Computation from Food Tables, *Canad. J. Res.*, **26**:367, 1948.

Issekutz, B., Jr., N. C. Birkhead, and K. Rodahl: Effect of Diet on Work Metabolism, *J. Nutr.*, **79**:109, 1963.

Kagan, E. M., P. Dolgin, P. M. Kaplan, C. O. Linetzkaja, J. L. Lubarsky, M. F. Neumann, J. J. Semernin, J. S. Starch, and P. Spilberg: Physiologische Vergleichsuntersuchung der Hand-und Fleiss-(Conveyer) Arbeit, *Arch. Hyg. (Berlin)*, **100**:335, 1928.

Kraut, H., and E. A. Müller: Caloric Intake and Industrial Output, *Science*, **104**:2709, 1946.

Lehmann, G., E. A. Müller, and H. Spitzer: Der Calorienbedarf bei gewerblicher Arbeit, *Arbeitsphysiol.*, **14**:166, 1950.

Lundgren, N.: Physiological Effects of Time Schedule Work on Lumber Workers, *Acta Physiol. Scand.*, **13**:(Suppl. 41), 1946.

Mattice, M. R.: "Bridge's Food and Beverage Analysis," Lea & Febiger, Philadelphia, 1950.

McCance, R. A., and E. M. Widdowson: "The Chemical Composition of Foods," Chemical Publishing Co., Inc., Brooklyn, N.Y., 1947.

McCance, R. A., and E. M. Widdowson: "The Composition of Foods," Spec. Rep. Ser. Med. Res. Coun. (London), no. 297, 1960.

Müller, E. A., and H. Franz: Energieverbrauchsmessungen bei beruflicher Arbeit mit einer verbesserten Respirations-Gasuhr, *Arbeitsphysiol.*, **14**:499, 1952.

Nelson, N. A., W. B. Shelley, S. M. Horvath, L. W. Eichna, and T. F. Hatch: Influence of Clothing, Work and Air Movement on the Thermal Exchanges of Acclimatized Men in Various Hot Environments, *J. Clin. Invest.*, **27**:209, 1948.

Passmore, R., and J. V. G. A. Durnin: Human Energy Expenditure, *Physiol. Rev.*, **35**:801, 1955.

Rodahl, K.: "Nutritional Requirements under Arctic Conditions," Norsk Polarinstitutt, Skrifter Nr. 118, Oslo University Press, Oslo, 1960.

Spitzer, H., and Th. Hettinger: "Tafeln für Kalorienumsatz bei körperlicher Arbeit," REFA publication, Darmstadt, 1958.

Taylor, C. M., M. W. Lamb, M. E. Robertson, and G. MacLeod: Energy Expenditure for Quiet Playing and Cycling of Boys 7 to 15 Years of Age, *J. Nutr.*, **35**:511, 1948.

Taylor, C. M., O. F. Pye, A. B. Caldwell, and E. R. Sostman: Energy Expenditure of Boys and Girls 9 to 11 Years of Age (1) Sitting Listening to the Radio (Phonograph), (2) Sitting Singing and (3) Standing Singing, *J. Nutr.*, **38**:1, 1949.

U.S. War Department: "Hospital Diets," Technical Manual TM8-500, March, 1945.

Watt, B. K., and A. L. Merrill: "Composition of Foods," U.S. Department of Agriculture, Handbook no. 8, Washington, D.C., 1963.

Wolff, H. S.: The Integrating Motor Pneumotachograph: A New Instrument for the Measurement of Energy Expenditure by Indirect Calorimetry, *Quart. J. Exper. Physiol.*, **43**:270, 1958.

Wooster, H. A., Jr., and F. C. Blanck: "Nutritional Data," Heinz Nutritional Research Division, Mellon Institute, Pittsburgh, 1949.

14

Nutrition and Physical Performance

contents

chapter fourteen

Nutrition and Physical Performance

INTRODUCTION

The effect of nutrition on physical performance capacity has been a subject of considerable interest for at least 2,000 years. Many of the food taboos and dietary superstitions of primitive tribes are related to this question. Even some of our present-day athletes and coaches may at times be affected by similar conceptions. Often a clue to the success of a champion athlete is sought in the manner in which he trains or carries out his warm-up exercises, or in what he eats. Such clues are often used as a basis for some sort of theory as to what components of the program are important. The athlete's subsequent success may then be taken as proof of the validity of the particular theory, while the fact may be that the athlete won his championship in spite of his practices. As the athlete now has an almost limitless number of food items or dietary combinations to choose from, he has naturally been tempted to see if he can become stronger, develop longer legs, achieve greater endurance, or become more alert by eating certain foods.

The question of what an athlete should profitably eat in order to achieve superior performance is as old as the recorded history of organized sports. According to Christophe and Mayer (1958), the practice of consuming large quantities of meat to replenish the supposed loss of muscular substances during heavy muscular work was first recorded in Greece during the fifth century B.C. Two athletes, instead of the predominantly vegetarian diet of the time, adopted a regimen consisting of large quantities of meat, resulting in increased body bulk and weight. In the nineteenth century, such drastic measures as fluid restriction, bloodletting, and the use of laxatives were practiced by some athletes. More detrimental practices could hardly have been suggested, even by the athlete's worst enemies. The discovery of vitamins offered a new and promising area of experimentation by the more imaginative minds on the basis of the assumption that if a little of it is good, a great deal of it must be much better. More recently, protein tablets have figured prominently in dietary discussion, especially among weight lifters. Since the muscles consist mainly of protein, it was assumed that ingestion of excess protein might stimulate muscle growth and improve strength, as did the Greek pioneers some 2,000 years ago.

Repeated claims have been made by different workers over the last several decades to the effect that physical work performance of normal men can be significantly improved by special diets or dietary supplements (Simonson, 1951), and even more so in the case of athletic performance (Mayer and Bullen, 1960). And yet, Mayer and Bullen conclude: "The concept that any well-balanced diet is all that athletes actually require for peak performance has not been superseded."

(It should be noted that the body's water exchange will be discussed in Chap. 15.)

FUEL FOR MUSCULAR WORK

An intelligent analysis of this question would require a rather complete understanding of the fuel utilized during physical work of different types and intensity to provide the energy required for muscular work.

Protein

It is well established that *protein* is not used as a fuel to any appreciable extent when the caloric supply is adequate (Pettenkofer and Voit, 1866; Chauveau, 1896; Krogh and Lindhard, 1920). Nitrogen excretion does not rise significantly following muscular work (Crittenden, 1904; Cathcart and Burnett, 1926). Similar findings have been reported by Hedman (1957) in skiers during $2\frac{1}{2}$ hr of skiing at a rate corresponding to about 3.8 liters O_2/min, amounting to a total expenditure of about 3000 kcal. Even after exhaustion of the glycogen depots, continued exercise does not raise the nitrogen excretion significantly. The choice of fuel for the working muscle, therefore, is actually limited to *carbohydrate* and *fat*.

Carbohydrate versus Fat

The percentage participation of these two fuels in the energy metabolism is usually assessed by the determination of the nonprotein respiratory quotient (RQ) which is the ratio CO_2 volume produced/O_2 volume utilized. An approximate estimation of the amount of O_2 (ml/min) used for the oxidation of fat can be obtained from the formula: $(1 - RQ)/0.3 \times O_2$. In this calculation the RQ is handled as a "nonprotein RQ." The fact that the caloric value of O_2 varies with RQ is not taken into consideration, but this may at the most represent a 7 percent difference.

In the fasting animal, cellular fuel obviously must be obtained from available energy in the form of stored material in the body. In the development of the ideal energy stores for the mobile animal, such as man, nature has to meet a number of requirements. For reasons of portability, each molecule should carry a large amount of energy per unit weight. The material should be fitted into various oddly shaped spaces and compartments of the body. It should possess a great storage stability and, at the same time, be readily available, capable of being rapidly converted into oxidizable substrate when needed, without being spontaneously explosive. As in all other efficient operations, the overhead cost should be low, i.e., the handling expenses including cost of storage and transport should be minimal.

As pointed out by Dole (1964), fat meets these requirements remarkably well. It is high in energy content, and it is stable, yet readily mobilized. The amount of energy held per unit weight of any molecule depends on its content of oxidizable carbon and hydrogen. Oxygen in a molecule of stored energy merely adds dead weight, since oxygen atoms can be easily obtained from the air as needed. Fat, therefore, approaches maximal storage efficiency since it contains as much as 90 percent carbon and hydrogen and has an energy density of about 9 kcal/g. It is much superior in storage efficiency to carbohydrates, containing only 46 percent carbon and hydrogen and having an energy density of only 4 kcal/g. The difference is even greater when one considers that fat is deposited in droplets, while carbohydrate is deposited together with an appreciable amount of water, in mammals 2.7 g per gram dried glycogen (Weis-Fogh, 1967). In other words, hydration dilutes the energy density of glycogen to about 1 kcal/g. Thus, carbohydrate is a rather inferior material as an energy store in terms of portability. The caloric value of 1 g of adipose tissue, which does not consist of pure fat, is about 6 to 7 kcal.

The different animal species have through adaptation met their own peculiar needs. As A. V. Hill (1956) put it: "In most animals the energy for muscular effort is obtained largely by burning carbohydrate: but hydrated glycogen is a very heavy fuel compared with fat and, in the locust which flies great distances (over water or desert), a sufficient load of glycogen would severely limit range and endurance. The animal therefore rapidly transforms the carbohydrate of its diet into fat, which alone is used during flight."

In migrating fishes, like the salmon and the eel, and in birds, fat also constitutes the main source of energy. Weis-Fogh (1967) points out that due to impaired weight economy, carbohydrate cannot sustain flight for more than a few hours, and the recorded endurance of 1 to 3 days observed in some typical migrants therefore depends almost exclusively on the utilization of fat mobilized from stored triglycerides.

In the past, however, it has been generally assumed that the main and primary energy source during muscular work or exercise in mammals was carbohydrate, and that fat was only reserve fuel used mainly at rest and during recovery. This is in accordance with the Hill-Meyerhof theory. One of the strongest arguments for this theory was the decreased work performance of men on a high-fat diet. It is now clearly established that fat also is an important source of energy fuel for the working muscle, and that the percentage participation of fat and carbohydrate in the metabolic mixture depends on a number of factors, including the severity and duration of the work in relation to the subject's maximal aerobic power and his diet (Zunts, 1911; Krogh and Lindhard, 1920; Bock et al., 1928).

Christensen and Hansen (1939), in their classical experiments, examined the participation of fat and carbohydrate in energy metabolism on the basis of the RQ during physical work of different intensities. In subjects on a *normal diet* engaged in exercise of such an intensity that the metabolic processes were essentially aerobic, they found that about 50 to 60 percent of the energy was supplied by fat. In prolonged, standardized aerobic work of up to 3-hr duration, an increased participation of fat was observed, supplying up to 70 percent of the energy. In heavy work, on the other hand, where anaerobic metabolic processes were involved, their findings indicated a major participation of carbohydrates.

Subjects on an extremely *high-fat diet* for several days, in which less than 5 percent of the caloric intake was derived from carbohydrates, were only able to work at the given intensity for a period of about 1 hr. Throughout the entire work period, 70 to 99 percent of the energy was obtained from fat combustion. Although there was a significantly reduced capacity for prolonged heavy exercise, it is noteworthy that the subjects nevertheless were able to carry on this work for 1 hr while utilizing fat almost exclusively as a source of fuel (Fig. 14-1).

In subjects on a very *high-carbohydrate diet* where 90 percent of the food calories were derived from carbohydrates, the standard load could be performed for a much longer time, up to 4 hr. Initially, fat contributed only 25 to 30 percent to the metabolic fuel, compared with more than 70 percent when the diet was rich in fat. Gradually the contribution from fat combustion increased, and it was about 60 percent at the end of the exercise.

The subjects were able to work about three times as long on this very high carbohydrate diet than on the very high fat diet.

In these basic experiments of Christensen and Hansen (1939), the exercise was carried out until exhaustion of the subjects. At the point of exhaustion, both subjective symptoms and laboratory findings indicated reduced blood-sugar levels. It is of interest to note that at this point the evidence suggests the muscles were still utilizing carbohydrate in spite of the fact that the blood sugar was approaching critically low levels. It should be pointed out that the glycogen stores in the muscles cannot supply the blood with sugar directly, but only through the formation of lactic acid, which is then converted to sugar in the liver and released to the blood. Ingestion of 200 g glucose at the time of exhaustion enabled the subject to continue the work for another hour. Following the glucose administration, the blood-sugar level rose and the subjective symptoms disappeared within

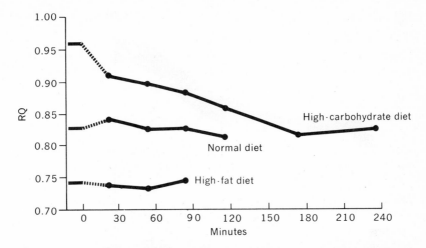

Fig. 14-1   Respiratory quotient (RQ) in a subject working at 1080 kpm/min, while on three different diets. The subject was exhausted in 90 min on the high-fat diet, but he could carry on for 240 min on the high-carbohydrate diet. While he was on normal diet, the experiment was discontinued at 120 min. (From Christensen and Hansen, 1939.)

15 min. The RQ, however, was not influenced but remained relatively low. It thus appears that the primary limiting factor in exhausting work of this kind may be the fall in blood sugar. This has been confirmed in more recent experiments by Rodahl et al. (1964).

The central nervous system with its low glycogen content depends to a very great extent on the blood sugar. It has been stated that in man approximately 60 percent of the hepatic sugar output serves the brain metabolism (Shreeve et al., 1956; Reichard et al., 1961). Christensen and Hansen's findings seem to suggest that exhaustion may in many cases be a central nervous system phenomenon and not the result of a lack of muscle glycogen per se.

With a sugar content of 1 g/liter the total glucose content in the blood is 5 to 6 g. Since some 3 g of carbohydrate may be utilized during every minute of heavy exercise in an individual with a high aerobic power, this blood glucose can cover only 2 min work. With a reduction to 0.5 g/liter of blood (a reduction of about 3 g), severe symptoms of hypoglycemia would develop. To prevent a marked decrease in the blood glucose concentration, some sort of a barrier must exist to stop the glucose from entering the muscle cells and being used up in their metabolism.

The permeability of the cell membrane for glucose depends on the plasma insulin concentration which falls during heavy exercise (Hunter and Sukkar, 1968). Secondly phosphorylating enzymes are necessary for the uptake of glucose across the membrane, and at least one such enzyme (hexokinase) is inhibited by products from the breakdown of

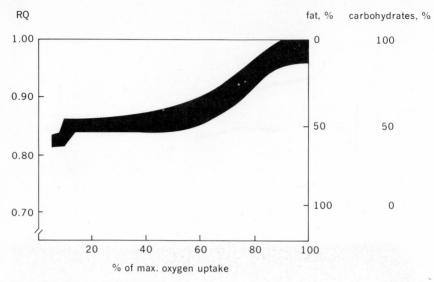

Fig. 14-2   Schematic illustration of nonprotein respiratory quotient at rest and during exercise, related to the oxygen uptake in percent of the subject's maximal oxygen uptake. To the right, the percentage contribution to the energy yield of fat and carbohydrate. Prolonged exercise and diet can markedly modify the metabolic response.

glycogen (Hultman, 1967). Therefore glycogen is a more readily available substrate for energy production in the working muscle cell than exogenous glucose. This fact is an advantage for the central nervous system which might otherwise suffer from lack of glucose. It should be recalled that the net energy yield is higher when glycogen is utilized than is the case with glucose (Chap. 2).

Summary   On the basis of the early experiments of Christensen and Hansen, it may be concluded that the following factors affect the participation of fat versus carbohydrates in the work metabolism: (1) the diet; (2) the duration of the work; (3) intensity of work in relation to the individual's total work capacity, i.e., whether or not the work can be accomplished under aerobic metabolic conditions (Fig. 14-2). Thus, the utilization of carbohydrates depends on the oxygen supply of the working muscles; the more inadequate the oxygen supply, the higher the carbohydrate utilization. If a certain aerobic power is attained with the arms only, more carbohydrate is utilized than when the same caloric output is accomplished with the larger leg muscles. It should be pointed out that there is an almost 10 percent higher energy yield per liter O_2 used when carbohydrate is combusted than when protein and fat are burned.

The findings of Christensen and Hansen (1939) have been confirmed by more recent studies applying the more sophisticated and advanced techniques available today. Such techniques have also facilitated more detailed studies concerning the storage, release, transportation, transformation, and stepwise oxidation of various substances.

Muscle Glycogen

A technique has been developed by which small pieces of muscle tissue (10 to 20 mg) can be sampled in man, and their content of glycogen and other substances analyzed. By obtaining samples during various stages of different dietary regimen, during exercise, etc., the variation in glycogen content of the muscles can be followed (Bergström et al., 1967; Hermansen et al., 1967; Hultman, 1967; Saltin and Hermansen, 1967).

On a normal mixed diet, the glycogen content in the quadriceps femoris muscle is about 1.4 g/100 g wet muscle with a range from about 1.0 to 2.0 g (Hultman, 1967). In the deltoid muscle the glycogen content is significantly lower, or about 1 g/100 g wet muscle. Figure 14-3 presents data from experiments in which biopsies were taken every twentieth min during exercise on a bicycle ergometer with an average oxygen uptake of 77 percent of the individuals' maximal aerobic power. Two groups of subjects participated in the study; 10 were trained, 10 were untrained and had a lower maximal oxygen uptake. Therefore, the "77 percent load" corresponds to a mean oxygen uptake of 3.4 liters/min for the trained and only 2.8 liters/min for the untrained subjects. The heart rates

Fig. 14-3   Average values for glycogen content in needle-biopsy specimens from the lateral portion of the quadriceps muscle taken before and at intervals during exercise until exhaustion, in a group of 10 trained and 10 untrained subjects. Oxygen uptake averaged 77 percent of the maximal aerobic power (3.4 and 2.8 liters/min, respectively). (From Hermansen et al., 1967.)

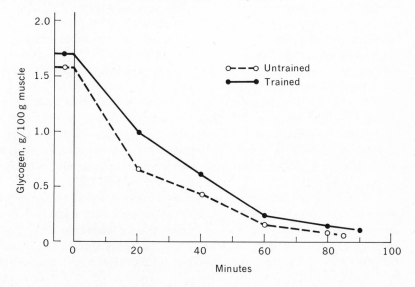

after about 10 min of exercise were 164 and 172 beats/min respectively. After about 90 min, the work had to be terminated due to the subjects being exhausted. The glycogen content was then on the order of 0.1 g/100 g wet muscle. The RQ was higher in the untrained group (about 0.95) than for the trained subjects (about 0.90). The fat metabolism evidently played a somewhat larger role in the trained men. The actual combustion of glycogen was about 2.8 g/min in both groups despite the difference in energy expenditure. At the end of exercise the blood sugar level was still 80 mg/100 ml.

With these high work loads the RQ only showed a small decrease during the work period in contrast to the experimental finding discussed above. A high glycogen combustion was apparently essential for a high work load, higher than in Christensen's and Hansen's studies. With the glycogen depots emptied, the work had to be stopped, or the subject had to reduce the rate of work. In such a case the RQ dropped, indicating an increased energy yield from free fatty acids.

The high carbohydrate metabolism which was maintained even at the end of exercise, despite the lower rate of glycogen utilization (Fig. 14-3), can be explained by an increased transfer of glucose from the liver to the working muscle. Hultman (1967) has noticed a glucose output from the liver of about 1 g/min during heavy exercise. It should also be emphasized that the biopsies were only taken from the lateral portion of the quadriceps femoris muscle, and the analyses do not reveal events in other portions of the working muscles.

Figure 14-4 shows a very good relation between the carbohydrate utilization, calculated from oxygen uptake and RQ on the one hand, and the decrease in the muscle glycogen content on the other. The results support the concept that the RQ can be used to estimate the proportional contribution from fat and carbo-hydrates in total energy yield even during heavy exercise. For a group of subjects it could be calculated that increasing the work load from 29 to 78 percent of the maximal aerobic power increased the combustion of fat from 2.5 to 3.6 kcal/min compared to a change from 3.4 to 12.3 kcal/min for carbohydrates. With high work loads, demanding about 75 or more of the max $\dot{V}_{O_2}$, it seems that the glyco-gen stores in the exercising muscles are an important determinant of maximal work time.

In other words, assuming a glycogen concentration of 1.5 to 2.0 percent at the beginning of exercise, the glycogen utilization at a work level of 25 to 30 per-cent of the individual's maximal aerobic power would be sufficient to allow him to continue for 8 to 10 hr before depletion of the glycogen depots. At a work load demanding 75 to 85 percent of the maximum the same amount of glycogen would be combusted within $1\frac{1}{2}$ hr (Saltin and Hermansen, 1967).

These experiments indicate that the initial glycogen content in the skeletal muscles is of very decisive importance for the individual's ability to sustain pro-longed heavy exercise. The question arises then to what extent the glycogen con-tent can be modified by the diet, as suggested by the studies by Christensen and Hansen (1939). Bergström et al. (1967) found that after a normal mixed diet giving

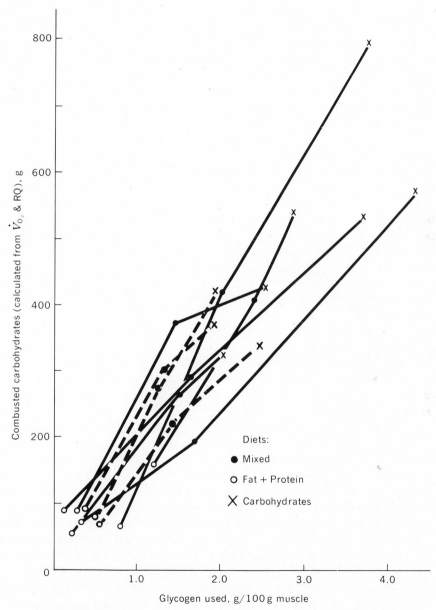

Fig. 14-4   Individual values for the total amount of glycogen utilized during exercise at different work levels, in relation to the amount of combusted carbohydrates calculated from data on RQ and oxygen uptake. Utilized glycogen could be calculated since the glycogen content of muscle specimens was analyzed before and after the exercise. (From Hermansen et al., 1967.)

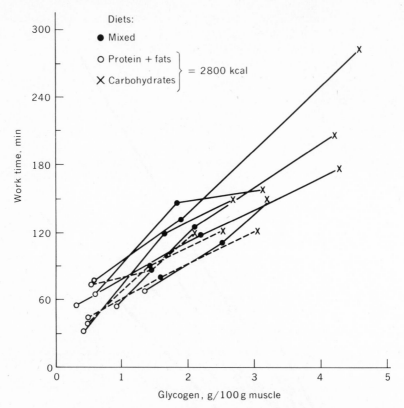

Fig. 14-5   Relation between initial glycogen content in the quadriceps muscle in nine subjects who had been on different diets, and maximal work time when working at a given load demanding 75 percent of maximal aerobic power. Dotted lines denote subjects who had been on a carbohydrate diet prior to the fat-plus-protein diet. (From Bergström et al., 1967.)

an initial glycogen content of 1.75 g/100 g wet muscle, the subject could tolerate a work load demanding an average of 75 percent of the maximal oxygen uptake for 115 min (see Fig. 14-5). After the subject spent 3 days on an extreme fat and protein diet, the glycogen concentration was reduced to about 0.6 g/100 g wet muscle and the standard load could only be performed for about 60 min. After 3 days on a carbohydrate-rich diet, the subject's glycogen content became higher, 3.5 g/100 g wet muscle, and the time on the "75 percent load" could now be prolonged to about 170 min (average figures).

It was further observed that the most pronounced effect was obtained if the glycogen depots were first emptied by heavy prolonged exercise and then maintained low by giving the subject a diet low in carbohydrate, followed by a few days with a diet rich in carbohydrates. With this procedure the glycogen content could exceed 4 g/100 g wet muscle and the heavy load could be tolerated for longer periods, in some subjects for more than 4 hr. The total muscle glycogen content under these conditions could exceed 700 g.

Figure 14-6 illustrates an experiment in which one subject worked with his left leg and the other subject simultaneously worked with his right leg on the same bicycle ergometer. After several hours' work the exercising leg was emptied of glycogen while the resting leg still had a normal glycogen content. It should be recalled that there are no enzymes present in the skeletal muscles that can transform glycogen to glucose (Chap. 2; Fig. 2-2). Feeding the subjects carbohydrate-rich diet the following days did not markedly influence the depots of the resting limb, but in the previously exercised leg the glycogen content increased

Fig. 14-6   Two subjects were exercising on the same bicycle ergometer, one on each side working with one leg, while the other leg was resting (dashed line). After working to exhaustion, the subjects' glycogen content was analyzed in specimens from the lateral portion of the quadriceps muscle. Thereafter, a carbohydrate-rich diet was followed for 3 days. Note that the glycogen content increased markedly in the leg that had been previously emptied of its glycogen content. (Modified from Bergström and Hultman, 1966.)

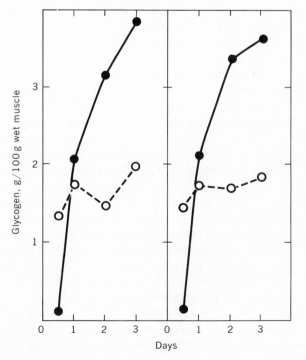

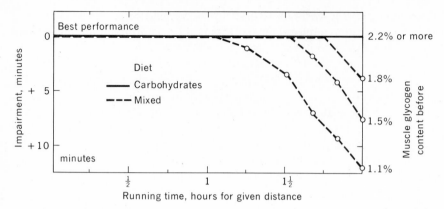

Fig. 14-7   Schematic illustration of the importance of a high glycogen content in the muscle before a 30-km race (running). The lower the initial glycogen store, the slower became the speed at the end of the race compared with the race performed when the muscle glycogen content was 2.2 g or more per 100 g muscle at the start of the race. For the first hour, however, no difference in speed was observed. (By courtesy of B. Saltin.)

rapidly until the values were about twice as high as those in the nonexercised leg. Again, exercise with glycogen depletion enhances the resynthesis of glycogen and the factor (presently unknown) must be operating locally in the exercising muscle (Lamb et al., 1968).

We may, therefore, conclude that different diets can markedly influence the glycogen stores in the muscles. The ability to perform heavy, prolonged exercise is correspondingly affected and the higher the initial glycogen content the better the performance. This is schematically illustrated in Fig. 14-7. A group of subjects participated twice in a 30-km race, cross-country running, on the first occasion after their normal mixed diet, and on the second occasion after a few days on an extremely rich carbohydrate diet after previously emptying the glycogen depots. At several points of the track the running time was recorded. It was calculated that the rate of glycogen utilization was 0.8 to 1.0 g/100 g muscle/ hr.

Two conclusions can be drawn from this study: (1) a high glycogen content in the muscles did not enable the subject to attain a higher speed at the beginning of the race compared to the case when the initial glycogen level was low; (2) when the glycogen concentration approached zero, the speed was reduced. The lower the initial store the sooner this impairment in running occurred.

It should be pointed out that there is one drawback with the high glycogen storage: it was mentioned that each gram of glycogen is stored together with about 2.7 g of water. With a glycogen storage of 700 g there is therefore an increase in body water amounting to about 2 kg. In activities in which the body weight has to be lifted an excessive glycogen store should therefore be avoided.

After prolonged heavy exercise a maximal effort does not produce the same high blood lactic acid concentration as is normally found in connection with short-term exercise (Fig. 14-8). This cannot be explained simply by emptied glycogen stores; there is, for some other reason, a gradual decrease in the capacity of the muscle cells to produce a tension high enough to initiate the anaerobic processes to maximal power (Saltin and Hermansen, 1967; Karlsson et al., 1968).

If glucose is infused continuously during exercise, bringing the blood sugar level well above 200 mg/100 ml blood, the fall in muscle glycogen is significantly smaller than under normal conditions (Hultman, 1967). The difference in this reduction is, however, not large. It seems as if glucose administered during exercise will to some extent replace the utilization of free fatty acids (Havel et al., 1963).

The beneficial effect of the sugar supply may be more pronounced when the glycogen depots are depleted. Various factors may then facilitate the transport of glucose across the cell membrane of the muscle cell.

The glycogen content of the liver is 50 to 100 g. This acts as the carbohydrate store to maintain the blood glucose level and indirectly to supply glucose to the nervous tissue which has no carbohydrate reserve of its own.

Summary The ability to perform heavy prolonged exercise is directly related to the initial glycogen stores of the working muscle. From about 1.5 g/100 g muscle after a mixed diet the glycogen content may, after about 2 hr of work with 75 percent of the maximal oxygen uptake, approach zero, and the subject becomes exhausted.

It is possible to increase the glycogen content of the muscles by a carbohydrate-rich diet (to about 2.5 g/100 g muscle). The effect will be still more

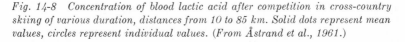

Fig. 14-8   Concentration of blood lactic acid after competition in cross-country skiing of various duration, distances from 10 to 85 km. Solid dots represent mean values, circles represent individual values. (From Åstrand et al., 1961.)

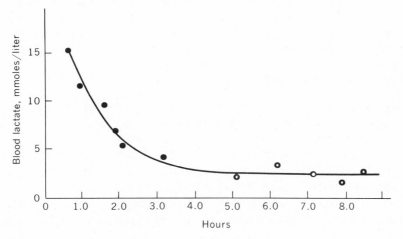

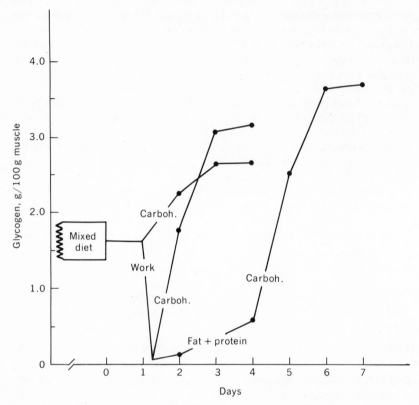

Fig. 14-9 Different possibilities of increasing the muscle glycogen content. For further explanation see text. (From Saltin and Hermansen, 1967.)

pronounced if prolonged exercise to exhaustion precedes this special diet (up to 4 g/100 g muscle). Even higher values (up to 5 g/100 g muscle) can be obtained if the low glycogen content of the muscle is maintained low for a few days by eating fat and protein with no carbohydrates included in the diet (see Fig. 14-9).

Free Fatty Acids (FFA)

At low work loads, even the starving man can work for long periods of time as long as he has fat depots. One practical example: Two men participated in a ski trip in the mountains, covering a distance of about 65 km in 3 days. The calculated total caloric output was about 18,000 kcal but only 1000 kcal were consumed, almost exclusively in the form of carbohydrate. Some 14,000 to 15,000 kcal were derived from body fat, and the RQ determined at rest and during a standardized step test did in fact become very low. With the few calories in the form of sugar taken at appropriate intervals and by avoiding peak loads, the 3 days of heavy work could be completed with hypoglycemic symptoms occurring only on a few occasions.

The role of free fatty acids as an important fuel for muscular work is directly revealed by modern techniques, including the use of isotopes (Andreas et al., 1956; Fritz et al., 1958; Issekutz and Spitzer, 1960; Fritz, 1961; Friedberg and Estes, 1962; Carlson, 1967). Carbohydrates start, as emphasized above, to play a major role as a source of fuel in heavy exercise, particularly when the oxygen supply to the muscles become insufficient. Issekutz et al. (1963a) found that it is the actual carbohydrate intake rather than the amount of fat in the diet which determines whether fatty acids or glycogen is the preferred fuel.

In the adipose tissue we have a substantial amount of energy stored, normally more than 50,000 kcal. At the beginning of exercise there is a drop in the plasma FFA due to an increased uptake in the active muscles. After some time there is a subsequent rise in the plasma FFA concentration, provided that the work is not extremely heavy. This rise is due to an enhanced mobilization of FFA from the adipose tissues, which is probably induced by an increased sympathetic activity, which in turn leads to a proportionately increased turnover of the FFA, as shown by Armstrong et al. (1961) (Fig. 14-10). Furthermore, local fat stores in the muscle cells may also play an important role as a readily available fuel source for the energy yield in the working muscle (Issekutz et al., 1963b; Carlson, 1967).

FFA can be mobilized from the adipose tissue. An FFA-albumin complex is the most important system for transport of endogenous fatty acids from the storage sites. At rest FFA is only about 2 to 3 percent of the fat content of the plasma, but about 25 percent of it disappears every minute, so that the "life time" is very short (Carlson, 1967). Thus up to several hundred grams of fatty acids can be transported during a 24-hr period from adipose tissue to various organs in the form of the albumin-bound FFA. A mobilization of FFA is caused directly or indirectly by epinephrine, norepinephrine, growth hormone, and hypoglycemia. The following factors may inhibit the FFA mobilization: insulin, glucose, lactic acid, and a reduction of blood pH (Issekutz, 1964; Carlson, 1967).

The relatively moderate role played by FFA as a fuel with increasing work load can at least partly be due to an inhibiting effect by produced lactic acid (Issekutz and Miller, 1962). In experiments on dogs Issekutz et al. (1965) have shown that the fitter the dog, the greater is his ability to utilize fatty acids in his metabolism. At a given work load, the fit dog produces less lactic acid than one which is less fit.

By intravenous administration of nicotinic acid the release of free fatty acids from the adipose tissue can be blocked. Bergström et al. (1969) have noted that such an administration did not interfere with the subject's ability to perform brief work or prolonged submaximal work. The reduced supply of FFA was compensated for by an increased metabolism of glycogen. After intravenous infusion of nicotinic acid, the glycogen content of the working muscles, as revealed in needle-biopsy specimens, was significantly smaller than that found in control studies. Also, the RQ was higher after administration of nicotinic acid.

We may now try to tie together some other relations which have been mentioned earlier. During prolonged exercise, the elevation of blood lactate content at the earlier stage may be followed by a gradual decline—hypoglycemia may develop; the sympathetic

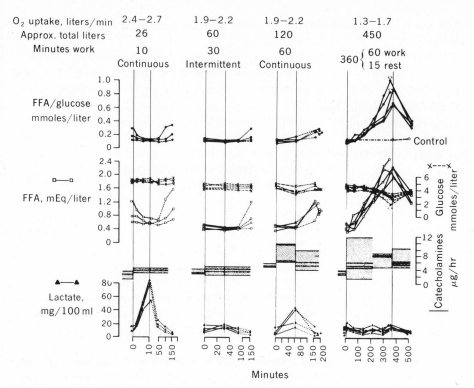

Fig. 14-10 Effects of work of different intensity and duration (heavy exercise on the bicycle ergometer for 10 min; intermittent run: 5 sec work, 5 sec rest, on the treadmill, slope 8.6 percent, speed 7.5 mph for 30 min; continuous work on the bicycle ergometer at 900 kpm/min for 60 min; and alternating work on the bicycle ergometer at 600 kpm/min for 60 min and work on the treadmill with a slope 8.6 percent at a speed 3.5 mph for 60 min for a total period of 360 min, or until exhaustion) in normal fasting subjects in relation to blood lactate levels, urinary cate- cholamine excretion, plasma FFA, and blood glucose. (From Rodahl et al., 1964.) Note the rise in plasma FFA and the drop in blood glucose during prolonged work, and the marked rise in blood lactate level during heavy short exercise, while it is essentially unchanged during the prolonged work. Note also the rise in plasma FFA following the cessation of short, heavy work.

activity may increase. All these factors will promote a mobilization of FFA from the adipose tissue and bring more fuel to the working muscle. For the transport of fatty acids from the sarcoplasma into the mitochondria or sarcosomes, a cofactor, carnitine, is essential. It also "assists" in the oxidation of the long-chain fatty acids in the skeletal muscles (Weis-Fogh, 1967).

Summary

Biologically there are "two extreme metabolic types, those which utilize only carbohydrates, and those who depend almost exclusively if not entirely on fat" (Weis-Fogh, 1967). Man and other mammals depend on both fuels for their metabolism. During *light* or *moderate* muscular work, the energy is supplied by fat and by carbohydrate, mainly glycogen, in about equal amounts. As the work progresses, fat contributes in an increasing amount to the energy fuel; the control is probably mediated through an increased production of norepinephrine, mobilizing FFA from the adipose tissue. During *heavy* muscular work the major source of energy fuel is glycogen.

Various enzymes and other factors can modify the permeability of cell membranes (and sarcosomal membranes) as well as the pathways for the oxidation of the substrates. Training and diet may affect both sets of factors. A low-carbohydrate diet tends to favor the utilization of fat as a source of fuel in the work metabolism, but this will reduce the capacity for prolonged heavy work; a carbohydrate-rich diet enhances the storage of glycogen, at least temporarily, which improves the endurance.

OPTIMAL SUPPLY OF NUTRIENTS

With the exception of children during growth, a period of pregnancy, and sometimes convalescence, the caloric intake should not exceed the caloric expenditure. The caloric requirements vary naturally with the individual's physical activity. On the other hand the need for most of the nutrients is comparatively independent of the individual's activity level; therefore the less active the individual, the higher is the content of the essential nutrients required per calorie in order to obtain the desired optimal nutritional level (Wretlind, 1967). In a homogeneous population a dietary tradition is usually developed which is practically the same for everyone. Blix (1965) found that a linear relation exists between caloric supply and supply of many nutrients (protein, calcium, vitamin A, thiamin, iron). This is illustrated in Fig. 14-11. Through the centuries man obtained his choice of food geared to a caloric output of 3000 kcal or more, which also gave all the nutrients he needed. For many of the nutrients mentioned the caloric intake should actually exceed about 2500 to ensure an adequate supply. In other words the diet in Sweden, and no doubt many other countries, seems to be adjusted to persons with a caloric requirement of at least 2500 to 3000 kcal (Wretlind, 1967). This is, however, not suitable for the large number of "low caloric consumers" which actually exist, as exemplified in Fig. 14-11. This may explain the rather common disturbances in the state of health and well-being associated with malnutrition even in countries with plenty of food available (lowered resistance to infections, dental caries, iron-deficiency anemia, atherosclerosis, osteoporosis, obesity, constipation). It has been noted that in Sweden, as well as in many other countries, the percentage of calories from protein has remained remarkably constant from the end of the nineteenth century to the present day, or between 11 and 12 percent; the proportion of fat in the caloric supply has, however, greatly increased from about 20 to about 40 calories percent. *Per capita,* the amount of calories of

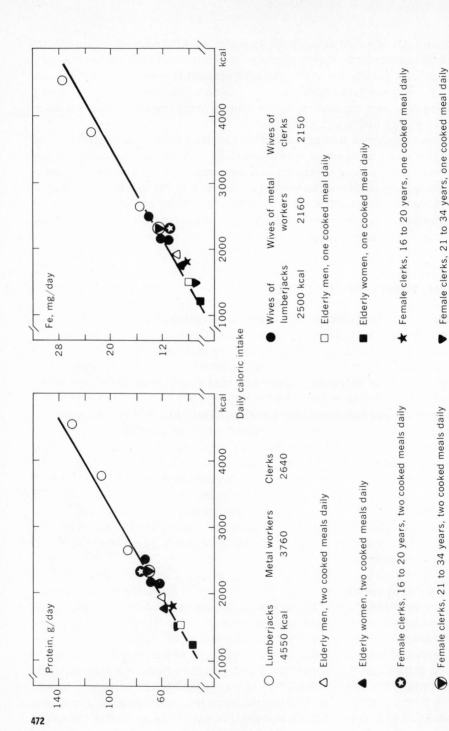

Daily caloric intake

| Lumberjacks 4550 kcal | Metal workers 3760 | Clerks 2640 |
| Wives of lumberjacks 2500 kcal | Wives of metal workers 2160 | Wives of clerks 2150 |

○ Lumberjacks 4550 kcal

△ Elderly men, two cooked meals daily

◢ Elderly women, two cooked meals daily

★ Female clerks, 16 to 20 years, two cooked meals daily

✪ Female clerks, 21 to 34 years, two cooked meals daily

● Wives of lumberjacks 2500 kcal

□ Elderly men, one cooked meal daily

■ Elderly women, one cooked meal daily

★ Female clerks, 16 to 20 years, one cooked meal daily

▼ Female clerks, 21 to 34 years, one cooked meal daily

Fig. 14-11 Relation between total caloric intake per day and the supply of protein and iron respectively. With the traditional food composition, some 2500 kcal should be consumed in order to ensure a sufficient supply of these and other nutrients. (From Blix, 1965.)

protein and carbohydrates has decreased, but the amount of fat has increased during this period of time (Wretlind, 1967).

There are two general ways of improving the nutritional conditions of the low caloric consumers: (1) Change their food habits so that their diet consists of a higher content of essential nutrients per calorie than is now the case. A dietary habit should be developed so that the requirements of those who consume only 1500 to 2000 kcal/day can be satisfied. Wretlind (1967) has pointed out as an example that an increase in the iron content of the diet of 20 percent (from 10 to 12 mg/day) would reduce the risk of anemia from 25 to 3 percent. (2) Low caloric consumers should be stimulated to become high caloric consumers by taking part in regular physical activity in one form or another. By an increase in caloric output they can, without the risk of obesity, eat more and automatically get a greater supply of essential nutrients.

These two measures are of equal importance. There are sections of the population which, for various reasons, are inevitably "forced" to physical inactivity and hence a low caloric consumption.

It should be emphasized that the "easy" and most convenient way to spend calories is in activities involving large muscle groups. Furthermore, the distance covered in jogging and running is the decisive factor, not the speed of movement. Walking briskly or jogging 10 km demands about as many calories as running the same distance (Chap. 16).

Two km/day (1¼ mile) requires about 150 kcal for a person weighing 70 kg. This distance covered every day corresponds to an expenditure of 55,000 kcal in a year; this amount of energy can be stored in about 9 kg of adipose tissue!

Summary An improved diet in a modern society should include the following: (1) The fat content should be no more than between 25 and 35 percent of the calories, partly in the form of polysaturated fatty acids. (2) The amount of refined sugar in the diet should be low. (3) The consumption of vegetables, fruit, berries, skimmed milk, fish, lean meat, and cereal product should be relatively high (Wretlind, 1967). The content of protein, iron, calcium, and certain vitamins in the diet will then be high enough to satisfy the requirements of low caloric consumers.

Low consumers should be stimulated to become more physically active. This will allow them to eat more, which will automatically furnish them with more of the essential nutrients. The need for such nutrients does not vary significantly with the level of physical activity.

REGULATION OF FOOD INTAKE

The exact mechanism by which man regulates the amounts and kinds of food he needs is not completely understood. As in many control systems one has to extrapolate from animal studies (Hamilton, 1965; Mayer and Thomas, 1967). There is " . . . a powerful and complex central regulatory system, one capable of closely matching energy intake to energy expenditure in the face of marked variations in the energy requirements of the organism and the nutritional value of the diet. Although transient factors may momentarily override its control this central system normally prevails maintaining the energy balance of the organism with remarkable accuracy" (Mayer and Thomas, 1967). In the hypothalamus there is a *feeding center* responsible for the urge to eat or the initiation of feeding, and a *satiety center* capable of exerting inhibitory control over this feeding center.

Mayer and Thomas point out that there is some sort of *short-term feedback* (1) concerning the nutritive value of ingested foods relayed from gastric sensors to the hypothalamic regulatory system by way of neural or humoral pathways or both. (2) Food intake is also intimately related to the regulation of blood glucose. The information concerning the availability of glucose to the cells mediated via glucose receptors is more important than the actual blood-glucose concentration. Hypoglycemia is, however, almost always followed by hunger sensations. (3) The control of heat exchange is in a complex manner connected with the center regulating food intake. An elevation of the body temperature inhibits the sensation of hunger (Andersson, 1967). This explains the poor appetite experienced during periods of intense physical activity, and the poor appetite often noticed in patients with fever. It is also known that homeotherms increase their food intake in a cold environment. All this information is integrated with myriads of other exteroceptive and interoceptive inputs at a particular moment, determining the balance of appetite and satiety and the initiation, continuance, or termination of the feeding response.

Cumulative errors in this system could, in turn, be corrected through a *long-term feedback*, or a lipostatic regulation, i.e., an inhibition of food intake whenever sufficient energy is derived from the mobilization of surplus body fat. How the adjustment of feeding is modified by the state of the fat stores is not clear. It is, however, noticed that after a period of weight gain there is often a reduction in food intake and the body weight becomes stabilized at a new level.

It is evident that a number of factors are involved in short-term reactions controlling the initiation and cessation of single meals and the actual rate of eating. The balance between food intake and body weight over longer periods of time somehow modifies the food intake as a long-term feedback.

Mayer and Thomas (1967) point out that there are a great variety of sensations recognized as signaling hunger, and that the frequency with which they are experienced varies appreciably between age groups, between sexes, and between individuals.

PHYSICAL ACTIVITY, FOOD INTAKE, AND BODY WEIGHT

The level of physical activity largely affects the food intake. Any energy expenditure will inevitably be reflected in the energy storehouse of the body. It has been observed that animals whose activities are restricted by confinement in small cages consume more calories than they require and therefore accumulate fat (Gasnier and Mayer, 1939; Ingle and Nezamis, 1947). This is the basis for the practice of fattening cattle, pigs, and geese by restricting their activity.

The relationship between food intake, body weight, and physical activity has been systematically studied in laboratory animals. Mayer et al. (1954) exercised mature rats, accustomed to a "caged" sedentary existence, on a treadmill for increasing daily periods. They observed that for moderate exercise of 20- to 60-min duration there was no corresponding increase in food intake. As a matter of fact, there was a significant decrease in food intake, and consequently body weight also decreased. When the exercise was extended from 1 to 6 hr, food intake increased linearly with energy expenditure and body weight was maintained. When the daily exercise was extended beyond 6 hr, the animals lost weight, their food intake decreased, and their appearance deteriorated. It has also been shown that rats which are overfed voluntarily reduce their activity. This, in turn, increases the weight gain, and a vicious circle is created. On the other hand, when fat rats are fed a calorically restricted diet, their spontaneous activity increases as the weight loss progresses. Thus, both slight and exhausting activities do not appear to be directly related to corresponding changes in food intake.

The regulation of food intake is, in a way, adapted to an energy expenditure exceeding the basal metabolic rate and even the demands in an habitually sedentary individual. In the past the very act of seeking food required much more energy output than it does now in many species for whom food is readily available. Figure 14-12 illustrates schematically how at low energy expenditure the appetite may be adapted for a higher energy output than is actually attained in a sedentary man. At 2000 kcal/day and higher caloric output, there may be a balance between spontaneous caloric intake and caloric output; at a very high caloric output (e.g., during hiking) the individual is satiated before he has covered his daily expenditure and he uses part of the energy stored in his adipose tissue. Mayer has concluded from his experiments that there are two ways to maintain a normal body weight: by being physically inactive but often hungry, or by being physically fairly active and eating as much as one desires.

Summary   Within a wide range of caloric expenditure through various degrees of physical activity there is an accurate balance between caloric output and intake so that the body weight is maintained constant. If the daily activity is very intensive and prolonged, the spontaneous caloric intake is often less than the output, with a reduction in body weight as a result. A daily energy expenditure below a threshold level often leads to an energy surplus and consequent obesity. In this case satiety is not reached until more calories have been consumed than have been expended.

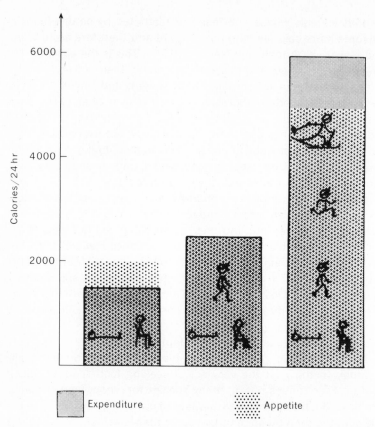

Fig. 14-12   The appetite center is set at a level of caloric intake which is well above the amount of calories consumed at rest. With increasing physical activity a balance is attained between appetite and caloric expenditure. At extremely high caloric expenditure the feeling of hunger no longer ensures an adequate caloric intake.

OBESITY

If caloric intake exceeds caloric output the excess energy will be stored mainly as adipose tissue. If this state of affairs is maintained over a period of time, it will lead to obesity. Mayer and Thomas (1967) point out that "while below-average muscular activity on the part of the obese is a matter of popular and immemorial record, there is a physiologic basis for reassessing the possibility that inactivity plays a primary role in the etiology of their disorder."

Greene (1939) studied more than 200 overweight adult patients in whom the onset of obesity could be traced to a sudden decrease in activity. Mayer et al. (1956) made a study on an industrial population in India engaged in a very wide range of physical activity, from tailors and clerks to coolies carrying heavy loads for several hours a day. The diet was quite uniform for each individual and showed little variety within groups and from group to group. It was observed that the sedentary individuals had the highest caloric intakes and the highest body weight. The light workers had lower caloric intakes and lower body weight. The groups engaged in medium, heavy, and very heavy work had increasing caloric intakes, but their body weights were the same. It thus appears that sedentary individuals, like the animals, are apt to eat more than they need and become obese. The remarkable thing is that similar to the rats on light treadmill work, the group engaged in light work did not eat more than the sedentary workers; they did, in fact, eat less.

In a study concerning the relationship of body weight and physical activity in children, Bruch (1940) found that inactivity was characteristic of the majority of 160 obese children examined; 76 percent of the boys and 88 percent of the girls were physically inactive. It has been suggested by Rony (1940) that laziness or decreased tendency to muscular activity is a primary characteristic of obese subjects, and Bronstein et al. (1942) found that the majority of the obese children they studied spent most of their leisure time in sedentary activities. Similar observations have been made by Simonson (1951), Juel-Nielsen (1953), and others. Other studies (Peckos, 1953; Fry, 1953) indicate that obese children do not have higher average caloric intakes than do control children of the same height and age.

In a study by Johnson et al. (1956), caloric intake and activity were systematically compared in paired groups of obese and normal-weight school girls. Their findings were that suburban high school girls were generally not very active, but nevertheless there was a marked difference between the obese and nonobese groups, the obese group being much less active than the nonobese. Generally speaking, the time spent by the obese groups in sports or any other sort of exercise was less than half of that spent by the lean girls. Caloric intakes were generally larger in the nonobese girls than in the obese, and it was concluded that inactivity was of greater importance than overeating in the development of obesity. It is interesting to note that when these high school girls attended summer camp, they all, both obese and nonobese, almost without exception lost weight under a program of enforced strenuous activity in spite of simultaneous increased food intake.

Stefanik et al. (1959) in a summer camp study found that obese boys had significantly smaller caloric intakes both during the school year and at the summer camp than the nonobese controls.

By employing a motion picture technique, Bullen et al. (1964) compared the activity of obese and nonobese adolescent girls while they were engaged in various sports at a summer camp. With this method it became possible to demonstrate that the average obese girl expends far less energy during scheduled exercise periods than her nonobese counterpart. The obese girls were spending much more time just sitting and standing than the nonobese girls when going for a swim or when playing tennis. On the average only 9 percent of the obese group, as compared to 55 percent of the nonobese group, were actually swimming at one time when attending the swimming pool. The obese youngsters were aware that they were inactive, though they had no conception of the degree of their inactivity. They seemed totally unaware that they might not like physical activity. Bullen et al. concludes that (1) inactivity is a significant factor in perpetuating obesity, and (2) more attention should be devoted to the prevention of inactivity in obese adolescent individuals. "It would appear essential to attempt to teach these girls to enjoy more intensive exertion in sport or dance activities which are of social significance to them."

Chirico and Stunkard (1960) made a rough estimate of the degree of physical activity of obese and normal subjects of similar occupation and social status. They were asked to wear a pedometer for recording the number of steps throughout the day. The nonobese subjects were about twice as active as the obese ones. Durnin (1967) has also noticed that overweight individuals are less physically active than those who are nonobese.

Statistics show that people who are greatly overweight are more susceptible to certain diseases, such as diabetes and cardiovascular diseases (Bogert, 1957; Mayer, 1967*a* and *b*). Larsson (1967) has observed that the obese state in mice is not only accompanied by a decreased ability to mobilize fat, but also by a decreased ability to utilize FFA in the skeletal muscles. If the obese animals are forced to exert muscular activity, they slowly regain the fat-mobilizing capacity, and the utilization of FFA in skeletal muscles increases. Similar observations on man, however, are not reported.

Mayer and Thomas (1967) remark that comparison of the hunger and satiety pictures in obese and nonobese individuals suggests that abnormalities of satiety may be more prevalent in the obese than abnormalities of hunger. Nisbett (1968) made observations which fit this assumption. He told his subjects, normal as well as overweight ones, to eat as many sandwiches as they wanted, but a limited number were served at the table, the rest being left in a refrigerator in the same room. He found that the overweight individuals confronted with three sandwiches ate 57 percent more than those confronted with only one sandwich. In contrast, normal and underweight subjects were completely unaffected by the difference between experimental conditions; both groups ate as many sandwiches whether they were initially offered one or three. The obese individual will habitually eat everything he is served in a typical meal, but he may not necessarily refill from the refrigerator. Schachter et al. (quoted from Nisbett, 1968) found that obese subjects ate no more food after being deprived for several hours than they did after being recently fed, while normal subjects ate much more food after they had been deprived.

Summary It appears, conclusively, that obesity is to a large extent the result of reduced activity with a maintenance of an "old-fashioned" appetite center set for an energy expenditure well above the one typical for a sedentary individual. This is true for children as well as for adults. The reason for their different attitudes toward physical exertion is not clear. When studying very obese individuals on the bicycle ergometer, the authors noted that these obese subjects complained of fatigue and felt exhausted at low work loads, when the blood lactic acid levels were not noticeably elevated and would be easily tolerated by normal individuals. Obese individuals are characterized by their response to external rather than to internal cues in their eating.

It is particularly important to stimulate young individuals to regular physical activity, for such activity will in the long run effectively counteract obesity by keeping the individual within the range where spontaneous caloric intake is properly regulated by the caloric output. The amount of work (e.g., distance moved) is more important for the caloric expenditure than the intensity of work. Walking 3 km/day (about 2 miles) during a 10-year period demands an amount of energy contained in about 80 kg (180 lb) of adipose tissue.

"IDEAL" BODY WEIGHT

The body size and shape are largely determined by the skeletal size, for there is usually a certain amount of muscle and other tissues that go along with a certain amount of bone. Therefore, the ideal body weight, the body weight which includes only a minimal amount of body fat, is largely dependent upon the skeletal size. This "ideal" body weight may be modified to some extent by enlargement of the muscles by training, especially such training as weight lifting; a person may therefore be overweight without being obese. However, for all practical purposes the excess weight, i.e., weight over and above the ideal body weight, represents accumulated body fat.

There are graphs and tables available usually based on height and sex, giving the so-called ideal body weight. One of the better known of these norms is the Metropolitan Life Insurance table (1963). Such norms are, however, rather inaccurate and only meant as a general guide. Thus the need for a relatively simple but meaningful method of assessing body composition, including the amount of adipose tissue, is apparent. There are several methods for a quantitative classification of body build based on a series of anthropometric measurements, including various diameters (e.g., the radial-ulnar diameter), skin-fold thickness, body densitometry, total body potassium determination (von Döbeln, 1959; Behnke, 1961; Mayer and Thomas, 1967). Body composition, particularly in athletes, is a better guide for determining the desirable weight than the standard height-weight-age tables because of the high proportion of muscular content in their total body composition. This was clearly shown by Behnke et al. (1942) and Welham and Behnke (1942). They measured the body density of 25 professional athletes, 17 of whom were overweight according to the standard height-weight tables, and found that 11 of the 17 had excessively high lean body mass as measured by specific gravity determinations. One athlete, for example, 182 cm high (72.6 inches) and weighing 93 kg (205.7 lb) had a specific gravity of 1.086, indicating considerable leanness. This man would, according to accepted World War II

Standards for Military Service, have been classified as overweight and unfit for military service since his ratio of weight over height was greater than 2.65, the upper allowable limit.

There are observations on animals as well as on man showing that an increase in physical activity may have a profound effect on the body composition, increasing the protein/fat ratio but the body weight may remain the same (Larsson, 1967; see Chap. 13).

As a general guide the body weight at the age of about twenty is usually not far from the ideal body weight in most adults.

PRACTICAL APPLICATIONS

Slimming Diets

There are numerous slimming diets available, but most of them are very unphysiological. It should be emphasized that the most lenient way to reduce weight involves allowing plenty of time for the measures to take effect. The diet should be critically examined and an attempt made to eliminate some hundred kcal/day, for instance by substituting artificial sweetenings for sugar and skim milk for whole milk. Furthermore, 2 km ($1\frac{1}{4}$ miles) of walking per day would add 100 kcal to the expenditure, the result being a total net reduction in fatty tissue equivalent to 200 kcal/day. Therefore after a month, everything else being equal, the body will be storing 6000 kcal less than before, equivalent to 1 kg of adipose tissue. After a year the body weight would be reduced by 12 kg.

It is a common experience that when a slimming diet is instituted, there is often a sudden drop in body weight, followed by a more gradual decline. This initial drop, which may indeed have a gratifying and encouraging effect, may largely be due to loss of body water which is regained later on. During the first day or two of caloric restriction, stored glycogen is mobilized to cover the energy requirement. Since each gram of glycogen is stored together with almost 3 g of water, the mobilized glycogen liberates this water, which is eliminated and may thus account for the initial drop in body weight, a fact which is often skillfully exploited in the advertising of different slimming preparations. When returning to an adequate diet, glycogen is again stored, together with the required amount of water. This, then, would account for the rapid transient weight gain when changing from caloric restriction to adequate diet.

It is a mistake to avoid carbohydrates completely, as muscle and nerve cells need carbohydrates in their metabolism. This is particularly important for anyone who is physically active.

Athletes' Diet

It should be emphasized that the term "athlete" may also refer to a sportsman with a relatively modest daily energy expenditure, especially if his sporting activity involves highly technical events of short duration. In sharp contrast to this are the participants in endurance events. Hedman (1957) has calculated that "Vasaloppet," a cross-country skiing competition covering 85 km (53 miles), demanded the expenditure of 6000 to 7000 kcal. Repeated measurements of oxygen uptake during a 300-km (about 190 miles) bicycle race revealed a total caloric output of up to 10,500 kcal during the 14 hr it lasted. This brings the 24-hr metabolism to about 12,000 kcal. A simulation of the same race in the laboratory with some of the participants working on bicycle ergometers confirmed these figures (K.-E. Olsson, personal communication).

In heavy physical exertions or athletic events lasting *less than* 1 *hour*, the available supply of stored energy fuel is generally ample to cover the need. Under such conditions the diet is of less importance. Because the digestion of a meal causes a redistribution of blood from the muscles to the guts, physical exercise shortly after a meal will result in a competition between the guts and the working muscles for the available blood supply. Heavy muscular work should therefore be avoided immediately following a heavy meal. As a general rule, a meal should not be ingested later than $2\frac{1}{2}$ hr prior to an athletic event. Furthermore, the last meal prior to the event should be light. It should consist only of ingredients which the individual knows from experience he can tolerate well. Excessive quantities of carbohydrates should not be ingested prior to the event, for Christensen and Hansen (1939) showed that the ingestion of large quantities of sugar a few hours prior to strenuous muscular work may drastically impair the maximal work capacity.

In the case of athletic events involving large muscle groups at high work loads for periods *exceeding an hour or so*, the available evidence suggests that it would be advisable for the competitor to ingest ample quantities of carbohydrates the days preceding the event in order to fill the glycogen depots (Fig. 14-9). He should avoid heavy physical work which might deplete the existing glycogen depots prior to the event. On the other hand, it is not advisable to live on a high carbohydrate diet regularly, since this would condition the metabolic processes to a high utilization of carbohydrate fuel, rather than FFA. If the period of exertion during the event is very long, it might be advantageous to consume moderate amounts of sugar, preferably in the form of a lemon-flavored glucose solution, before warm-up. During a 50-km race, the cross-country skier may ingest as much as 1 liter of such a solution (corresponding to about 400 g sugar), divided into 7 to 8 helpings, ingested at the various control posts some 5 to 6 km apart. The ideal arrangement would be to issue the sugar solution some minutes before or after an uphill section of the track, preferably just before a slack downhill slope which would enable the skier to drink the solution while gliding downhill. The sweet sugar solution should be rinsed down with water; this would also serve to replace some of the fluid loss. In view of the fact that the sugar solution is rather concentrated, it is important that the competitor becomes accustomed to it during his training in order to avoid any unpleasant surprises during the actual event. This regimen has been well tried and has survived the test of time. It may therefore be recommended for other competitive events such as long-distance bicycle racing, walking, and marathon running. In the case of marathon running (42 km), the existing regulations concerning the service facilities violate established physiological principles, in that no nourishments are permitted during the first 10 km, and after that only at the official control points every fifth kilometer. The ingestion of sugar during games such as soccer and football might be physiologically justified.

It should be emphasized that the rate of absorption of glucose, water, and various minerals in the gastrointestinal tract is not affected by exercise, at least not up to loads demanding about 70 percent of the maximal oxygen uptake. The data of Fordtran and Saltin (1967) suggest that gastric emptying and intestinal absorption of saline solutions would be rapid enough to replace all of the losses of sweat incurred during heavy exercise, even in hot environments. An addition of glucose to water causes, however, a rather marked inhibition of gastric emptying, and high concentrations of sugar in orally ingested water may cause large amounts of fluid to be maintained in the stomach. This may produce abdominal discomfort during exercise. It is a critical and yet unsolved question how strong a sugar solution should be ingested. In our experience with cross-country skiers we have noticed that some of them prefer a weak (about 10 percent) glucose solution, while others tolerate and actually prefer a 40% glucose solution.

In Fordtran and Saltin's experiments (1967) at least 50-g glucose was emptied from the stomach during 1 hr of heavy exercise. This amount corresponded to one-fourth to one-half of the carbohydrates required by the body during this period.

With the exception of the days preceding the event and on the day of the actual event, the athlete should consume a regular well-balanced diet. His greatly increased caloric expenditure automatically increases his appetite, with the result that he ingests more food. If his diet is well balanced to start with, this will also automatically cover his increased requirement for protein, vitamins, and minerals. It has been shown that during a 10-day period physical work capacity is not affected by a reduction of the protein intake to 4 g/day (Rodahl et al., 1962), nor is the performance improved by an increased ingestion of protein, up to 160 g/day (Darling et al., 1944). If an extra supply of protein seems advisable in connection with muscle training or "muscle building," during convalescence or during growth, there is no reason to resort to costly preparations, since all the required protein and all the essential amino acids can be obtained from meat, fish, and milk. However, the old rule of thumb that 1 g of protein/kg body weight/day covers the demand is also valid for the hard-training athlete.

With regard to the distribution and the frequency of meals throughout the day, there appears to be no valid physiological basis for any particular regimen, with the possible exception of breakfast. Certain studies have indicated that breakfast improves performance (Tuttle et al., 1949, 1950, 1951; Daum et al., 1950; Coleman et al., 1953). It appears that rather frequent, moderate meals are more effective in yielding maximal performance than fewer larger meals, at least for psychological reasons. Hutchinson (1952) and Mayer and Bullen (1960) recommend that athletes should have at least three meals a day.

It is well known that vitamin deficiency causes an impairment of the performance capacity, but it takes a very long time to develop such vitamin deficiency in an individual living on a deficient diet (Rodahl, 1960). For this reason, an individual may be without any vitamin intake for a week or so without any detectable detrimental effect on the work capacity. As long as the vitamin intake is adequate, additional vitamin intake does not improve the performance (Keys, 1943; Simonson, 1951; Mayer and Bullen, 1960). Particular attention has been focused on the possible beneficial effect of the B-vitamins on performance capacity, but all studies almost without exception have been negative. In the case of the water-soluble vitamins, the ingestion of large quantities of vitamin pills is a rather expensive way of increasing the vitamin content of the urine, which serves no useful purpose in the first place.

The iron intake is at present subject to considerable interest because low serum-iron values have been observed in many top athletes. It has been pointed out that inasmuch as the oxygen is transported by hemoglobin, which contains iron, an iron deficiency may cause a reduced capacity for oxygen transport. It is hard to conceive how reduced serum-iron levels may affect the oxygen-transporting system as long as the hemoglobin content of the blood is normal, as is usually the case in the athletes in question. However, it is conceivable that the observed reduction in the serum iron may be interpreted as a forerunner for a reduction in the hemoglobin content of the blood. On the basis of the available evidence, it appears that there is no obvious justification for excessive iron intake in the athlete.

In certain sports, such as boxing and wrestling, athletes at times subject themselves to stringent dietary regimens combined with dehydration in an endeavor to lose weight in order to make lower weight classifications, where they do not properly belong. This allows them the advantage of competing with contestants who are actually lighter than they themselves. As Mayer and Bullen (1960) point out, such weight classifications have been established to provide competition on an equitable basis. The violation of these standards by sudden self-inflicted starvation and dehydration not only violates sportsmanship ethics, but may also perhaps harm the health of the individual concerned, and such practices have been condemned by the American Medical Association (1959).

During competitive cross-country orientation, a mean fluid loss of 3.0 liters during the 90-min duration of the competition has been observed (Saltin, 1964). Even greater fluid loss of up to 5 or 6 liters has been observed during ski racing and in bicycle racing in a hot environment. Sweat loss of a similar magnitude has been reported for many industrial operations.

In the army it used to be advocated that during strenuous field maneuvers, the soldier should drink as little as possible in order to reduce the amount of sweating. This advice is in frank violation of the physiological fact that sweating is a vital process which during heavy muscular work or in the heat cools the body and prevents overheating. It is now generally recognized that the lost water has to be replaced, preferably at the same rate at which it is lost (Adolph, 1947; see Chap. 15).

As part of the dietary advice for the athlete, emphasis should be placed on the fact that weight loss through dehydration should be avoided. If the training is intense, and especially if it takes place in a hot climate, ample fluid should be taken, particularly on the day preceding the event. It is also advisable to drink water a few hours prior to the event. Too much water is far better than too little; the kidneys eliminate any excess anyhow. In the case of prolonged efforts in a hot climate, adequate fluid intake is essential. It appears that the death of an athlete during the bicycle race in Rome in 1960 may at least in part have been due to extreme dehydration. A dehydration corresponding to a loss of body water in excess of 1 to 2 percent of the body weight should in any case be avoided (Adolph, 1947; Ladell, 1955; Saltin, 1964).

The sweat is hypotonic compared to the body fluid (Robinson and Robinson, 1954), so that relatively more fluid than salts is lost from the body during sweating. Sweating, therefore, causes an increase in the NaCl concentration in the body. Under these conditions, ingestion of extra NaCl is contraindicated (Ladell, 1955). Only in the case of prolonged activity associated with intense sweating lasting more than a week or so is the ingestion of additional salt indicated.

It should be kept in mind that whatever the physiological principles for an optimal diet, the practical considerations dictate that the diet has to be acceptable to the individual. If an athlete believes in a food fad or in a miracle pill, the fad or the pill may cause him to win, providing of course that his diet otherwise is fully adequate.

REFERENCES

Adolph, E. F.: "Physiology of Man in the Desert," Interscience Publishers, New York, 1947.

American Medical Association (editorial): Crash Diets for Athletes Termed Dangerous, Unfair, *AMA News*, **2:**2, 1959.

Andersson, B.: The Thirst Mechanism as a Link in the Regulation of the "Milieu Intérieur," in "Les Concepts de Claude Bernard sur le Milieu Intérieur," p. 13, Masson et Cie, Paris, 1967.

Andreas, R., G. Cader, and K. L. Ziegler: The Quantitatively Minor Role of Carbohydrate in Oxidative Metabolism by Skeletal Muscle in Intact Man in the Basal State: Measurement of Oxygen and Glucose Uptake and Carbon Dioxide and Lactate Production in the Forearm, *J. Clin. Invest.*, **35:**671, 1956.

Armstrong, D. T., R. Steele, N. Altszuler, A. Dunn, J. S. Bishop, and C. DeBodo: Regulation of Plasma Free Fatty Acid Turnover, *Amer. J. Physiol.*, **201:**9, 1961.

Åstrand, P.-O., I. Hallbäck, R. Hedman, and B. Saltin: Blood Lactates after Prolonged Severe Exercise, *J. Appl. Physiol.*, **18**:619, 1963.

Behnke, A. R., Jr.: Quantitative Assessment of Body Build, *J. Appl. Physiol.*, **16**:960, 1961.

Behnke, A. R., Jr., B. G. Feen, and W. C. Welham: The Specific Gravity of Healthy Men: Body Weight Divided by Volume as Index of Obesity, *J. Am. Med. Asso.*, **118**:495, 1942.

Bergström, J., L. Hermansen, E. Hultman, and B. Saltin: Diet, Muscle Glycogen and Physical Performance, *Acta Physiol. Scand.*, **71**:140, 1967.

Bergström, J., and E. Hultman: Muscle Glycogen Synthesis after Exercise: An Enhancing Factor Localized to the Muscle Cells in Man, *Nature*, **210**:309, 1966.

Bergström, J., E. Hultman, L. Jorfeldt, B. Pernow, and J. Wahren: Effect of Nicotinic Acid on Physical Working Capacity and on Metabolism of Muscle Glycogen in Man, *J. Appl. Physiol.*, **26**:170, 1969.

Blix, G.: A Study on the Relation between Total Calories and Single Nutrients in Swedish Food, *Acta Soc. Med. Upsal.*, **70**:117, 1965.

Bock, A. V., C. Vancaulaert, D. B. Dill, A. Fölling, and L. Hurxthal: Studies in Muscular Activity, part IV, The "Steady State" and the Respiratory Quotient during Work, *J. Physiol.*, **66**:162, 1928.

Bogert, L. J.: "Nutrition and Physical Fitness," 6th ed., W. B. Saunders Company, Philadelphia, 1957.

Bronstein, I. P., S. Wexler, A. W. Brown, and L. J. Halpern: Obesity in Childhood, *Am. J. Diseases Children*, **63**:238, 1942.

Bruch, H.: Energy Expenditure of Obese Children, *Am. J. Diseases Children*, **60**:1082, 1940.

Bullen, B. A., R. B. Reed, and J. Mayer: Physical Activity of Obese and Nonobese Adolescent Girls Appraised by Motion Picture Sampling, *Amer. J. Clin. Nutr.*, **14**:211, 1964.

Carlson, L. A.: Plasma Lipids and Lipoproteins and Tissue Lipids during Exercise, in G. Blix (ed.), "Nutrition and Physical Activity," p. 16. Almqvist & Wiksell, Uppsala, 1967.

Cathcart, E. P., and W. A. Burnett: Influence of Muscle Work on Metabolism in Varying Conditions of Diet, *Proc. Roy. Soc. (Biol.)*, **99**:405, 1926.

Chaveau, A.: Source et Nature du Potentiel Directement Utilise dans le Travail Musculaire d'après les Exchanges Respiratoires, Chez l'homme en Etat d'abstinence, *C. R. A. Sci. (Paris)*, **122**:1163, 1896.

Chirico, A. M., and A. J. Stunkard: Physical Activity and Human Obesity, *New Engl. J. Med.*, **263**:935, 1960.

Christensen, E. H., and O. Hansen: Arbeitsfähigkeit und Ehrnährung, *Skand. Arch. Physiol.*, **81**:160, 1939.

Christophe, J., and J. Mayer: Effect of Exercise on Glucose Uptake in Rats and Men, *J. Appl. Physiol.*, **13**:269, 1958.

Coleman, M. C., W. W. Tuttle, and K. Daum: Effect of Protein Source on Maintaining Blood Sugar Levels after Breakfast, *J. Am. Dietet. Ass.*, **29**:239, 1953.

Crittenden, R. H.: "Physiological Economy in Nutrition," Frederick A. Stokes Company, New York, 1904.

Darling, R. C., R. E. Johnson, G. C. Pitts, R. C. Consolazio, and P. F. Robinson: Effects of Variations in Dietary Protein on the Physical Well-being of Men Doing Manual Work, *J. Nutr.*, **28**:273, 1944.

Daum, K., W. W. Tuttle, C. Martin, and L. Myers: Effect of Various Types of Breakfast on Physiologic Response, *J. Am. Dietet. Ass.*, **26**:503, 1950.

Döbeln, W. von: Anthropometric Determination of Fat-free Body Weight, *Acta Med. Scand.*, **165**:37,1959.

Dole, V. P.: Fat as an Energy Source, in K. Rodahl and B. Issekutz, Jr. (eds.), "Fat as a Tissue," McGraw-Hill Book Company, New York, 1964.

Durnin, J. V. G. A.: Activity Patterns in the Community, *Canad. Med. Ass. J.*, **96**:882, 1967.

Fordtran, J. S., and B. Saltin: Gastric Emptying and Intestinal Absorption during Prolonged Severe Exercise, *J. Appl. Physiol.*, **23**:331, 1967.

Friedberg, S. J., and E. H. Estes, Jr.: Direct Evidence for Oxidation of Free Fatty Acids by Peripheral Tissues, *J. Clin. Invest.*, **41**:677, 1962.

Fritz, I. B.: Factors Influencing the Rates of Long-chain Fatty Acid Oxidation and Synthesis in Mammalian Systems, *Physiol. Rev.*, **41**:52, 1961.

Fritz, I. B., D. G. Davis, R. H. Holtrop, and H. Dundee: Fatty Acid Oxidation by Skeletal Muscle during Rest and Activity, *Am. J. Physiol.*, **194**:379, 1958.

Fry, R. C.: A Comparative Study of "Obese" Children Selected on the Basis of Fat Pads, *Am. J. Clin. Nutr.*, **1**:453, 1953.

Gasnier, A., and A. Mayer: Recherches sur la Régulation de la Nutrition, I, Qualités et Cotes des Mécanismes Régulateurs Généraux; III, Mécanismes Régulateurs de la Nutrition et Intensité du Métabolisme, *Ann. Physiol.*, **15**:145, 1939.

Greene, J. A.: Clinical Study of the Etiology of Obesity, *Ann. Intern. Med.*, **12**:1797, 1939.

Hamilton, C. L.: Control of Food Intake, in W. S. Yamamoto and J. R. Brobeck (eds.), "Physiological Controls and Regulations," p. 274, W. B. Saunders Company, Philadelphia, 1965.

Havel, R. J., A. Naimark, and C. F. Borchgrevink: Turnover Rate and Oxidation of Free Fatty Acids of Blood Plasma in Man during Exercise: Studies during Continuous Infusion of Palmitate-1-C^{14}, *J. Clin. Invest.*, **42**:1054, 1963.

Hedman, R.: The Available Glycogen in Man and the Connection between Rate of Oxygen Intake and Carbohydrate Usage, *Acta Physiol. Scand.*, **40**:305, 1957.

Hermansen, L., E. Hultman, and B. Saltin: Muscle Glycogen during Prolonged Severe Exercise, *Acta Physiol. Scand.*, **71**:129, 1967.

Hill, A. V.: The Design of Muscles, *Brit. Med. Bull.*, **12**:165, 1956.

Hultman, E.: Studies on Muscle Metabolism of Glycogen and Active Phosphate in Man with Special Reference to Exercise and Diet, *Scand. J. Clin. Lab. Invest.*, **19**(Suppl. 94), 1967.

Hunter, W. M., and M. Y. Sukkar: Changes in Plasma Insulin Levels during Muscular Exercise, *J. Physiol.*, **196**:110P, 1968.

Hutchinson, R. C.: Meal Habits and Their Effects on Performance, *Nutr. Abstr. Rev.*, **22**:283, 1952.

Ingle, D. J., and J. E. Nezamis: The Effect of Insulin on the Tolerance of Normal Male Rats to the Overfeeding of a High Carbohydrate Diet, *Endocrinology*, **40**:353, 1947.

Issekutz, B., Jr.: Effect of Exercise on the Metabolism of Plasma Free Fatty Acids, in K. Rodahl and B. Issekutz, Jr. (eds.), "Fat as a Tissue," chap. 11, McGraw-Hill Book Company, New York, 1964.

Issekutz, B., Jr., and J. J. Spitzer: Uptake of Free Fatty Acids by Skeletal Muscle during Stimulation, *Proc. Soc. Exp. Biol. Med.*, **105**:21, 1960.

Issekutz, B., Jr., and H. Miller: Plasma Free Fatty Acids during Exercise and the Effect of Lactic Acid, *Proc. Soc. Exp. Biol. Med.*, **110**:237, 1962.

Issekutz, B., Jr., N. C. Birkhead, and K. Rodahl: Effect of Diet on Work Metabolism, *J. Nutr.*, **79**:109, 1963a.

Issekutz, B., Jr., H. I. Miller, and K. Rodahl: Effect of Exercise on FFA Metabolism of Pancreatectomized Dogs, *A. J. Physiol.*, **205**:645, 1963b.

Issekutz, B., Jr., H. I. Miller, P. Paul, and K. Rodahl: Aerobic Work Capacity and Plasma FFA Turnover, *J. Appl. Physiol.*, **20**:293, 1965.

Johnson, M. L., B. S. Burke, and J. Mayer: Relative Importance of Inactivity and Over-eating in the Energy Balance of Obese High School Girls, *Am. J. Clin. Nutr.*, **4**:37, 1956.

Juel-Nielsen, N.: On Psychogenic Obesity in Children, *Acta Paediat. (Stockholm)*, **42**:130, 1953.

Karlsson, J., B. Diamant, and B. Saltin: Lactate Dehydrogenase Activity in Muscle after Prolonged Severe Exercise in Man, *J. Appl. Physiol.*, **25**:88, 1968.

Keys, A.: Physical Performance in Relation to Diet, *Fed. Proc.*, **2**:164, 1943.

Krogh, A., and J. Lindhard: Relative Value of Fat and Carbohydrate as Source of Muscular Energy, *Biochem. J.*, **14**:290, 1920.

Ladell, W. S. S.: Effects of Water and Salt Intake upon Performance of Men Working in Hot and Humid Environments, *J. Physiol. (London)*, **127**:11, 1955.

Lamb, D. R., H. Wallace, R. Jeffress, and J. Peter: "Postexercise Glycogen Supercom-pensation in Skeletal Muscle," paper presented Apr. 26, 1968, at the S. W. District AAHPER Convention, Albuquerque, N. Mex.

Larsson, S.: Diet, Exercise, and Body Composition, in G. Blix (ed.), "Nutrition and Physical Activity," p. 132, Almqvist & Wiksell, Uppsala, 1967.

Mayer, J.: Nutrition, Exercise and Cardiovascular Disease, *Fed. Proc.*, **26**:1768, 1967a.

Mayer, J.: Inactivity, an Etiological Factor in Obesity and Heart Disease, in G. Blix (ed.), "Nutrition and Physical Activity," p. 98, Almqvist & Wiksell, Uppsala, 1967b.

Mayer, J., N. B. Marshall, J. J. Vitale, J. H. Christensen, M. B. Mashayekhi, and F. J. Stare: Exercise, Food Intake and Body Weight in Normal Rats and Genetically Obese Adult Mice, *Am. J. Physiol.*, **177**:544, 1954.

Mayer, J., P. Roy, and K. P. Mitra: Relation between Caloric Intake, Body Weight and Physical Work in an Industrial Male Population in West Bengal, *Am. J. Clin. Nutr.*, **4**:169, 1956.

Mayer, J., and B. Bullen: Nutrition and Athletic Performance, *Physiol. Rev.*, **40**:369, 1960.

Mayer, J., and D. W. Thomas: Regulation of Food Intake and Obesity, *Science*, **156**:328, 1967.

Metropolitan Life Insurance Company: "How to Control Your Weight, ' p. 4, 1963.

Nisbett, R. E.: Determinants of Food Intake in Obesity, *Science*, **159**:1254, 1968.

Peckos, P. S.: Caloric Intake in Relation to Physique in Children, *Science*, **117**:631, 1953.

Pettenkofer, M. von, and C. Voit: Untersuchungen Über dem Stoffverbrauch des nor-malen Menschen, *Z. Biol.*, **2**:459, 1866.

Reichard, G. A., B. Issekutz, Jr., P. Kimbel, R. C. Putnam, N. J. Hochella, and S. Wein-house: Blood Glucose Metabolism in Man during Muscular Work, *J. Appl. Physiol.*, **16**:1001, 1961.

Robinson, S., and A. H. Robinson: Chemical Composition of Sweat, *Physiol. Rev.*, **34**:202, 1954.

Rodahl, K.: "Nutritional Requirements under Arctic Conditions," Norsk Polarinstitutt, Skrifter no. 118, Oslo University Press, Oslo, 1960.

Rodahl, K., S. M. Horvath, N. C. Birkhead, and B. Issekutz, Jr.: Effects of Dietary Protein on Physical Work Capacity during Severe Cold Stress, *J. Appl. Physiol.*, **17**:763, 1962.

Rodahl, K., H. I. Miller, and B. Issekutz, Jr.: Plasma Free Fatty Acids in Exercise, *J. Appl. Physiol.*, **19**:459, 1964.

Rony, H. R.: "Obesity and Leanness," Lea & Febiger, Philadelphia, 1940.

Saltin, B.: Aerobic Work Capacity and Circulation at Exercise in Man: With Special Reference to the Effect of Prolonged Exercise and/or Heat Exposure, *Acta Physiol. Scand.*, **62**(Suppl. 230), 1964.

Saltin, B., and L. Hermansen: Glycogen Stores and Prolonged Severe Exercise, in G. Blix (ed.), "Nutrition and Physical Activity," p. 32, Almqvist and Wiksell, Uppsala, 1967.

Shreeve, W. W., N. Baker, M. Miller, R. A. Shipley, G. E. Ingefy, and J. W. Craig: C[14] Studies in Carbohydrate Metabolism: Oxidation of Glucose in Diabetic Human Subjects, *Metabolism*, **5**:22, 1956.

Simonson, E.: Influence of Nutrition on Work Performance, Nutrition Fronts in Public Health, Nutrition Symposium Series no. 3, p. 72, National Vitamin Foundation, New York, 1951.

Stefanik, P. A., F. P. Heald, Jr., and J. Mayer: Caloric Intake in Relation to Energy Output of Obese and Non-obese Adolescent Boys, *Am. J. Clin. Nutr.*, **7**:55, 1959.

Tuttle, W. W., M. Wilson, and K. Daum: Effect of Altered Breakfast Habits on Physiologic Response, *J. Appl. Physiol.*, **1**:545, 1949.

Tuttle, W. W., K. Daum, L. Myers, and C. Martin: Effect of Omitting Breakfast on Physiologic Response of Men, *J. Am. Dietet. Ass.*, **26**:332, 1950.

Tuttle, W. W., K. Daum, C. J. Imig, C. Martin, and R. Kisgen: Effects of Breakfasts of Different Sizes and Content on Physiologic Response of Men, *J. Am. Dietet. Ass.*, **27**:190, 1951.

Weis-Fogh, T.: Metabolism and Weight Economy in Migrating Animals, Particularly Birds and Insects, in G. Blix (ed.), "Nutrition and Physical Activity," p. 84, Almqvist and Wiksell, Uppsala, 1967.

Welham, W. C., and A. R. Behnke, Jr.: The Specific Gravity of Healthy Men: Body Weight Divided by Volume and Other Physical Characteristics of Exceptional Athletes and of Naval Personnel, *J. Am. Med. Ass.*, **118**:498, 1942.

Wretlind, A.: Nutrition Problems in Healthy Adults with Low Activity and Low Caloric Consumption, in G. Blix (ed.), "Nutrition and Physical Activity," p. 114, Almqvist & Wiksell, Uppsala, 1967.

Zunts, N.: Betrachtungen Über Die Beziehungen Zwischen Nährstoffen und Leistungen Des Körpers, *Oppenheimers Handbuch der Biochemie*, **4**:826, 1911.

15

Temperature Regulation

contents

chapter fifteen

Temperature Regulation

Certain plants may grow at temperatures below freezing; others may survive temperatures in excess of 80°C. But the warm-blooded animals can only live within a very narrow range of body temperature. "In all climates and everywhere on the earth mammals maintain a body temperature of about 38°C. It looks as if evolution has settled this temperature as an optimum for the mammalian class" (Irving, 1966). The warm-blooded species can operate regularly in almost all types of weather and climate, which makes them superior to cold-blooded forms of life. However, they have to pay for this advantage in the currency of food. The metabolic rate at 37 to 38°C is rather high and may be even higher when the animal is exposed to cold environments.

The protected man may well tolerate variations in environmental temperature between −50°C and 100°C. But he can only tolerate a variation of about 4°C in his own deep body temperature without impairment of his optimal physical and mental work capacity. Changes in body temperature affect cellular structures, enzyme systems, and numerous chemical reactions and physical processes that take place in the body. The maximal limits which the living cell can tolerate range from about −1°C at one end of the scale, when the ice crystals formed during freezing break the cell apart, to thermal heat coagulation of vital proteins in the cell at about 45°C at the other end of the scale. Only for shorter periods of time can they tolerate an internal temperature exceeding 41°C. In fact many animals, including man, live their entire lives only a few degrees removed from their thermal death point.

The hot end of the scale is more of a problem than the cold end, for man can more easily protect himself against overcooling than against overheating. Consequently, the controlling mechanism for temperature regulation is particularly geared to protect the body tissues against overheating (Hardy, 1961, 1967).

HEAT BALANCE

If the heat content of the body is to remain constant, heat production and heat gain must equal heat loss, according to the equation: $M \pm R \pm C - E = 0$ where M = metabolic heat production, R = radiant heat exchange (positive if the environment is hotter than the skin temperature, but negative if the temperature of the environment is lower than that of the skin), C = convective heat exchange (positive if the air temperature is higher than that of the skin, negative if the other way around), E = evaporative heat loss. This equation is valid only for conditions when the body temperature is constant. If the body temperature varies, a correction has to be introduced, and the following equation is applicable: $M \pm S \pm R \pm C - E = 0$, where S = storage of body heat (Winslow, Gagge, and Herrington, 1939). S is positive if the body heat content is falling, negative if the heat content increases. The specific heat of most tissues is about 0.83. [Conductive heat exchange (K) is in most conditions negligible but increases in importance during such activities as swimming, since water has a heat-removing capacity which is some twenty times that of air.]

One very important function of the blood circulation is to transport heat: to cool or to heat various tissues as may be needed, and to carry excess body heat from the interior of the body to the body surface, or skin. In this function the blood is very effective, for it has a high heat capacity (0.9), which means that the blood may carry a great deal of heat with only a moderate increase in temperature. Conductance of the tissue [kcal/(m²)(hr)(°C)] is the amount of heat given off per square meter body surface per hour and per degree temperature difference between the interior of the body and its surroundings. When the skin blood flow increases, there is a rise in skin temperature, and the conductance increases. When the skin blood flow is reduced, the temperature difference between arterial and venous blood becomes greater, the conductance is reduced, and the insulating value of the skin is increased.

This control of body temperature, the balance between overcooling and overheating, is the role of temperature regulation. This regulation endeavors to keep the temperature of certain tissues such as the brain, heart, and guts relatively constant. Within the body, the temperature is by no means uniform. The greatest gradient is found between the "shell" (the skin) and the "core" (deep central areas including heart, lungs, abdominal organs, and brain). The temperature of the core may be as much as 20°C higher than that of the shell; the ideal difference between shell and core is about 4°C at rest. But even within the core the temperature varies from one place to another. This complicates the calculation of the heat content of the body and makes it difficult to study temperature regulation. Evidently the term *body temperature* is a misnomer. The maintenance of a normal body temperature is actually quite compatible with considerable gains or losses of heat. The problem is, which temperatures are being regulated.

METHODS OF ASSESSING HEAT BALANCE

Measurements of the *deep body temperature* (core temperature) may be accomplished with the aid of mercury thermometers, thermocouples, or thermistors. The classical site of measurement is the rectum. Since the temperature in the rectum (T_r) varies with distance from the anus, it is customarily measured at a depth of 5 to 8 cm. This rectal temperature is, in a resting individual, slightly higher than the temperature of the arterial blood; it is about the same as liver temperature, but slightly lower (0.2 to 0.5°C) than those parts of the brain where the thermal regulatory center is located. During physical exertion or during exposure to heat, the temperature of this part of the brain increases more rapidly than does the rectal temperature, and the time interval until a new temperature equilibrium is established has been found to be about 30 min (Nielsen, 1938). The temperature increase or decline in the brain and in the rectum is of the same magnitude, however. The rectal temperature is therefore a representative indicator for the purpose of assessing changes in the deep body temperature, providing the measurement is made under steady-state conditions, i.e., after some 30 to 40 min. It has been found that the eardrum temperature is a good indication of the actual brain temperature. This may be obtained by placing a thermocouple, introduced through the ear, against the eardrum (Benzinger and Taylor, 1963). However, the eardrum temperature is not quite identical with the temperature in the thermoregulatory center. Another alternative is to measure the temperature in the esophagus, which is relatively accessible for such measurements. While the temperature of the esophagus is not identical with any of the above-mentioned core temperatures, it generally changes parallel with these temperatures (Nielsen and Nielsen, 1962; Saltin and Hermansen, 1966). In work and heat studies measurements of the oral temperature has its limitations (Strydom et al., 1965).

The skin temperature is measured with the aid of a radiometer, or by placing thermocouples or thermistors on the skin at certain locations, and the mean skin temperature (T_s) is calculated by assigning certain factors to each of the measurements, as follows (Newburgh, 1949):

Head	0.07
Arms	0.14
Hands	0.05
Feet	0.07
Legs	0.13
Thighs	0.19
Trunk	0.35
	1.00

For the calculation of the heat content of the body the following equation may be applied:

Heat content $= 0.83W \, (0.65T_r + 0.35T_s)$

where $W =$ body weight, 0.83 is the specific heat of the body, and 0.65 and 0.35 are the factors assigned to the rectal and the mean skin temperatures, respectively.

The *metabolic rate*, or the magnitude of heat production, is assessed by the measurement of oxygen uptake. The volume of 1 liter oxygen consumed corresponds to approximately 4.9 kcal (from 4.7 to 5.05 kcal, depending on the relative amounts of fat and carbohydrates combusted). Human calorimeters large enough to measure the rate of heat production by direct calorimetry have been constructed and are in use in certain laboratories.

The *evaporative heat loss* (E) plays a major role in the cooling of the skin and the blood during exposure to heat. At normal skin temperature the evaporation of 1 liter of sweat requires 580 kcal. The magnitude of the sweat loss may be estimated simply by weighing the subject nude or dressed in dry clothing, before and after the experiment, and by weighing food and fluid ingested and stools and urine voided during the period of observation. Such measurements are only valid for the calculation of heat balance as long as all the sweat produced during the experiment is actually evaporated. On the other hand, whether the produced sweat is evaporated or part of it has run off the body, the sweat rate is an indication of the magnitude of the heat stress.

The *air temperature* affects convective heat loss or gain (C) and is most conveniently measured with the aid of the usual mercury thermometer. If the thermometer is exposed to radiation, it should be shielded. A piece of tinfoil may be used, but care should be taken to allow free air passage around the thermometer.

The *humidity of the air* may be measured with the aid of a sling psychrometer or an electronic device for measuring humidity. The rate of evaporation of the produced sweat is greatly dependent on the humidity of the air.

The *air movement*, which may be measured by a hot-wire animometer, affects both convective heat exchange (C) and evaporative heat loss (E).

The *radiant heat exchange* (R) depends on the temperature difference between the individual and his surroundings. This may be assessed by the values obtained from a mercury thermometer placed in a hollow spheric black copper container with a diameter of 15 cm (a globe thermometer). Because of rapidly changing radiation intensities, typical for many industrial operations, it is often almost impossible to obtain a true picture of the intensity of the radiation.

Several attempts have been made to measure and evaluate the relative importance of the different climatic factors in the individual's feeling of well being and his physical and intellectual work capacity. The concept *effective temperature* has been introduced; it takes into account the air temperature, wind velocity, and the humidity, or *corrected effective temperature*, where instead of the air temperature, the globe thermometer reading is introduced (Leithead and Lind, 1964). The main objection to any of these assessments of the effects of climate on the individual is that the metabolic rate is not taken into account, i.e., the assessment is valid for resting conditions only. It would be far more meaningful to determine to what extent the various physiological functions are stressed by climatic conditions.

MAGNITUDE OF METABOLIC RATE

Man may be considered to be a tropical animal inasmuch as he requires an ambient temperature of 28°C if he, in a nude state, is to remain in thermal balance, maintain a resting metabolic rate, and at the same time be in the so-called comfort zone. The oxygen uptake under these conditions is about 0.20 to 0.30 liter/min. It is slightly higher when the body size is larger. This corresponds to a production of 60 to 90 kcal/hr, or 70 to 100 watts. This energy is the byproduct of metabolic processes which are essential for the maintenance of life. This produced heat makes up for the heat lost through convection (C), radiation (R), and evaporation (E). Under these conditions, $C + R$ accounts for about 75 percent of the heat loss, and E accounts for only 25 percent. Heat loss through the lungs, through the saturation of the air with water vapor during respiration, accounts for about two-fifths of E. Not all of the rest of E is due to the evaporation of sweat, for part of the water loss through the skin occurs without the involvement of the sweat glands, the so-called perspiratio insensibalis. The total water loss through the skin amounts to a minimum of 0.5 liter/day.

Muscular work is associated with an increase in metabolic rate. Since the mechanical efficiency (the ratio of mechanical energy to chemical work) may vary from 0 to 25 percent depending on the kind of work, at least 75 percent of the energy used is converted into heat. Well-trained athletes may, during short work periods of 5- to 10-min duration, attain an oxygen uptake of up to about 6 liters/min (2,000 watts), and during more prolonged work as much as 4 to 5 liters/min. The amount of heat thus produced during 1 hr could theoretically increase the body temperature of a 70 kg individual from 37°C to about 60°C if the excess heat were not dissipated. With fever, or during shivering due to cold exposure, the heat production may be increased two- to fourfold.

EFFECT OF CLIMATE

Cold

For a nude resting individual, the ideal ambient temperature is about 28°C. Under such conditions the mean skin temperature is about 33°C, and the temperature of the core is about 37°C. The temperature gradient from core to skin is then adequate to facilitate the transfer of the excess heat from the metabolically active tissues to the surroundings. Of the total amount of blood (the circulating blood volume) pumped by the heart, about 5 percent flows through the blood vessels of the skin. If the ambient temperature drops, the temperature difference between the skin and the environment is increased; this causes an increased heat loss through convection and radiation. A reduced heat flow to the skin would result in a gradual lowering of the skin temperature. This would result in a reduced temperature gradient between the skin and the environment. Actually, a reduction in the "conductance of the tissue" occurs, partly because of *vasoconstriction of the skin's blood vessels* causing a reduction in blood flow, partly because the blood in the veins of the extremities is deviated from the superficial to the deep veins. Because of the proximity of the deep veins to the arteries, a heat exchange occurs (Fig. 15-1). Because of this system of countercurrent, in a subject exposed to an ambient temperature of 9°C, the blood leaving his heart will have a temperature of about 37°C. As it is flowing through his arm it will be gradually cooled so that by the time it reaches his hand it may have dropped to about 21°C. The venous blood absorbs a considerable part of the heat as the blood flows through the arm. In other words, cooling of arterial blood flowing through the arteries of the limbs is dependent on the rewarming of cold blood returning in adjacent veins from more distal areas. Thus a cooling of the body core is prevented (Bazett et al., 1948; Schmidt-Nielsen, 1963). In a hot environment, on the other hand, the blood from the limbs returns primarily through superficial veins, which facilitates further cooling of the blood.

The heat exchange between arterial and venous blood is still more important for arctic animals. Sea birds swim in icy water, and the large surfaces of their bare feet are exposed to the cooling effect of the water. However, little body heat is lost because their feet cool down to the temperature of the water, and this reduces the loss of metabolic heat (Irving, 1967). Warm feet of a gull or a duck standing on snow or ice would cause it to melt, and soon the feet would be frozen solidly to the ground where they stood! Hogs, naked as a man in a cold environment, and narwhal, walrus, or seals in arctic waters can prevent the escape of heat by having a very cold skin. These animals have a considerable layer of subcutaneous fatty tissue as an effective insulation when their blood vessels constrict. It is an interesting observation that fats in the peripheral regions have a lower melting point than those in the warmer internal tissues; if this were not the case, the peripheral tissues and legs would become too inflexible in cold weather (Irving, 1967).

Thickly furred animals can use their bare extremities to release excess heat from the body (heat can also be dissipated by evaporation from the mouth and tongue during panting).

Through peripheral vasoconstriction, a sixfold increase in the insulating capacity of the skin and subcutaneous tissues is possible (Burton, 1963). This vascular constriction is particularly active in the fingers and toes. It has been estimated that the blood flow through the fingers may vary a hundredfold or more [from 0.2 to 120 ml blood/(min)(100 g tissue)] (Robinson, 1963). The disadvantage of this vasoconstriction is that the temperature in the peripheral tissues may approach that of the environment. For this reason one is apt to

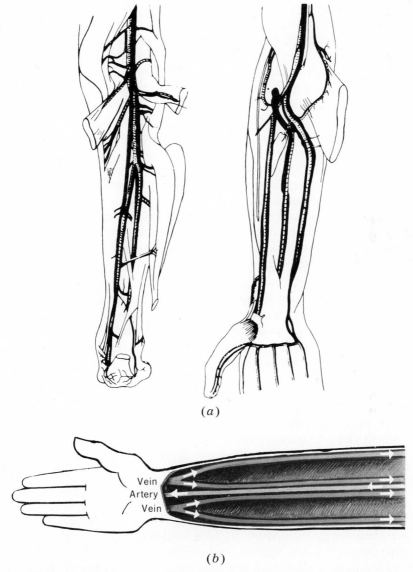

(a)

(b)

Fig. 15-1   (a) The anatomic relationship between the arteries and the deep veins in the forearm which constitute a heat exchange system in the extremity. (From Todt, 1919.)
(b) Schematic illustration of the two ways possible for venous blood flow from the hand; the superficial veins or the deeper ones anatomically close to the arteries making a heat exchange possible.

suffer cold fingers and toes. The blood vessels of the head are far less subject to active vasoconstriction.

Another protective mechanism to maintain heat balance is an *increase of the metabolic rate*, mediated through muscle activity in the form of shivering by a reflex mechanism. Shivering consists of a synchronous activation of practically all muscle groups; antagonists are made to contract against each other. Since the mechanical efficiency of shivering is 0 percent, the heat production is relatively high, and the metabolic rate may increase to two to four times that of resting metabolic rates. Through "active" muscular work the metabolic rate may be further increased in the cold.

Even though the resources for maintaining the core temperature may be quite effective, this maintenance does to some extent take place at the expense of the peripheral tissues, the shell. Local cold injury may be the result in extreme conditions. During prolonged severe cold exposure even the core temperature may drop. Under certain circumstances, especially when exposed to cold water, obese individuals may be better off in the cold than lean individuals because of the insulating value of the adipose tissue.

Summary   When man is exposed to a cold environment, there are two main mechanisms by which a lowering of the body temperature can be prevented: (1) a reduction in the peripheral blood flow with a secondary drop in the skin temperature (reducing the heat loss by radiation and convection); (2) an increase in the metabolic heat production by shivering.

Heat

When the nude resting body is exposed to heat (when the ambient temperature exceeds 28°C), or during muscular work, the heat content of the body tends to increase. Under such conditions the blood vessels of the skin dilate, venous return in the extremities takes place through superficial veins, and the conductance of the tissue increases. In the comfort zone, the skin blood flow amounts, as mentioned, to about 5 percent of the cardiac minute volume; in extreme heat it may increase to 20 percent or more. The increased heat flow to the skin increases the skin temperature. If the temperature of the surroundings is lower than that of the skin, heat loss is facilitated through $C + R$. If the heat load is sufficiently large, the sweat glands are activated, and as the produced sweat is evaporated, the skin is cooled. It has been calculated that there are a total of at least 2 million sweat glands in the skin. It is claimed that the sweating starts on the legs, next the trunk, and finally head and arms are involved (Randall, 1963). Thus, the increase in sweat gland activity follows a certain pattern. The activity of individual sweat glands follows a cyclic pattern. The sweat starts to drop off the skin when the sweat intensity has reached about one-third of the maximal evaporative capacity (Kerslake, 1963).

The individual difference in the capacity for sweating is quite large; some people have no sweat glands at all. As a person becomes accustomed to heat, the amount of sweat produced in response to a standard heat stress increases. A person may produce several liters of sweat per hour. Workers exposed to intense heat may lose as much as 6 to 7 liters of sweat in the course of the working day. Sweat loss up to 10 to 12 kg in 24 hr has been reported (Leithead and Lind, 1964).

During prolonged exposure to a hot environment there is a gradual reduction in the sweat rate, even if the body water loss is replaced at the same rate. This decline in sweat rate is greater in humid than in dry heat, greater "when the men wore Army tropical uniforms than when they wore only broadcloth shorts" (Gerking and Robinson, 1946). The explanation of this "fatigue" of the sweat mechanism is presently not clear. The decline has been attributed simply to a soaking of the skin with sweat (or water) with blockage of the sweat ducts by swelling of the surrounding tissue (Hertig et al., 1961; Brebner and Kerslake, 1964, 1968). Ahlman and Karvonen (1961) report that exercise could again induce sweating after the sweating had ceased during repeated thermal stimuli in a "sauna bath." Wyndham et al. (1966b; Peter and Wyndham, 1966), on the other hand, interpret their experimental findings as unequivocal evidences of fatigue of the sweat glands.

The sweat contains different salts, notably NaCl, in varying concentrations, and excessive sweating may therefore cause a considerable salt loss.

Summary   When man is exposed to a hot environment, he experiences (1) a vasodilation in the skin, making an increased heat transfer from "core" to "shell" possible; (2) and he may experience activation of the sweat glands, and the evaporation of the sweat takes heat from the body, causing an evaporative heat loss.

EFFECT OF WORK

Figure 15-2 shows an example of how the thermal balance is maintained during muscular work of different intensity over a 1-hr period. The body temperature increases during work, and this temperature elevation may be interpreted as the result of an active regulation (Christensen, 1931; Nielsen, 1938; Berggren and Christensen, 1950). The difference between "energy output" and "heat production" in Fig. 15-2 is an expression of the mechanical efficiency (about 23 percent), and the difference between "heat production" and "total heat loss" is a consequence of the elevated body temperature. Note that the convective and radiative heat losses are almost constant despite the large variations in heat production. Evaporation takes care of the extra heat loss as the work load increases.

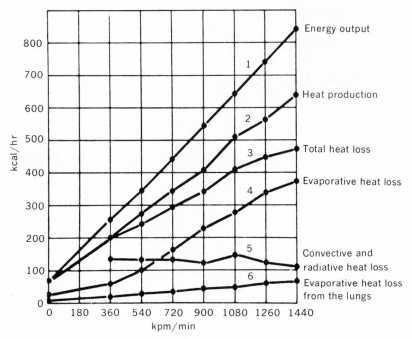

Fig. 15-2   Heat exchange at rest and during increasing work intensities (expressed in kilopond meters per minute along the abscissa) in a nude subject at a room temperature of 21°C. Further explanation in text. (From Nielsen, 1938.)

Figure 15-3 shows how the rectal temperature measured after about 45 min of work on the bicycle ergometer increases linearly with the O_2 uptake. This end temperature does not depend on the absolute magnitude of the energy output but on the level of metabolism relative to the individual's maximal aerobic power. A subject with a maximal O_2 uptake of 2.0 liters/min attains a body temperature of about 38°C during a work load which demands an O_2 uptake of 1.0 liter/min, i.e., 50 percent of his maximal aerobic power. A subject with a maximum of 5.0 liters O_2/min may expend $2\frac{1}{2}$ times more energy (O_2 uptake of 2.5 liters/min) without his body temperature exceeding 38°C (I. Åstrand, 1960).

Saltin and Hermansen (1966) have further studied the relationship between body temperature (esophageal temperature) and oxygen uptake. Figure 15-4 illustrates their data. There is a large scatter of the individual curves when temperature is related to the oxygen uptake (left panel), but the curves come closer together when the oxygen uptake is expressed in percent of the individual's maximal oxygen uptake (right panel).

With an oxygen uptake of about 25 percent of the maximal aerobic power the esophageal temperature was, on an average, 37.3°C; with 50 percent oxygen uptake,

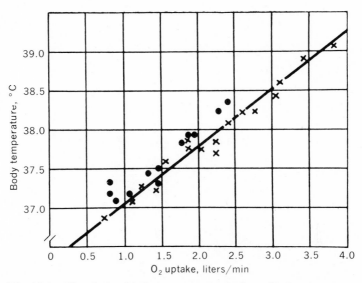

Fig. 15-3 *The relationship between oxygen uptake and body temperature in work with the legs (×) and work with the arms (●). (From Berggren and Christensen, 1950.)*

38.0°; and with 70 percent of the maximal oxygen uptake involved in the work, the temperature rose to 38.5°C. This is illustrated in Fig. 15-5, which also shows the above-mentioned variability in temperature in the core, depending on where the temperature is measured. Naturally, it is highest in working muscles producing the heat. The difference in the temperature measured in the esophagus and in the rectum is small (on an average

Fig. 15-4 *Individual values for esophageal temperature in relation to oxygen uptake or external work load (left panel) and to oxygen uptake in percent of the individual's maximal oxygen uptake (right panel). (From Saltin and Hermansen, 1966.)*

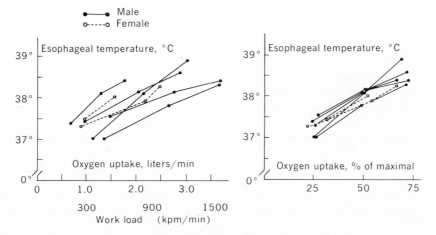

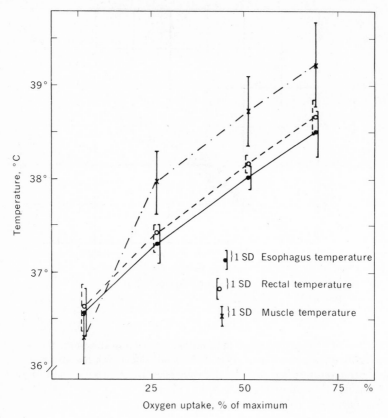

*Fig. 15-5   Average temperature measured simultaneously in the esophagus,
the rectum, and the working muscle in relation to the oxygen uptake in percent
of the individual's maximal oxygen uptake. Seven subjects were working for
60 min on a bicycle ergometer. To the left, data obtained at rest. (SD =
standard deviation.) (From Saltin and Hermansen, 1966.)*

0.14°C) and any of these temperatures are probably a good index of the core temperature
during work.

A further example of the importance of the relative work load rather than the
absolute load for the core and skin temperatures is illustrated in Fig. 15-6. Two subjects
with identical body size were working simultaneously in the same room, side by side, for
60 min on bicycle ergometers with an oxygen uptake which was about 50 percent of the
maximal aerobic power. Since this maximum was 5.35 liters/min for one of the subjects
but "only" 3.54 liters/min for the other subject (open circles in the figure), the actual
work load, bringing the aerobic energy output up to the 50 percent level, was different.
It can be calculated that the stronger subject produced about 640 kcal of heat during the
60-min work period and the weaker subject, 440 kcal. Despite this difference in heat production, the respective temperatures were almost identical. Therefore the stronger man

must have dissipated more heat, and since heat exchange via radiation and convection could not differ, the two subjects must have differed with regard to the sweating rate. The noted weight loss due to water loss was 825 g for the stronger man but only 500 g for the less stronger man during their 60 min of work.

Nielsen (1938) had a subject perform a certain amount of work, 900 kpm/min, at different ambient temperatures varying between 5°C and 36°C. After 30 to 40 min of work, the rectal temperature was the same, regardless of the room temperature. Since the mechanical efficiency was constant, the heat dissipation had to be the same in all these experiments. Figure 15-7 shows how radiation and convection $(R + C)$ accounted for about 70 percent of the heat loss at the lower ambient temperatures. The skin temperature was then 21°C. In the experiment

Fig. 15-6 Temperature of skin (weighed mean value) and esophagus in two subjects at rest and during 60 min of work. The work load is 750 kpm/min ($\dot{V}_{O_2}$ = 1.87 liters/min = 52.8 percent of max $\dot{V}_{O_2}$) for one subject (open circles) and 1200 kpm/min ($\dot{V}_{O_2}$ = 2.73 liters/min = 51.0 percent of max $\dot{V}_{O_2}$) for the other subject (filled circles). The subjects' body surface area according to DuBois formula was 1.86 and 1.90 m², respectively. The weight loss during the 60 min of work was 500 and 825 g, respectively. Squares denote the temperature in the lateral portion of quadriceps femoris. (From Saltin and Hermansen, 1966.)

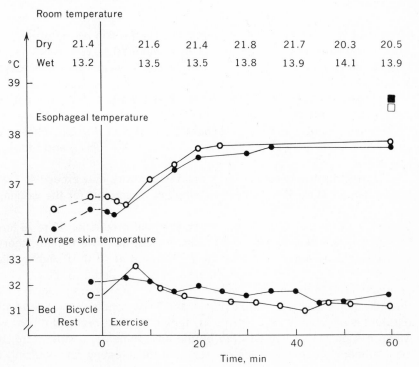

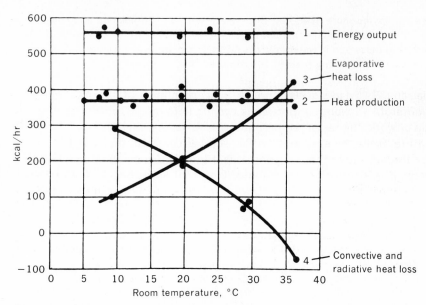

Fig. 15-7   Heat exchange during work (900 kpm/min) at different room temperatures in a nude subject. (Modified from Nielsen, 1938.)

carried out at the highest ambient temperature, the skin temperature was 35°C, and the body absorbed heat from the environment (i.e., 36°C air temperature). This was completely counteracted by an increasing evaporative heat loss. In the cold, the subject evaporated 150 g sweat; in the heat he evaporated 700 g. The greater variation in the rectal temperature at the end of the work period in all these experiments was only 0.11°C. This difference in body temperature may be brought about merely by evaporating about 11 g sweat. These findings have been confirmed by Nielsen and Nielsen (1962, 1965a) and Stolwijk et al. (1968).

During maximal exercise the rectal temperature may exceed 40°C and the muscle temperature 41°C without causing any discomfort for the working man.

Summary   Muscular work can increase the heat production 10 to 20 times the heat production at rest. During work in a "neutral" environment, there is an increase in body temperature up to a maximum of 40°C or slightly higher at maximal work loads. The body temperature is not related to the absolute heat production but to the relative work load, i.e., actual oxygen uptake in relation to the individual's maximal aerobic power; at a 50 percent load the deep body temperature is about 38°C. The deep body temperature at rest and during work is, within a wide range, not affected by the environmental temperature; but the skin temperature is. In a given environment the sweating rate is mainly dependent on the actual heat production and not primarily on the skin or rectal temperature.

TEMPERATURE REGULATION

In the hypothalamus and the adjacent preoptic region there are nerve cells which by local heating and cooling in animal experiments may elicit the same reactions which occur during exposure to heat or cold (Hammel, 1965; Hardy, 1967). The temperature regulatory center is connected via nervous pathways with receptors in the skin. These receptors consist of a net of fine nerve endings which are specifically activated by heat or cold stimuli (Zotterman, 1959; Hensel, 1963). These temperature receptors are especially sensitive to rapid changes in temperature and are highly susceptible to adaptation. In the case of the heat receptors, the maximal frequency of the impulses occurs in a steady-state condition between 38 to 43°C; in the case of the cold receptors, the maximal impulse frequency occurs at 15 to 34°C (Zotterman, 1959). At temperatures above 45°C, the cold receptors may again be activated. This may explain the paradoxical cold sensation experienced at the first contact with very hot water. The receptors not only register temperature changes but also temperature levels, especially if the skin temperature is below 32°C in the case of the cold receptors, and above 37°C in the case of the heat receptors (Kenshalo et al., 1961). The number of active receptors determine to some extent the sensation of temperature. The fact that temperature sensation is a relative matter is best illustrated by the simple experiment of putting one finger in warm water, another finger in cold water. When both fingers are then simultaneously put into lukewarm water, the finger which previously was exposed to cold water will sense the lukewarm water as warm; the other finger will interpret it to be cold.

For the regulation of body temperature, the hypothalamic temperature regulatory center and the temperature-sensitive receptors in the skin play a dominating role. Precisely how the temperature regulatory mechanism works is unknown. This is a particularly interesting question for, as we have just learned, the change in the setting of the body thermostat during work is not related to the absolute but to the relative work load. It may conceivably operate as expressed by the following equation

$$R = aT_h + bT_s$$

where R = response, T_h = hypothalamic temperature, T_s = skin temperature; a and b are constants.

According to Benzinger et al. (1963), the metabolic rate at rest increases when the hypothalamic temperature drops below 37.0°C. Below this temperature the cold receptors of the skin will modify R, and the lower the skin temperature, the greater the increase

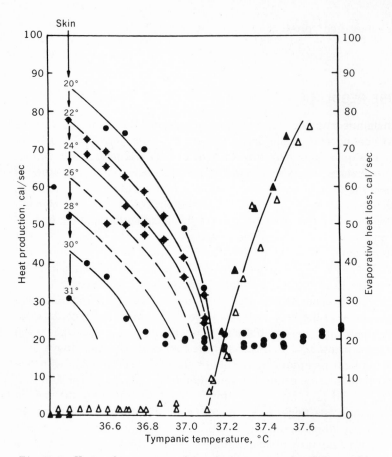

Fig. 15-8   Heat exchange expressed in calories per second at different skin temperatures (20 to 31°C) and eardrum temperatures (abscissa). When the central temperature is below 37.1°C (set point of thermostat), the heat production increases more with the lowering of the skin temperature. When the central temperature is above 37.1°C, the subject begins to sweat. In this experiment the sweat rate was not affected by the skin temperature (up to 37.5°C). (From Benzinger et al., 1963b.)

in metabolic rate. The maximal effect is assumed to occur at about 18°C, at which point the cold receptors show maximal firing frequency. Figure 15-8 shows data from experiments in man. Via sympathetic pathways, impulses are transmitted to the smooth muscles in the blood vessels in the skin, which cause vasoconstriction.

According to Benzinger et al. (1963), an increase of the brain temperature above 37°C, considered to be the normal "set point of the thermostat," will elicit sweating, and the vasoconstrictor impulses to the cutaneous blood vessels decline in frequency or are totally absent. He finds that the sweat intensity diminishes at a certain skin temperature below 33°C. This is interpreted as being the result of cold receptors inhibiting the heat loss center. If, on the other hand, the skin temperature is above 33°C, the sweat production is independent of whether the skin temperature is 33°C or 39°C. The heat receptors, according to Benzinger, do not affect the sweating. Other investigators do not support Benzinger on this point (Hardy, 1967).

The *anterior hypothalamus* and the preoptic region are sensitive to changes in the local temperature. Many neurons have been observed to increase their discharge rate when heated; only a few cells increase their discharge frequency when the local tissue temperature is lowered (Hardy, 1967). Preoptic heating in animals exposed to a neutral environment can induce vasodilation in the skin and, eventually, panting and sweating. With the animal in a cool environment such a local heating can inhibit the normal response of shivering and vasoconstriction of peripheral blood vessels. The depression of the metabolic rate causes the body temperature to drop.

An intact *posterior hypothalamus* is required to induce the reactions to a cold environment, namely shivering and an increase of metabolic rate, and to restrict flow of heat to the skin by a vasoconstriction of skin blood vessels. This posterior center is, however, temperature blind: it is essentially insensitive to local temperature changes. The function of this area is largely coordinating and it receives rather than generates temperature signals. Afferent impulses from cold receptors in the skin seem to be the main drive for this center.

The two thermoregulatory centers are in a way connected so that the response to stimulation of the anterior center includes stimulation of sweating but inhibition of shivering and vasoconstriction. Conversely, the action of a stimulation of the posterior center involves a stimulation of vasoconstriction and an increase of heat production, but a simultaneous inhibition of responses to heat. The final common pathways not only include the motor pathways of the synaptic and somatic systems, but also blood humoral transmissions. Cold may, via the thermoregulatory center in the hypothalamus, affect the pituitary gland and the release of hormones which in turn act on their target organs to release thyrotropic and adrenal hormones, increasing the heat production in the tissues. According to Hardy (1967) many large animals, however, show relatively little of the neuroendocrine response to cold. There are functional connections between the central control of body temperature and the areas regulating water and food intake (Andersson, 1967).

We may conclude that cold acts primarily on the periphery, stimulating cold receptors which signal to the central nervous system. Heat has a direct central effect, but peripheral signals from thermosensitive receptors can modify the response. In other words "chemical" thermoregulation (increased metabolic rate) originates mainly from cold reception in the skin, but "physical" thermoregulation (sweating, increased heat conductance of the skin) is to a high degree elicited by central warm reception. [In a warm environment of 33°C only a small increase in oxygen uptake is observed in dogs when cooling the hypothalamic region to 33.5°C. The same degree of cooling in a neutral environment of 23°C caused, however, a fourfold increase in the heat production (Hammel et al., 1963). This experiment illustrates that the peripheral rather than the central sensing receptors determine the response to a cold environment. Benzinger (1967) points out that shivering ceases immediately when one takes a warm shower, even if the rectal temperature is low. On the other hand, it is difficult to suppress sweating completely even during light exercise by any manipulation of the ambient air or skin temperature and still keep the subject acceptably comfortable (Stolwijk et al., 1968).]

To simplify the understanding of the thermoregulation, we can assume that the center for thermoregulation has a "set point," and adjustments are made to minimize the deviation of the actual body temperature from this set point (Hammel, 1965; Hardy, 1965). The set point is not constant but may change with many physiological conditions. If the body temperature exceeds the set point, the thermoregulating center switches on the cooling actions; if the body temperature is below the set point, the metabolic rate increases and heat conservation mechanisms are switched on. For example, when the skin temperature falls in a cold environment, the afferent nerve impulses from the cold receptors elevate the set point so that the hypothalamic temperature will be below it, and therefore will start driving the heat conserving mechanisms. In fact, the internal body temperature in man may *rise* during mild exposure to cold and still cause shivering and cutaneous vasoconstriction (Hardy, 1965, 1967). Conversely, in a hot environment the set point goes below the hypothalamic temperature so as to drive the mechanisms which promote heat loss.

The diurnal variations in body temperature may be due to variations in the set point; sleep decreases the set point and gradually lowers the body temperature, which upon awakening increases the set point again.

From this point of view exercise should adjust the set point by a decrease. Actually, within seconds after the onset of work, an increase in sweat secretion has been observed (Meyer et al., 1962; Beaumont and Bullard, 1963). On the other hand, another factor enters into the picture: the body temperature increases during exercise (and does not go down to an assumed lowered set point). In other words, the body regulates at a higher temperature during exercise than at rest. The optimal temperature may range from just above 37°C up to 40°C, depending on the severity of work. The knowledge of the thermoregulation during exercise is deficient. The facts are: (1) Deep body temperature is mainly a function of the energy output and is within a wide range independent of the ambient air temperature (Nielsen, 1938; Nielsen and Nielsen, 1965*a*; Stolwijk et al., 1968). Actually the total aerobic energy production seems to be more decisive for the final body temperature than the heat production (Nielsen, 1966). (2) Skin temperature is principally related to the ambient air temperature but not to metabolic rate (and body temperature) [for references see (1); Saltin et al., 1968]. (3) At a *constant work load* the sweat rate increases with the ambient temperature (and skin temperature) and is then unrelated to the body temperature (Fig. 15-7); in a *constant environment* the skin sweating is a linear function of the heat production but unrelated to the skin temperature [for references see (1); Nielsen, 1966]. Therefore skin temperature and hypothalamic temperature may independently modify the sweating response to exercise. This is illustrated in Fig. 15-9.

Nielsen and Nielsen (1965*b*) noticed a remarkable similarity in the thermoregulatory responses to passive heating by diathermia and to active heating by muscular exercise. Thus, at the same level of heat production, rectal and skin temperatures and estimated skin blood flow were increased to the same level in the two kinds of experiments. Therefore, they conclude, a work factor of nervous origin (from mechanoreceptors or cortical irradiation) may operate on the thermoregulatory centers at the start of exercise but hardly in the later phase of exercise. In intermittent work with the same heat production over a given period of time as in continuous work (with different intensity of work during the periods of activity), Nielsen (1968) found the same body temperature and sweat rate, which should also exclude nervous impulses related to the severity of work as the "work factor." She proposes that a chemical factor liberated during work in proportion to the

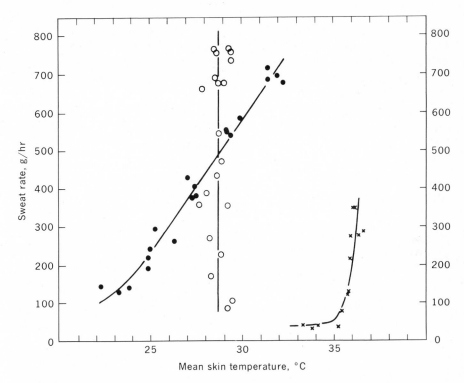

Fig. 15-9   Steady-state values of sweat rate plotted against the corresponding values of mean skin temperature: (○) *work intensity from 540 kpm/min to 1440 kpm/min at constant environmental temperature of 20°C;* (●) *constant work intensity (900 kpm/min) at environmental temperatures from 5 to 30°C;* (×) *experiments at rest at environmental temperatures from 25 to 44°C. (From Nielsen and Nielsen, 1965a.)*

engagement of the aerobic processes may be responsible for the temperature resetting. Since the work load in relation to the individual's maximal aerobic power dominates the adjustment of the body thermostat more than the absolute aerobic power (I. Åstrand, 1960), the attention should be focused at some stress factor, e.g., epinephrine and norepinephrine.

Summary   The temperature-regulating center is mainly located in the hypothalamus. It behaves like a thermostat, and its set point may change during different physiological conditions. Thermosensitive receptors, particularly in the skin, contribute to the regulation of the set point. In a cold environment stimulation of the cold receptors may elevate the set point, and the heat-conserving mechanisms are switched on. In a hot environment the set point becomes lowered, and the heat loss can increase by means of vasodilation in the skin and sweating. The factors involved in the regulation of the body temperature during exercise are largely unknown.

The sweating and the dilation of the skin blood vessels do not run parallel. The produced sweat may in itself cause a vasodilation; on the other hand, the local skin temperature may affect the diameter of the blood vessels in that area. At high evaporative sweat rates the skin may cool, causing vasoconstriction and reduced blood flow.

It is probable that other impulses to the temperature regulatory center exist. Fever is caused by the "thermostat" being set at a higher level. The reason for the diurnal variations in the body temperature is unknown.

ACCLIMATIZATION

Continuous or repeated exposure to heat, and possibly also to cold, causes a gradual adjustment or acclimatization resulting in a better tolerance of the temperature stress in question. Many plants prepare for the winter by increasing their carbohydrate content, and certain types of apple trees may, during the winter, tolerate air temperatures below −40°C, but during the month of July, they fail to survive air temperatures below −3°C. Certain insects accumulate the "antifreeze," glycerol, in the fall, which enables them to survive cold. (Literature, see Dill, 1964.)

Heat

After a few days' exposure to a hot environment, the individual is able to tolerate the heat much better than was the case when he was first exposed. This improvement in heat tolerance is associated with increased sweat production and a lowered skin and body temperature (Robinson et al., 1943; Kuno, 1956; Bass, 1963; Leithead and Lind, 1964; Collins and Weiner, 1965; Peter and Wyndham, 1966; Wyndham, 1967). An example is illustrated in Fig. 15-10. Usually the skin blood flow is reduced; in one experiment it declined from 2.6 to 1.5 liters/(m²)(min), i.e., to about 60 percent of the original value (Bass, 1963). The increased sweat rate provides the possibility for a more effective cooling of the skin through the evaporative heat loss, and the resultant lowered skin temperature provides for a better cooling of the blood flowing through the skin. Thus, the body can afford to cut down on the skin blood flow. In acute experiments, the sweat glands have the capacity to produce more sweat than they actually do under ordinary circumstances. The reason why this capacity is not fully utilized until after several days' exposure to a hot environment is not known. The increase in sweat production may go as high as 100 percent (Leithead and Lind, 1964). Some investigators have found an increase in the blood volume in connection with heat acclimatization, but apparently this is transitory. It appears that heat acclimatization may also occur without changes in blood volume (Bass, 1963).

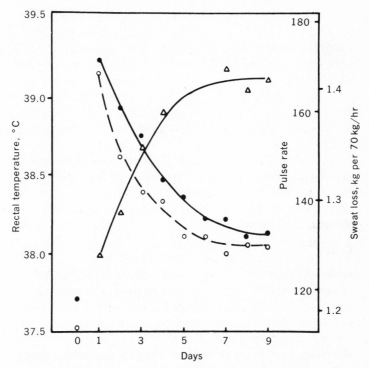

Fig. 15-10   Mean rectal temperature (●), heart rates (○), and sweat losses (△) in a group of men during a 9-day acclimatization to heat. On day 0 they worked for 100 min at a rate of energy expenditure of 300 kcal/hr in a cool climate. On the following day they worked in a hot climate (48.9°C dry-bulb and 26.7°C wet-bulb temperature). (Modified from Lind and Bass, 1963.)

It is possible, then, to demonstrate objectively physiological alterations in response to prolonged exposure to heat, and it is also possible to explain why the heat tolerance gradually increases. It is found that the skin blood flow in acute experiments increases at the expense of the blood flow through other tissues (Williams et al., 1962; Rowell et al., 1965). As the individual becomes acclimatized, normal distribution of the blood flow is once more established. Within 4 to 7 days' exposure to a hot environment most of the changes have taken place, and at the end of 12 to 14 days, the acclimatization is complete. Even a relatively short daily heat exposure will have some effect.

In one series of experiments, daily exposures for 100 min consisting of walking on a treadmill at 5.7 km/hr in a room temperature of 48.9°C dry bulb (26.7°C wet bulb) resulted in satisfactory acclimatization. This effect was better than when the individual worked for 50 min or in two 50-min periods each day. The effect was not improved if the exposure was increased to two 100-min periods per day (Leithead and Lind, 1964).

A well-trained individual adjusts better to heat than one who is in poor physical condition, but training cannot replace acclimatization (Strydom et al., 1966). If muscular work is involved in the exposure to the hot climate, physical work should be included in the acclimatization period. The effect is the same whether the climate is hot and dry or less hot but damp (high humidity). Similarly, the response seems to be the same whether the work in question is heavy and of short duration or less heavy but carried out for a longer period. During the period of acclimatization and during heat exposure, it is important that fluid and salt losses be replaced.

The effect of heat acclimatization persists several weeks following heat exposure, although some impairment in heat tolerance may be detected after a few days following cessation of exposure, such as after a long weekend, especially if the individual is fatigued and alcohol has been consumed (Bass, 1963; Williams et al., 1967).

Cold

Acclimatization to heat may be conveniently studied in subjects living in a hot climate or in laboratory experiments. Studies of adaptation to cold, on the other hand, require climatic chamber experiments inasmuch as man normally when moving to a cold climate brings his semitropical climate along with him. He has the ability to protect himself against the cold by clothing or adequately insulated dwellings, even in arctic regions. Animals habitually exposed to cold develop an effective protection in the form of fur and an effective heat exchange system in their peripheral blood vessels. In certain types of seals, the skin may be cooled to 0°C without an increase in oxygen uptake. In man, a lowering of the skin temperature a few degrees may cause a doubling of his metabolic rate. Man, then, has, with respect to his reaction to cold, taken a path somewhat different from that followed by arctic animals.

According to some investigators, the metabolic rate of "cold acclimatized" individuals is often elevated when they are exposed nude to a standardized cold stress, even though shivering is said to be less pronounced. Nevertheless, the metabolic rate is unchanged at normal room temperature. It is possible that hormones, notably norepinephrine, play a role in the elevation of the metabolic rate, but the mechanism is unclear. It should be noted that the body temperature is reported to be lower in the cold-acclimatized individual during standardized cold stress (Andersen et al., 1960; Hammel, 1963; Wyndham et al., 1964). Some investigators have reported a somewhat lowered skin temperature; others, that it is unchanged or even elevated (Burton, 1963; Davis, 1963).

While it is difficult to demonstrate definite evidence of general physiological acclimatization to cold in man, such acclimatization can be produced in animals exposed to severe cold. If rats are placed in a refrigerator kept at 5°C for as long as 6 weeks, the metabolic rate of the animals will increase twofold. They will double their food intake, and their thyroid function will increase. The main feature of the cold-acclimatized rat is its ability to maintain a high rate of heat production. This ability is absent in the non-acclimatized rat, which is unable to survive in the cold. If these findings in the cold-acclimatized rats are compared with findings in human beings, such as Eskimo or Caucasians habitually exposed to cold, it is observed that the changes found in the rat do not necessarily occur in man. His food intake is not materially increased in a cold environment, nor is his rate of heat production. It is true that the Eskimo's rate of heat production is higher than in Caucasians, but much of this is due to the specific dynamic effect of his diet, for Eskimo living on the white man's diet do not as a rule have any higher rate of heat production than they do. When whites move to the Arctic, their metabolic rate is no higher than it was at home. Normal Eskimo do not show increased thyroid function compared to normal Caucasians (Rodahl and Bang, 1957).

When man, on the other hand, is exposed to cold stress which is far more severe than what is normally encountered by the clothed individual living in the Arctic, certain physiological changes do occur which may be interpreted as hormonally induced adjustments to cold (Rodahl et al., 1962). When young men, dressed in shorts and sneakers only, were confined continuously to cold chamber temperatures of 8°C for 3 to 10 days, they responded by violent shivering which lasted more or less continuously night and day throughout the experiment, even at night when they slept under a blanket. They soon learned to continue to sleep in spite of the shivering. As a result, the increased heat production due to the shivering was maintained even during sleep, so that the body temperature remained normal and the subjects slept relatively comfortably. Occasionally, however, the same subjects, for some reason, failed to keep up the vigorous shivering during the night while they slept. Consequently their body temperature continued to drop and approached dangerously low levels. Under these conditions the subjects occasionally objected to being disturbed, and wanted to be left in peace to continue to sleep. If this were allowed, the subject might continue to cool and, conceivably, eventually die from hypothermia. It thus appears that an exposed person sleeping in the cold may actually freeze to death if his rate of cooling is slow and gradual so that violent shivering is not produced which would wake him up.

These subjects in the climatic chamber doubled their metabolic rates because of their constant shivering. Their resting heart rate was markedly elevated, and the nitrogen loss in the urine was greatly increased. These changes were interpreted as being the result of hormonal changes brought about by the cold stress (Issekutz et al., 1962). It should be borne in mind that the cold exposure of these subjects in the climatic chamber (8°C) is far greater than the cold exposure of the Eskimo or any clothed group of individuals living in the Arctic. Consequently, all the cold-room subjects suffered ischemic cold injury of their feet, although the temperature never approached the freezing temperature of the tissue.

However, any physiological adaptation to cold in man is of little practical value compared to the importance of know-how, experience, and state of physical fitness. There is definitely a limited capacity of the automatic thermostatic system. In the case of the Eskimo, his success in getting along in the cold depends on his ability to avoid the extreme cold. Since most arctic clothing (except fur clothing) is inadequate in terms of insulation to maintain thermal balance when an individual is exposed without a shelter to extreme prolonged cold, the only way to survive is to remain active in order to increase the heat production. The fitter an Eskimo is, the longer he can do this. Thus, survival in the Arctic is a matter of survival of the fittest.

When a person, whether he is an Eskimo, a Caucasian, or a Negro, allows his hands to be repeatedly exposed to cold for about $\frac{1}{2}$ hr daily for a few weeks, this cold stress will cause an increased blood flow through the hand, so that his hand will remain warmer and is not so apt to become numb when exposed to cold. This may be termed local acclimatization to cold. While this inevitably will cause a greater amount of heat to be lost from the hands, it will improve the ability of the hand and fingers to perform work of a precise nature in the cold. To this extent this local cold acclimatization is beneficial (LeBlanc, 1962; Nelms and Soper, 1962; Strömme et al., 1963).

LIMITS OF TOLERANCE

Normal Climate

Optimal function requires that the body temperature be maintained between 36.5 to 39.5°C. The ideal room temperature is about 20°C for clothed individuals who are sitting or standing still. The more active the individual, the lower the room temperature should be. Thus, when performing heavy physical work, a person may prefer a room temperature of 15°C or even lower. Not only may acclimatization play a role, but habits and established traditions may affect the so-called comfort temperature. This may explain the fact that the preferred room temperature is higher in the United States than in England and higher in the summer than in winter in both the United States and England. The preferred room temperature in Singapore is higher than in the United States (Pepler, 1963).

Failure to Tolerate Heat

The most serious consequence of exposure to intense heat is heat stroke, which may be fatal. The heat stroke victim may have a body temperature of 41°C or higher, have ceased sweating, and be confused or unconscious. This form of temperature-regulatory failure is rare and may occur in only one in a million persons exposed to extreme heat. The risk is higher in nonacclimatized than in acclimatized individuals. Obese persons and older individuals are most susceptible. The treatment is rapid cooling of the patient.

Another type of temperature-regulation failure is the so-called anhidrotic heat exhaustion. The victim may have a body temperature of 38 to 40°C; he may sweat very little or not at all. He feels very tired, may be out of breath, and has tachycardia. His main trouble is reduced sweat production. When the patient stops working and is removed to a cool place, his condition rapidly improves, but it may take him a long time to regain full tolerance to heat.

A third type of serious disturbance due to heat exposure is excessive loss of fluid and salt, usually because of failure to replace fluid and salts lost through sweating. After several weeks' exposure, the patient may eventually experience cramps, the so-called miner's cramps, which in rare cases may be fatal. Intravenous administration of NaCl will promptly relieve the cramps.

Heat syncope is a less serious affliction due to heat exposure. This is primarily caused by an unfavorable blood distribution. A large proportion of the blood volume is distributed to the peripheral vessels, especially in the lower extremities as the result of prolonged standing, by a reduction in blood volume due to dehydration. The result is a fall in blood pressure and inadequate oxygen supply to the brain, which may lead to unconsciousness. If he is placed in the horizontal position, preferably with the legs elevated, he quickly regains consciousness. This type of heat collapse is a form of built-in safety mechanism of the body.

Certain individuals exhibit an untoward reaction to heat in the form of heat rash. This condition may make the individual unsuited for work in a hot environment (Minard and Copman, 1963).

Upper Limit of Temperature Tolerance

It is not feasible to quote exact permissible limitations for the working environment. These limitations depend on the combination of the different climatic factors, on the nature and type of work in question, on the severity of the work load, and on the duration of the work. Finally, there are wide individual variations in the tolerance of climatic stress, apart from the effect of acclimatization.

Leithead and Lind (1964) have suggested the following categories: (1) intolerable conditions; (2) just tolerable conditions, i.e., tolerable for intermittent exposure only; (3) easily tolerable conditions. The first two conditions may be encountered in cases of emergency, accidents, fire, and military operations. At air temperatures above 120°C, the heat pain may be the limiting factor. If an individual can tolerate 120°C for about 10 min, he may tolerate 200°C for about 2 min, if the air is dry. Under favorable conditions, temperatures of 50 to 60°C may be tolerated for hours. At higher air temperatures the sweating may not be able to prevent a continuous gradual increase of the body temperature. Under such conditions the body's capacity to store heat may be the limiting factor.

Wyndham et al. (1965) suggest the following criteria for the strain on man at work in hot conditions, based on the measurements of rectal temperature: conditions of work and heat should be judged to be "easy" when the T_r does not exceed 38°C; it should be considered to be "excessive" when the T_r exceeds 39.2°C; with the T_r between these two limits the conditions should be graded as increasingly "difficult" as T_r approaches 39.2°C. It is evident that the man who is acclimatized to heat can maintain a higher energy output at a given body temperature than one who is unacclimatized.

The sweat rate is a good indication of the total heat load. As a method of assessment, the sweat production during a 4-hr period may be calculated. This is known as the predicted 4-hourly sweat rate, or P4SR. As a limit for just tolerable conditions, 4.5 liters has been suggested. This figure provides for a certain margin of safety (maximum 5 to 6 liters in 4 hr). If the subject is over forty-five years old, untrained, unacclimatized, and obese, it has been suggested that the limit be lowered to 3.0 liters P4SR. If industrial workers are to be exposed to heat continuously for more than 4 hr, the criteria have to be changed accordingly (Leithead and Lind, 1964).

The margin of safety is far more limited when the humidity is high, in which case an individual may tolerate an air temperature of 30.8°C for 60 min, 33.5°C for 42 min, or 37.8°C for 29 min (Leithead and Lind, 1964). Wyndham et al. (1960) have found that in loading operations in mines, the performance is reduced by 4 percent when the air temperature is increased from 27.2 to 28.9°C, but by 50 percent when it is increased to 33.9°C when the air is saturated with water vapor.

It is simpler to state the limit of tolerance for heat exposure than to state what kind of climate may be tolerated for an 8-hr working day. Leithead and Lind (1964) suggest that a room temperature of 30°C may be acceptable if the caloric expenditure is below 180 kcal/hr, 28°C for 300 kcal/hr, and 26.5°C for 420 kcal/hr when the humidity is low. They point out that the solution is not to reduce the duration of the work day, from 8 to 6 hr for instance, in a hot working environment, but rather to reduce the physical work load or to include rest periods or to provide the opportunity at intervals to work in a cooler environment.

Age

Although the experimental data are still limited, the available evidence suggests that heat tolerance is reduced in older individuals (Leithead and Lind, 1964; Robinson, 1963). Older individuals start to sweat later than do young individuals. Following heat exposure it takes longer for the body temperature to return to normal levels in older individuals. Older people react with a higher peripheral blood flow, but their maximal capacity is probably lower. In one study it was found that 70 percent of all individuals who suffered heat stroke were over sixty years of age (Minard and Copman, 1963).

Sex

Women have a lower tissue conductance in cold and a higher tissue conductance in heat than do men. This fact suggests a greater variation in the peripheral reaction to climatic stress in women. It appears that this fact is of no importance for the performance of work.

State of Training

As mentioned above, it appears that a trained individual is better able to adjust to heat than one who is untrained. A convalescent patient during recovery from illness is particularly sensitive to heat stress. It may be wise in such cases to subject the patient to a period of acclimatization prior to the assumption of full duties during an 8-hr working day in a hot environment.

COORDINATED MOVEMENTS

The speed of nervous impulses and the sensitivity of the receptors are affected by the temperature of the tissues. At about 5°C the skin receptors for pressure and touch do not react on stimulation. The execution of coordinated motions depends upon the inflow from these receptors to the central nervous system. The numbness in the cold is the result of this lack of sensitivity of the skin receptors. Irving (1966) reports that the skin at a temperature of 20°C was only one-sixth as sensitive as at 35°C; i.e., an impact on the skin had to be six times greater to be felt at the lower skin temperature. The muscle spindles show an increased sensitivity at moderately lowered muscle temperatures, but at 27°C the activity in response to a standardized stimulus is reduced to 50 percent; at a temperature of 15 to 20°C it is completely abolished (Stuart et al., 1963). This phenomenon also contributes to the difficulty of performing fine coordinated movements in the cold.

MENTAL WORK CAPACITY

An evaluation of the mental or intellectual work capacity during exposure to heat or cold is hampered by subjective variations and lack of suitable objective testing methods (Pepler, 1963). As a rule, a deterioration is observed when the room temperature exceeds 30 to 35°C if the individual is acclimatized to heat. For the unacclimatized, clothed individual, the upper limit for optimal function is about 25°C.

The observed deterioration in performance capacity refers to precise manipulation requiring dexterity and coordination, ability to observe irregular, faint optical signs, the ability to remain alert during prolonged, monotonous tasks, and the ability to make quick decisions. During a 3-hr drilling operation, the best results were achieved at 29°C, but at a room temperature of 33°C the performance was reduced to 75 percent; at 35.5°C, to 50 percent; and at 37°C, to 25 percent. A high level of motivation may to some extent counteract the detrimental effect of the climate.

WATER BALANCE

Normal Water Loss

Reasonable figures for the daily water loss are as follows: from gastrointestinal tract, 200 ml; respiratory tract, 400 ml; skin, 500 ml; kidneys, 1,500 ml = 2,600 ml. This loss is balanced by an intake as follows: as fluid, 1,300 ml; as water in the food, 1,000 ml; water liberated during the oxidation in the cells, 300 ml = 2,600 ml. As mentioned the water loss can increase considerably when the individual exercises or is exposed to a hot environment.

Water loss through the respiratory tract varies roughly with the pulmonary ventilation (dryness and temperature of the inspired air have some influence). The ventilation varies within a wide range directly with the production of CO_2, and this production is, in turn, proportional to the metabolic rate. Therefore the water volume from oxidation, proportional to the metabolic rate, equals, by coincidence, roughly the water loss through the respiratory tract.

During heavy exercise, glycogen is the preferred fuel. About 2.7 g of water is stored together with each gram of glycogen (Chap. 14) and this water becomes free as the glycogen is combusted. If during heavy exercise 1200 kcal is totally consumed, 80 percent, or 960 kcal, may be derived from glycogen. The liberated volume of water (including the water of oxidation) will be close to 800 ml. Assuming a mechanical efficiency of about 25 percent, 900 kcal of the 1200 kcal should be dissipated as heat if the body temperature should be maintained unchanged. An exclusive evaporative heat loss demands the evaporation of about 1,500 ml of water to eliminate 900 kcal. Under these conditions only approximately half of the necessary water volume must be taken from body "stores," for the rest is apparently liberated in the processes producing the heat. It should be emphasized that more sweat may be secreted than is evaporated from the skin. On the other hand, radiative and convective heat exchange may reduce the demand on the evaporative heat loss. When the glycogen depots are again restored, extra water is certainly needed.

Thirst

In the adult man about 70 percent of the lean body weight is water, so there is a substantial buffer to cover water losses over limited periods of time. However, in the long run, water intake must balance water loss by the several routes mentioned. Hypothalamus and adjacent preoptic regions play the essential role in the thirst mechanism (Stevenson, 1965; Andersson, 1967). There are some sort of osmoreceptors reacting on an increase in the osmolarity of the intracellular fluid. Any change in the internal environment leading to cellular hypohydration (dehydration) will elicit thirst. A second effect of a rise in body fluid osmolarity is an increased secretion of antidiuretic hormone (ADH) from the neurohypophysis, an effect mediated from the same center (Verney, 1947). The kidneys excrete a minimal amount of water as a vehicle for the elimination of other materials. When water is in excess in the body, little or no ADH is brought to the kidneys and more water is excreted. A water deficit will, as stated, stimulate ADH secretion, causing an increased reabsorption of water by increasing the water permeability of the wall of the collecting ducts in the kidneys. It is a common observation that the volume of urine is reduced when sweating is profuse.

It has been shown that injection of minute amounts of hypertonic saline into, and electrical stimulation within, the anterior parts of the hypothalamus may elicit excessive drinking in the goat (Andersson, 1967). Similar stimulations will not only induce drinking but inhibit feeding. Certainly during hot conditions this mechanism helps to ensure extra supply of water essential to cover the loss by the evaporative heat exchange. A local cooling markedly reduces water intake but influences the food intake in a reverse manner.

The intracellular fluid volume in the specific cells of the hypothalamus may be the crucial factor in thirst (Andersson, 1967). The osmotic pressure across the cell membrane will of course influence this volume. There are volume receptors (stretch receptors) in the walls of the atria and great veins which can detect and reflexly adjust variations in the volume of intravascular fluid (Gauer, 1967).

Sensation of oral-pharyngeal dryness can elicit the urge to drink, but this reflex is not essential for the maintenance of a normal water intake. When one drinks, there is a temporary relief of thirst. This negative feedback operates partly from the oral-pharyngeal level but is also induced from stomach distension. The rapid relief of thirst after water intake is also explained by a normalization of the osmolarity of the blood. Not only does water move out of the gastrointestinal tract but salts diffuse in the opposite direction along a concentration gradient.

Satiety is to a large extent a matter of behavior. Stevenson (1965) points out that man usually waits until he takes food to replace the last part of a water deficit.

The salt content of the sweat is less than that of the blood, and sweat loss therefore causes an increase in the salt concentration of the blood. As discussed above, increased salt concentration of the body fluids leads to the sensation of thirst and reduced urine volume. Certain studies have indicated that the salt content of sweat may be reduced as the result of acclimatization to heat, which should result in increased osmolarity and increased thirst at a given sweat loss (Robinson, 1963). These findings have not been confirmed, at any rate not in experiments in which the salt intake was adequate (Leithead and Lind, 1964). The fact remains, however, that the acclimatized individual is better able to maintain his fluid balance than one who is not acclimatized. Here again, experience may be an important factor, for a water deficit will negatively influence the physical condition.

It is a common observation that a voluntary water intake does not necessarily cover the water loss induced by excessive sweating (Pitts et al., 1944; Adolph, 1947; Leithead and Lind, 1964). The risk of a voluntary hypohydration is greatest in an individual unaccustomed to heat. The risk is also greater when a large portion of the food consists of dried or dehydrated rations, since a considerable volume of the daily water intake comes normally with the regular meals.

Summary   Osmometric, volumetric, and thermal excitations appear to be nature signals which feed information into the control system for water intake, located mainly in the hypothalamus. Sweat loss increases the osmotic pressure of the body fluids and thereby elicits the urge to drink. However, the sensation of thirst

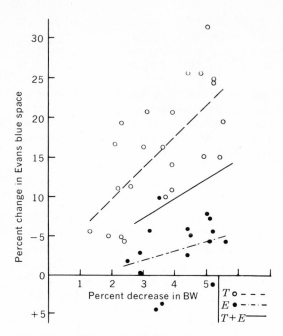

Fig. 15-11   Relationship between changes in Evans blue space (plasma volume) and changes in body weight (BW) during thermal hypohydration (○) and during exercise hypohydration (●). Full line represents average data from experiment with less severe exercise performed in a hot environment. (From Kozlowski and Saltin, 1964.)

does not always "force" the individual to cover the water loss, particularly not when this loss is pronounced due to profuse sweating or if the individual does not eat normal meals containing a large amount of water.

Water Deficit

We have concluded that high sweat rates with excessive loss of body fluids may cause a deficit of body water (hypohydration or dehydration). The regulation of body temperature has priority over the regulation of body water. Therefore a hypohydration can be driven very far, and may in fact be a threat to life if the environment is very hot and water is not available.

Sweating during heat exposure, without muscular work, causes a relatively marked reduction in blood volume; during sweating associated with heavy muscular work, the fluid is primarily taken from the intracellular space, causing a lesser reduction in blood volume (P.-O. Åstrand and Saltin, 1964; Saltin, 1964). Figure 15-11 shows how the plasma volume (Evans blue space) is reduced at various degrees of dehydration. Sweating was induced by exposure to heat, prolonged exercise, or a combination of both. With a 4 percent reduction in body weight, the plasma volume was on an average 20 percent less than normal after exposure to heat. When the same water loss was due to heavy exercise, the plasma volume was only reduced by a few percent, indicating that relatively more water was lost from the intracellular space. Probably part of this water became available as glycogen was consumed. Irrespective of the cause of sweating, hypohydration is associated with a decrease in stroke volume during exercise and a concomitant increase in heart rate during submaximal work. It is remarkable, however, that during maximal exercise, oxygen uptake, cardiac output, and stroke volume are not modified by a sweat loss of up to 5 percent of body weight. However, the work time which can be tolerated on a standardized maximal work load is definitely reduced after dehydration (Fig. 15-12).

The explanation for the gradual decrease in physical performance as a hypohydration develops is presently not available. It apparently cannot be primarily a modification of the aerobic energy yield, because the maximal aerobic power was not impaired in Saltin's experiments, despite a pronounced hypohydration. The explanation should be sought at the cellular level where changes may occur during a hypohydration. The maximal isometric strength after a water deficit is reported to be unaffected (Saltin, 1964) or slightly decreased in connec-

Fig. 15-12   (a) Oxygen uptake (liters/min) and peak blood lactate concentration (HLa) at a work load which could be tolerated for 5½ min during normal conditions (unfilled symbols) but only for 3½ min after hypohydration (filled symbols). Arrows indicate maximal work time.
(b) Oxygen uptake after about 5 min of work at submaximal and maximal work loads as well as peak blood lactate concentration under normal conditions (unfilled symbols) and after hypohydration (filled symbols). Arrows indicate maximal work load. (From Saltin, 1964.)

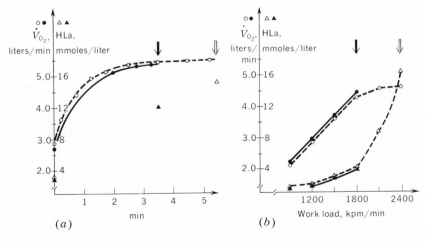

tion with progressive hypohydration (Bosco et al., 1968). In any case, a reduced water content within the muscle cell and a disturbed electrolyte balance can easily influence the muscle cell's ability to contract and its susceptibility to metabolites. The reduction in work performance is more marked if a water deficit is caused by extended heavy work than after exposure to hot environment without exercise being involved.

Dehydration causes a reduced tilt-table tolerance. A person who normally could tolerate prolonged 45° head-up tilt with a heart rate of 90 beats/min fainted within 7.5 min after a fluid loss corresponding to 3 percent of his body weight. Following a fluid loss corresponding to 6 percent of his body weight, he fainted within 1.5 min. His heart rate before fainting was 115 and 135 respectively. A low stroke volume was a characteristic finding (Adolph, 1947).

During prolonged and physically heavy training or during participation in certain competitive sports, the sweat rate may be very high. In some cases it may be as high as 2 liters/hr. An adequate water balance plays an important role in maintaining optimal performance capacity. It is unfortunate that in certain athletic events such as marathon running or walking, the established rules may actually limit the available fluid supply to the athletes. Similarly, the current practice of weighing in wrestlers at least 2 hr before the start of the match of the day may permit those who have purposely become dehydrated, in order to qualify for a lower weight class, to replace their lost body water before the match if they are to compete late, while this may not be possible in the case of those who are to compete first. In any case, such dehydration is not only harmful to the individual, but also poor sportmanship.

Well-trained subjects are less affected in their performance by a hypohydration than untrained subjects (Buskirk et al., 1958; Saltin, 1964). Acclimatization to heat does not seem to protect from the deteriorating effect of a hypohydration (but the water balance may be better maintained).

The simplest method of determining whether the fluid intake has been adequate is by weighing the individual under standard conditions. Even a reduction in body weight of 1 to 2 percent may represent a deterioration in work capacity (Pitts et al., 1944; Adolph, 1947; Ladell, 1955; Saltin, 1964).

The fluid loss during prolonged heat exposure should preferably be replaced by drinking 100 to 150 ml water several times per hour. The water temperature should be about 15°C. In the case of heat exposure lasting for several weeks, the ingestion of salt tablets is advisable, 5 to 15 g per day depending on diet, climate, and degree of physical activity.

Summary   An individual tolerates heavy physical work less well if subjected to a water deficit, even if the water loss is only about 1 percent of the body weight. At a submaximal work load the heart rate is increased, the stroke volume reduced, and the body temperature is higher than normal. Drinking water to satiety may not fully compensate for a water loss. We should also consider that "There are some who claim that sweat drips less profusely from their faces and into their eyes when they have deliberately restricted their fluid intake. These people should not be disbelieved; they can be told instead that they are trading safety for comfort and that the effects of voluntary dehydration (increased temperatures and heart rates and reduced sweating) predispose to water-depletion heat exhaustion and, in the right circumstances, to heatstroke" (Leithead and Lind, 1964).

PRACTICAL APPLICATION

Physical Work

It should be borne in mind that heat exposure in itself represents an extra load on the blood circulation. Exhaustion occurs much sooner during heavy physical work in the heat because the blood, in addition to carrying oxygen to the working muscle, also has to carry heat from the interior of the body to the skin. This represents an extra burden on the heart, which has to pump that much harder. This is convincingly demonstrated in Table 15-1, which shows the difference in work pulse in a subject performing the same work in the heat and in a cool environment. The stress of heat and the hydrostatic factors in prolonged standing work may be added to the stress of work itself.

Williams et al. (1962) observed no difference in maximal O_2 uptake in subjects working in the heat and at comfort temperature. At submaximal work loads, however, they found that the major change in hemodynamics in the heat was an increase in heart rate and a fall in stroke volume. Neither cardiac output nor arteriovenous difference was significantly altered compared to comfortable

TABLE 15-1

Effect of environmental temperature on human response to standard work on a bicycle ergometer for 45 min

Environmental temp.	Heart rate, beats/min	Rectal temp., °C T_r	Mean skin temp., °C T_s	O_2 uptake, liters/min	Weight loss	
					kg	% of body weight
Cool	104	37.7	32.8	1.5	0.25	0.3
Hot steel mill, air temp. 40 to 50°C + radiation	166	38.8	37.6	1.5	1.15	1.6

At the same work load, the temperature difference between core and shell is 4.9°C in the cool environment, but only 0.9°C in the heat. This necessitates a much greater skin blood flow in the heat. Hence, the markedly elevated heart rate in the heat.

conditions (Rowell et al., 1965). Williams et al. (1962) also demonstrated a larger lactate production in the subject who worked in the heat as compared with work in a neutral environment. This finding can be explained to be a result of a reduced muscle blood flow.

In order to avoid incidents of heat stroke and to improve the work conditions in the South African mining industry the workers are, prior to the actual work, adapted to work in a hot and humid environment. During about a week they perform daily a modified "step test" for 4 hr with gradually increasing intensity, and the progress in acclimatization is checked with frequent measurements of oral temperature (Wyndham et al., 1954). Wyndham et al. (1966a) have also developed a standard heat-stress test to evaluate the effectiveness of the acclimatization procedure.

Warm-up

The benefit of the higher temperature during work lies in the fact that the metabolic processes in the cell can proceed at a higher rate, since these processes are temperature-dependent. For each degree of temperature increase, the metabolic rate of the cell increases by about 13 percent. At the higher temperature, the exchange of oxygen from the blood to the tissues is also much more rapid. Physical work capacity is increased following warm-up (Simonson et al., 1936; Asmussen and Böje, 1945). Furthermore, the nerve messages travel faster at higher temperatures. At the temperature of the human body, which is much higher than that of a frog, our nerve messages go up to eight times as fast as those of the frog (Hill, 1927). Thus, there is a very good reason for man to keep his body temperature up as he does, even at considerable expense, in order that he may move more quickly. This is also the reason why athletes have discovered that it pays to warm up before an athletic event. This warming up may make a difference of 3 sec in a 400-yard dash. The warming up may profitably consist of rather vigorous exercise, such as running at a rate of about 12 km/hr for 15 to 30 min just before the event (Högberg and Ljunggren, 1947). In the case of ordinary exercise, a 5-min warm-up consisting of light to moderate exercise is usually adequate.

Högberg and Ljunggren (1947) examined the effect of warm-up in the form of running at moderate speed combined with calisthenics on the speed of running 100, 400, or 800 m in well-trained athletes. They compared this effect with the effect of heating the body passively in a sauna bath for a period of 20 min prior to the race and found that the beneficial effect of passively elevating the body temperature by a sauna bath was much less than that of elevating the body temperature by a warm-up through physical exercise. In the 100-m dash the improvement after a proper warm-up was in the order of 0.5 to 0.6 sec, corresponding to 3 to 4 percent compared with the results without any warm-up. In the 400-m race the improvement amounted to 1.5 to 3.0 sec, corresponding to 3 to 6 percent. In the 800-m race the improvement was 4 to 6 sec, or 2.5 to 5.0 percent.

Thus, the percentage improvement was roughly the same at all distances examined. Similar results have been obtained in swimming (Muido, 1946).

With regard to the duration of the warm-up, Högberg and Ljunggren (1947) observed better results after 15-min than after 5-min warm-up, but no further significant improvement occurred in the 100-m race when the warm-up was extended from 15 to 30 min. The authors observed no deterioration in performance attributable to fatigue as a consequence of rather vigorous warm-up. They recommend a warm-up period of 15 to 30 min, at a relatively high rate of energy expenditure (in their experiments about 3.0 to 3.4 liter O_2 uptake/min, equivalent to running at a speed of 12 to 14 km/hr). The duration and intensity of warm-up should be adjusted according to the environmental temperature and amount of clothing. The higher the environmental temperature and the greater the amount of clothing, the sooner the desired body temperature of about 38.5°C is attained (muscle temperature 39°C or higher). Ideally the rest period between warm-up and the start of the race should be no more than a few minutes, in any case no more than 15 min. After 45 min rest the beneficial effect of the warm-up is abolished, at which time the muscle temperature has also returned to pre-warm-up levels. The use of warm clothing is recommended during warm-up, and this warm clothing should be worn until the athlete is ready to start the race.

Figure 15-13 presents a summary of this discussion of the beneficial effect of warm-up for the physical performance. The improvement is particularly related to the increase in muscle temperature. The higher the muscle temperature (i.e., the heavier the preceding warm-up exercise), the better is the performance.

Radiation

Figure 15-14 illustrates how the radiant heat may be reduced from 1300 kcal/hr to about 15 kcal/hr by placing an aluminum shield between the worker and the heat source, which in this particular case had a temperature of 188°C. An inexpensive protection against radiation may be provided by a sheet of masonite covered with tinfoil. It is important that the surface be kept clean. If the shield has to be transparent, substances which will reflect infrared light, such as glass, should be used.

Air Motion

Air motion increases the evaporation of sweat. However, if the air temperature is higher than that of the skin, the air motion may serve to increase the heat load in that it will cause the skin to pick up more heat through convection.

Clothing

In a moist, hot climate where the temperature of the environment is lower than that of the skin, it is advisable to use as little clothing as possible. If the ambient temperature is higher than that of the skin, the clothing may protect the individual from the radiant heat of the environment. Loose-fitting clothing which permits free

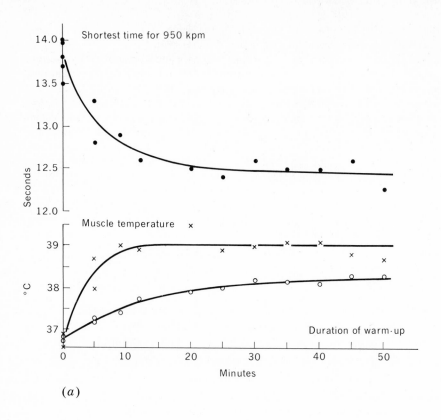

(a)

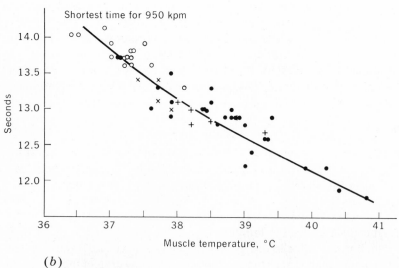

(b)

(See legend on opposite page)

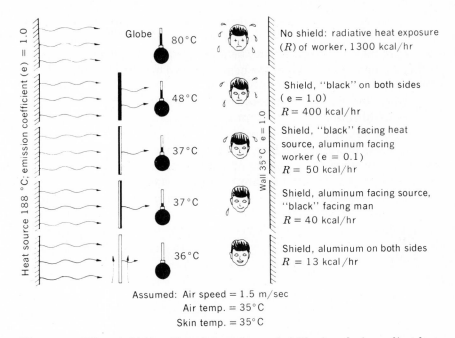

Fig. 15-14 Effect of shields with various surface emissivities in reducing radiant heat load. (Modified from Hertig and Belding, 1963.)

circulation of air between the skin and the clothing is preferable. Workers habitually exposed to intense heat in their work have learned to dress in heavy clothing for protection. This allows some of the radiant heat to be absorbed in the clothing a distance away from the skin. However, it also impairs the facility for evaporative heat loss.

The problem of clothing in the cold when heavy physical work is alternated with rest periods is even more complicated, since there is no single item of clothing capable of protection against cold at rest, yet capable of facilitating heat dissipation during heavy work. The conventional method is to unbutton the coat during work and to button it up during inactivity.

Fig. 15-13 (a) Two lower curves show the temperature in the lateral vastus muscle (upper curve) and in the rectum (lower curve) after warm-up at a work load of 985 kpm/min of different durations (abscissa). The top curve shows the shortest period of time required for the completion of an energy output corresponding to 950 kpm on a bicycle ergometer following warm-up of different durations as shown on the abscissa.
(b) The relationship between the time required for a spurt on the bicycle ergometer (950 kpm) and the temperature measured in the lateral vastus muscle immediately prior to the spurt. O = no warm-up; $\bullet$ = 30-min warm-up of different intensity; $\times$ = warm-up in the form of warm showers; $+$ = warm-up with the aid of diathermy. (From Böje, 1945.)

It should also be pointed out that the surface of the hands represents about 5 percent of the total surface area of the body. In a nude individual, about 10 percent of the heat produced may be eliminated through the hands. In a clothed individual, up to 20 percent of the heat produced may be eliminated through the hands (Day, 1949).

When the Eskimo is exposed to the elements, his simple but practical clothing of fur offers him excellent protection under almost every conceivable kind of weather. In order to remain in heat balance, a man sleeping outdoors at −40°C needs protective clothing with an insulation value of about 12 Clo units. (A "Clo" unit equals the amount of insulation provided by the clothing a man usually wears at room temperature.) However, when the same man is physically active, moving about or walking along, he will only need the equivalent of 4 Clo units because his body heat production is now at least three times greater than it was when he was sleeping due to the increased metabolic rate associated with the increased physical activity (Fig. 15-15). This requirement is adequately met

Fig. 15-15   Insulating requirements in the cold when protected from the wind during different rates of heat production. A Clo unit is the thermal insulation which will maintain a resting man indefinitely comfortable in an environment of 21°C, relative humidity less than 50 percent, and air movement 6 m/min. The unit referred to as met. is the metabolic rate of a resting man. (From Burton and Edholm, 1955.)

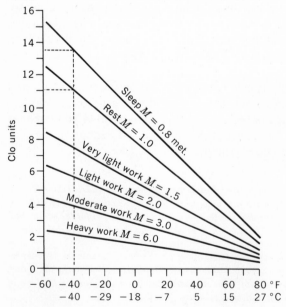

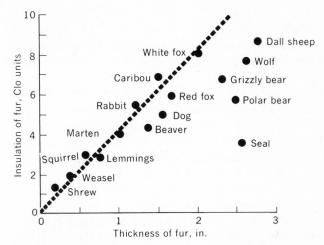

Fig. 15-16   Insulating value of different furs. (Redrawn from Scholander et al., 1950.)

by the double-layer caribou clothing of the Eskimo. Two layers of caribou fur, amounting to a thickness of 3 in., has a total insulation value of about 12 Clo units (Fig. 15-16). This is adequate to maintain heat balance under practically any condition likely to be encountered by the Eskimo. Temperature measurements inside the Eskimo clothing carried out in the field have confirmed that the Eskimo's body inside the clothing is indeed comfortably warm. There is, therefore, some truth to the old statement that the Eskimo, by virtue of his clothing, is really surrounded by a tropical climate. The ordinary uniform usually worn by airmen and soldiers in the Arctic, on the other hand, has an insulation value of only 4 Clo units, which is only one-third that of the Eskimo's clothing. The arctic uniform offers adequate protection for an active man at temperatures as low as −40°C, but he would be in negative heat balance if inactive (Fig. 15-15). Temperatures below −40°C occur on an average of about 2 days per month in the winter in the interior of Alaska.

The insulation value of most materials is proportional to the amount of air which is trapped within the material itself, since air is such a superb insulation. In the case of fur, air is trapped in the space between each hair, but the superior insulation quality of caribou fur, over and above ordinary fur, lies in the fact that the caribou hair is hollow and contains trapped air inside each hair as well as in the spaces between them.

Microclimate

The solution to the problem of providing optimal working environment may be to create local microclimates by cooling or heating the clothing, by providing environmental suits to be worn under special circumstances, or by enclosing the work area in a suitable man-made environment which will facilitate an effective climatic control. Radiant heaters may be used, and exposure suits may be applicable. In any case there will be certain complications. In the ideal climate the temperature of the skin is about 33°C, but not uniformly so. The feet are normally colder than the trunk, and man may accept a greater lowering of the temperature of the feet without feeling cold. A bath, with a water temperature of 33°C, feels cold. In order to feel comfortable, the water temperature has to be about 35°C. However, such a water temperature does not produce temperature equilibrium; it will cause the body temperature to rise. As Burton (1963) puts it: man is not constructed to spend much time in water.

A local heating of the floor may represent an unphysiological manner of regulating room temperature. Burton (1963) interprets this as depending on the fact that the receptors in the skin of the feet exert a relatively dominating influence upon the temperature regulating center. An induced vasodilation of the feet is effective in increasing heat loss. In spite of warm feet, the subjects eventually became cold. With this background, it might appear unphysiological to heat our dwellings, factories, etc., by keeping the floor hotter than the room air.

It is thus possible by improper clothing, or by local heating or cooling of limited skin areas, to upset the normal physiological temperature regulation. Local heating of hands and feet may, for instance, bring about shivering and sweating at the same time. Even if the air temperature is high, heat loss by outgoing radiation to cold surfaces, like a cold window, may cause a most unpleasant sensation, commonly referred to as draft. Because of radiative heat loss to the night sky, a person exposed, unshielded, in the arctic environment may actually be exposed to a cold stress 10 to 20° colder than that which the air thermometer might indicate. In a small enclosed area such as the cabin of an aircraft or a car, it is difficult to satisfy the requirement for adequate ventilation, heat or cold, without causing certain areas of the body to be too hot or too cold. It is evident that much money as well as many heart beats and much sweating could be saved by proper planning of lecture halls, office and factory buildings, and machinery, taking into account all the factors which constitute the optimal climate.

The secret of the success of an experienced arctic traveler or hunter lies in his ability to avoid the extreme cold. The Eskimo's dwelling, whether it be a skin tent, a peat-covered house, or a log cabin, is comfortably warm at all times. The temperature is kept around 21°C (70°F) in the day and may drop to about 10°C (50°F) during the night when the sleeping Eskimo is well covered by fur. While in the summer he may spend as much as 9 hr or more out of doors, the average amount of time spent outside in the winter is only 1 to 4 hr. Furthermore, experience has taught him to take advantage of the characteristic temperature distribution in the arctic environment, produced by the so-called temperature inversion during the winter (Fig. 15-17). The coldest spot is at the surface of the

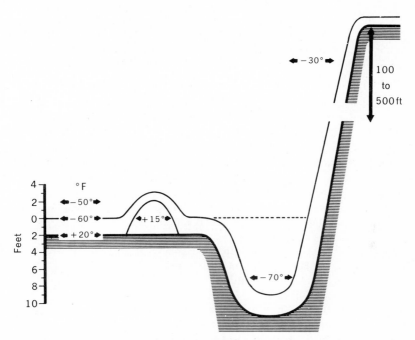

Fig. 15-17   The characteristic temperature distribution (°F) during the Arctic winter.

snow, and especially in a depression in the terrain, such as a river-bed where cold, heavy air is trapped. A few feet up the hillside the temperature is usually many degrees warmer. While the temperature at the actual snow surface may be as cold as $-57°C$ ($-70°F$), the temperature under the snow cover, in the narrow air space between the ground and the snow crust, is usually maintained at -9 to $-6°C$ (15 to 20°F) throughout the winter. It is here that the field mice and other arctic rodents which do not hibernate run around perfectly comfortable all winter long. Actually smaller animals like weasels and mice are not able to carry a fur thick enough for insulation and must therefore spend the winter mostly underneath the snow. By taking advantage of this characteristic temperature distribution, the experienced traveler may successfully escape the extreme degrees of cold stress.

REFERENCES

Adolph, E. F., and Members of the Rochester Desert Unit: "Physiology of Man in the Desert," Interscience Publishers, Inc., New York, 1947.

Ahlman, K., and M. J. Karvonen: Stimulating of Sweating by Exercise after Heat Induced "Fatigue" of the Sweating Mechanism, *Acta Physiol. Scand.*, **53**:381, 1961.

Andersen, K. L., Y. Löyning, J. D. Nelms, O. Wilson, R. H. Fox, and A. Bolstad: Metabolic and Thermal Response to a Moderate Cold Exposure in Nomadic Lapps, *J. Appl. Physiol.*, **15**:649, 1960.

Andersson, B.: The Thirst Mechanism as a Link in the Regulation of the "Milieu Intérieur," in "Les Concepts de Claude Bernard sur le Milieu Intérieur," p. 13, Masson et Cie, Paris, 1967.

Asmussen, E., and O. Böje: Body Temperature and Capacity for Work, *Acta Physiol. Scand.*, **10**:1, 1945.

Åstrand, I.: Aerobic Work Capacity in Men and Women with Special Reference to Age, *Acta Physiol. Scand.*, **49**(Suppl. 169): 67, 1960.

Åstrand, P.-O., and B. Saltin: Plasma and Red Cell Volume after Prolonged Severe Exercise, *J. Appl. Physiol.*, **19**:829, 1964.

Bass, E. E.: Thermoregulatory and Circulatory Adjustments during Acclimatization to Heat in Man, in J. D. Hardy (ed.), "Temperature: Its Measurement and Control in Science and Industry," vol. 3, part 3, p. 299, Reinhold Book Corporation, New York, 1963.

Bazett, H. C., L. Love, M. Newton, L. Eisenberg, R. Day, and R. Forster: Temperature Changes in Blood Flowing in Arteries and Veins in Man, *J. Appl. Physiol.*, **1**:3, 1948.

Beaumont, W. van, and R. W. Bullard: Sweating: Its Rapid Response to Muscular Work, *Science*, **141**:643, 1963.

Benzinger, T. H.: The Thermal Homeostasis of Man, in "Les Concepts de Claude Bernard sur le Milieu Intérieur," p. 325, Masson et Cie, Paris, 1967.

Benzinger, T. H., C. Kitzinger, and A. W. Pratt: The Human Thermostat, in J. D. Hardy (ed.), "Temperature: Its Measurement and Control in Science and Industry," vol. 3, part 3, p. 637, Reinhold Book Corporation, New York, 1963.

Benzinger, T. H., and G. W. Taylor: Cranial Measurements of Internal Temperature in Man, in J. D. Hardy (ed.), "Temperature: Its Measurement and Control in Science and Industry," vol. 3, part 3, p. 111, Reinhold Book Corporation, New York, 1963.

Berggren, G., and E. H. Christensen: Heart Rate and Body Temperature as Indices of Metabolic Rate during Work, *Arbeitsphysiol.*, **14**:255, 1950.

Bosco, J. S., R. L. Terjung, and J. E. Greenleaf: Effects of Progressive Hypohydration on Maximal Isometric Muscular Strength, *J. Sports Med.*, **8**:81, 1968.

Brebner, D. F., and D. McK. Kerslake: The Time Course of the Decline in Sweating Produced by Wetting the Skin, *J. Physiol.*, **175**:295, 1964.

Brebner, D. F., and D. McK. Kerslake: The Effects of Soaking the Skin in Water at Various Temperatures on the Subsequent Ability to Sweat, *J. Physiol.*, **194**:1, 1968.

Burton, A. C.: The Pattern of Response to Cold in Animals and the Evolution of Homeothermy, in J. D. Hardy (ed.), "Temperature: Its Measurement and Control in Science and Industry," vol. 3, part 3, p. 363, Reinhold Book Corporation, New York, 1963.

Burton, A. C., and O. G. Edholm, "Man in a Cold Environment," Edward Arnold, Publishers, London, 1955.

Buskirk, E. R., P. F. Iampietro, and D. E. Bass: Work Performance after Dehydration: Effects of Physical Conditioning and Heat Acclimatization, *J. Appl. Physiol.*, **12**:189, 1958.

Christensen, E. H.: Beiträge zur Physiologie schwerer Körperlicher Arbeit, Die Körpertemperatur während und unmittelbar nach schwerer körperlicher Arbeit, *Arbeitsphysiol.*, **4**:154, 1931.

Collins, K. J., and J. S. Weiner: The Effect of Heat Acclimatization on the Activity and Number of Sweat Glands: A Study on Indians and Europeans, *J. Physiol.*, **177**:16P, 1965.

Davis, T. R. A.: Acclimatization to Cold in Man, in J. D. Hardy (ed.), "Temperature: Its Measurement and Control in Science and Industry," vol. 3, part 3, p. 443, Reinhold Book Corporation, New York, 1963.

Day, R.: Regional Heat Loss, in L. W. Newburgh (ed.), "1, Physiology of Heat Regulation," p. 240, W. B. Saunders Company, Philadelphia, 1949.

Dill, D. B. (ed.): "Handbook of Physiology," Sec. 4, Adaptation to the Environment, American Physiological Society, Washington, D.C., 1964.

Gauer, O. H.: Étude des Rapports Entre la Régulation Volumétrique et la Régulation Osmotique, in "Les Concepts de Claude Bernard sur le Milieu Intérieur," p. 29, Masson et Cie, Paris, 1967.

Gerking, S. D., and S. Robinson: Decline in the Rates of Sweating of Men Working in Severe Heat, *Am. J. Physiol.*, **147**:370, 1946.

Hammel, H. T.: Summary of Comparative Thermal Patterns in Man, *Fed. Proc.*, **22**:846, 1963.

Hammel, H. T.: Neurons and Temperature Regulation, in W. S. Yamamoto and J. R. Brobeck (eds.), "Physiological Controls and Regulations," p. 71, W. B. Saunders Company, Philadelphia, 1965.

Hammel, H. T., S. Strömme, and R. W. Cornew: Proportionality Constant for Hypothalamic Proportional Control and of Metabolism in Unanesthetized Dogs, *Life Sciences*, **12**:933, 1963.

Hardy, J. D.: Physiology of Temperature Regulation, *Physiol. Rev.*, **41**:521, 1961.

Hardy, J. D.: The "Set-Point" Concept in Physiological Temperature Regulation, in W. S. Yamamoto and J. R. Brobeck (eds.), "Physiological Controls and Regulation," p. 98, W. B. Saunders Company, Philadelphia, 1965.

Hardy, J. D.: Central and Peripheral Factors in Physiological Temperature Regulation, in "Les Concepts de Claude Bernard sur le Milieu Intérieur," p. 247, Masson et Cie, Paris, 1967.

Hensel, H.: Electrophysiology of Thermosensitive Nerve Endings, in J. D. Hardy (ed.), "Temperature: Its Measurement and Control in Science and Industry," vol. 3, part 3, p. 191, Reinhold Book Corporation, New York, 1963.

Hertig, B. A., M. L. Riedesel, and H. S. Belding: Sweating in Hot Baths, *J. Appl. Physiol.*, **16**:647, 1961.

Hertig, B. A., and H. S. Belding: Evaluation and Control of Heat Hazards, in J. D. Hardy (ed.), "Temperature: Its Measurements and Control in Science and Industry," vol. 3, part 3, p. 347, Reinhold Book Corporation, New York, 1963.

Hill, A. V.: "Living Machinery," Harcourt, Brace & World, Inc., New York, 1927.

Högberg, P., and O. Ljunggren: Uppvärmningens inverkan på löpprestationerna, *Svensk Idrott*, **40**, 1947.

Irving, L.: Adaptations to Cold, *Sci. Am.*, **214**(1):94, 1966.

Irving, L.: Ecology and Thermoregulation, in "Les Concepts de Claude Bernard sur le Milieu Intérieur," p. 381, Masson et Cie, Paris, 1967.

Issekutz, B., Jr., K. Rodahl, and N. C. Birkhead: Effect of Severe Cold Stress on the Nitrogen Balance of Men under Different Dietary Conditions, *J. Nutr.*, **78**:189, 1962.

Kenshalo, D. R., J. P. Nafe, and B. Brooks: Variations in Thermal Sensitivity, *Science*, **134**:104, 1961.

Kerslake, D. McK.: Errors Arising from the Use of Mean Heat Exchange Coefficients in Calculation of the Heat Exchanges of a Cylindrical Body in a Transverse Wind, in J. D. Hardy (ed.), "Temperature: Its Measurement and Control in Science and Industry," vol. 3, part 3, p. 183, Reinhold Book Corporation, New York, 1963.

Kozlowski, S., and B. Saltin: Effect of Sweat Loss on Body Fluids, *J. Appl. Physiol.*, **19:**1119, 1964.

Kuno, Y.: "Human Perspiration," Charles C Thomas, Publisher, Springfield, Ill., 1956.

Ladell, W. S. S.: The Effects of Water and Salt Intake upon the Performance of Men Working in Hot and Humid Environments, *J. Physiol.*, **127:**11, 1955.

LeBlanc, J.: Local Adaptation to Cold of Gaspé Fishermen, *J. Appl. Physiol.*, **17:**950, 1962.

Leithead, C. S., and A. R. Lind: "Heat Stress and Heat Disorders," Cassell & Co., Ltd., London, 1964.

Lind, A. R., and D. E. Bass: Optimal Exposure Time for Development of Acclimatization to Heat, *Fed. Proc.*, **22:**704, 1963.

Meyer, F. R., S. Robinson, J. L. Newton, C. H. Ts'ao, and L. O. Holgersen: The Regulation of the Sweating Response to Work in Man, *The Physiologist*, **5:**182, 1962.

Minard, D., and L. Copman: Elevation of Body Temperature in Disease, in J. D. Hardy (ed.), "Temperature: Its Measurement and Control in Science and Industry," vol. 3, part 3, p. 253, Reinhold Book Corporation, New York, 1963.

Muido, L.: The Influence of Body Temperature on Performances in Swimming, *Acta Physiol. Scand.*, **12:**102, 1946.

Nelms, J. D., and J. G. Soper: Cold Vasodilatation and Cold Acclimatization in the Hands of British Fish Filleters, *J. Appl. Physiol.*, **17:**444, 1962.

Newburgh, L. H.: "Physiology of Heat Regulation and the Science of Clothing," W. B. Saunders Company, Philadelphia, 1949.

Nielsen, B.: Regulation of Body Temperature and Heat Dissipation at Different Levels of Energy and Heat Production in Man, *Acta Physiol. Scand.*, **68:**215, 1966.

Nielsen, B.: Thermoregulatory Responses to Arm Work, Leg Work and Intermittent Leg Work, *Acta Physiol. Scand.*, **72:**25, 1968.

Nielsen, B., and M. Nielsen: Body Temperature during Work at Different Environmental Temperatures, *Acta Physiol. Scand.*, **56:**120, 1962.

Nielsen, B., and M. Nielsen: On the Regulation of Sweat Secretion in Exercise, *Acta Physiol. Scand.*, **64:**314, 1965a.

Nielsen, B., and M. Nielsen: Influence of Passive and Active Heating on the Temperature Regulation of Man, *Acta Physiol. Scand.*, **64:**323, 1965b.

Nielsen, M.: Die Regulation der Körpertemperatur bei Muskelarbeit, *Skand. Arch. Physiol.*, **79:**193, 1938.

Pepler, R. D.: Performance and Well-being in Heat, in J. D. Hardy (ed.), "Temperature: Its Measurement and Control in Science and Industry," vol. 3, part 3, p. 319, Reinhold Book Corporation, New York, 1963.

Peter, J., and C. H. Wyndham: Activity of the Human Eccrine Sweat Gland during Exercise in a Hot Humid Environment before and after Acclimatization, *J. Physiol.*, **187:**583, 1966.

Pitts, G. C., R. E. Johnson, and F. C. Consolazio: Work in the Heat as Affected by Intake of Water, Salt and Glucose, *Amer. J. Physiol.*, **142:**253, 1944.

Randall, W. C.: Sweating and Its Neural Control, in J. D. Hardy (ed.), "Temperature: Its Measurement and Control in Science and Industry," vol. 3, part 3, p. 275, Reinhold Book Corporation, New York, 1963.

Robinson, S.: Circulatory Adjustments of Men in Hot Environments, in J. D. Hardy (ed.), "Temperature: Its Measurement and Control in Science and Industry," vol. 3, part 3, p. 287, Reinhold Book Corporation, New York, 1963.

Robinson, S., E. S. Turrell, H. S. Belding, and S. M. Horvath: Rapid Acclimatization to Work in Hot Climates, *Am. J. Physiol.*, **140**:168, 1943.

Rodahl, K., and G. Bang: "Thyroid Activity in Men Exposed to Cold," *Arctic Aeromed. Lab. Tech. report* 57-36, 1957.

Rodahl, K., S. M. Horvath, N. C. Birkhead, and B. Issekutz, Jr.: Effects of Dietary Protein on Physical Work Capacity during Severe Cold Stress, *J. Appl. Physiol.*, **17**:763, 1962.

Rowell, L. B., H. J. Marx, R. A. Bruce, R. D. Conn, and F. Kusumi: Reduction in Cardiac Output, Central Blood Volume and Stroke Volume with Thermal Skin in Normal Man during Exercise, *J. Clin. Invest.*, **45**:1801, 1965.

Saltin, B.: Aerobic Work Capacity and Circulation at Exercise in Man, *Acta Physiol. Scand.*, **62**:(Suppl. 230), 1964.

Saltin, B., and L. Hermansen: Esophageal, Rectal and Muscle Temperature during Exercise, *J. Appl. Physiol.*, **21**:1757, 1966.

Saltin, B., A. P. Gagge, and J. A. J. Stolwijk: Muscle Temperature during Submaximal Exercise in Man, *J. Appl. Physiol.*, **25**:679, 1968.

Schmidt-Nielsen, K.: Heat Conservation in Counter-current Systems, in J. D. Hardy (ed.), "Temperature: Its Measurement and Control in Science and Industry," vol. 3, part 3, p. 143, Reinhold Book Corporation, New York, 1963.

Scholander, P. F., V. Walters, R. Hook, and L. Irving, Body Insulation of Some Arctic and Tropical Mammals and Birds, *Biol. Bull.*, **99**:225, 1950.

Simonson, E., N. Teslenko, and M. Gorkin: Einfluss von Vorübungen auf die Leistung beim 100 m.Lauf, *Arbeitsphysiol.*, **9**:152, 1936.

Stevenson, J. A. F.: Control of Water Exchange, in W. S. Yamamoto and J. R. Brobeck (eds.), "Physiological Controls and Regulations," p. 253, W. B. Saunders Company, Philadelphia, 1965.

Stolwijk, J. A. J., B. Saltin and A. P. Gagge: Physiological Factors Associated with Sweating during Exercise, *J. Aerospace Med.*, **39**:1101, 1968.

Strömme, S., K. L. Andersen, and R. W. Elsner: Metabolic and Thermal Responses to Muscular Exertion in the Cold, *J. Appl. Physiol.*, **18**:756, 1963.

Strydom, N. B., C. H. Wyndham, C. G. Williams, J. F. Morrison, G. A. G. Bredell, and A. Joffe: Oral/Rectal Temperature Differences during Work and Heat Stress, *J. Appl. Physiol.*, **20**:283, 1965.

Strydom, N. B., C. H. Wyndham, C. G. Williams, J. F. Morrison, G. A. G. Bredell, A. J. S. Benade, and M. von Rahden: Acclimatization to Humid Heat and the Role of Physical Conditioning, *J. Appl. Physiol.*, **21**:636, 1966.

Stuart, D. G., E. Eldred, A. Hemingway, and Y. Kawamura: Neural Regulation of the Rhythm of Shivering, in J. D. Hardy (ed.), "Temperature: Its Measurement and Control in Science and Industry," vol. 3, part 3, p. 545, Reinhold Book Corporation, New York, 1963.

Todt, C.: "An Atlas of Human Anatomy," The Macmillian Company, New York, 1919.

Verney, E. B.: The Antidiuretic Hormone and the Factors Which Determine Its Release, *Proc. Roy. Soc.*, ser. B., **135**:25, 1947.

Williams, C. G., G. A. G. Bredell, C. H. Wyndham, W. B. Strydom, J. F. Morrison, J. Peter, P. W. Fleming, and J. S. Ward: Circulatory and Metabolic Reactions to Work in Heat, *J. Appl. Physiol.*, **17**:625, 1962.

Williams, C. G., C. H. Wyndham, and J. F. Morrison: Rate of Loss of Acclimatization in
    Summer and Winter, *J. Appl. Physiol.*, **22:**21, 1967.
Winslow, C.-E. A., A. P. Gagge, and L. P. Herrington: Influence of Air Movement upon
    Heat Losses from Clothed Human Body, *Amer. J. Physiol.*, **127:**505, 1939.
Wyndham, C. H.: Effect of Acclimatization on the Sweat Rate/Rectal Temperature
    Relationship, *J. Appl. Physiol.*, **22:**27, 1967.
Wyndham, C. H., N. B. Strydom, J. F. Morrison, F. D. Du Toit, and J. G. Kraan: A New
    Method of Acclimatization to Heat, *Arbeitsphysiol.*, **15:**375, 1954.
Wyndham, C. H., N. B. Strydom, H. B. Cooke, and J. S. Maritz: The Temperature Re-
    sponses of Men after Two Methods of Acclimatization, *Intern. Z. Angew. Physiol.*,
    18:112, 1960.
Wyndham, C. H., J. F. Morrison, J. S. Ward, G. A. G. Bredell, M. J. E. von Rahden,
    L. D. Holdsworth, H. G. Wenzel, and A. Munro: Physiological Reactions to Cold of
    Bushmen, Bantu and Caucasian Males, *J. Appl. Physiol.*, **19:**868, 1964.
Wyndham, C. H., N. B. Strydom, J. F. Morrison, C. G. Williams, G. A. G. Bredell, J. S.
    Maritz, and A. Munro: Criteria for Physiological Limits for Work in Heat, *J. Appl.
    Physiol.*, **20:**37, 1965.
Wyndham, C. H., N. B. Strydom, J. F. Morrison, G. A. G. Bredell, C. H. van Graan, L. Holds-
    worth, A. van Rensburg, A. Munro, and A. Levin: A Test of the Effectiveness of
    Acclimatization Procedures in the Gold Mining Industry, *J. Appl. Physiol.*, **21:**1586,
    1966a.
Wyndham, C. H., N. B. Strydom, J. F. Morrison, C. G. Williams, G. A. G. Bredell, and
    J. Peter: Fatigue of the Sweat Gland Response, *J. Appl. Physiol.*, **21:**107, 1966b.
Zotterman, Y.: Thermal Sensations, in J. Field (ed.), "1, Handbook of Physiology: Neuro-
    physiology," vol. I, p. 431, American Physiological Society, Washington, D.C., 1959.

16

Physiology of Various Sport Activities

contents

chapter sixteen

Physiology of Various Sport Activities

INTRODUCTION

The development of light-weight electronic instruments and devices capable of transmitting impulses by telemetry has made it possible to study a variety of physiological functions in man exposed to different types of work stress, including athletic events. This development is of recent date, however. For this reason few data of this kind have as yet been published.

In this chapter we shall merely present a few examples of various types of analysis of certain athletic events. It is hoped that a more comprehensive discussion may be included in future editions.

In the evaluation of the physiological requirements of different types of physical work it may be useful to consult Table 9-1 in Chap. 9. The discussions in Chaps. 9 and 12 lead us to the following conclusions.

Energy Yield

Work which engages large muscle groups continuously during 1 min or more may often tax the *aerobic power* to a maximal degree and thereby also impose a maximal load on the circulation (see Fig. 9-4). In many types of exercise (walking, running, swimming, rowing, cycling, cross-country skiing, skating, or calisthenics) the individual may set a pace which calls for an aerobic power which may vary from low to maximal (Fig. 9-3b). These types of activities are therefore excellent examples of exercises suitable to develop general physical fitness.

However, the same types of activities may also tax the *anaerobic processes* to a great extent when performed at maximal intensity. The feeling of exertion depends largely on the rate at which glycogen is broken down to lactic acid. Naturally, maximal work in which small muscle groups are engaged brings about the accumulation of lactic acid in the muscles in question and a feeling of exertion, although the total amount of energy applied may be quite small.

In work involving great intensity lasting for a few seconds, interrupted by periods of rest or light work, the peak load is moderate both with respect to aerobic processes and to anaerobic energy yield, as evidenced by the formation of small amounts of lactic acid (Fig. 9-5; Tables 12-2, 12-3). Various kinds of ball games are typical examples of this type of activity.

In athletic events which require heavy work of 1-hr duration or more, the availability of glycogen will eventually determine the level of achievement (Chap. 14). Special measures may therefore be necessary in order to attain improvement. Such events carried out in a hot climate place an additional demand on the heat-dissipating systems and on continuous replacement of lost water (Chap. 15).

Figure 12-7 shows how athletes representing certain athletic events usually have a high maximal aerobic power, while others, although ranking among the elite in their particular field, have a relatively low maximal aerobic power. The requirements of the respective competitive athletic events are evident from this figure.

Neuromuscular Function

Any evaluation of the engagement of individual muscle groups is impossible without the use of *electromyography*. Electromyography furnishes information pertaining to (1) which muscle or which parts of a muscle are activated; (2) the chronological order of the participation of the respective muscles in the activity; (3) the degree and duration of the contraction of the respective muscles in each movement. Such studies may also facilitate the development of an individual muscle training program. As a general rule, strength and technique are best trained through the utilization of the respective athletic event.

TABLE 16-1

Factor	Evaluation of functional demand	"Capacity"
Energy yield		
Aerobic power	Oxygen uptake	Maximal aerobic power
Anaerobic power	Oxygen debt	Maximal oxygen debt
	Blood lactates	Maximal blood lactate concentration
Neuromuscular	Electromyography	Muscle strength
function	Kinematic analysis	?*
	Kinetic analysis	?*
	Oxygen uptake at a given task	?*
Psychological	?	?

* Comparisons with top athletes may be useful, but the sources of error are considerable. Reproducibility may be a valuable criterion.

The *kinematic* analysis describes the geometrical form of a movement. In order to arrive at an idea of the forces which produce movement, i.e., a *kinetic* analysis, force platforms with strain gauges as force-sensitive devices may be used. It is typical for an elite athlete, such as a champion golf player, to be able to repeat again and again in a precise manner a certain motion or force, while the path of movement, the force developed, and the electromyogram are practically identical each time (Carlsöö, 1967). A discussion of the neuromuscular function as applied to sport activities falls primarily within the area of kinesiology and is therefore beyond the scope of this book.

Summary   It may be stated that the requirement of a particular activity and the ability of the individual to meet these requirements can be judged as indicated in Table 16-1.

The relationship between the duration of work and the participation of aerobic or anaerobic processes in the energy yield during maximal work was discussed in connection with Fig. 9-11 and Table 9-1.

ANALYSIS OF VARIOUS SPORTS

Walking

All active individuals have to walk, and occasionally even run, in order to move about in the course of their normal daily life and activity, whether they otherwise engage in any recreational physical activity or not. The energy cost of walking may vary within wide limits, not only between individuals but also in the same individual, depending on the circumstances. It certainly depends on total body weight including clothing, speed of walking, type of surface, and gradient.

There is a fair amount of data accumulated from different countries on the energy cost of walking on the level, and generally these data are in good agreement. Figure 16-1, based on data from Passmore and Durnin (1955), shows the combined effects of varying speed and varying body weight on the energy expenditure of walking.

A comprehensive treadmill-grade walking study was carried out by Margaria (1938), who found that going down a slope of 1 in 10 at varying speeds involved an energy expenditure of up to 25 percent less than walking on the level. However, on very steep declines, particularly at low speeds, energy expenditure may be considerably higher than when walking on the level.

The type of surface may affect the energy cost of walking from 5.5 kcal/min on an asphalt road to 7.5 kcal/min on a ploughed field for a 70-kg man walking at a speed of about 5.5 km/hr (Granati and Busca, 1945). Walking up and down stairs may represent an energy expenditure of as much as 10 kcal/min for a 75-kg person (Passmore et al., 1952). But since stair-climbing for most people does not represent an extensive activity, the total amount of calories spent that

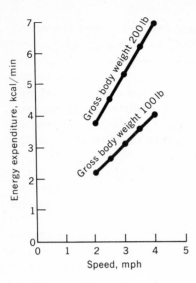

Fig. 16-1 Effect of speed (mph) and gross body weight (lb) on energy expenditure (kcal/min) of walking. (Data from Passmore and Durnin, 1955.)

way may not add a significant amount to the total energy metabolism. Going down stairs involves only about one-third of the energy used in going up stairs. Going, and especially running, up stairs nevertheless represents fairly heavy work, and it is therefore not surprising that so many housewives find this activity very tiring, especially when their care of sick children on the second floor necessitates frequent ascents from their normal occupation in the kitchen or elsewhere on the first floor. On the other hand, stair-climbing may be effective in improving the physical fitness of the housewife, or anyone else for that matter.

After the Olympic Games in London, 1948, the winner of the 10-km walking event, John Mikaelsson, was studied when walking on a treadmill. When simulating the race by adjusting the speed of the treadmill so that the walking speed during the race was attained (13.3 km/hr = 10 km in 45 min 13.2 sec), his measured oxygen uptake was 4.0 liters/min or 58 ml/kg × min (unpublished results). Unfortunately, his maximal aerobic power was not measured.

Running

The caloric expenditure of running varies tremendously, as is to be expected. In adults individual variations are fairly small at submaximal speeds. Under these conditions O_2 uptake per kilogram body weight is the same regardless of sex or athletic rating (P.-O. Åstrand, 1956). On the other hand the oxygen uptake per kilogram body weight is higher for children than for the adults when they both run at a certain speed (P.-O. Åstrand, 1952). It is not clear whether the reason for this inferior efficiency of running depends on inferior technique or is due to different dimensions.

As long as the speed is kept below a certain level, which varies individually, it is more economical to walk than to run (Böje, 1944, Fig. 16-2). In the case of elite walkers this level is higher than in the case of less expert walkers. It may cost the same amount of energy to run at a rate of 14 km/hr (8.7 mph) as it does to walk at a rate of only 10 km/hr (6.2 mph).

The O_2 uptake is dependent upon the stride length, as is evident from Fig. 16-3. The subject ran at a given speed paced by a metronome, and the oxygen uptake was measured at steady state. In some experiments he was free to choose the stride frequency. In general, the stride length which is natural for the individual is also the most economical one. The energy cost of running is greatly increased with a further increase of the stride lengths. This fact is of considerable practical significance. In long-distance running, the factor of economy and energy efficiency is of importance, so that it becomes essential to maintain the stride length most efficient for the individual. In short-distance running, such as the 100-m dash, on the other hand, where speed is more important than economy, the individual who can run with rapid long strides is at an advantage, and he can afford to disregard energy cost for the limited period of time involved.

The well-known runner, Nurmi, had an unusually long stride length. It was thought that this was the key to his success, and efforts were made by others to copy his style. This, however, resulted in reduced efficiency and inferior results. It appears that the long stride was natural for Nurmi, and for him this represented the most economical style. This example clearly shows the danger

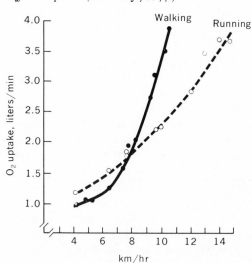

Fig. 16-2 Energy cost of walking and running at different speeds. (From Böje, 1944.)

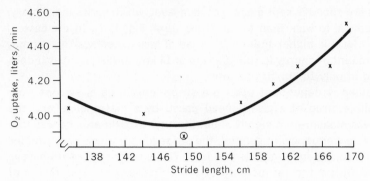

Fig. 16-3 Oxygen uptake during running at a speed of 16 km/hr with different lengths of stride. The encircled cross represents the freely chosen length of stride. (From Högberg, 1952.)

of generalization on the basis of single cases and emphasizes the need for objective physiological studies.

An increase in speed of running is primarily brought about by an increase in the stride length. A good 800-m runner ran on the treadmill at different speeds from 8 up to 30 km/hr. The length of the stride increased more or less rectilinearly from about 80 to 220 cm, while the stride frequency only increased from about 170 to 230 steps/min (Högberg, 1952). It should be emphasized that when running within a wide range of speeds the energy demand per kilogram body weight is practically the same (jogging at lower speeds; running at higher speeds). Walking at *low* speed is less costly, but at high speeds the energy cost approaches and even exceeds that of jogging or running (Zuntz and Schumburg, 1901; Böje, 1944; Margaria et al., 1963). For total energy cost when running the distance covered is more important than the speed. For a 70-kg individual the cost above the basal metabolic rate is about 75 kcal/km when jogging or running on the level (120 kcal/mile).

In order to attain a level of achievement equivalent to the world elite in middle- and long-distance running, a maximal oxygen uptake close to, or preferably above, 80 ml/kg per min is necessary in the case of men (Saltin and P.-O. Åstrand, 1967). In the case of women runners of 400 and 800 m a maximal aerobic power of 65 ml/kg per min or higher is necessary in order to attain results qualifying for the elite class. An athlete with a high maximal oxygen uptake has the advantage of being better able to tolerate bouts of forced tempo than does his competitor with a lower aerobic power in that he does not have to utilize the anaerobic energy yield to the same extent during the bouts of increased tempo. Ideally, the pace should be selected corresponding to the intensity when the oxygen uptake, following a linear increase with the increasing intensity, begins to level off (Fig. 9-3*b*).

Swimming

Swimming engages practically all muscle groups. It is therefore not surprising that very high oxygen uptakes have been obtained on swimmers (P.-O. Åstrand et al., 1963). Since the specific gravity of the body is not much different from that of water, the weight of the body submerged in water is reduced to a few kilograms. The fat individual especially may keep his body floating with very little energy expenditure. Swimming may therefore be an easy task when performed at a low level of intensity. For this reason swimming and different exercises performed in water are a very common form of training for physically handicapped individuals.

A maximal O_2 uptake of 3.75 liters/min was attained by the female silver medal winner in the 400-m freestyle in the Olympic Games in Rome, 1960. In male swimmers of good European standard a maximum of 5.5 liters/min has been measured.

The functional demands in hard swimming were evaluated in 22 girl swimmers on the basis of the relationship between oxygen uptake during swimming at competitive speed and maximal oxygen uptake during work on a bicycle ergometer (P.-O. Åstrand et al., 1963). A very high correlation was observed, but on the average, the oxygen uptake during swimming was only 92.5 percent of the maximal oxygen uptake reached during cycling. However, 5 girls reached a higher value in the former case. The good correlation is also evident from the fact that the blood lactate concentration after swimming was of the same order of magnitude as after maximal cycling (10.3 and 10.5 mmoles/liter respectively). Similar results were reported by P.-O. Åstrand and Saltin (1961). The quotient of pulmonary ventilation to oxygen uptake ($\dot{V}_E/\dot{V}_{O_2}$) was significantly lower during maximal swimming than during cycling (27.7 and 35.5 respectively). The reason for this relative hypoventilation may be the different mechanical conditions of breathing. The water pressure on the thorax makes the respiration more difficult. Furthermore, the breathing is not as free during swimming as in the case of most other types of work, in that the respiration during competitive swimming is synchronized with the swimming strokes. A reduction in the oxygen content of the arterial blood, due to hyperventilation, may in any case partly explain the somewhat lower maximal oxygen uptake observed during swimming compared to cycling. It should be noted that elite swimmers have a large oxygen debt following a good race, as well as high blood lactate levels. If the oxygen transport is impaired, it is obviously important to be able to rely on anaerobic processes for support.

In recent years world records have been attained in swimming by girls at an increasingly younger age. In Chap. 9 an analysis of their physiological possibilities was presented. It was concluded that the girls already at an age of thirteen to fourteen years had almost reached the maximal effect of their aerobic processes. During puberty the organism responds more strongly to the training. In the mentioned study of girl swimmers (P.-O. Åstrand et al., 1963) it was found that they had significantly greater functional dimensions than ordinary girls who had not taken part in competitive sport or undergone any special physical training. Vital capacity and heart volume (Fig. 6-24) were highly correlated to the maximal oxygen uptake.

Thus young girls may exhibit a very high motor effect. It has been shown that women during breaststroke swimming at a certain speed have a lower oxygen uptake (greater mechanical efficiency) than men. This may be explained by the fact that the lower specific gravity in the case of women, due to their greater fat content, reduces the effort required to keep the body floating. However, considerable individual variations in technique are typical for swimming. Thus at a speed of 40 m/min the oxygen uptake varied from 460 to 710 ml/kg body weight $\times$ min (the body weight being that measured in water) (Hemmingsen, 1957).

The energy requirement of different types of swimming is best mirrored in the record tables, which show that crawling is the most economical type of swimming as long as the swimmer masters the proper technique. It is possible to study the different swimming techniques in detail at a specially constructed "swim mill" which has recently been installed at the College of Physical Education in Stockholm. In this swim mill the water can be made to flow at different speeds through a channel, which provides opportunity for studies similar to those on walking and running on a treadmill.

Figure 16-4 represents an example of the application of electromyography in the evaluation of the involvement of different muscle groups in various types of work. Ikai et al. (1964) have studied top swimmers of the Japanese team selected for the Olympics in Rome. They used surface electrodes attached to the skin of the arms, legs, and trunk. The figure presents the integrated voltage of the electrical discharge arranged in chronological order of participation in one full stroke. This integrated voltage is apparently linearly related to the developed muscle strength. There is a definite difference in the EMG pattern obtained on a top swimmer (right panel) compared to a less successful swimmer of a university swimming club (left panel). Whether the difference is only due to different techniques or due to different constitutional factors is not clear (the maximal aerobic power of the swimmers was not reported). The technique appears promising, however, in the mapping of the engagement of different muscle groups in the performance of the work effort.

Various aspects of swimming are discussed in "Medical Research on Swimming" (Firsov and Jokl, 1968).

Speed Skating

Ekblom et al. (1967) have made a study of Swedish speed-skating athletes including the 1968 Olympic champion in 10,000 m, J. Höglin, the 1968 bronze medal winner, Ö. Sandler, and the 1964 World champion and Olympic champion in

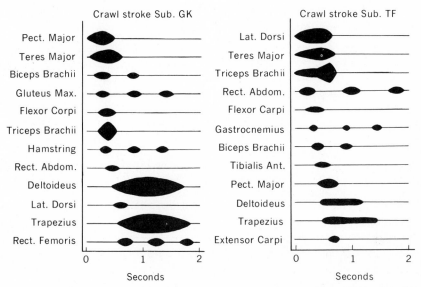

Fig. 16-4   Diagrams of the pattern of the muscular activities during a full crawl stroke evaluated from the recorded electromyograms. To the left, the diagram of one swimmer not particularly outstanding; to the right, the diagram of a Japanese top swimmer. The amplitude of the black area indicates the integrated voltage which is related to the developed strength. Abscissa shows the time course, revealing the beginning and duration of the activity of the muscle in question. (From Ikai et al., 1964.)

10,000 m, J. Nilsson. They were studied when running on a treadmill as well as when skating at submaximal and maximal speeds.

Table 16-2 summarizes some of the maximal data attained in the two types of activities. The average value for maximal oxygen uptake was 5.49 liters/min during treadmill running, as against 4.85 liters/min when skating, i.e., a difference of about 12 percent. Otherwise, the achieved maximal values were rather uniform. It should be pointed out that the extra burden imposed by the equipment which the skaters had to carry for the collection of the expired air might have caused the oxygen uptake values to be higher than would be the case when skating without equipment. Most of the determinations were made on Sandler who, relatively speaking, had the highest oxygen uptake when skating (95 percent of that attained during treadmill running). The values for blood lactate and heart rate which were attained in connection with skating competitions were similar to those obtained during the determination of maximal oxygen uptake.

When the ice condition is excellent, the maximal oxygen uptake expressed in liters per min is probably of greater practical importance than the maximal oxygen uptake expressed as ml/O_2 per kg body weight per min. During speed skating the center of gravity of the body is moving relatively parallel to the ice

TABLE 16-2

		Maximal values									
Subject	Best time 10,000 m	Oxygen uptake				Pulmonary ventilation, liters/min		Blood lactates, mmoles/liter		Heart rate, beats/min	
		liters/min		ml/kg × min							
		R*	S*	R	S	R	S	R	S	R	S
Sandler, Ö.	15,20.6	5.77	5.48	79.0	75.1	184	183	20.3	19.9	186	186
Nilsson, J.	15,47.0	5.70	4.80	79.2	66.7	172	139	17.0	18.0	188	186
Höglin, J.	15,23.6	5.39	4.69	71.9	62.5	137	159	15.4	15.4	185	185
Claesson	17,08.0	5.39	4.89	64.9	58.9	141	141	17.4	17.4	194	190
Nilsson, I.	16,19.4	5.20	4.38	76.5	64.4	138	121	16.8	16.6	171	171
Mean value		5.49	4.85	74.3	65.5	154	149	17.4	17.4	185	184

* R = running; S = speed skating.

surface without marked vertical movements as is the case during running. When the ice surface is porous, it is a drawback for the skater to be heavy, since this increases the friction. The results from the mentioned study show that the elite long-distance skaters have a maximal aerobic power of about 5.5 liters/min. In most speed-skating competitions, such as the World championship, the final results are based on the combined results of the 500-m, 1,500-m, 5,000-m, and 10,000-m distances. The 500-m race and especially the 1,500-m race (skating time, a little over 2.0 min) impose particularly high demands on the anaerobic power (the peak blood lactate concentration for three of the skaters at the end of the race at the Swedish championship competition in 1965 was on the average as follows: 500 m, 13.6 mmoles/liter; 1,500 m, 17.3 mmoles/liter; 5,000 m, 15.1 mmoles/liter; 10,000 m, 13.3 mmoles/liter). The technique is also different in the shorter distances compared to the longer ones. The skaters who come first in the 500-m race, and to some extent also in the 1,500-m race, rarely win the 10,000-m race, which illustrates that the physiological requirements are different in the two types of races. The skaters referred to in Table 16-2 are all typical long-distance skaters.

Figure 16-5 presents oxygen uptakes (Douglas bag method) during skating at different speeds at two different skating rinks. As is evident, the curve is not linear. An increase in the speed from 4 to 6 m/sec requires an additional 0.7 liter/min, while the increase from 8 to 10 m/sec necessitates an increase in the aerobic power of 2.0 liters/min. (A possible contribution from anaerobic processes was not taken into consideration in this study.) The explanation must primarily be sought in the increased air resistance at the higher speed. It increases with the speed raised to the second power, and accounts for a large part of the energy expenditure at high speeds. Thoman (personal communication) has, from experiments with Sandler in a wind tunnel, calculated that at a speed of 10 m/sec, 70 percent of the external work is devoted to overcoming the air resistance, while the remaining 30 percent is required to overcome the friction of the ice. Thus a more ideal aerodynamic profile would to a large extent reduce the air resistance. The speed-skating style, with the arms held on the back, is undoubtedly the result of experience. It is also known that the type of clothing is of great importance in regard to the air resistance. If the conventional speed-skating suit of a competitive speed skater were replaced with a suit made of a type of material similar to that used by many skin divers, the time on a 5,000-m speed-skating race might theoretically speaking be improved by 10 to 20 sec.

The style of the speed skating appears rather rigorous, with many muscle groups involved in static contraction. This may explain the fact that the maximal oxygen uptake is lower during skating than during running. Figure 16-6 also shows that the blood lactate concentration at a given oxygen uptake is con-

Fig. 16-5   Oxygen uptake during speed skating at different speeds. The different symbols represent experiments performed at different speed-skating rinks. In excellent ice conditions as well as at high altitude (reduced atmospheric pressure = reduced air resistance), the curve is shifted to the right; i.e., the same expenditure of energy gives greater speed. Vertical pillars indicate speed during the skaters' best performances on 500 and 10,000 m respectively. Subjects J. Höglin, J. Nilsson, and Ö. Sandler. (From Ekblom et al., 1967.)

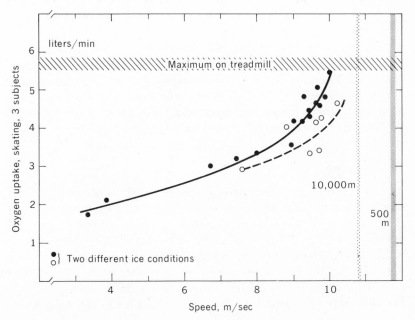

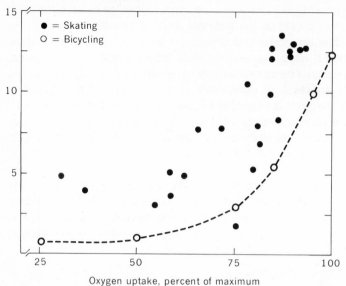

Fig. 16-6 *Blood lactate concentration at different work loads expressed in percentage of maximal oxygen uptake during speed skating and bicycling. The bicycling data are from well-trained noncyclists. The speed skating data are obtained from well-trained speed skaters. At corresponding oxygen uptakes, more lactic acid is produced during speed skating than during bicycling. (From Ekblom et al., 1967.)*

siderably higher during skating than when cycling. From a training point of view it appears essential that the speed skater allow the muscle groups engaged in skating to become accustomed to tolerating a high lactic acid concentration.

The maximal isometric and dynamic muscle strength of skaters is not particularly great. It has been observed, however, that the leg muscles of the skaters who are the best sprinters and specialize in 500-m skating are much stronger than in the case of long-distance skaters.

Alpine Skiing

Agnevik et al. (1969) have studied the Swedish elite athletes in this event, both in the laboratory and in connection with regular international competitions (i.e., "Årebragden" in Sweden; Holmenkollen in Norway; Polarnight Cup in Narvik in Norway, 1965). One of the foremost skiers in the world in special slalom, Bengt-Erik Grahn, was included in this study. His maximal oxygen uptake measured on the bicycle ergometer was 3.9 liters/min or 66 ml/(kg)(min); his maximal heart rate was 207 beats/min (see Fig. 16-7). With the aid of telemetry his heart rate was recorded before and during a competitive slalom race. Due to technical difficulties, his heart rate could only be followed during the end of the race in the case of giant slalom and downhill skiing. Figure 16-7 shows a heart rate of more than 160 beats/min at the start. This high heart rate no doubt is due to emotional factors and some degree of nervousness at the start of the race. The heart rate quickly rose to over 200 beats/min, to the same maximal heart rate which was obtained at a maximal load on the bicycle ergometer. The same heart rate was recorded at the end of the other competitive events. The figure also shows how the heart rate increases at "supermaximal" work loads on the bicycle ergometer. In this case the increase in heart rate occurs more slowly than is the case during competitions in Alpine skiing, a difference which must be attributed to psychic factors. (It should be pointed out that the skier prior to the start of the race skis the track uphill in order to familiarize himself with the track. This physical effort does not explain the high heart rate, since an ample period of rest precedes the actual start of the race.) The day following the giant slalom race, some of the participants covered exactly the same track, but in this case the oxygen uptake was determined by the Douglas bag method as well. The subjects completed the run in almost exactly the same time as was the case during the actual race. Figure 16-7 presents an example of the data collected. The oxygen uptake in this particular subject was 3.3 liters/min, or 87 percent of his maximal O_2 uptake measured in laboratory experiments. In another subject the oxygen uptake was 3.9 liters/min, or 78 percent of his maximal aerobic power. On this occasion the recorded heart rates were submaximal. Figure 16-7 shows that the heart rate measured lies on the regression line which was obtained for the heart rate–oxygen uptake relationship during tests on the bicycle ergometer. This observation supports the assumption that the competition itself represents an extra psychic stress leading to an elevated heart rate. Figure 16-7 also presents mean values for peak blood lactic acid in connection with competitions, maximal work on the bicycle ergometer, and running. The high lactic acid levels during skiing, in spite of a relatively short work time, may be explained by the assumption that certain muscle groups are engaged in intense static work. The best skier, Grahn, attained in his special event the highest lactic acid levels of the entire group. Finally, it is evident from Fig. 16-7 that the Alpine skiers are characterized by a great isometric strength in the stretch muscles of the legs. A group of military recruits is shown for comparison (values taken as 100 percent). The skiers had even greater muscle strength than a group of weight lifters.

Summary   It may be pointed out that competitive Alpine skiing places heavy demands on both aerobic and anaerobic motor power. Great strength in the muscle groups involved is required. In addition to these basic requirements the technique will obviously determine the level of the achievement. From the viewpoint of training it is important to learn to master a good technique in spite of a high lactic acid concentration in the muscles.

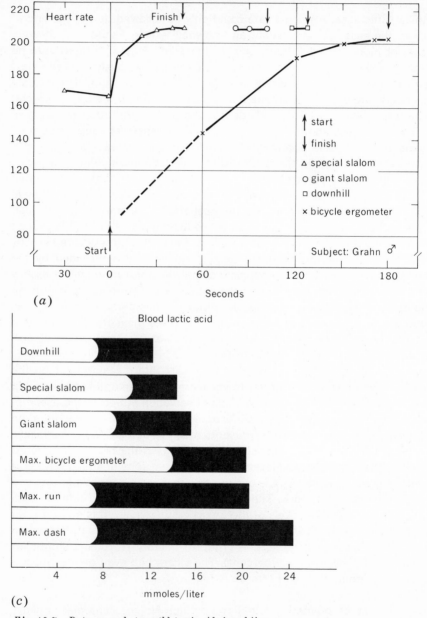

Fig. 16-7   Data on male top athletes in Alpine skiing.
(a) *Heart rate before and during competitions in Alpine skiing and during maximal work on a bicycle ergometer. Note the high heart rate before start and the rapid increase in heart rate after start.*
(c) *Peak blood lactate concentration after various activities; mean values of three to five determinations. "Max. run" = 3 × 1,000 m at maximum speed with a few minutes' rest in between; "Max. dash" = 5 × 50 sec in a similar manner.*

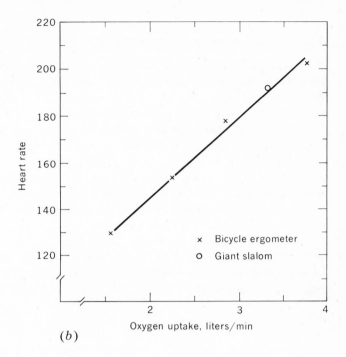

(b)

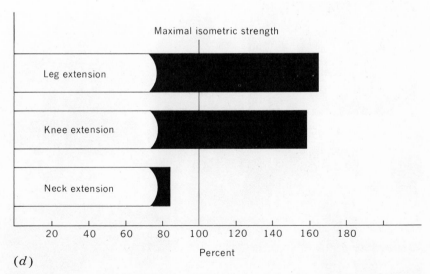

(d)

(b) Heart rate in relation to oxygen uptake during cycling and giant slalom the day after actual race on the same course.

(d) Maximal isometric muscle strength; 100 percent = data obtained on a group of recruits. (From Agnevik et al., 1968.)

Cross-country Motorcycling (Moto Cross)

In this competitive event the motorcycle (250 or 500 ml) is driven across a field track, which may be very difficult, for up to 45 min (Fig. 16-8). Agnevik and Saltin (1967) have studied a group of Swedish motorcyclists including four world champions both in the laboratory and in the field during actual motorcycle events. Heart rates and blood samples were obtained in connection with important competitions. Furthermore, following the actual race, the oxygen uptake was measured while the driver covered the same track at as near as possible the same speed which was achieved during the actual race (Fig. 16-8). Table 16-3 presents some of the data. Considering the fact that the motor effect required to move the body and the motorcycle is supplied by the motorcycle engine, the demand on the driver is surprisingly great. The oxygen uptake varied from 2.09 to 2.61 liters/min, with a mean value of 2.1 liters/min for the two 250-ml drivers and 2.5 liters/min for the four 500-ml drivers. The maximal oxygen uptake was 3.8 and 3.9 liters/min for the two 250-ml drivers, which means that they taxed their oxygen-transporting capacity by 55 percent. The corresponding values for

Fig. 16-8   Cross-country motorcycle racing (moto cross). Insert at upper left shows the arrangement of noseclip, mouthpiece, respiratory valve, and three-way stopcock with automatic timing for the determination of oxygen uptake during a simulated motorcycle race (insert at lower right).

TABLE 16-3

Subject	Maximal values, bicycle ergometer					Simulated competition				Competition	
	Oxygen uptake		Pulmonary ventilation, liters/min	Heart rate, beats/min	Blood lactates, mmoles/liter	Oxygen uptake, liters/min	Pulmonary ventilation, liters/min	Heart rate, beats/min	Blood lactates, mmoles/liter	Heart rate, beats/min	Blood lactates, mmoles/liter
	liters/min	ml/kg × min									
Blomqvist (250*)	3.81	52	153	194	16.9	2.13	55	174	4.3		
Eneqvist (250)	3.90	56	132	180	16.3	2.09	61	161	4.6	180	7.0
Jonsson (250)	4.07	58	129	200	14.5	....	....	....	....	200	4.8
Lundin, B. (500)	3.54	45	150	192	16.6	2.10	60	155	11.6		
Lundin, S. (500)	3.74	52	151	180	18.2	2.61	76	169	3.5		
Persson (500)	4.40	56	147	193	15.1	2.57	56	164		193	7.8
Tibblin (500)	4.65	56	139	180	14.3	2.57	71	158	2.0	180	7.9

* Cylinder volume of motorcycle in milliliters.

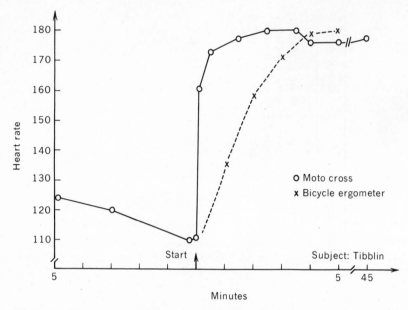

Fig. 16-9   Heart rate recorded during a competition in cross-country motorcycle racing and during maximal work on the bicycle ergometer. (From Agnevik and Saltin, 1967.)

500-ml drivers were 4.1 liters/min (3.54 to 4.65) and 65 (55 to 70) percent. The speed during this simulated race was identical to the speed during the actual event, so that the values for the aerobic power may be considered to correspond to the requirement during the actual competitive motorcycle race. The difference in the blood lactic acid concentration (see Table 16-3) probably depends on the fact that the simulated race lasted only 5 to 10 min while the actual race lasted 45 min. A large number of determinations of the blood lactate concentration during three major races gave as a mean result: 7.7 mmoles/liter for the winners, 7.4 mmoles/liter for those who came second, and 6.0 mmoles/liter for those who came third. These values are relatively high considering the fact that the oxygen uptakes were not very high. Presumably the leg and gluteal muscles were stressed statically during the race. The drivers themselves stated that they were exceedingly tired in their legs at the end of the race. In cross-country motorcycling, as in several other events, such as riding and alpine skiing, involving prolonged static muscle contractions, it is important to utilize every opportunity to relax the tensed muscle groups or to engage them in dynamic work to facilitate the removal of local metabolites and the supply of oxygen.

When the heart rate during work on the bicycle ergometer and the heart rate recorded during cross-country motorcycling are compared, there is a marked difference. As an example it may be mentioned that Eneqvist at an oxygen uptake of 2.1 liters/min on the bicycle ergometer had a heart rate of 130 beats/min. During a simulated cross-country motorcycle race with the same oxygen uptake of 2.1 liters/min the heart rate was on an average 161 beats/min, and during the actual race it was 180 (Table 16-3). The heart rate should be evaluated on the basis of (1) the energy load; (2) static muscular work; (3) psychic stress. Figure 16-9 presents an example of telemetric heart rate recordings on one of the most successful cross-country motorcyclists. The heart rate rose very rapidly to the maximal level which was recorded during laboratory tests on this subject and remained at this maximal level throughout the race, which lasted 45 min.

The isometric muscle strength measured as hand grip strength was no greater in these motorcycle drivers than in the case of relatively untrained persons. The ability to maintain a muscle contraction corresponding to 50 percent of the maximal strength, however, was about double that of the control group. The difference was even greater when they had to contract with a force corresponding to 50 percent of the maximal strength for periods of 5 sec interspersed with 5-sec rest periods. Similar differences were also observed for isometric work with the leg muscles. Evidently cross-country motorcycle racing requires static endurance, and it is therefore natural that the elite motorcyclists exhibit a superior performance capacity in this type of work.

REFERENCES

Agnevik, G., and B. Saltin: Moto Cross, *Idrottsfysiologi*, Rapport nr. 3, Framtiden, Stockholm, 1967.

Agnevik, G., B. Wallström, and B. Saltin: A Physiological Analysis of Alpine Skiing, *J. Appl. Physiol.*, in press.

Åstrand, P.-O.: "Experimental Studies of Physical Working Capacity in Relation to Sex and Age," Munksgaard, Copenhagen, 1952.

Åstrand, P.-O.: Human Physical Fitness with Special Reference to Sex and Age, *Physiol. Rev.*, 36:307, 1956.

Åstrand, P.-O., and B. Saltin: Maximal Oxygen Uptake and Heart Rate in Various Types of Muscular Activity, *J. Appl. Physiol.*, 16:977, 1961.

Åstrand, P.-O., L. Engström, B. Eriksson, P. Karlberg, I. Nylander, B. Saltin, and C. Thorén: Girl Swimmers, *Acta Paediat.*(Suppl.147), 1963.

Böje, O.: Energy Production, Pulmonary Ventilation, and Length of Steps in Well-trained Runners Working on a Treadmill, *Acta Physiol. Scand.*, 7:362, 1944.

Carlsöö, S.: A Kinetic Analysis of the Golf Swing, *J. Sports Med.*, 7:76, 1967.

Ekblom, B., L. Hermansen, and B. Saltin: Hastighetsåkning på Skridsko, *Idrottsfysiologi*, Rapport nr. 5, Framtiden, Stockholm, 1967.

Firsov, S., and E. Jokl (eds.): "Medical Research on Swimming," All-American Productions and Publishers, 1968.

Granati, A., and L. Busca: Il lavoro della tribiatura, *Boll. Soc. Ital. Biol. Sper.*, 20:51, 1945.

Hemmingsen, I.: Energiomsättningen under svoemning hos maend og kvinder, *Tidskrift for Legemsövelser*, **22**:53, 1957.

Högberg, P.: How Do Stride Length and Stride Frequency Influence the Energy Output during Running? *Arbeitsphysiol.*, **14**:437, 1952.

Ikai, M., K. Ishii, and M. Miyashita: An Electromyographic Study of Swimming, *Res. J. Phys. Educ.*, **7**:47, 1964.

Margaria, R.: Sulla fisiologia e specialmente sul consumo energetico, della marcia c della corsa a varie velocita ed inclinazioni del terreno, *Atti dei Lincei*, **7**:299, 1938.

Margaria, R., P. Cerretelli, P. Aghema, and G. Sassi: Energy Cost of Running, *J. Appl. Physiol.*, **18**:367, 1963.

Passmore, R., J. G. Thomson, and G. M. Warnock: Balance Sheet of the Estimation of Energy Intake and Energy Expenditure as Measured by Indirect Calorimetry, *Brit. J. Nutr.*, **6**:253, 1952.

Passmore, R., and J. V. G. A. Durnin: Human Energy Expenditure, *Physiol. Rev.*, **35**:801, 1955.

Saltin, B., and P.-O. Åstrand: Maximal Oxygen Uptake in Athletes, *J. Appl. Physiol.*, **23**:353, 1967.

Zuntz, L., and W. Schumburg: "Studien zu einer Physiologie des Marsches," Bibliothek v. Coler, Band 6., Verlag von A. Hirschwald, Berlin, 1901.

17

Factors Affecting Performance

contents

chapter seventeen

Factors Affecting Performance

INTRODUCTION

Figure 9-1 is a schematic presentation of the components of the aerobic work capacity and the factors which affect this capacity. Most of these factors have been discussed earlier rather extensively from various viewpoints and in different connections. Thus the effect of the type of exercise was discussed in Chap. 9, oxygen-transporting functions were discussed in Chaps. 6 and 9, metabolic aspects in Chap. 14, and training and deconditioning in Chap. 12. As far as the effect of the environment is concerned, the effect of temperature and water balance was discussed in Chap. 15. The present chapter will be devoted to a combined discussion of the remaining factors mentioned in Fig. 9-1, including the effect of altitude.

HIGH ALTITUDE

The climbing of high mountains has always fascinated man. His quest for new discoveries has taken him beyond the earth's atmosphere into outer space. Permanent human habitation is encountered up to an altitude of over 4,500 m. It is becoming increasingly popular to spend summer and winter vacations in high mountain areas undertaking the fairly heavy physical work involved in hiking, mountaineering, or skiing. In spite of pressure cabins, passenger flights at high altitudes involve a certain lack of oxygen. The air pressure in the airplane cabin usually corresponds to an altitude of 1,000 to 1,500 m. The decision to hold the 1968 Olympic Games in Mexico City at an altitude of 2,300 m created a special interest in the problems concerning the effects of altitude on physical performance. Several international symposia have recently been arranged to discuss problems related to physical performance at high altitude (for reports see

Weihe, 1964; Dill, 1964; Luft, 1964*b*; *Schw. Zschr. Sportmed.*, **14**:1–329, 1966; Margaria, 1967; Goddard, 1967; Jokl and Jokl, 1968; Roskamm et al., 1968).

It is thus apparent that the interest in the effect of altitude on human performance is considerable. Luft (1964*a*) has reviewed some historic events in the exploration of altitude and the facilities available in various parts of the world for research in this field.

Physics

In the nineteenth century Bert (1878) recognized that the detrimental effects of high altitude were due to the *diminished partial pressure of oxygen* at reduced barometric pressure. The barometric pressure at a given altitude is dependent on the weight of the air column over the point in question. The atmosphere is compressed under its weight. Its pressure and density are therefore highest at the surface of the earth and decrease exponentially with altitude. Due to temperature differences and turbulence, any sedimentation of the gas molecules of different molecular weight is avoided, and the chemical composition of the atmosphere is practically uniform up to an altitude of more than 20,000 m.

Table 17-1 presents barometric pressure and the oxygen pressure of the inspired air (tracheal air) at various altitudes. With a constant oxygen concentration of 20.94 percent of the dry air, the oxygen pressure of the inspired air in the trachea, saturated with water vapor, can easily be calculated from the formula $P_{O_2} = (P_{Bar} - 47)20.94/100$. This means that at an altitude of about

TABLE 17-1

Barometric pressure (standard atmosphere) at various altitudes and the pressure of oxygen after the inspired gas has been saturated with water vapor at 37°C (tracheal air)*

Altitude		Pressure, mm Hg	P_{O_2} tracheal air, mm Hg	Altitude		Pressure, mm Hg	P_{O_2} tracheal air, mm Hg
m	ft			m	ft		
0	0	760	149	5,500	18,050	379	69
500	1,640	716	140	6,000	19,690	354	64
1,000	3,280	674	131	6,500	21,330	330	59
1,500	4,920	634	123	7,000	22,970	308	55
2,000	6,560	596	115	7,500	24,610	287	50
2,500	8,200	560	107	8,000	26,250	267	46
3,000	9,840	526	100	8,500	27,890	248	42
3,500	11,840	493	93	9,000	29,530	230	38
4,000	13,120	462	87	9,500	31,170	214	35
4,500	14,650	433	81	10,000	32,800	198	32
5,000	16,400	405	75	19,215	63,000	47	0

* Based on dry conditions for average temperature at altitude when the temperature at sea level is 15°C and the barometric pressure is 760 mm Hg.

19,000 m where the barometric pressure is 47 mm Hg there should be nothing but water molecules in the trachea!

The oxygen tension of the alveolar air, and thereby also the oxygen tension of the arterial blood, is determined by the magnitude of the pulmonary ventilation in addition to the composition and pressure of the inspired air. The more frequently the air is exchanged, the more closely the composition of the pulmonary air resembles the inspired air (water vapor subtracted). This will be discussed below.

The reduced density of the air at high altitudes affects the mechanics of breathing. Part of the work of breathing is expended in moving the air against the resistance of the airways. The resistance is relatively high when the flow is turbulent, as is the case during exercise. Therefore, the influence of reduced density is more noticeable at high rates of air flow as in hyperpnea, during heavy exercise or in flow-dependent pulmonary function tests. The maximal breathing capacity is considerably higher at high altitude than at sea level (Cotes, 1954; Miles, 1957; Ulvedal et al., 1963). The respiratory muscles' ability to exert pressure seems to be reduced at low barometric pressure; however, the net effect of the reduced resistance to air flow at lowered barometric pressure is a *diminished respiratory work* to move a given volume of air in and out of the lungs (Fenn, 1954). On several occasions pulmonary ventilations of 200 liters/min have been measured during maximal exercise at high altitude (see below). Another effect of the reduced density of the air at a low barometric pressure is a diminished air resistance. The air resistance changes with the wind speed raised to the second power. The external work is therefore reduced at high altitude in sprint-type activities, speed skating, cycling, and alpine skiing with high velocities. The *air temperature* is on the whole lower, the higher the altitude. With a mean annual temperature of 15°C at sea level, the air temperature decreases linearly by 6.5°C/1,000 m to about 11,000 m. The air also becomes increasingly *dry* with increasing altitude. Therefore the water loss via the respiratory tract is higher at high altitudes than at sea level. If much work is performed this may give rise to a hypohydration at high altitude and a sensation of soreness and dryness in the throat.

The *solar radiation* is more intense at high altitude and the ultraviolet radiation may cause difficulties in the form of sunburn or snow blindness. Finally, *the force of gravity* is reduced with the distance from the earth's center. This may have some favorable effect in the case of athletic events involving jumping or throwing at high altitudes.

Physical Performance

The reduced work capacity in many physical activities performed at high altitude is well established. It is already evident at an altitude of about 1,200 m in the case of heavy exercise engaging large muscle groups for about 2 min or longer. Henderson (1938) remarked that men had come near to the summit of Mount

Everest in distance, but they had been far from it in terms of time. As an example of the physical strain experienced during the conquest of high mountain peaks Somervell (1925) may be quoted: "It may be of interest to record one or two personal observations which I made while climbing in the neighborhood of 27,000 to 28,000 feet. *Pulse:* The heart rate during the actual motion upwards was found to be beating 160–180 per minute, sometimes even more, regular in rhythm and of good volume . . . . *Respiration:* About 50 to 55 per minute while climbing. Approaching 28,000 feet I found that for every single step forward and upward, seven to ten complete respirations were required. Breathing quickly and deeply is very easy at a great height owing to the low density of the air." Norton (1925) writes that at an altitude of 8,500 m he required 1 hr to climb a distance of 35 m even though the climb was not particularly difficult.

Athletic competitions at different altitudes provide an experimental condition where highly motivated, well-trained athletes subject themselves to all-out tests. Leary and Wyndham (1966) report observations from South Africa where important track meetings take place at altitudes of 1,500 m or above as well as at sea level. They conclude that the best performances recorded in middle- and long-distance events were on the coast, whereas sprinters recorded better times at medium altitude. South African championships held at medium altitude have been consistently dominated by athletes domiciled at such altitudes and those who acclimatized themselves by training for medium altitude for 3 to 4 weeks before the championship competitions. In competitions in the mile event in Johannesburg (1,760 m or 5,780 feet) five International-class athletes all performed better at sea level (average time 4 min and 0.5 sec in the best coastal performance and 4 min and 15.2 sec in Johannesburg).

In competitions in Mexico City (altitude 2,300 m) the same or better performance in running distances up to 400 m has been reported. In 800-m running there is an impairment of about 3 percent, and in 5,000- and 10,000-m an impairment of roughly 10 percent compared to sea level. In swimming one considers an impairment of the time for 100 m of about 2 to 3 percent and an impairment of about 10 percent in the case of 400-m or longer distances when comparing Mexico City altitude with that of sea level (Goddard and Favour, 1967). In jumping and throwing no marked difference has been recorded when competing at different altitudes. It has frequently been stated that recovery rates in Mexico City are considerably longer than at low altitudes. The effect of altitude on running performance in races from 100 m to 42 km is summarized in Fig. 17-1.

This brief summary shows that in types of work requiring an intense activity of no more than one minute's duration, and especially events of this nature where the technique is of primary importance, there is no noticeable difference in performance between sea level and high altitude, at least up to an altitude of 2,500 m. The capacity to perform heavy work for 2 min or more is, on the other hand, definitely reduced at high altitude, with the exception of types of activities in which the air resistance plays a major role. [For further details

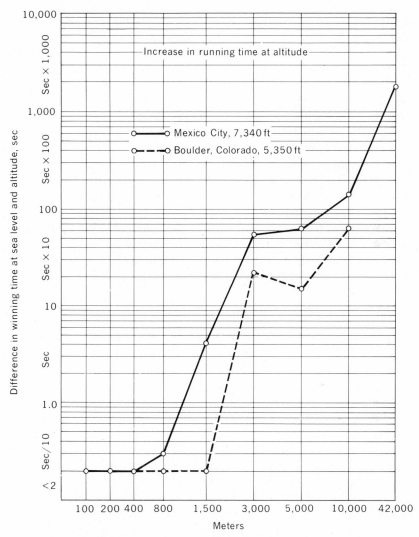

Fig. 17-1 The effect of altitude on running performance in races from 100 m to 42 km, shown on logarithmic paper. The impairment due to lowered oxygen pressure becomes statistically significant at 5,350 ft only for distances of 1,500 m and longer; and at 7,340 ft for distances of 800 m and longer. The two ascending lines indicate differences of running times between sea level, Boulder, and Mexico City. (From Jokl et al., 1966.)

see data presented in *Schw. Zschr. Sportmed.*, **14:** 1966, and Goddard (1967),
mentioned at the beginning of this chapter.] What, then, is the physiological
explanation for this phenomenon?

Limiting Factors

We may base our analysis on the factors listed in the table of factors affecting
physical work capacity at the beginning of Chap. 9. If one disregards altitudes
above 3,000 m, which may cause a disturbance of psychological functions, it is
evident that it is the aerobic power which is affected by a reduced oxygen pressure in the inspiratory air. Studies have shown that work during acute exposure
to high altitude causes a rise in the blood lactate concentration at lighter work
loads than is the case at sea level. At a fixed work load the lactate level is higher,
but the maximal concentration attained during exhaustive work is roughly the
same as at sea level (Edwards, 1936; Asmussen et al., 1948; P.-O. Åstrand, 1954;
Stenberg et al., 1966; Buskirk et al., 1967; Hermansen and Saltin, 1967). Figure
17-2 gives an example of how the blood lactate concentration is affected during
work at sea level, 2,300 m, and 4,000 m simulated altitude. In this subject, how-

*Fig. 17-2   Blood lactate concentrations in one subject during exercise at sea level (filled
dots) and acute exposure to 2,300 (crosses) and 4,000 m (triangles) (760, 580, and 462 mm
Hg respectively). Work times for all submaximal work loads were 10 min, and all
blood samples were taken between the ninth and tenth min. Work times for the maximal
work loads were 4 to 5 min, and peak values for lactic acid are given.*
*On the left panel the absolute oxygen uptake is on the abscissa; to the right, oxygen uptake
in percent of the maximum. (From Hermansen and Saltin, 1967.)*

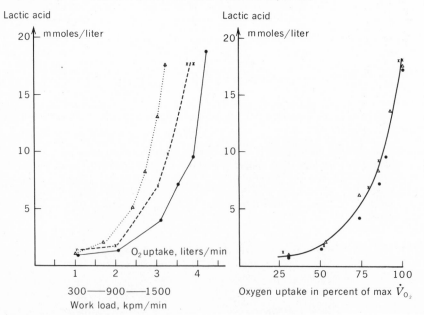

ever, the lactate concentration was the same if it was related to the relative instead of the absolute oxygen uptake. This must be interpreted as indicating that the anaerobic processes are brought into play at a relatively lower work load at high altitude. The maximal anaerobic power, in any case the part of it which is determined by the glycogenolysis, is on the other hand not affected. The fact that the maximal oxygen debt is the same following maximal work at sea level and at altitude also points in the same direction (acute exposure or limited acclimatization) (Saltin, 1966; Buskirk et al., 1967).

As an example, Saltin (1966) mentions that the sprinter Bodo Tümmler ran 1,500 m in Stockholm in 3.42 min and in Mexico City in 3.54 min. The O_2 uptake during the subsequent 60-min rest was 38 and 42 liters respectively; peak blood lactate concentration was 18.6 and 19.3 mmoles/liter respectively. After a 3,000-m steeplechase, the values obtained for Bengt Persson in Stockholm were 28 liters O_2, blood lactates 18.2 mmoles/liter, running time 8.34 min. Corresponding values in Mexico City were 33 liters O_2, 20.3 mmoles/liter, and 9.32 min.

With regard to the neuromuscular function, Christensen and Nielsen (1936) have shown that speed and strength are not affected by moderate hypoxia. In a test on a Hill's wheel, contraction time being less than 6 sec, their subjects attained the same maximum at sea level as at a barometric pressure of 440 and 390 mm Hg respectively.

The oxygen uptake at a given submaximal task, i.e., on a bicycle ergometer, is the same at different altitudes (Christensen, 1937; Asmussen and Chiodi, 1941; P.-O. Åstrand, 1954; Pugh et al., 1964).

It is difficult to explain the higher lactic acid level in the blood at a standard submaximal work load at high altitude in view of the fact that the oxygen uptake is the same as at sea level. We are not aware of any systematic studies showing that the increase in oxygen uptake is slower at the beginning of a given work load at high altitude; a slower rise in oxygen uptake up to a given steady-state level would mean a greater participation of anaerobic processes.

From a psychological point of view a stay at even moderate altitudes may represent a considerable stress, or in any case an unusual sensation. From experience one knows how a certain work intensity feels or affects the body under normal circumstances. The same physical effort at higher altitudes produces a higher pulmonary ventilation, a higher heart rate, and possibly other symptoms of fatigue which under normal conditions would not be customary at this work load. Gradually an adaptation takes place to the new situation. For the athlete the tactics may be different in training and competitions at high altitude compared with the situation at sea level. This will be further discussed at the end of this chapter.

Summary It may be concluded that it is the aerobic power which is directly affected during work under conditions of reduced oxygen pressure in the inspired air. We shall therefore briefly summarize how the different steps in the oxygen transport from the air to the mitochondria of the exercising muscle cell are affected during exposure to high altitude.

Oxygen Transport

Figure 17-3 indicates the pressure levels at different distances along the transport chain from atmospheric air to the ultimate destination, the mitochondria. In the given example the pressure gradient for oxygen between air and mixed venous blood at rest at sea level is about 110 mm Hg (150 − 40 mm Hg). When one goes to an altitude of 5,400 m, $P_{O_2} = 70$ mm Hg, which is about the highest altitude to which man can become acclimatized and at which he can live and work for months and years (Rahn, 1966), the pressure drop for the same oxygen delivered is reduced to about 50 mm Hg. A reduction of the oxygen pressure in the inspired air by 80 mm Hg is associated with a decrement of only 20 mm Hg in the mixed venous blood. Close to the mitochondria the oxygen pressure may be 10 mm Hg in the sea-level situation and about 5 mm Hg at 5,400 m. This pressure is still adequate to provide optimal conditions for the oxidative enzyme

Fig. 17-3   A comparison of the oxygen cascade from inspired air to tissue in man at sea level and at 5,500 m (18,000 ft). (Modified from Rahn, 1966.)

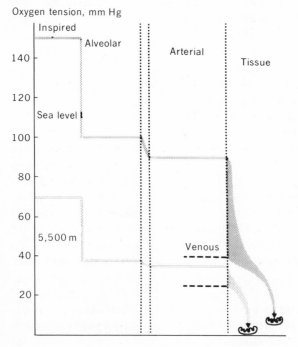

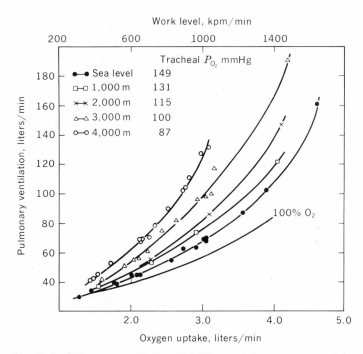

Fig. 17-4   Pulmonary ventilation (BTPS) in relation to oxygen uptake at different work loads when breathing oxygen or air during exposure to various simulated altitudes. Note the high pulmonary ventilation of 190 liters/min reached during work at 3,000-m altitude. (From P.-O. Åstrand, 1954.)

reactions (Chap. 2). There are many factors explaining the increased "conductance" to keep the tissue oxygen pressure at this almost constant level. We shall first consider the *acute hypoxia*.

1   The *pulmonary ventilation* at a given oxygen uptake is markedly elevated, see Fig. 17-4. In this subject the ventilation at an oxygen uptake of 4.0 liters/min was 80 liters/min when pure oxygen was inhaled, 105 liters/min at sea level when the subject was breathing air, 140 liters/min at 2,000 m, and 160 liters/min at 3,000 m, i.e., twice as high as when breathing oxygen. Even at sea-level conditions there exists a hypoxic drive which is evident when this subject's oxygen uptake exceeds 1.5 liters/min. This hypoxic hyperpnea is elicited through the reflex pathway originating in the chemoreceptors of the carotid body and aortic body (see Chap. 7). Since the production of carbon dioxide is roughly the same at a given oxygen uptake, this hyperventilation will inevitably wash out CO_2 from the blood into the inspired air, the dissolved CO_2

of the blood being more affected than the bicarbonate. The secondary effect of the hyperpnea will therefore be a rise in the pH of the blood, i.e., an uncompensated respiratory alkalosis. The reduced P_{CO_2} and elevated pH of the arterial blood exert an inhibitory influence on the respiratory center. On the other hand, the earlier accumulation of lactic acid in the blood must be considered to cause the pH to fall.

At sea level the alveolar P_{CO_2} was about 38 mm Hg in the experiment, bringing the pulmonary ventilation to 80 liters/min (Fig. 17-4). The reduction in ventilation when breathing oxygen brought the alveolar P_{CO_2} to a somewhat higher level; at a simulated altitude of 4,000 m the alveolar P_{CO_2} became reduced to 28 mm Hg during exercise with a similar pulmonary ventilation of 80 liters/min (cf. Dejours et al., 1963). The effect is that the hypoxic drive during work at high altitude must be stronger than reflected in the magnitude of the pulmonary ventilation. If the alveolar and arterial P_{CO_2} is maintained at about 40 mm Hg by the addition of a proper volume of CO_2 to the inspired air, the pulmonary ventilation during exercise in hypoxic conditions will be still higher than illustrated in Fig. 17-4. The ventilatory response must be viewed as a physiological compromise with the call for an adequate oxygen supply matched against the need to maintain the acid-base balance as normal as possible.

Due to the great increase in ventilation during work at high altitude and a given oxygen uptake, the alveolar P_{O_2} is higher than what it otherwise would be. This obviously facilitates the diffusion of oxygen to the blood in the pulmonary capillaries.

The maximal pulmonary ventilation during work at high altitude is the same or higher than at sea level (Stenberg et al., 1966; Saltin, 1966; Grover and Reeves, 1967; Roskamm et al., 1968).

2   The diffusing capacity is reported to be unchanged after arrival at higher altitudes in newcomers who have lived at sea level (West, 1962). However, the alveolar-arterial P_{O_2} gradient becomes less at altitude (Fig. 17-3).

   Normally, a certain amount of mixing of venous blood with the arterial blood takes place in the lungs. The effect of this venous admixture in producing an alveolar-arterial O_2 difference becomes less as a result of the shift in the alveolar gases down to the steep part of the oxygen dissociation curve (Rahn, 1966). (It is evident from Fig. 6-18 that a certain change in the oxygen saturation is brought about with a smaller difference in the O_2 pressure around $P_{O_2} = 50$ mm Hg compared to $P_{O_2} = 100$ mm Hg.)

3   During submaximal work with reduced O_2 pressure in the inspired air, the lower O_2 saturation is compensated with an increased cardiac output (Asmussen and Nielsen, 1955; Stenberg, 1966). This and other

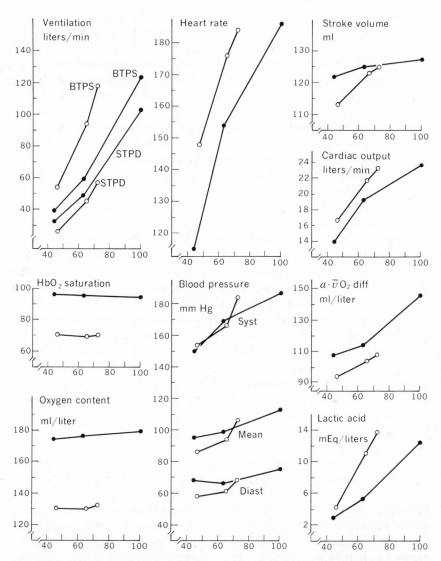

Fig. 17-5   Mean values on six subjects studied at two submaximal work loads and one maximal work load (bicycle ergometer) at sea level (filled dots) and when acutely exposed to simulated altitude of 4,000 m in a decompression chamber (open dots). Note that the maximal oxygen uptake was reduced to 72 percent of the value at sea level. Abscissa = oxygen uptake in percent of the maximum attained at sea level. (From Stenberg et al., 1966.)

effects of high altitude on the oxygen transport in man are illustrated by Fig. 17-5. The increase in the cardiac output is brought about by an increase in the heart rate; the stroke volume may even be reduced. The arterial blood pressure is largely unchanged. A vasodilation as the result of hypoxia causes a reduced peripheral resistance.

The previously mentioned hypocapnia during acute lack of oxygen produces a shift in the Hb dissociation curve (Fig. 6-18a), which means a net advantage for oxygen transport due to a higher arterial saturation. The arterial oxygen content is definitely reduced at high altitudes, however, and the arteriovenous O_2 difference drops.

It is rather interesting to note that the observed maximal values for heart rate, cardiac output, and stroke volume are the same at an altitude of 4,000 m (acute exposure) as at sea level. Apparently the lack of oxygen is not of such a magnitude that the pumping capacity of the heart muscle is reduced, in spite of the fact that the Pa_{O_2} is estimated to be lower than 50 mm Hg. In a comparable study Blomqvist and Stenberg (1965) showed that there were no ECG signs of myocardial ischemia when their subjects were performing maximal work at the same simulated altitude of 4,000 m. The conclusion of this study by Stenberg et al. was that the reduced maximal oxygen uptake in moderate acute hypoxia compared with normoxia was closely related to the reduction of the arterial oxygen content. During maximal work at hypoxia the oxygen uptake was on an average 72 percent, the arterial oxygen content was 74 percent, and the cardiac output was 100 percent of the values attained at sea level. In other words, the maximal oxygen uptake was highly correlated with the volume of oxygen offered to the tissue (arterial oxygen content $\times$ maximal cardiac output).

During maximal exercise at sea level almost all the oxygen is extracted from the blood passing the working muscles, so that there is nothing more to gain in this respect at acute exposure to high altitude.

Summary Acute exposure to a reduced oxygen pressure in the inspired air during exercise is associated with hyperpnea in excess of that at sea-level conditions for the same energy requirement. The cardiac output also rises out of proportion to the oxygen uptake. These factors, combined with the displacement of the physiological range of the O_2 dissociation curve to its steep part, enhance the oxygen transport. These compensatory responses cannot, however, fully compensate for the reduced oxygen pressure. The maximal oxygen uptake is reduced, and the importance of the anaerobic energy yield increases.

The quantitative effect on the oxygen transport during maximal work at different altitudes is illustrated in Fig. 17-6. At the altitude of Mexico City, 2,300 m, the reduction is in the order of 15 percent, at 4,000 m it is about 30 percent (from 4.24 to 3.07 liters/min in the study by Stenberg et al., 1966). The considerable scatter of the data observed when different studies are compared may be

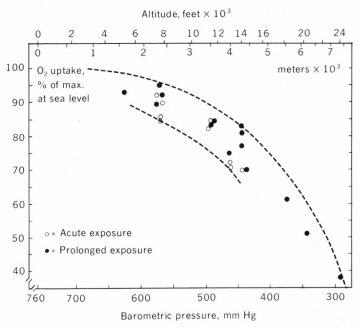

Fig. 17-6 Reduction in maximal oxygen uptake in relation to altitude barometric pressure. Unfilled dots denote experiment in the acute hypoxic stage; filled dots, data obtained after various periods of acclimatization. In principle the maximal aerobic power during acute exposure to reduced oxygen pressure falls at the lower part, within the dotted lines; during acclimatization it is shifted toward the upper part of the field. (Data from Balke, 1960; Pugh, 1964; Stenberg et al., 1966; Buskirk et al., 1967; Hansen et al., 1967b; Saltin, 1967; Roskamm et al., 1968.)

explained by the fact that (1) the effect of the reduced oxygen pressure on the physical work capacity is different in different individuals; (2) different techniques have been applied, especially concerning criteria for ascertaining that the maximal oxygen uptake has been reached; (3) conceivably persons with a high maximal aerobic power are more affected than persons with a lower maximal aerobic power in that the diffusing capacity may be more critical for the former. The progressive fall in arterial oxygen saturation as the work level is raised at high altitude in spite of an increasing alveolar tension, and the resulting large alveolar-arterial oxygen differences can be explained by diffusion limitations of the lung (West et al., 1962; Saltin, 1967; Grover and Reeves, 1967).

We shall now discuss the effect of a prolonged stay at high altitude, i.e., *acclimatization* to reduced oxygen pressure in the inspired air. It is customary to distinguish between short-term adaptation when it is a matter of days, weeks, or a few months at high altitude, and long-term adaptation when it is a matter of spending years at high altitude.

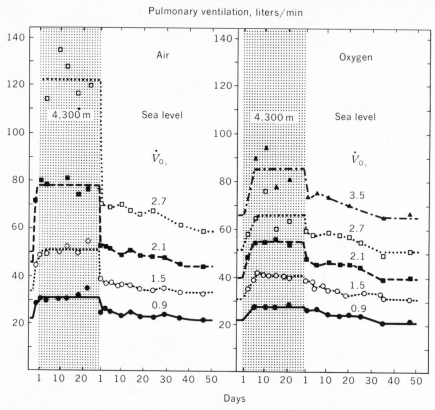

Pulmonary ventilation, liters/min

Fig. 17-7 Pulmonary ventilation in one subject working on a bicycle ergometer (1) at sea level (base line to the left), (2) during a 4-week sojourn at an altitude of 4,300 m (shaded area), and (3) again after return to sea level including almost 50 days' observation time. Oxygen uptake is indicated on respective line. Left panel: Subject breathing room air; data from experiments in a low-pressure chamber simulating the high altitude are included between base line and altitude studies. Right panel: Subject breathing oxygen during exercise. (Unpublished data obtained by I. Åstrand and P.-O. Åstrand at White Mountain Research Station, California, 1957.)

1   The first few days' exposure to reduced oxygen pressure entails a further increase in the pulmonary ventilation at a given work load. This hyperpnea will further raise the P_{O_2} and reduce the P_{CO_2} of the alveolar air. This ventilatory response is illustrated in Fig. 17-7. It shows data obtained at sea level, during acclimatization for 4 weeks to an altitude of 4,300 m (14,250 ft), and again at sea level during the reacclimatization. Four subjects were studied, but the figure only presents the data on one of the subjects. However, he represents reasonably well the normal reaction. Three comments should be made: (*a*) It is evident that within

a week at high altitude a new level for pulmonary ventilation is attained, exceeding the value noticed in acute exposure to the same degree of hypoxia. The prolonged exposure to this hypoxia caused a 40 to 100 percent increase in the pulmonary ventilation compared with the sea level controls, the increase being more pronounced the heavier the work loads. At the end of the stay the ventilation at an oxygen intake of 2.7 liters/min (200 watts) was about 120 liters/min when breathing air, compared with 60 liters/min at sea level. The alveolar P_{CO_2} was 24 mm Hg at 4,300 m, compared with 40 mm Hg at sea level. (*b*) Even when oxygen is inhaled during exercise at high altitude, blocking the peripheral chemoreceptor drive, there is a gradual increase in pulmonary ventilation (see Fig. 17-7). In the example chosen the ventilation in the control experiment at sea level was 50 liters/min, but at the end of the 4-week sojourn at high altitude it was raised to 65 liters/min during the standard exercise. If oxygen is used during work at high altitude it is certainly oxygen-saving if the individual is unacclimatized. Since on the other hand, oxygen can hardly be supplied continuously for long periods of time, an acclimatization is desirable, especially at very high altitudes. Furthermore, the oxygen-providing equipment may fail. (*c*) When the subjects returned to sea level following exposure to altitude, it took several weeks before the control level was attained (see Buskirk et al., 1967). The return of the alveolar P_{CO_2} to control levels paralleled the shift in pulmonary ventilation.

A calculation of the pulmonary ventilation at STPD gives practically the same volumes at a given oxygen uptake in subjects acclimatized to various altitudes for about 4 weeks (Christensen, 1937; Dejours et al., 1963; Pugh et al., 1964). In other words the number of oxygen molecules which are inhaled per unit of time is constant at the various altitudes. It should be remembered, however, that the alveolar oxygen pressure will still not be the same at high altitude as at sea level.

At high altitude, the energy cost for the respiratory muscles to move the greater volume of air may not be higher than at sea level because of the reduced density of the air. When sea level is again reached, the "abnormally" high ventilatory response at a given oxygen uptake must demand extra respiratory work.

It has been suggested that the extra increase in ventilation in connection with work at high altitude must be attributed to a hypoxic drive via the peripheral chemoreceptors. This hypoxic drive is prevalent even during chronic hypoxia, at least for a considerable period of time (P.-O. Åstrand, 1954; Dejours et al., 1963). The induced alkalosis on sudden exposure to low P_{O_2}, reducing the central chemoreceptor drive, becomes gradually compensated by a proportionate decrease in the blood bicarbonate, and restoration of a normal pH occurs in the acclimatized man.

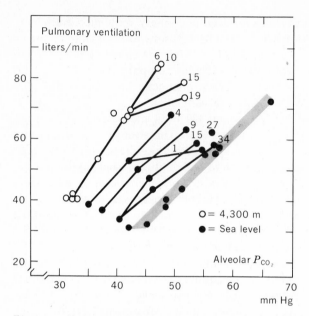

Fig. 17-8   Ventilatory response during a standard work load (100 watts) to inhalation of various CO_2-O_2 mixtures at sea level (shaded area), during a prolonged sojourn at altitude of 4,300 m (unfilled dots), and at various intervals during the reacclimatization to sea-level conditions (filled dots). Figures at top of lines denote the day for the experiment after arrival at altitude and sea level respectively. (Same subject as in Fig. 17-7.) (Unpublished data by I. Åstrand and P.-O. Åstrand.)

(Prolonged alteration in arterial P_{CO_2} in either direction tends to bring about alterations in the renal acid-base excretion that slowly tend to return the arterial H⁺ toward normalcy.) As the alkalosis is reduced (pH is lowered), there is a further increase in pulmonary ventilation, as illustrated in Fig. 17-7.

   Severinghaus et al. (1963, 1964) point out that the renal base excretion of bicarbonate is a rather slow process and that the changes of pH and HCO_3^- ions in the cerebrospinal fluid (CSF) are better correlated to the ventilatory increase during the first days of acclimatization. Thus, medullary respiratory chemoreceptor drive, initially reduced at altitude by hyperventilation alkalosis, is restored to normal during acclimatization by reduction in CSF HCO_3^-, the incremental ventilatory drive being supplied by peripheral chemoreceptors.

   Evidence of a real change in the regulation of the body P_{CO_2} developing during the first week after ascent from sea level to high altitudes is presented in Fig. 17-8. The respiratory response to CO_2 was tested by breathing CO_2-O_2

mixtures during work with 100 watts (oxygen uptake 1.5 liters/min). There was a marked shift to the left so that after acclimatization a given pulmonary ventilation was attained at 15 to 20 mm Hg lower alveolar P_{CO_2} compared with controls at sea level. The difference in the response to CO_2 can be illustrated by the following example: At sea level a pulmonary ventilation of 35 liters/min was obtained when the end-expiratory P_{CO_2} was 45 mm Hg; after 1 week at 4,300 m the same P_{CO_2} was recorded when ventilation was as high as 83 liters/min. (In this subject the high CO_2 mixture had an almost narcotic effect, with a reduced ventilatory response as a consequence; see days 15 and 19 at high altitude, and day 1 at sea level.)

After the return to sea level, there was a gradual return of the CO_2 response curve to the control level. This slow reacclimatization shows a pattern parallel with the one illustrated in Fig. 17-7. When the hypoxic drive was reduced by the increased oxygen pressure in the inspired air, the pulmonary ventilation was reduced. This produces a reduced wash-out of CO_2 and an uncompensated acidosis with increased central chemoreceptor drive as a consequence. (For a more detailed discussion of the regulation of respiration during hypoxia, see Kellogg, 1964.)

There is another effect of the *reduced alkaline reserve* in the blood of the man acclimatized to high altitude. He will have less ability to withstand acidosis from other acids which may arise in the course of metabolism (Roughton, 1964). During acute exposure of a few weeks' duration to high altitude he can apparently attain the same high blood lactate level as at sea level, but data by Edwards (1936) suggest a gradual decline in this maximum (Hansen et al., 1967a). The reduced alkaline reserve may be one factor behind this decrease in maximal anaerobic power.

Cerretelli (1967) reports that an increase in the blood lactate by 8.0 mmoles/liter normally should result in a lowering of the pH to 7.25. In the case of a person acclimatized to an altitude of 5,000 m, the same amount of lactate should lower the blood pH to below 7.0. Roughton (1964) points out that the protein buffer power towards H^+ ions arising from hydration of CO_2 to H_2CO_3 is increased in the acclimatized man, and accordingly, the change in pH during the respiratory cycle is about one-fourth less than at sea level.

Natives at high altitude apparently have a somewhat lower pulmonary ventilation than the short-term acclimatized individuals, although it is higher than the values typical at sea level (Chiodi, 1957; Mazess, 1966). The reason may be a reduced peripheral chemoreceptor drive; in other words, there may be a gradual adaptation of these receptors (Severinghaus et al., 1966; Milledge and Lahiri, 1967).

2   With regard to the transport of O_2 from the alveolar air to the pulmonary capillary blood, the situation is somewhat controversial. West (1962)

found no change in the diffusing capacity after a 6-month exposure to 5,800-m altitude. Natives and long-term residents at high altitude may have a greater diffusing capacity than comparable sea-level residents (Velasquez, 1959; DeGraff et al., 1965). Kreuzer et al. (1964) found that the values of the alveolar-arterial P_{O_2} gradient in the Andean natives was higher than in normal sea-level residents, which is in contrast to Hurtado's (1964) observation of a particularly small gradient in his native subjects in the same area.

3   Pugh et al. (1964; Pugh, 1964) have done hemodynamic studies including maximal work at very high altitudes. They report that a prolonged sojourn at various altitudes brought the *cardiac output* at a given work load down to the level typical for the same work load performed at sea level. However, the maximal cardiac output was markedly reduced, and after several months' stay at 5,800 m, the values were 16 to 17 liters/min compared with 22 to 25 liters/min at sea level. This reduction of cardiac output was a combined effect of a lowered stroke volume and maximal heart rate (reduced from 192 down to 135 beats/min). This study confirms the data by Christensen and Forbes (1937).

A number of measurements of cardiac output during work at altitudes between 3,000 and 4,300 m have been made with exposure up to a few weeks (Klausen, 1966; Alexander et al., 1967; Hartley et al., 1967; Vogel et al., 1967; Saltin et al., 1968). The results indicate that already after a few days the minute volume of the heart during submaximal work is reduced compared with the cardiac output during acute exposure to the hypoxic condition, and it returns gradually to values typical for sea-level conditions, or it may even become subnormal. During maximal work the cardiac output is reduced. A reduced stroke volume appears to be the primary reason for the reduced cardiac output; the lowering of the heart rate, in any case during maximal work, is a more inconsistent finding. (Figure 17-5 shows that the stroke volume during light work is reduced at acute exposure to the hypoxia.)

In residents at 3,100-m altitude the cardiac output actually increased somewhat (8 percent) after 10 days at sea level, and the stroke volume increased 15 percent (Hartley et al., 1967). In residents at sea level who stayed for 10 days at 3,100-m altitude the cardiac output decreased during exercise compared with the sea-level data. Reduced cardiac output was chiefly due to a decrease in stroke volume (Alexander et al., 1967). The authors advance the hypothesis that a depressant effect of chronic hypoxia upon the ventricular myocardium could result in reduced myocardial contractile force and stroke volume. Saltin et al. (1968) found that sea-level residents studied after two weeks at 4,300 m had a 20 percent reduction in maximal cardiac output due to a smaller stroke volume, but two of their four subjects had also a significantly lowered maximal heart rate. The ECGs were within normal limits.

An example of the heart rate response to fixed work loads was presented by P.-O. Åstrand and I. Åstrand (1958); see Fig. 17-9. During acute exposure to a tracheal oxygen tension of about 85 mm Hg (4,300 m) the heart rate was 15 to 30 beats higher per minute than at sea level. When the hypoxia was prolonged, there was a gradual decrease in heart rate at a given oxygen uptake. In the later stage of acclimatization the heart rates attained during lower levels of work fell in the same range as those recorded at sea level. At the heavier loads, however, the heart rate was even lower than in experiments with high tracheal P_{O_2}. In this subject the normal maximal heart rate was about 190 beats/min. At the high altitude it gradually declined to 135 (Christensen and Forbes, 1937; Cerretelli and Margaria, 1961; Pugh et al., 1964). When the subject was allowed to breath 100 percent oxygen during almost maximal work, the heart rate increased within seconds by as much as 25 beats/min. Pugh et al. (1964) noticed that a maximal heart rate of 130 to 150 was elevated to almost sea-level values when the subjects were breathing oxygen. [However, Hartley et al. (1967) report that the increase in cardiac output and stroke volume induced in their high-altitude residents (3,100 m) by chronic relief of the hypoxia could not be achieved by acute administration of oxygen at high altitude. The altitude may

Fig. 17-9 Pulmonary ventilation and heart rate during exercise on a bicycle ergometer, work load 1200 kpm/min at sea level, and during a 22-day sojourn at an altitude of 4,300 m. On some days room air was inhaled during work; on other days, pure oxygen. (Same subject as in Figs. 17-7 and 17-8.) (Modified from P.-O. Åstrand and I. Åstrand, 1958.)

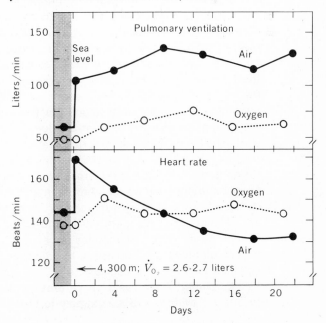

be critical for these reactions.] As already mentioned, the physiological basis
for the reported variations in cardiac output, stroke volume, and heart rate
is still unknown.

Within the first few days at high altitude the *hemoglobin concentration*
in the blood increases, but this increase is mainly due to a hemoconcentration secondary to a decrease in the plasma volume (Merino, 1950;
Surks et al., 1966; Buskirk et al., 1967). Gradually the increased erythro-
cytopoiesis brings hemoglobin content to high levels so that the oxygen
content per liter of arterial blood may be the same in the acclimatized
man at 4,500 m as it is in man at sea level (Christensen and Forbes,
1937; Hurtado et al., 1945; Chiodi, 1957; Reynafarje, 1967). At an altitude
of 4,500 m at Morococha in Peru the native residents had a hemoglobin
concentration which averaged 20.8 g/100 ml blood (Hurtado et al., 1945).

As a consequence of an increased hemoconcentration during acclimatization to high altitude, the oxygen offered to the tissues per liter
of arterial blood may be the same in the individual living at high altitude
as in the resident at sea level. However, these values of oxygen are
really not comparable. The oxygen pressure gradient between blood
and tissue is most important for the final transfer of oxygen to the
mitochondria, and the gradient is reduced during hypoxia. However,
the gradual decline in cardiac output during prolonged exposure to a
low oxygen pressure in the ambient air can partly be explained by the
concomitant rise in the oxygen-combining capacity of the blood. The
increased viscosity of the blood with the elevated hematocrit must
necessitate an increased cardiac work at a given cardiac output, but
the net effect of the hematologic response to prolonged hypoxia in
terms of work of the heart cannot be evaluated at present. In any case,
the increase in hemoglobin concentration and the mentioned shift in
the operational range to the steeper slope of the oxygen-dissociation
curve provide major contributions to the gradually increased oxygen
conductance within the body at altitude.

4   Barbashova (1964) has summarized various aspects of cellular adaptation to high altitude. It is concluded that the oxygen-utilization efficiency
for an aerobic energy yield increases at a low O_2 tension; this increase
must be interpreted as an adaptation at the enzyme level. The ability
to tolerate lack of oxygen increases, furthermore, in connection with
an acclimatization. Vannotti (1946) and Cassin et al. (1966) report that
an increased capillarization takes place after a period of acclimatization
to high altitudes. An increase in the number of capillaries reduces the
distance between the capillary and the most distant cells within its
tissue cylinder. A relatively low O_2 tension in the capillaries may therefore still provide the oxygen supply to these distant cells. Rahn (1966)

emphasizes that an increase in the number of open capillaries plays a most important role not only in daily life at sea level but particularly during acclimatization to high altitude by changing the O_2 conductance.

Reynafarje (1962) reports that the myoglobin content in the skeletal muscles increases during altitude adaptation; this will have a favorable effect on the O_2 transport.

The critical alveolar P_{O_2}, at which an unacclimatized person loses consciousness within a few minutes in acute exposure to hypoxia, is 30 mm Hg, with minimal individual variations (Christensen and Krogh, 1936). This limit is set at an altitude of slightly more than 7,000 m. Down to this low P_{O_2} the demand of the nerve cells for oxygen can apparently be maintained (Noell, 1944). The well-acclimatized individual can spend hours at an altitude above 8,000 m breathing the ambient air. This must be an example of adaptation on the cellular level. The comprehension of the last step in the oxygen transfer from air to tissue, i.e., from the capillary to the mitochondria, is however still very incomplete.

5　There is one more finding in high altitude dwellers that should be mentioned. Residents at altitudes of around 3,500 m or above develop a pulmonary hypertension with increased pulmonary vascular resistance and hypertrophy of the right ventricle of the heart (Penaloza et al., 1963; Hultgren et al., 1965). This phenomenon does not change rapidly when returning to lower altitudes. This type of hypoxia modification of the cardiopulmonary system, as well as other changes in connection with adaptation to hypoxia, is believed to occur primarily in the fetal and infant period and is retained in adult life. The physiological consequences and the significance of these findings are not revealed.

Summary　It may be stated that with prolonged exposure to reduced oxygen pressure in the inspired air the compensatory devices are more slowly acquired, such as (1) a further increase in pulmonary ventilation (in long-term dwellers eventually followed by a small but significant reduction in ventilation); (2) an increased hemoglobin concentration in the blood; and (3) morphological and functional changes in the tissues (increased capillarization, myoglobin content, modified enzyme activity). The initially observed increase in cardiac output during exercise is replaced by a gradual decline to or even below the values observed at sea level. During submaximal as well as maximal work the stroke volume becomes reduced. If the sojourn has been at very high altitude, 4,000 m or higher, the maximal heart rate may become reduced compared with sea-level values. All these adaptive changes are reversible, but it may take several weeks before the values return to sea-level values in the case of sea-level dwellers who have stayed at high altitudes for a month or more.

The net effect of this acclimatization to high altitude is a gradual improvement in the physical performance in endurance events or prolonged work. The

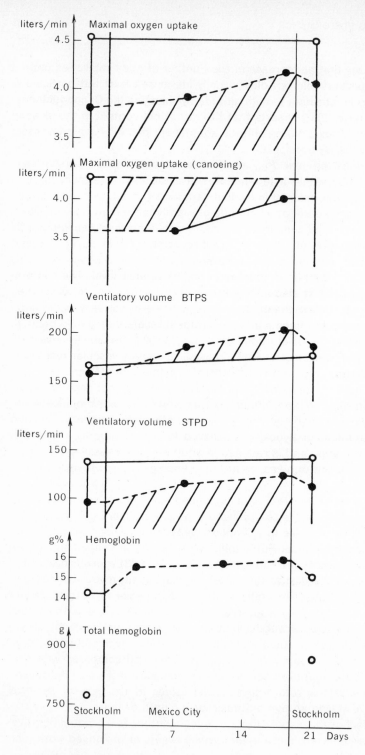

liters/min — Maximal oxygen uptake

4.5

4.0

3.5

liters/min — Maximal oxygen uptake (canoeing)

4.0

3.5

Ventilatory volume BTPS

liters/min

200

150

Ventilatory volume STPD

liters/min

150

100

g% — Hemoglobin

16

15

14

g — Total hemoglobin

900

750

Stockholm   Mexico City   Stockholm

7   14   21   Days

(See legend on opposite page)

maximal anaerobic power after a prolonged period of acclimatization has not been carefully analyzed. At altitudes up to 2,500 m there are objective measurements which show that the maximal aerobic power increases during the first few weeks' stay at the altitude in question (Pugh, 1965; Saltin, 1966; Roskamm et al., 1968).

It is also true that the oxygen content per liter of blood eventually increases, but the maximal cardiac output apparently is reduced at about the same rate. It should be pointed out that the well-trained individual is not acclimatized to high altitude any sooner or any more effectively than the untrained individual.

Figure 17-10 presents one example of data collected at the altitude of Mexico City with one top athlete serving as the subject. His initial decrease in maximal oxygen uptake of 14 percent became less during the stay at Mexico City but it was still 6 percent below the sea-level value after 19 days. The same trend was observed for the oxygen uptake measured during maximal efforts in a canoe. In a group of eight international top athletes the reduction in maximal oxygen uptake at the altitude of 2,300 m was on the average 16 percent (range 9 to 22 percent); after 19 days at this altitude the maximum was still 11 percent below the sea-level average (range 6 to 16 percent) (Saltin, 1967). This example illustrates the individual variations in response to hypoxic conditions.

Performance after Return to Sea Level

The opinions differ concerning the question whether or not the performance capacity at sea level is improved following exposure to high altitude, or whether a certain amount of training at high altitude is more effective than the same amount of training at sea level (Goddard, 1967). The subject presented in Fig. 17-10 was the best adapted athlete in the Swedish test group sent to Mexico City in 1965. When he returned to Stockholm after the 3-week sojourn in Mexico City, his maximal oxygen uptake was no higher than before the trip. Buskirk et al. (1967), as well as Consolazio (1967), state that their subjects who stayed at altitudes up to about 4,000 m for 4 weeks or more did not attain any better results than usual when they returned to sea level. The measured maximal O_2 uptake was not improved. Buskirk et al. conclude that there is little evidence to indicate that performance on return from altitude is better than before going to high altitude, if training remains relatively constant; one should view with caution any statement about the physical superiority of the indigenous resident

Fig. 17-10   Respiratory and cardiovascular reactions to a 19-day stay in Mexico City (2,300-m altitude) for one subject, Gunnar Utterberg, gold medal winner in Tokyo, 1964 (canoeing). Unfilled circles denote determinations at sea level in Stockholm. Filled circles represent the altitude values either in a low-pressure chamber at 580 mm Hg or in Mexico City. The ventilatory volumes (BTPS and STPD) are measured during the maximal work on the bicycle ergometer. (From Saltin, 1966.)

in any environment until he can be compared with individuals of comparable training and experience (and physical endowment).

Grover and Reeves (1967) also came to similar conclusions studying the exercise performance at altitudes of 300 and 3,100 m in men native to those two altitudes. The maximal oxygen uptake was 26 percent less at the higher altitude for both groups; the sea-level athletes won all track competitions at both low and high altitude; their performance after return to low altitude was not improved by their sojourn at medium altitude. (See also Hansen et al., 1967b; Kollias et al., 1968.)

It is true that both training and prolonged exposure to hypoxia produce similar changes with respect to increased vascularization in the skeletal muscles, increased myoglobin content, and possibly similar changes in the oxidative transport system. On the other hand, training at sea level produces no increase in the Hb concentration. Exposure of more than a few days' duration to high altitudes evidently hinders the attainment of a maximal stroke volume, and the increased pulmonary ventilation following return to sea level represents no advantage. The same is true for the reduced alkaline reserve in the blood, with reduced buffer capacity for lactic acid as a consequence. The training intensity has to be reduced at high altitude. For these reasons the improvements attained at high altitude cannot directly be transferred to sea-level conditions.

Practical Applications

In order to attain top achievement at altitudes of 2,000 m or higher in activities requiring the engagement of the maximal aerobic power, an acclimatization period of no less than 3 weeks is necessary. At lower altitudes the time required is probably less. A longer exposure to high altitude would probably be beneficial from a physiological point of view, but this must be considered against possible psychological, social, and economic factors. After the initial acclimatization the improvement per week is so small that it may easily be concealed by day to day variations in the physical fitness.

Athletes competing in events when the technique is of prime importance or events primarily involving anaerobic metabolic processes may more or less arrive at the time of the competition if the altitude is not so high that mountain sickness may be expected.

A theoretical consideration shows that the performance capacity in certain events ought to be better in competitions at high altitude (e.g., 100- to 400-m, bicycling). Dickinson et al. (1966) state on the basis of ballistic calculations that at an altitude corresponding to Mexico City one might expect an improvement of 6 cm in the case of shot putting, 53 cm in throwing the hammer, 69 cm in javelin throwing, and 162 cm in discus throwing. In view of the fact that throwers are used to mastering different wind conditions, the reduced density of the air at high altitude certainly does not require any special period of preparation.

There is no evidence to suggest that it is necessary to take it easy during the initial period of exposure to high altitude. It is necessary, however, to become accustomed to the fact that the subjective feeling of fatigue is different, which must be reflected in the choice of tactics. Experience has shown (especially in the case of cross-country skiers) that if the effort is too intense, a considerably longer recovery time will be required than is the case at sea level. (Perhaps the effect of accumulated lactic acid?)

One is forced to accept a slower tempo, and the intensity and duration of training activities must be reduced. The swimmer discovers that he can remain under water for a shorter period of time after turning than is normally the case, and he must adapt his swimming strokes to a different breathing rhythm. The ability to tolerate an intense tempo for long periods of time at high altitudes is different from one individual to the next. This complicates the selection of the athletes for a team. There are examples of outstanding athletes in long-distance events at sea level who consistently fail at high altitude. In competitions at high altitude, collapse from unknown causes occurs more frequently than is the case at sea level. This, however, does not appear to represent an increased health hazard. Ingestion of additional fluid and possibly also carbohydrates may be required during exposure to high altitude when it is combined with heavy physical work. There is no evidence to suggest that the preparation for competitions which place heavy demands on the aerobic power should include training at an altitude which is higher than that of the actual place of competition. Pugh (1964) states that after suitable acclimatization man can by his own efforts and without supplementary oxygen ascend to about 8,600 m (28,200 feet). It is, in a way, a pity that Mount Everest has an altitude of 8,848 m; it is true that the peak has been reached by men breathing oxygen, but from a sporting point of view Mount Everest is still unconquered!

HIGH GAS PRESSURES

While man can become acclimatized to low air pressures, there is no way to become biologically acclimatized to high air pressures, such as are encountered in deep sea diving and during escape from submarines when the survivor attempts to get out from the inside of the craft, where the pressure is normal, to the surface through the sea where the air pressure is higher.

Pressure Effects

For every 10 m (33 feet) of sea water the diver descends, an additional pressure of 1 atm is acting upon his body. Small changes in sea depth thus bring about great pressure changes. The effect of changing pressures on the blood P_{O_2} and P_{CO_2} and the consequence of this for underwater swimming and breath holding were discussed in Chap. 7. The body may tolerate high pressures as long as the pressure is the same inside and outside the body. When diving with a snorkel connected to the mouth, one maintains the atmospheric pressure in the lungs, while the surface of the thorax, in addition, is exposed to the pressure of the water. At a depth of about 1 m the pressure difference becomes so large that the inspiratory muscles no longer have the strength to overcome the external pressure, and normal breathing becomes impossible. For this reason a snorkel system does not permit diving to depths exceeding about 1 m. At greater depths breathing apparatus has to be used in which the pressure in the system corresponds to that prevailing at the depth in question. If there is an overpressure in the system, the lung tissues may be damaged, with hemorrhaging as a consequence.

As the pressure increases, more gases can be taken up by the diver's body and dissolved in the various tissues. At a depth of 10 m, twice as much gas will be dissolved in his blood and tissues as at sea surface. This is apt to give him trouble, mainly because of the nitrogen.

Nitrogen

The problem with nitrogen is that it diffuses into various tissues of the body very slowly, and once dissolved, it also leaves the body very slowly when the pressure is once more reduced to the normal atmospheric pressure. This is especially bad when the pressure is suddenly reduced from several atmospheres, as may be the case during submarine escape or deep sea diving. Then the nitrogen is released from the tissues in the form of insoluble gas bubbles. These bubbles congregate in the small blood vessels, where they obstruct the flow of blood. This, then, gives rise to symptoms such as pains in the muscles and joints, and even paralysis may develop if the bubbles become trapped in the brain. These symptoms are known as the *bends*. Obviously, the severity of the symptoms depends on the magnitude of the pressure, which means the depth to which the person has descended under water, the length of time the person has spent at that depth, and the speed with which he ascends to the surface.

The bends can be avoided to a large extent by a slow return to normal pressure so as to allow time for the tissues to get rid of their excess nitrogen without the formation of bubbles. Another way to avoid the bends is to prevent the formation of nitrogen bubbles by replacing atmospheric nitrogen by helium, which is less easily dissolved in the body. This is done by having the diver breathe a helium-oxygen gas mixture. Another advantage with this is that it is more apt to prevent the so-called nitrogen narcosis which occurs when air is breathed at 3 atm or more, when there is an onset of euphoria and impaired mental activity with lack of ability to concentrate. With increasing pressures, they progressively handicap the individual and may render him helpless at 10 atm. Diving to depths exceeding 100 m while breathing ordinary atmospheric air may thus be fatal.

Pilots flying high-altitude aircraft may also suffer from bends if there is a sudden loss of pressure in the pressurized cabin, but the symptoms in these cases are usually not so severe as in the divers, and they usually do not occur at altitudes lower than 10,000 ft in any case.

Oxygen

Prolonged breathing of 100 percent oxygen at 1 atm may also be quite harmful (Lambertsen, 1965); irritation of the respiratory tract may occur after 12 hr and frank bronchopneumonia after 24 hr and the peripheral blood flow (the flow through the brain) may be reduced. In most individuals, no harmful effects result from breathing mixtures with less than 60 percent oxygen, but newborn infants are particularly susceptible to oxygen poisoning and may suffer harmful effects with oxygen concentrations over 40 percent. The remarkable thing is that oxygen poisoning is apparently no problem when breathing 100 percent oxygen at altitudes over 6,000 m, no matter for how long. Oxygen poisoning, therefore, is not much of a problem in aviation medicine, but it is indeed an important problem in deep sea diving where it may even affect the brain function when pure oxygen is breathed under increased pressure. This latter form of oxygen toxicity is apt to occur in divers at depths greater than 10 m, but there are great individual variations in sensitivity to 100 percent oxygen. The onset of symptoms may be hastened by vigorous physical activity at great depths; it starts with muscular twitchings and a jerking type of breathing and ends in unconsciousness and convulsions. The exact cause of this is unknown, but it is assumed that it is a matter of interference with certain enzyme systems in the tissues.

When a person breathes pure oxygen at a pressure of 3 atm or higher, the dissolved oxygen covers the oxygen need of the body at rest. No oxygen would be removed from HbO_2 during its passage through the capillary bed. Therefore the hemoglobin of the venous blood would still be saturated with oxygen, and this would interfere with the amount of H^+ ions taken up by Hb, which is a weaker acid than HbO_2 (Chap. 5). Thus CO_2 entering the blood from metabolizing cells would raise the blood P_{CO_2}, and the H^+ concentration would be higher than under normoxic conditions when the desaturation of HbO_2 simultaneously favors the removal of H^+ ions. The end result would be a CO_2 retention in the tissues and an acidosis.

Carbon Dioxide

If the pressure of the carbon dioxide in the respiratory air is increased, as is the case when the absorption of CO_2 fails in a closed breathing system, CO_2 pressure may be reached (about 75 mm Hg or higher) which produces a narcotic effect.

Oxygen Inhalation in Sports

In athletic events oxygen breathing is of very limited value. While with a lung volume of 5 liters one normally has a supply of about 0.8 liter of oxygen, this oxygen volume may be increased to about 4.5 liters if 100 percent oxygen is inhaled. This volume may be consumed by the cells within less than a minute during heavy exercise. An increase of the oxygen content of the blood is of no practical significance, since the saturation is over 95 percent already when one breathes air. The increased volume of dissolved oxygen is in this connection also unimportant.

After a few breaths of atmospheric air, the oxygen is quickly diluted, since the nitrogen content of the air is as much as 79 percent. The elimination of anaerobic metabolites is not speeded up if pure oxygen is inhaled during recovery.

Oxygen inhalation may improve the performance capacity just prior to swimming under water and when there is a need to increase the supply of oxygen *during* heavy muscular work. (It increases the maximal oxygen uptake, reduces the anaerobic participation in the energy yield, and reduces the pulmonary ventilation.) The "sniffing" of oxygen practiced by football players during the rest periods may perhaps have a psychological effect. From a physiological point of view, however, the extra oxygen is definitely wasted.

TOBACCO SMOKING

Larson et al. (1961) have prepared a very extensive summary on this subject. In this connection only a few aspects of tobacco smoking and physical performance will be discussed.

Circulatory Effects

Tobacco smoke contains up to 4 percent by volume of carbon monoxide. By inhalation some of this carbon monoxide is absorbed. The affinity of the hemoglobin to carbon monoxide is 200 to 300 times greater than to oxygen. The presence of even small amounts of carbon monoxide may, therefore, noticeably reduce the oxygen-transporting capacity of the blood. Carbon monoxide also interferes in a negative way with the unloading of oxygen in the tissues by shifting the oxyhemoglobin dissociation curve to the left (Chap. 6). A study has shown that subjects who smoked 10 to 12 cigarettes a day had 4.9 percent carbon monoxide hemoglobin, those who smoked 15 to 25 cigarettes a day had 6.3 percent, and those who smoked 30 to 40 cigarettes per day had 9.3 percent. Other studies have confirmed these findings; it may take 1 day or more for the carbon monoxide content of the blood to return to normal (Larson et al., 1961, p. 108). This amount of carbon monoxide in the blood gives no subjective symptoms at rest, as long as the concentration in the blood is below 10 percent. The adverse effect is only noticeable during physical exertion.

In a study normal subjects worked on a bicycle ergometer without having smoked 12 hr prior to the experiment and immediately after smoking one or two cigarettes (Juurup and Muido, 1946). Oxygen uptake, pulmonary ventilation, and respiratory rates were unaffected by smoking. A slight rise in arterial blood pressure was noted, but the difference was not significant. At a fixed oxygen uptake, the heart rate was 10 to 20 beats/min higher when the work test was preceded by smoking. The difference in heart rate between smokers and non-smokers was greater the higher the work load. The effect of smoking could not be observed if the subject waited 10 to 45 min after smoking before the work test started.

All other factors being equal, a reduction in the oxygen-transporting capacity is associated with a corresponding reduction in physical performance capacity during heavy or maximal work. At a given state of training the maximal cardiac output is quite constant. At high altitudes and low oxygen pressures in inspired air the oxygen content of the arterial blood drops. The amount of blood the heart can pump per minute during maximal work effort is the same, however, as when the work is performed at sea level at normal barometric pressure (Stenberg et al., 1966). Since the oxygen-transporting capacity of the blood declines with increasing altitude, and the maximal aerobic power depends partly on the maximal cardiac output and partly on the oxygen-carrying capacity of the blood, this power declines with increasing altitude. The same reasoning is also applicable to a situation in which the oxygen-carrying capacity of the hemoglobin is blocked by carbon monoxide of 5 to 10 percent. The result is that the oxygen-transporting capacity of the blood is reduced by at least a corresponding amount. During the performance of strenuous physical work involving large muscle groups for more than a few minutes, this represents a definite handicap. The smoker cannot compensate for the carbon monoxide content of the blood during maximal work. Because a regular physical training program only increases the maximal oxygen uptake by some 10 to 20 percent, a 5 to 10 percent reduction in maximal aerobic power due to smoking may play a significant role in many types of athletic events and in very heavy work.

Respiratory Effects

In Chap. 7 (Airway Resistance) it was pointed out that inhalation of smoke from a cigarette could cause within seconds a two- to threefold rise in the airway resistance. In addition to this acute effect, smoking also causes a more chronic swelling of the mucous membranes of the airways, leading to an increased airway resistance. At rest when the pulmonary ventilation is less than 10 liters/min, the increased airway resistance is not noticeable, however. When the demand on respiration is elevated, the increased respiratory resistance caused by smoking may be noticeable. A reduced pulmonary ventilation capacity may cause a smaller volume of oxygen to reach the alveoli, resulting in an impaired gas exchange. The result may be subjective symptoms of distress.

Smoking Habits among Athletes

A study of 285 top athletes in Great Britain (Report in $JAMA$, **170:**1106, 1959) showed that 16 percent of them smoked, but only 4 percent of them smoked more than 11 cigarettes per day. These findings are in agreement with the results of similar studies of top athletes elsewhere. Most of the smokers are among those engaged in athletic events requiring skill rather than endurance. None of the middle- and long-distance runners or swimmers were smokers. Among skiers, some of the ski jumpers and downhill skiers may smoke, but none of the cross-country skiers of the elite class are smokers. This may indicate that under conditions where the demand on the oxygen-transporting system is very great, smoking has been found to impair performance. If the current trends continue, even athletes participating in events primarily requiring skill will be required to undergo strenuous physical training which taxes the oxygen-transporting system to a considerable degree. This may make it desirable for the athlete to refrain from smoking while training, even though tobacco smoking at the time of the competitive event may not affect the outcome.

ALCOHOL AND EXERCISE

Neuromuscular Function

It is well established that alcohol may temporarily cause impaired coordination. The performance of rather simple movements is used to test whether or not an individual is under the influence of alcohol (walking on a straight chalk line, touching the tip of the nose with the index finger with the eyes closed, etc.). A precise assessment in borderline cases is impossible, however. The tolerance of alcohol varies greatly from one individual to another. Ikai and Steinhaus (1961) noted that the maximal isometric muscle strength could actually be improved in some cases, especially with untrained subjects, after moderate alcohol consumption. They explain this on the basis of a depressing influence of alcohol on the central inhibition of the impulse traffic in the nerve fibers to the skeletal muscles during maximal effort. The result is an increased impulse activity and an increased strength (see Chap. 4). At times individuals taking part in competitive events, such as target-shooting, maintain that they achieve better results following moderate alcohol consumption. They feel more relaxed. It is conceivable that the depression of the inhibiting effect of certain CNS centers may cause routine procedures to progress normally without the disturbing effect caused by the anxiety of the actual competition.

Aerobic and Anaerobic Power

Blomqvist et al. (1969) studied eight young male subjects during bicycle exercise at 2 submaximal and 1 maximal work load before and after peroral ethanol intake producing blood levels of 90 to 200 mg%. Oxygen uptake and heart rate were determined. In four of the subjects cardiac output, stroke volume, and intra-arterial pressure were also measured. Table 17-2 summarizes the most important changes. During maximal work with an O_2 uptake of about 3 liters/min and a cardiac output of about 21 liters/min, no difference was observed when the results before and after alcohol consumption were compared. Maximal heart rate, stroke volume, $a\text{-}\bar{v}O_2$ difference, and calculated peripheral resistance were also unaffected. During submaximal work, on the other hand, the heart rate was on an average 12 to 14 beats higher per minute in the alcohol experiments ($p < 0.01$). In the latter case the cardiac output was greater while the stroke volume was unaffected. The O_2 uptake during submaximal work was slightly higher after alcohol, but the $a\text{-}\bar{v}O_2$ difference was nevertheless reduced (i.e., the cardiac output was more elevated than would be expected from the increased oxygen uptake). At rest and during submaximal work the calculated total peripheral resistance was reduced.

Asmussen and Böje (1948) studied the effect of alcohol on the ability to perform a total of 956 and 9,860 kpm respectively as fast as possible. The first type of work could be performed in 12 to 15 sec, simulating a 100-m sprint in its effect on the organism, whereas the latter, lasting about 5 min, simulated a 1,500-m run. A blood concentration of alcohol of up to 100 mg percent did not significantly affect the ability to perform this kind of maximal work.

Summary These studies have revealed an effect on the circulatory response to submaximal exercise in individuals with elevated blood alcohol level. Measurements during maximal work, however, showed no effect on oxygen uptake, cardiac output, heart rate, stroke volume, or total peripheral resistance. Nor

TABLE 17-2

Hemodynamic effects of alcohol

	$\dot{V}_{O_2}$	Heart rate	Cardiac output	Stroke volume	$a\text{-}\bar{v}O_2$ difference	Total peripheral resistance
Rest, sitting	⇑ *	⇑	⇑	O	⇓	⇓
Submaximal exercise	⇑	⇑	⇑	O	⇓	⇓
Maximal exercise	O	O	O	O	O	O

* The arrows denote the observed changes, which are significant in all cases except the O_2 uptake.

SOURCE: Blomqvist et al., 1969.

was the maximal time the subjects could tolerate a standard load affected by the alcohol intake.

It should be pointed out that the question of dosage in the case of alcohol consumption in connection with athletic performance is a most difficult one. The tolerance varies greatly from individual to individual, and probably also from time to time in the same individual. Furthermore, the use of alcohol is definitely to be considered as a form of doping.

REFERENCES

Alexander, J. K., L. H. Hartley, M. Modelski, and R. F. Grover: Reduction of Stroke Volume during Exercise in Man Following Ascent to 3,100 m Altitude, *J. Appl. Physiol.*, **23**:849, 1967.

Asmussen, E., and O. Böje: The Effect of Alcohol and Some Drugs on the Capacity for Work, *Acta Phys ol. Scand.*, **15**:109, 1948.

Asmussen, E., and H. Chiodi: The Effect of Hypoxemia on Ventilation and Circulation in Man, *Am. J. Physiol.*, **132**:426, 1941.

Asmussen, E., W. von Döbeln, and M. Nielsen: Blood Lactate and Oxygen Debt after Exhaustive Work at Different Oxygen Tensions, *Acta Physiol. Scand.*, **15**:57, 1948.

Asmussen, E., and M. Nielsen: Cardiac Output during Muscular Work and Its Regulation, *Physiol. Rev.*, **35**:778, 1955.

Åstrand, P.-O.: The Respiratory Activity in Man Exposed to Prolonged Hypoxia, *Acta Physiol. Scand.*, **30**:343, 1954.

Åstrand, P.-O., and I. Åstrand: Heart Rate during Muscular Work in Man Exposed to Prolonged Hypoxia, *J. Appl. Physiol.*, **13**:75, 1958.

Balke, B.: Work Capacity at Altitude, in W. R. Johnson (ed.), "Science and Medicine of Exercise and Sports," p. 339, Harper & Row, Publishers, Incorporated, New York, 1960.

Barbashova, Z. I.: Cellular Level of Adaptation, in D. B. Dill (ed.), "Handbook of Physiology," sec. 4, Adaptation to the Environment, p. 37, American Physiological Society, Washington, D.C., 1964.

Bert, P.: "La Pression Barométrique," Masson et Cie, Paris, 1878.

Blomqvist, G., B. Saltin, and J. H. Mitchell: Acute Effects of Ethanol Ingestion on the Response to Submaximal and Maximal Exercise in Man, *Circulation*, **42**:463, 1970.

Blomqvist, G., and J. Stenberg: The ECG Response to Submaximal and Maximal Exercise during Acute Hypoxia, in G. Blomqvist, The Frank Lead Exercise Electrocardiogram, *Acta Med. Scand.*, **178**(Suppl. 440):82, 1965.

Buskirk, E. R., J. Kollias, R. F. Akers, B. K. Prokop, and E. Picón-Reátegui: Maximal Performance at Altitude and on Return from Altitude in Conditioned Runners, *J. Appl Physiol.*, **23**:259, 1967.

Carter, E. T.: Effect of Altitude on Maximum Breathing Capacity, *J. Aviat. Med.*, **28**:195, 1957.

Cassin, S., R. D. Gilbert, and E. M. Johnson: Capillary Development during Exposure to Chronic Hypoxia, Report SAM-TR-66-16, USAF School of Aviation Medicine, Randolph Field, Tex., 1966.

Cerretelli, P.: Lactacid O_2 Debt in Acute and Chronic Hypoxia, in R. Margaria (ed.), "Exercise at Altitude," p. 58, Excerpta Medica Foundation, Amsterdam, 1967.

Cerretelli, P., and R. Margaria: Maximum Oxygen Consumption at Altitude, *Intern. Z. Angew. Physiol.*, **18**:460, 1961.

Cervantes, J., and P. V. Karpovich: Effect of Altitude on Athletic Performance, *Res. Quart.*, **35**:446, 1964.

Chiodi, H.: Respiratory Adaptations to Chronic High Altitude Hypoxia, *J. Appl. Physiol.*, **10**:81, 1957.

Christensen, E. H.: Sauerstoffaufnahme und Respiratorische Funktionen in Grossen Höhen, *Skand. Arch. Physiol.*, **76**:88, 1937.

Christensen, E. H., and A. Krogh: Fliegerundersuchungen; die Wirkung niedriger O_2-Spannung auf Höhenflieger, *Skand. Arch. Physiol.*, **73**:145, 1936.

Christensen, E. H., and H. E. Nielsen: Die Leistungsfähigkeit der menschlichen Skelettmuskeln bei niedrigen Sauerstoffdruck, *Skand. Arch. Physiol.*, **74**:272, 1936.

Christensen, E. H., and W. H. Forbes: Der Kreislauf in grossen Höhen, *Skand. Arch. Physiol.*, **76**:75, 1937.

Consolazio, C. F.: Submaximal and Maximal Performance at High Altitude, in R. F. Goddard (ed.), "The International Symposium on the Effects of Altitude on Physical Performance," p. 91, The Athletic Institute, Chicago, 1967.

Cotes, J. E.: Ventilatory Capacity at Altitude and Its Relation to Mask Design, *Proc. Roy. Soc. (London)*, ser. B., **143**:32, 1954.

DeGraff, A. C., Jr., R. F. Grover, J. W. Hammond, Jr., J. M. Miller, and R. L. Johnson, Jr.: Pulmonary Diffusing Capacity in Persons Native to High Altitude, *Clin. Res.*, **13**:74, 1965.

Dejours, P., R. H. Kellogg, and N. Pace: Regulation of Respiration and Heart Rate Response in Exercise during Altitude Acclimatization, *J. Appl. Physiol.*, **18**:10, 1963.

Dickinson, E. R., M. J. Piddington, and T. Brain: Project Olympics, *Schw. Zschr. Sportmed.*, **14**:305, 1966.

Dill, D. B. (ed.): "Handbook of Physiology," sec. 4, Adaptation to the Environment, American Physiological Society, Washington, D.C., 1964.

Edwards, H. T.: Lactic Acid in Rest and Work at High Altitude, *Am. J. Physiol.*, **116**:367, 1936.

Fenn, W. O.: The Pressure Volume Diagram of the Breathing Mechanism, in W. M. Boothby (ed.), "Handbook of Respiratory Physiology," USAF School of Aviation Medicine, Randolph Field, Tex., 1954.

Goddard, R. F. (ed.): "The International Symposium on the Effects of Altitude on Physical Performance," The Athletic Institute, Chicago, 1967.

Goddard, R. F., and C. B. Favour: United States Olympic Committee Swimming Team Performance in International Sports Week, Mexico City, Oct., 1965, in R. F. Goddard (ed.), "The International Symposium on the Effects of Altitude on Physical Performance," p. 135, The Athletic Institute, Chicago, 1967.

Grover, R. F., and J. T. Reeves: Exercise Performance of Athletes at Sea Level and 3,100 Meters Altitude, in R. F. Goddard (ed.), "The International Symposium on the Effects of Altitude on Physical Performance," p. 80, The Athletic Institute, Chicago, 1967.

Hansen, J. E., G. P. Stelter, and J. A. Vogel: Arterial Pyruvate, Lactate, pH, and P_{CO_2} during Work at Sea Level and High Altitude, *J. Appl. Physiol.*, **23**:523, 1967a.

Hansen, J. E., J. A. Vogel, G. P. Stelter, and C. F. Consolazio: Oxygen Uptake in Man during Exhaustive Work at Sea Level and High Altitude, *J. Appl. Physiol.*, **23**:511, 1967b.

Hartley, L. H., J. K. Alexander, M. Modelski, and R. F. Grover: Subnormal Cardiac Output at Rest and during Exercise in Residents at 3,100 m Altitude, *J. Appl. Physiol.*, **23**:839, 1967.

Henderson, Y.: "Adventures in Respiration," The Williams & Wilkins Company, Baltimore, 1938.

Hermansen, L., and B. Saltin: Blood Lactate Concentration during Exercise at Acute Exposure to Altitude, in R. Margaria (ed.), "Exercise at Altitude," p. 48, Excerpta Medica Foundation, Amsterdam, 1967.

Hultgren, H. N., J. Kelly, and H. Miller: Pulmonary Circulation in Acclimatized Man at High Altitude, *J. Appl. Physiol.*, **20**:233, 1965.

Hurtado, A.: Animals in High Altitudes: Resident Man, in D. B. Dill (ed.). "Handbook of Physiology," sec. 4, p. 843, Adaptation to the Environment, American Physiological Society, Washington, D.C., 1964.

Hurtado, A., C. Merino, and E. Delgado: Influence of Anoxemia on the Hemopoietic Activity, *Arch. Int. Med.*, **75**:284, 1945.

Ikai, M., and A. H. Steinhaus: Some Factors Modifying the Expression of Strength, *J. Appl. Physiol.*, **16**:157, 1961.

Jokl, E., and P. Jokl (eds.): "Exercise and Altitude," S. Karger, New York, 1968.

Jokl, Ernst, A. H. Frucht, M. J. Karvonen, D. C. Seaton, E. Simon, and Peter Jokl: Sports Medicine, *Ann. N. Y. Acad. Sci.*, **134**:908, 1966.

Juurup, A., and L. Muido: On Acute Effects of Cigarette Smoking on Oxygen Consumption, Pulse Rate, Breathing Rate and Blood Pressure in Working Organisms, *Acta Physiol. Scand.*, **11**:48, 1946.

Kellogg, R. H.: Central Chemical Regulation of Respiration, in W. O. Fenn and H. Rahn (eds.), "Handbook of Physiology," sec. 3, Respiration, vol. 1, p. 507, American Physiological Society, Washington, D.C., 1964.

Klausen, K.: Cardiac Output in Man in Rest and Work during and after Acclimatization to 3,800 m, *J. Appl. Physiol.*, **21**:609, 1966.

Kollias, J., E. R. Buskirk, R. F. Akers, E. K. Prokop, P. T. Baker, and E. Picón-Reátegui: Work Capacity of Long-time Residents and Newcomers to Altitude, *J. Appl. Physiol.*, **24**:792, 1968.

Kreuzer, F., S. M. Tenney, J. C. Mithoefer, and J. Remmers: Alveolar-arterial Oxygen Gradient in Andean Natives at High Altitude, *J. Appl. Physiol.*, **19**:13, 1964.

Lambertsen, C. J.: Effects of Oxygen at High Partial Pressure, in W. O. Fenn and H. Rahn (eds.), "Handbook of Physiology," sec. 3, Respiration, vol. 2, p. 1027, American Physiological Society, Washington, D.C., 1965.

Larson, P. S., H. B. Haag, and H. Silvette: "Tobacco," The Williams & Wilkins Company, Baltimore, 1961.

Leary, W. P., and C. H. Wyndham: The Possible Effect on Athletic Performance of Mexico City's Altitude, *S. Afr. Med. J.*, **40**:984, 1966.

Luft, U. C.: Laboratory Facilities for Adaptation Research: Low Pressures, in D. B. Dill (ed.), "Handbook of Physiology," sec. 4, p. 329, Adaptation to the Environment, American Physiological Society, Washington, D.C., 1964a.

Luft, U. C.: Aviation Physiology: The Effect of Altitude, in W. O. Fenn and H. Rahn (eds.), "Handbook of Physiology," sec. 3, Respiration, vol. 2, p. 1099, American Physiological Society, Washington, D.C., 1964b.

Margaria, R. (ed.): "Exercise at Altitude," Excerpta Med. Found., Amsterdam, 1967.

Mazess, R. B.: "Exercise Performance and Altitude Adaptation," paper read at International Congress of Americanists, Sept., 1966.

Merino, C.: Studies on Blood Formation and Destruction in the Polycythemia of High Altitude, *Blood*, **5**:1, 1950.

Miles, S.: The Effect of Changes in Barometric Pressure on Maximum Breathing Capacity, *J. Physiol.*, **137**:85P, 1957.

Milledge, J. S., and S. Lahiri: Respiratory Control in Lowlanders and Sherpa Highlanders at Altitude, *Respir. Physiol.*, **2**:310, 1967.

Noell, W.: Über die Durchblutung und die Sauerstoffversorgung des Gehirns, VI, Einfluss der Hypoxämie und Anämie, *Arch. Ges. Physiol.*, **247**:553, 1944.

Norton, E. F.: "The Fight for Everest," Edward Arnold (Publishers) Ltd., London, 1925.

Penaloza, D., F. Sime, N. Banchero, R. Gamboa, J. Cruz, and E. Marticorena: Pulmonary Hypertension in Healthy Men Born and Living at High Altitudes, *Am. J. Cardiol.*, **11**:150, 1963.

Pugh, L. G.: Animals in High Altitude: Man above 5,000 Meters-Mountain Exploration, in D. B. Dill (ed.), "Handbook of Physiology," sec. 4, p. 861, Adaptation to the Environment, American Physiological Society, Washington, D.C., 1964.

Pugh, L. G.: "Report of Medical Research Project into Effects of Altitude in Mexico City," report to the British Olympic Committee, 1965.

Pugh, L. G., M. B. Gill, S. Lahiri, J. S. Milledge, M. P. Ward, and J. B. West: Muscular Exercise at Great Altitudes, *J. Appl. Physiol.*, **19**:431, 1964.

Rahn, H.: Introduction to the Study of Man at High Altitudes: Conductance of O_2 from the Environment to the Tissues, in "Life at High Altitudes," Scientific Publ. 140, p. 2, Pan-American Health Organization, WHO, Washington, D.C., September, 1966.

Reynafarje, B.: Myoglobin Content and Enzymatic Activity of Muscle and Altitude Adaptation, *J. Appl. Physiol.*, **17**:301, 1962.

Reynafarje, C.: Humoral Control of Erythropoiesis at Altitude, in R. Margaria (ed.), "Exercise at Altitude," p. 165, Excerpta Medica Foundation, Amsterdam, 1967.

Roskamm, H., L. Samek, H. Weidemann, and H. Reindell: "Leistung und Höhe," Knoll AG, Ludwigshafen am Rhein, 1968.

Roughton, F. J. W.: Transport of Oxygen and Carbon Dioxide, in W. O. Fenn and H. Rahn (eds.), "Handbook of Physiology," sec. 3, Respiration, vol. 1, p. 767, American Physiological Society, Washington, D.C., 1964.

Saltin, B.: Aerobic and Anaerobic Work Capacity at 2,300 Meters, *Schw. Zschr. Sportmed.*, **14**:81, 1966.

Saltin, B.: Aerobic and Anaerobic Work Capacity at an Altitude of 2,250 Meters, in R. F. Goddard (ed.), "The International Symposium on the Effects of Altitude on Physical Performance," p. 97, The Athletic Institute, Chicago, 1967.

Saltin, B., R. F. Grover, C. G. Blomqvist, L. H. Hartley, and R. L. Johnson, Jr.: Maximal Oxygen Uptake and Cardiac Output after Two Weeks at 4,300 Meters, *J. Appl. Physiol.*, in press.

Schönholzer, G.: "Sport in mittleren Höhen," D. Haupt, Bern, 1967.

Severinghaus, J. W., R. A. Mitchell, B. W. Richardson, and M. M. Singer: Respiratory Control at High Altitude Suggesting Active Transport Regulation of CSF pH, *J. Appl. Physiol.*, **18:**1155, 1963.

Severinghaus, J. W., and A. Carcelen: Cerebrospinal Fluid in Man Native to High Altitude, *J. Appl. Physiol.*, **19:**319, 1964.

Severinghaus, J. W., C. R. Bainton, and A. Carcelen: Respiratory Insensitivity to Hypoxia in Chronically Hypoxic Man, *Respir. Physiol.*, **1:**308, 1966.

Somervell, T. H.: Note on the Composition of Alveolar Air at Extreme Heights, *J. Physiol.*, **60:**282, 1925.

Stenberg, J., B. Ekblom, and R. Messin: Hemodynamic Response to Work at Simulated Altitude, *J. Appl. Physiol.*, **21:**1589, 1966.

Surks, M. I., K. S. Chinn, and L. O. Matoush: Alterations in Body Composition in Man after Acute Exposure to High Altitude, *J. Appl. Physiol.*, **21:**1741, 1966.

Ulvedal, F., T. E. Morgan, Jr., R. G. Cutler, and B. E. Welch: Ventilatory Capacity during Prolonged Exposure to Simulated Altitude without Hypoxia, *J. Appl. Physiol.*, **18:**904, 1963.

Vannotti, A.: The Adaptation of the Cell to Effort, Altitude and to Pathological Oxygen Deficiency, *Schweiz. Med. Wschr.*, **76:**899, 1946.

Velasquez, T.: Tolerance to Acute Anoxia in High Altitude Natives, *J. Appl. Physiol.*, **14:**357, 1959.

Vogel, J. A., J. E. Hansen, and C. W. Harris: Cardiovascular Responses in Man during Exhaustive Work at Sea Level and High Altitude, *J. Appl. Physiol.*, **23:**531, 1967.

Weihe, W. H. (ed.): "The Physiological Effects of High Altitude," The Macmillan Company, New York, 1964.

West, J. B.: Diffusing Capacity of the Lung for Carbon Monoxide at High Altitude, *J. Appl. Physiol.*, **17:**421, 1962.

West, J. B., S. Lahiri, M. B. Gill, J. S. Milledge, L. G. Pugh, and M. P. Ward: Arterial Oxygen Saturation during Exercise at High Altitude, *J. Appl. Physiol.*, **17:**617, 1962.

18

Health and Fitness

contents

chapter eighteen

Health and Fitness

INTRODUCTION

Man is built for movement, for physical activity. The last century, and particularly the more recent decades, have witnessed a revolution in the lives of millions of people, whereby mechanical implements have taken over tasks earlier performed by human power. In all sections of life there is a breach between old and new customs. The human body is a product of millions of years of evolution, but for a very long time this development has been at a standstill. Man has created a new world, with a different environment and different living conditions from those of his ancestors. By and large these changes have been to our benefit, but they have also created problems that are difficult to manage. Many of these problems stem from the fact that man was built for the Stone Age but must now fit into the present extremely technical world. Infants born today have the same physical equipment as the children of our prehistoric ancestors. The vast changes have occurred in the world, not in them (Bortz, 1963). It has been emphasized in the preceding chapters that we are "constructed" for activity and that regular activity is essential for our optimal functioning and health. Therefore a portion of our leisure time should be devoted to active recreation and training.

In this chapter some concrete physiological and medical motives will be presented showing the need for regular physical activity. The basis for this need has largely been explained in the earlier chapters. Perhaps the scientific basis is not always adequate to substantiate the motives given, but it appears justifiable to base, to a certain extent, some conclusions on indirect evidence and extrapolations. (For further details see Rodahl, 1966.)

PHYSICAL FITNESS FOR EVERYDAY LIFE

The various organs and organ functions adjust to the demands placed upon them. Some feeling of fatigue is inevitable after hours of physically heavy work.

With the elimination of the heavier loads the job becomes easier, at least it does to start with. The well-trained individual, therefore, has a broad margin of safety between his maximal power or capacity on the one hand and what is being demanded of him physically on the other. As time goes by, however, he will become adapted to the reduced demand; he will to some extent deteriorate, as did the bedridden subjects discussed in Chap. 12. Consequently, his safety margin between capacity and demand will progressively narrow, and subjectively the job will become heavier. He may once again find the same feeling of tiredness at the end of a day's work, despite the reduced energy demand.

One important aim of regular physical training is to achieve a physical condition and degree of fitness that are well above that required for the routine job. If the performance of the job occasionally requires the heart to pump out 10 liters of blood/min at a rate of 120 beats/min, it is an evident advantage if it has been trained to attain an output of 15 liters/min at a rate of, say, 150. This is essentially what is meant by being fit for every day life. From this point of view occasional brisk walks may be sufficient for a teacher or a clerk, but a lumberjack, steelworker, or housewife should exercise somewhat harder a couple of times a week.

In many types of recreational activities, like sailing, bathing, mountaineering, and hiking, life or death can actually depend on the individual's fitness.

In conclusion it may be stated that if the work becomes physically less demanding, some physical activity must be included during leisure time to provide the body with the stimuli it needs to function at its best.

CARDIOVASCULAR DISEASES

We quote from Peterson (1965): "It is obvious that cardiovascular diseases are prevalent because they kill more persons than the next three most common causes of death combined. Their socioeconomic importance is illustrated by the statement that 'the 365,000 Americans between the ages of 25 and 65 who died of these diseases in 1962 would have earned wages totaling more than 1.5 billion dollars and paid close to 200 million dollars in federal income taxes had they lived *one* more healthy working year' (President's Commission on Heart Disease, Cancer and Stroke)."

In many countries cardiovascular diseases often account for more than 50 percent of all deaths. Naturally this fact motivates the intensive research presently being carried out to discover the genesis and treatment of such diseases. They certainly cause personal suffering and their social and economic consequences are enormous. In many countries medical care is actually one of the largest industries, with a direct budget that has increased enormously during the past decades. In the United States it has increased from 12 billion dollars in 1949 to 44 billion dollars in 1966. In Sweden, with a population of 8 million, the health budget covered by direct tax revenue was close to 1.4 billion dollars in 1967. About one million patients were treated in hospitals and 20 million ambulatory cases were treated in open wards. However, more money does not inevitably equal better health. Forbes (1967) points out that longevity, measured as the average remaining lifetime, increased markedly in the United States in each decade from 1900 to 1950 but appeared to have reached a plateau about 1954. Since that time there has been no increase for males and only a slight increase for females. In other countries, however, life-span for both sexes is still increasing, so that the United States has dropped from the fourteenth place among the nations of the world for males in 1949 to the thirty-first place in 1964. For females it has dropped from ninth to eleventh place. The average remaining lifetime at birth for males in the United States was 46.3 years in 1900, 66.7 years in 1954, and 66.9 years in 1964. The noticeable progress at the first half of this century is mostly due to a markedly reduced infant mortality. Actually, the remaining lifetime of a 50-year-old man in 1969 is only a few years longer than it was for the man who celebrated his fiftieth anniversary in 1900. Many diseases, particularly infectious diseases, have effectively been conquered, but diseases of a degenerative nature have increased extensively, particularly cardiovascular disease. One cannot avoid the thought that our modern way of life could be one important factor and that the main determinants of longevity are now more cultural than medical. Here let it merely be pointed out that several factors such as heredity, diet, and way of life seem to be of importance in the development of cardiovascular diseases. Individuals showing high blood pressure, obesity, a high concentration of cholesterol in the blood, or a combination of these run a higher risk of death from cardiovascular diseases than the nonobese with normal blood pressure and a low cholesterol level. In recent years interest has been focused on the possible role of physical inactivity in the genesis of these diseases.

The studies of Morris and collaborators (1958, 1959) have contributed greatly to focusing attention on the relationship between physical activity or physical training and coronary heart disease. On the basis of their large series of observations, they advance the hypothesis that "men in physically active jobs have a lower incidence of coronary (ischaemic) heart disease in middle-age than men in physically inactive jobs. More important, the disease is not so severe in physically active workers, tending to present in them in relatively benign forms" (Morris, 1958). Their general hypothesis is stated in causal terms that "physical activity of work is a protection against coronary (ischaemic) heart disease. Men in physically active jobs have less coronary heart disease during middle-age, what disease they have is less severe, and they develop it later than men in physically inactive jobs." They go on to advance the speculation that "habitual physical activity is a general factor of cardiovascular health in middle age, and that coronary heart disease is in some respect a deprivation syndrome, a deficiency disease." In their material, the hearts of sedentary and light workers show the pathology of the hearts of heavy workers 10 to 15 years older (Morris, 1958).

Since the time of this pioneer study by Morris and coworkers, a series of studies and discussions concerning the problem have been published. As a guide for the interested reader, the following may be listed: Burgess et al., 1963; Likoff and Mayer, 1963; Fox and Skinner, 1964; Frank et al., 1965; Hollmann, 1965; Barry, 1966; Evang and Andersen, 1966; Roskamm et al., 1966; Karvonen and Barry, 1967; Kellermann et al., 1967; Keys et al., 1967; "Proceedings of the International Symposium on Physical Activity and Cardiovascular Health" (Toronto, 1966), 1967.

In studies published so far it has been shown that the risk run by inactive individuals of death from a cardiovascular disease is two to three times greater than that run by the active. The probability of surviving the first heart attack is statistically two to three times greater for those who have previously been physically active than for those who have been inactive. These are statistical correlations and do not prove that the degree of physical activity per se has actually been the sole and decisive factor. The studies were carried out on selected groups of individuals, and it is possible that certain factors that determined choice of profession or degree of activity during leisure time may also have independently given rise to some sort of prevention against cardiovascular diseases. However, there are physiological explanations as to how physical activity could be beneficial. Investigations on animals and observations on man have revealed that physical training can open up more blood vessels in the arterial tree in the heart muscle, that is, collaterals may develop in the coronary arteries (Eckstein 1957; Chap. 12). Similarly, collaterals may develop in peripheral arteries. A narrowing or occlusion of a vessel due to arteriosclerosis will not mean the same catastrophe if there are other vessels that can take over the transport of blood with its oxygen and nutriments to the tissue peripheral to the damaged vessel.

Research in this area is very complicated, and it may take a hundred years or more of intensive studies, particularly longitudinal and intervention-type studies, to demonstrate with certainty that there is or is not a connection between cardiovascular disease and inactivity. The question is then critical whether we should wait so long for final proof one way or the other. In our opinion there is so much indirect evidence that regular physical activity, or training, has a beneficial effect on the functioning of the heart that the opportunity must be seized now actively to affect health in a positive way through a systematic improvement in physical fitness by training.

Training of the oxygen-transport system, including the heart and circulation, is particularly important then both as a prophylactic measure and as treatment.

The not uncommon belief that competitive athletic exertion may be harmful to the heart and frequently leads to permanent damage is an old misconception that dates back to Hippocrates. That this is not the case has been shown conclusively (Dublin et al., 1949; Rook, 1954; Montoye, 1962; Pyörälä et al., 1967).

CALORIC EXPENDITURE

People in most European countries and in North America show a definite tendency to put on weight after the age of thirty. At the same time, there is some decrease in muscle mass, and there will, therefore, be a more or less marked increase in body fat. Obesity may lead to esthetic problems, but it is also a drawback from the medical point of view. It is still an open question whether there is any direct connection between obesity and various diseases, but so far the incidence of cardiovascular diseases has been higher among the obese than the nonobese.

To prevent obesity and to ensure an adequate supply of essential nutriments, the individual should attain a caloric output that is not too low. Muscular activity is the most efficient way to increase the metabolism. The larger the active muscle mass, the higher is the caloric output at a given subjected strain. Within wide limits the work load or speed makes little difference to the energy demand, the determining factor being the total work done, the distance. A slow walk from A to B demands no fewer extra calories than a fast run over the same distance.

A 2-km (1¼-mile) walk per day would consume about 100 kcal/day, or 365,000 kcal after 10 years of daily trips. About 60 kg (130 lb) of fatty tissue can store this amount of calories. Therefore walking a few kilometers a day will in the course of time be very effective in keeping the body weight down. One can easily leave the bus or subway a few stops from the office in the morning, and walk back a similar distance in the afternoon.

BACK TROUBLES

Diseases in the spinal column rank very high on the list of common diseases. They are responsible for many days of sick leave and thus give rise to economic problems and cause much suffering, When a load is lifted or carried, a reflex mechanism calls the trunk muscles into action to fix the rib cage and to compress the abdominal contents. The intracavitary pressures are thereby increased and aid the support of the spine (Chap. 8). Such observations emphasize the important role that the trunk muscles have in supporting the spine. While flabby abdominal muscles may leave the spine exposed to injurious stress, well-developed abdominal muscles, on the other hand, are an important protective device which can prevent damage to the spinal column and its resultant backache, etc. To be sure, the trunk muscles probably have no influence on the inevitable changes in the spinal column that come with age, but if they are well developed and trained these muscles can, to a great extent, prevent the symptoms caused by the occurrence of a weak back. Walking or running upstairs or uphill will train the leg and trunk muscles. Simple, but if possible daily, exercises will also strengthen these muscles. As back troubles are so common, it is important to encourage workers to keep the trunk muscles fit. It is also important that everyone know how to lift and carry loads in a way that reduces the force on the spinal column.

JOINTS

There should be movement, even to the extreme, in many joints to counteract stiffening and impairment of the metabolism of the cartilage. A prolonged static load on a joint will compress the cartilage articular; movement in the joint on the other hand increases the thickness of the cartilage and improves the exchange of nutriments. Therefore, the limbs should be moved as far as they will go, gently and with no force applied. No pain should be caused by these movements. Aged or immobile individuals should do the exercises several times a day, unless doctor or physiotherapist recommends otherwise.

POSTURE

It has been emphasized that a sustained load has a detrimental effect on the functioning of muscles and joints. In the erect position moreover there is a tendency to blood stagnation in the legs. Patients whose cardiac function is impaired are particularly prone to such adverse blood distribution, resulting in edema of the lower leg. The contraction of muscles during movement compresses the veins and thus promotes the flow of blood toward the heart. This beneficial effect of dynamic muscular activity on the blood flow leads to the important conclusion that sitting or standing postures should not be too fixed but should permit movement. When driving long distances, it is important to stop regularly, get out of the car, and stretch the legs. No place of work should, in fact, be so well organized that everything is too comfortably within reach; it is good from a physiological point of view to get up from time to time, to walk a few steps, to climb the odd flight of stairs and so on.

In essence, well-developed muscles can promote the circulation of the blood. Training will increase the volume of blood, which is also a definite advantage. However, it should be pointed out that a period of rest in the middle of the day, lying down with the legs raised, may also work miracles, especially for the aged and those suffering from heart disease.

ACTIVE RECREATION

By active recreation is meant a stimulating hobby or activity with some muscular activity involved as opposed to passive recreation which, while it may still be stimulating, lacks any marked demand on the circulation or locomotor organs. Watching television, reading a book, playing cards or chess, stamp collecting, and listening to music are all examples of passive recreation. It is extremely important to have something exciting to look forward to every week, and possibly every day, and passive recreations are in this respect very valuable. They should, however, be supplemented by active recreation. A good example is ornithology. It is an outdoor activity. The bird being watched may move to the next tree, and in order to be identified must be followed. Further examples of active recreation are gardening, botany, and of course, sports, including fishing, hunting, and mountaineering. In the northern countries the climate does not favor outdoor hobbies all the year around; it is therefore necessary to substitute regular "contrived" training such as cross-country walking and running.

An important beneficial effect of active recreation is thought to be that impulses generated in the muscle spindles of the working muscles are probably essential to the optimal functioning of the central nervous system to which they are conveyed.

There are unfortunately many jobs for which it may be difficult to create conditions in which the worker can feel real satisfaction from the job itself. It is then essential that leisure time can be spent in a meaningful way.

This will focus attention on the proper utilization of leisure time for wholesome constructive activity and recreation, including exercise, sports, and outdoor activities. There will be a need for not only long-range planning and acquisition of community recreational facilities and recreational areas, but also the encouragement of physical recreational activities from childhood, especially during the habit-forming years of children.

We have to face the fact that most individuals are physically lazy from the age of puberty on. Therefore, the various types of facility for training and active recreation must be offered in an appetizing manner. The cost of simple sports grounds, including the track illustrated in Fig. 18-1, possibly with changing rooms, showers, and sauna, is low when compared with that of building and running a hospital, or of building a stadium for ice hockey, football, soccer, athletics, or similar sports.

AIMS OF PHYSICAL EDUCATION IN SCHOOLS

It is very important that children and youth in general be introduced to the principles of training and active recreation, in theory as well as practice. Good habits and motivations must be developed early in life. It is a pity that the time set aside

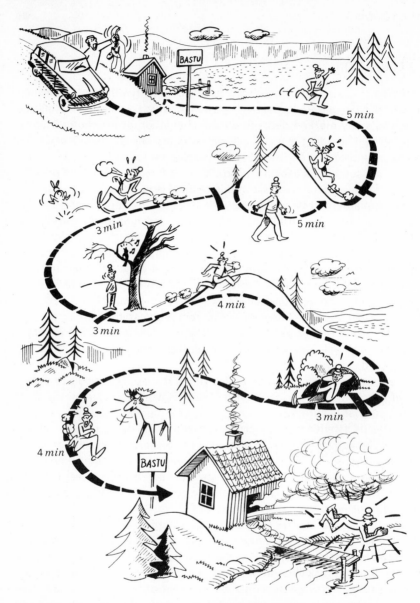

Fig. 18-1   Example of a training track and of an effective training program.
1. Start with a 5-min warm-up by walking and slow running (jogging).
2. Run uphill at top speed (or according to state of fitness) a distance of about 25 steps.
Walk downhill and repeat the uphill sprint. Repeat about 5 times in all. This takes
about 5 min and will develop strength in the leg and trunk muscles.
3. Run on the level at about 80 percent of top speed for 3 to 4 min. Then rest for a couple
of minutes. Repeat 3 or 4 times. This will condition the oxygen-transporting system.
This entire training program takes about 30 to 40 min. It should be repeated 2 or 3 times
a week. Untrained and older individuals may start by walking instead of running.
(Bastu means sauna.)

for compulsory physical education in schools is gradually being reduced in many countries. (There are essentially two main subjects in the school curriculum: physical education and all other theoretical subjects. If more time should be considered necessary for mathematics or physics, it should not be provided at the expense of that devoted to physical education, as has often been the case.) There is evidence that neglect of regular physical activity during adolescence cannot fully be compensated later on in life.

From a physiological-medical viewpoint, physical education in schools should

1 Train the oxygen-transporting system (respiration and circulation).
2 Generally train the locomotive organs (especially the muscles of the back and abdomen).
3 Give instruction on how to lie, sit, stand, walk, lift, carry, etc.
4 Give instruction in technique, tactics, rules, etc., in games and sports in order to reduce or eliminate accidents (the events are eventually practiced in the students' leisure time).
5 Provide physical and psychological recreation and variety.
6 Arouse interest in regular physical activity after schooling has been finished.

It is essential that we explain to the older student why we recommend regular physical training. The variety of events that can be taught and practiced within the frame of physical education in school is immense. In our opinion it is not necessary to teach any special sports, games, or athletic techniques (with the exception of swimming) for physiological-medical reasons. Actually track and field events, football, soccer, ice hockey, and other games are not so tempting to adults and less easily organized outside the context of the school. For this reason simple individual activities such as walking, running, swimming, and skiing should be stressed over group sports.

If, on the other hand, the teacher can "sell" these events so that the pupils become interested and stimulated, they are justifiable parts of the program. The teacher's personality and pedagogic ability are actually of more importance than the number of events and exercises he teaches. Thus he should not be rigidly confined by a curriculum as are teachers in other subjects. However, a limited section of the program should be so constructed that it effectively fulfills the requirements established by the school authorities. A major part of the activities should be developed according to the interest of the teacher and his pupils, depending on local facilities available, etc., and the choice should primarily be made on the basis of pedagogic principles.

A simple bicycle ergometer (for instance of the type made by Monark, available at Varberg, Sweden, or Quinton Instruments, 3051 44th Ave., West Seattle, Wash. 98199, U.S.A.) may be an excellent aid in teaching as an instrument for recording progress during training (see Fig. 11-7).

CONCLUDING REMARKS

It is more important to add life to years than years to life; and a continuous process of balance and compromise is involved in the choice of ways of living, comfort or resignation of comfort, stimulating agents, and so on. Fitness is a definite advantage to the enjoyment of meaningful and satisfying work and leisure time, and 2 to 3 half hours a week spent on efficient and rational training can achieve and maintain good physical condition.

Thanks to all our comforts, thanks to cars and other modern means of transport, we can really enjoy recreation and train under more pleasant conditions than when struggling for existence.

The question is frequently raised whether a medical examination is advisable before commencing a training program. Certainly anyone who is doubtful about his state of health should consult a physician. In principle, however, there is less risk in activity than in continuous inactivity. In a nutshell, our opinion is that it is more advisable to pass a careful medical examination if one intends to be sedentary in order to establish whether one's state of health is good enough to stand the inactivity! This question has another, practical aspect. If the beneficial effects of regular physical activity were generally recognized, there would not be enough physicians to examine all the individuals who would want to start a training program. Furthermore, there does not exist a health examination which is foolproof. Even very advanced examinations including work tests, ECG-recordings, etc., give too many false positive and false negative diagnoses to be absolutely reliable in this connection. On the other hand, since many patient categories may benefit from exercise, it is hoped that the physician of the future will prescribe specific physical exercise programs for his patients in the same way he prescribes drugs or any other form of treatment today. In any case, it is exceedingly important to start the program gently and to increase the intensity very gradually over a period of weeks.

REFERENCES

Barry, A. J.: Effects of Physical Training in Patients Who Have Had Myocardial Infarction, *Am. J. Cardiol.*, **17**:1, 1966.

Bortz, E. L.: "Creative Aging," The Macmillan Company, New York, 1963.

Burgess, A. M., Jr., Z. Fejfar, and A. Kagan: "Arterial Hypertension and Ischaemic Heart Disease: Comparison in Epidemiological Studies," WHO, Geneva, 1963.

Dublin, L. I., A. J. Lotka, and M. Spiegelman: "Length of Life: A Study of the Life Table," The Ronald Press Company, New York, 1949.

Eckstein, R. W.: Effect of Exercise and Coronary Artery Narrowing on Coronary Collateral Circulation, *Circulation Res.*, **5**:230, 1957.

Evang, K., and K. L. Andersen (eds.): "Physical Activity in Health and Disease," Universitetsforlaget, Oslo, 1966.

Forbes, W. H.: Longevity and Medical Costs, *New England J. Med.*, **277**:71, 1967.

Fox, S. M., and J. Skinner: Physical Activity and Cardiovascular Health, *Am. J. Cardiol.*, **14**:731, 1964.

Frank, C. W., E. Weinblatt, S. Shapiro, and R. V. Sager: Physical Activity as a Lethal Factor in Myocardial Infarctions among Men, *Circulation*, **32**(Suppl. II):87, 1965.

Hollmann, W.: "Körperliches Training als Prävention von Herz-Kreislaufkrankheiten," Hippokrates-Verlag, Stuttgart, 1965.

Karvonen, M. J., and A. I. Barry (eds.): "Physical Activity and the Heart," Charles C Thomas, Publisher, Springfield, Ill., 1967.

Kellermann, J. J., M. Levy, S. Feldman, and I. Kariv: Rehabilitation of Coronary Patients, *J. Chron. Dis.*, **20**:815, 1967.

Keys, A., et al.: Epidemiological Studies Related to Coronary Heart Disease: Characteristics of Men Aged 40–59 in Seven Countries, *Acta Med. Scand.*(Suppl. 460), 1967.

Likoff, W., and J. H. Mayer (eds.): "Coronary Heart Disease," Grune & Stratton, Inc., New York, 1963.

Montoye, H. J.: Physiology of Training Including Age and Sex Differences, *J. Sport Med.*, **2**:35, 1962.

Morris, J. N.: Occupation and Coronary Heart Disease, *Arch. Intern. Med. (Chicago)*, **104**:903, 1959.

Morris, J. N., and M. D. Crawford: Coronary Heart Disease and Physical Activity of Work; Evidence of a National Necropsy Survey, *Brit. Med. J.*, **2**:1485, 1958.

Peterson, L. H.: Control and Regulation of the Cardiovascular System, in W. S. Yamamoto and J. R. Brobeck (eds.): "Physiological Controls and Regulations," p. 308, W. B. Saunders Company, Philadelphia, 1965.

Proceedings of the International Symposium on Physical Activity and Cardiovascular Health, *Canad. Med. Ass. J.*, **96**, 1967.

Pyörälä, K., M. I. Karvonen, P. Taskinen, J. Takkunen, H. Kyrönseppä, and P. Peltokallio: Cardiovascular Studies on Former Endurance Athletes, *Am. J. Cardiol.*, **20**:191, 1967.

Rodahl, K.: "Be Fit for Life," Harper & Row, Publishers, Incorporated, New York, 1966.

Rook, A.: An Investigation into the Longevity of Cambridge Sportsmen, *Brit. Med. J.*, **1**:773, 1954.

Roskamm, H., H. Reindell, and K. König: "Körperliche Aktivität und Herz und Kreislauferkrankungen," Johann Ambrosius Barth, Munich, 1966.

Appendix

contents

Appendix

MEASUREMENT OF OXYGEN UPTAKE

The Douglas Bag Method

Oxygen uptake is a basic parameter in most studies pertaining to work physiology. The ability to measure oxygen uptake is therefore of fundamental importance to the work physiologist.

The classical method for the determination of oxygen uptake, i.e., the Douglas bag method, rests on a very secure foundation. It is theoretically sound, and it is well tested under a wide variety of circumstances. In all its relative simplicity it is unsurpassable in accuracy.

A disadvantage with the method is that the subject is somewhat hampered by the equipment required for the collection of the expired air. This limits the subject's freedom of movement. Furthermore, it merely provides a mean figure for the oxygen uptake of, say, 30 sec, depending on the length of the time in which the expired air is collected.

Figure A-1 shows the adaptation of this method for experiments using the bicycle ergometer (right) and the treadmill (left). The inner area of the mouthpiece is 400 mm², the inner diameters of the tubes in the valve, stopcock, and bag are 28 mm, and of the connecting tube (smooth, not corrugated), 35 mm. The tubes should be as short as possible. The resistance in this system with a connecting tube of 0.5-m length with a flow rate of 100 liters/min is 1 cm H_2O; of 200 liters/min, 3 cm H_2O; of 300 liters/min, 6 cm H_2O; and of 400 liters/min, 10 cm H_2O. A valve which is still better than this has recently been developed.

Respiratory rate is registered via a Marey capsule. Two Douglas bags are connected to the four-way stopcock to enable continuous collection of air. The subject is connected to the bag during an inspiration. The turning of the stopcock starts the stopwatch, which can be read down to .01 min (see the photograph inserted in the middle of Fig. A-1). When no less than about 50 liters of expired

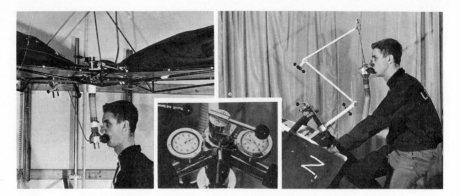

Fig. A-1   Arrangements of respiratory valves, stopcocks, and Douglas bags for the collection of expired air during experiments using a treadmill (left) or a bicycle ergometer (right).
With the aid of the arrangement shown in the center of the figure, the handle opens the stopcock for one of the two Douglas bags. At the same time the corresponding stopwatch is automatically started. When the handle is moved to close the opening to the first bag, the stopwatch is automatically stopped. With the simultaneous opening of the stopcock aperture, leading the expired air into the second bag, the corresponding stopwatch is automatically started. In this manner the time for the collection of the expired air is taken automatically.

air has been collected in the bag, the stopcock is turned to the second bag or back to room air. The stopcock is always turned during inspiration; the second turning of the stopcock automatically stops the stopwatch. The volume of expired air is measured in a balanced spirometer and the composition of the air is analyzed by the Haldane or Micro-Scholander or by electronic gas analyzers checked against the aforementioned manometric methods. Figures 9-2 and 16-8 give examples of the adaptation of the same method in experiments carried out in the field.

WORK TESTS

Different types of work and work tests were discussed in Chaps. 9 and 11. The current methods include the upright bicycle ergometer, the step test, or the treadmill. The procedure may vary from a single-level load to an intermittent series of increasing loads with intermittent rest periods, an almost continuous increase in load, or a continuous series of increasing loads with a steady state at each level. This latter procedure is only possible when applying loads below 50 to 70 percent of the subject's maximal oxygen uptake.

In the case of single-level loads the heart rate should reach a level above 120 beats/min and the work period should be about 6 min. In the case of multiple-level tests we prefer an intermittent series of test loads, or a continuous increase in load but with work periods at each load level which exceed 5 min.

If the oxygen uptake is to be measured and steady-state or "apparent" steady-state values are to be obtained, the noseclip, mouthpiece, and respiratory valve are placed on the subject roughly 4 min after the beginning of the test load and the collection of the expiratory air is initiated after about 5 min from the start of the test. (If the metabolic respiratory quotient is to be determined, a much longer work period is required before the air sample is collected.)

When the maximal aerobic power is to be assessed, the tests should preferably be extended to 2 days or more. On the first day two submaximal and one predicted maximal load are performed. On the second day, an additional one or two submaximal loads are performed, followed by a maximal load, which is determined on the basis of the results from the previous day's test. (The necessary criteria for ascertaining that the maximal oxygen uptake actually has been reached are discussed in Chap. 9.)

If for any reason only one test can be performed in each subject the following procedure may prove useful.

Bicycle Ergometer

The subject is exposed to one or two submaximal loads. The load should be adjusted so that the heart rate is at least 140 beats/min in the case of subjects below fifty years of age, and 120 beats/min in the case of subjects above fifty years of age. Then a "supermaximal" load should be tried. If the subject can tolerate this supermaximal load for at least 2 min, even with considerable difficulty, measurements carried out at the end of the work period are likely to give an oxygen uptake equal to or close to the individual's maximal oxygen uptake. In any case this will require a well-motivated subject. Usually this supermaximal load is selected on the basis of the individual's predicted maximal oxygen uptake from his heart rate at the submaximal work loads, using the nomogram of Åstrand and Åstrand (Fig. 11-6). From this a load is selected which would require an oxygen uptake which is about 10 to 20 percent higher than the predicted maximal oxygen uptake (see Fig. 11-9 and Table A-1). If the subject at the end

TABLE A-1

Work load		Oxygen uptake, liters/min
watts	kpm/min	
50	300	0.9
100	600	1.5
150	900	2.1
200	1200	2.8
250	1500	3.5
300	1800	4.2
350	2100	5.0
400	2400	5.7

of the first min on the selected load has difficulty in keeping up the pedaling rate and starts to hyperventilate markedly, the load is lowered slightly so as to allow the subject to continue for a total of about 3 min. If on the other hand the subject, after a minute or two, appears to have more strength left than originally predicted, the load is slightly increased. In such a case the individual's maximal oxygen uptake obviously has been underestimated when computing the predicted maximal oxygen uptake.

The objective is to be able to collect at least two Douglas bags of expiratory air. This requires a collection time of no less than 1 min. It is a matter of experience to decide when to put the mouthpiece and respiratory valve in place on the subject and when to start the collection of the air sample. It is always better, however, to start too early than too late.

This "quick method" requires a considerable amount of experience on the part of the investigator in order to attain satisfactory results. In our opinion the peak value of blood lactate mirrors fairly well the severity of the load on the aerobic and anaerobic processes and the degree of exhaustion, as does the respiratory quotient as well. Blood for the determination of lactate concentration can easily be obtained from the finger tip. (The hand should be prewarmed and washed in hot water.) Three to four blood samples should be secured within 1 min after the end of the work and again after about 3, 6, and 10 min. The most suitable pedaling rate has been found to be about 60 rpm.

Treadmill

In order to select a suitable starting speed and incline of the treadmill for a maximal run on the treadmill, the subject's maximal oxygen uptake (ml/min per kg body weight) is first predicted from the heart rate at a submaximal work load on the bicycle ergometer, as described above. Then Table A-2 is applied. Starting at this initial speed and incline, the incline is increased by 1.5 degrees (2.67 percent) every third min, keeping the speed constant. Under these conditions not even top athletes are able to run for more than 7 min. Experience has shown that it is advisable to give females a somewhat lower relative starting load than in the case of males (Table A-2). Immediately preceding the maximal run, the subject is given a 10-min warm-up on the treadmill at a load corresponding to 50 percent of the selected starting load. The subjects are not allowed to hold on to the treadmill railing during the run.

ENVIRONMENTAL CONDITIONS OF THE TEST LABORATORY

The room temperature should be between 19 and 21°C, and the relative humidity between 40 and 60 percent. The oxygen content of the room air should not be below 20.90 percent.

TABLE A-2
Starting work load used for the maximal run on the treadmill

Predicted max $\dot{V}_{O_2}$, ml/kg × min	Starting speed and inclination for max run on treadmill*							
	♂ Speed uphill				♀ Speed uphill			
	km/hr	Degree	mph	%	km/hr	Degree	mph	%
<40	10.0	3.0	6.2	5.25	10.0	1.5	6.2	2.67
40–50	12.5	3.0	7.8	5.25	10.0	3.0	6.2	5.25
55–75	15.0	3.0	9.3	5.25	12.5	3.0	7.8	5.25
>75								
A	15.0	4.5	9.3	7.0				
B	17.5	3.0	10.9	5.25				
C	20.0	1.5	12.5	2.67				

* A = cross-country skiers, skaters, etc., B = cross-country runners, orientation runners, etc., C = track runners.
SOURCE: Saltin and Åstrand: *J. Appl. Physiol.*, **23**:353, 1967.

PREDICTION OF MAXIMAL OXYGEN UPTAKE

The various sources of error associated with every type of prediction have already been emphasized. Having pointed out these sources of error, one may be justified in summarizing a principle which is based on personal experience and which requires a minimum of equipment and facilities. The most important thing is to keep in mind that not just any predicted value can be used as a substitute for the actual measurement of precise values when scientific exactness is required.

Estimation of the Maximal Oxygen Uptake on the Basis of the Heart Rate Response to Submaximal Work Loads

Under normal conditions there is in any given individual a roughly linear relationship between oxygen uptake and heart rate during submaximal work (Fig. A-2). The slope of the line changes with the state of physical training or physical fitness; a fit person is able to transport the same amount of oxygen at a lower heart rate than an unfit person. This relationship in general is independent of sex and age, although females acquire higher heart rates to transport the same amount of oxygen than males.

On the basis of this oxygen uptake–heart rate relationship, it is possible to predict an individual's maximal oxygen uptake by the heart rate response to two submaximal loads, as is evident from Fig. A-2, assuming a maximal heart rate of about 190 to 200 beats/min. However, since the maximal heart rate declines with age after about age 20, it is necessary to make corrections for age. It should be pointed out that at a given submaximal work load a trained 65-year-old subject

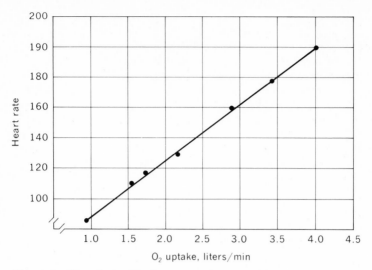

Fig. A-2   The relationship between work rate and O_2 uptake on the bicycle ergometer.

may have about the same heart rate as a trained 30-year-old subject. However, the 30-year-old subject, having a maximal heart rate of about 190, may have a maximal oxygen uptake of about 4.0 liters per min, while the older subject, having a maximal heart rate of about 160, can only take up 3.0 liters of oxygen at a maximum, using the example illustrated in Fig. A-2.

These considerations form the basis for the nomogram for the estimation of maximal O_2 uptake on the basis of heart rate or pulse response to submaximal work load developed by Åstrand and Åstrand (Table A-3). This nomogram was recently modified by I. Åstrand to include correction for the age factor (Table A-4). The error of the method is about 10 to 15 percent. The test can be performed on a bicycle ergometer such as that produced by Monark at Varberg, Sweden. A detailed description of the performance of the test accompanies each bicycle ergometer.

Performance of the Test

From the above it is evident that the pulse rate as such during a work load under standard conditions can be used as an indication of the state of circulatory fitness of an individual. This is especially valuable in following the effect of a training program when the pulse response in the same individual is compared before and after a training program and the subject serves as his own control.

Such tests should be done under steady-state conditions, i.e., during continuous steady work lasting at least 5 to 6 min. Under such steady-state conditions, O_2 equals oxygen demand, so that the metabolic processes are

TABLE A-3

Prediction of maximal oxygen uptake from heart rate and work load on a bicycle ergometer. The value should be corrected for age, using the factor given in Table A-4

Men

Heart rate	Maximal oxygen uptake, liters/min				
	300 kpm/min	600 kpm/min	900 kpm/min	1200 kpm/min	1500 kpm/min
120	2.2	3.5	4.8		
121	2.2	3.4	4.7		
122	2.2	3.4	4.6		
123	2.1	3.4	4.6		
124	2.1	3.3	4.5		
125	2.0	3.2	4.4	6.0	
126	2.0	3.2	4.4	5.9	
127	2.0	3.1	4.3	5.8	
128	2.0	3.1	4.2	5.7	
129	1.9	3.0	4.2	5.6	
130	1.9	3.0	4.1	5.6	
131	1.9	2.9	4.0	5.5	
132	1.8	2.9	4.0	5.4	
133	1.8	2.8	3.9	5.3	
134	1.8	2.8	3.9	5.2	
135	1.7	2.8	3.8	5.1	
136	1.7	2.7	3.8	5.0	
137	1.7	2.7	3.7	5.0	
138	1.6	2.7	3.7	4.9	
139	1.6	2.6	3.6	4.8	
140	1.6	2.6	3.6	4.8	6.0
141	1.6	2.6	3.5	4.7	5.9
142		2.5	3.5	4.6	5.8
143		2.5	3.4	4.6	5.7
144		2.4	3.4	4.5	5.7
145		2.4	3.4	4.5	5.6
146		2.4	3.3	4.4	5.6
147		2.4	3.3	4.4	5.5

Heart rate	Maximal oxygen uptake, liters/min				
	300 kpm/min	600 kpm/min	900 kpm/min	1200 kpm/min	1500 kpm/min
148		2.4	3.2	4.3	5.4
149		2.3	3.2	4.3	5.4
150		2.3	3.2	4.2	5.3
151		2.3	3.1	4.2	5.2
152		2.3	3.1	4.1	5.2
153		2.2	3.0	4.1	5.1
154		2.2	3.0	4.0	5.1
155		2.2	3.0	4.0	5.0
156		2.2	2.9	4.0	5.0
157		2.1	2.9	3.9	4.9
158		2.1	2.9	3.9	4.9
159		2.1	2.8	3.8	4.8
160		2.1	2.8	3.8	4.8
161		2.0	2.8	3.7	4.7
162		2.0	2.8	3.7	4.6
163		2.0	2.7	3.7	4.6
164		2.0	2.7	3.6	4.5
165		2.0	2.7	3.6	4.5
166		1.9	2.6	3.6	4.5
167		1.9	2.6	3.5	4.4
168		1.9	2.6	3.5	4.4
169		1.9	2.6	3.5	4.3
170		1.8	2.6	3.4	4.3

Women

Heart rate	Maximal oxygen uptake, liters/min				
	300 kpm/min	450 kpm/min	600 kpm/min	750 kpm/min	900 kpm/min
120	2.6	3.4	4.1	4.8	
121	2.5	3.3	4.0	4.8	
122	2.5	3.2	3.9	4.7	
123	2.4	3.1	3.9	4.6	
124	2.4	3.1	3.8	4.5	
125	2.3	3.0	3.7	4.4	
126	2.3	3.0	3.6	4.3	
127	2.2	2.9	3.5	4.2	
128	2.2	2.8	3.5	4.2	4.8
129	2.2	2.8	3.4	4.1	4.8
130	2.1	2.7	3.4	4.0	4.7
131	2.1	2.7	3.4	4.0	4.6
132	2.0	2.7	3.3	3.9	4.5
133	2.0	2.6	3.2	3.8	4.4
134	2.0	2.6	3.2	3.8	4.4
135	2.0	2.6	3.1	3.7	4.3
136	1.9	2.5	3.1	3.6	4.2
137	1.9	2.5	3.0	3.6	4.2
138	1.8	2.4	3.0	3.5	4.1
139	1.8	2.4	2.9	3.5	4.0
140	1.8	2.4	2.8	3.4	4.0
141	1.8	2.3	2.8	3.4	3.9
142	1.7	2.3	2.8	3.3	3.9
143	1.7	2.2	2.7	3.3	3.8
144	1.7	2.2	2.7	3.2	3.8
145	1.6	2.2	2.7	3.2	3.7
146	1.6	2.2	2.6	3.2	3.7
147	1.6	2.1	2.6	3.1	3.6

Heart rate	Maximal oxygen uptake, liters/min				
	300 kpm/min	450 kpm/min	600 kpm/min	750 kpm/min	900 kpm/min
148	1.6	2.1	2.6	3.1	3.6
149		2.1	2.6	3.0	3.5
150		2.0	2.5	3.0	3.5
151		2.0	2.5	3.0	3.4
152		2.0	2.5	2.9	3.4
153		2.0	2.4	2.9	3.3
154		2.0	2.4	2.8	3.3
155		1.9	2.4	2.8	3.2
156		1.9	2.3	2.8	3.2
157		1.9	2.3	2.7	3.2
158		1.8	2.3	2.7	3.1
159		1.8	2.2	2.7	3.1
160		1.8	2.2	2.6	3.0
161		1.8	2.2	2.6	3.0
162		1.8	2.2	2.6	3.0
163		1.7	2.2	2.6	2.9
164		1.7	2.1	2.5	2.9
165		1.7	2.1	2.5	2.9
166		1.7	2.1	2.5	2.8
167		1.6	2.1	2.4	2.8
168		1.6	2.0	2.4	2.8
169		1.6	2.0	2.4	2.8
170		1.6	2.0	2.4	2.7

SOURCE: From a nomogram by I. Åstrand: Acta Physiol. Scand. 49 (Suppl. 169):45–60, 1960.

TABLE A-4

Factor to be used for correction of predicted maximal oxygen uptake (1) when the subject is over 30 to 35 years of age or (2) when the subject's maximal heart rate is known. The actual factor should be multiplied by the value that is obtained from Table A-3

Age	Factor	Max heart rate	Factor
15	1.10	210	1.12
25	1.00	200	1.00
35	0.87	190	0.93
40	0.83	180	0.83
45	0.78	170	0.75
50	0.75	160	0.69
55	0.71	150	0.64
60	0.68		
65	0.65		

essentially aerobic, with no significant increase in blood lactate. Under these conditions pulmonary ventilation, heart rate, and cardiodynamic parameters, such as cardiac output, are essentially constant during the last 2 to 3 min of the test lasting 5 to 6 min. The advantage with this is that one may well compare the results from one test to another.

The subject should refrain from energetic physical activity 2 hr preceding the work test. In addition, the test should not be performed earlier than about 1 hr after a light meal, or after 2 to 3 hours if a heavier meal has been taken. Furthermore, the test subject should not smoke for the last 30 min prior to the commencement of the test.

Experience has shown that the resting heart rate does not normally give any information over and above that provided by the work test. It is therefore not necessary to record the resting heart rate.

The saddle and handlebars of the bicycle ergometer are adjusted to suit the test subject. Tests have shown that the mechanical efficiency does not vary with the height of the handlebars and saddle, providing that this is kept within reasonable limits. The most comfortable position, and in the case of very heavy loads the most effective one, is a saddle height that produces a slight bend of the knee joint when the ball of the foot rests on the pedal and the leg is stretched.

Provided that the work is not too heavy, respiration and circulation increase during the first 2 to 3 min of work and then attain a steady state. The increase in pulse rate can easily be established by counting the pulse once every minute. In any case, after 4 to 5 minutes' work the pulse rate has generally reached the steady state. As a rule, a working time of about 5 to 6 min is thus sufficient to adapt the pulse rate to the task being performed. The pulse rate should be taken every minute, the mean value of the pulse rate at the fifth and sixth minutes being designated the working pulse for the work in question. If the difference between these last two pulse rates exceeds 5 beats/min, the working time should be prolonged one or more minutes until a constant level is reached. The pulse rate is most easily felt over the carotid artery just below the mandible angle, and the most exact value is obtained by taking the time for 30 pulse beats. (The stopwatch showing tenths of a second is started at the "0" pulse beat.) Using Table A-5, the time recorded for 30 beats can be converted into the pulse rate per minute. (Example: If it takes 13.4 sec for the heart to beat 30 times, the heart rate is 134 beats/min.) For the inexperienced, it is rather difficult to count the pulse rate: the metronome confuses, the subject is in motion, and the pulse may vary. Practice under experienced supervision is important. The pulse rate may be suitably measured during the last 15 to 20 sec of every working minute.

A table similar to Table A-5 may be constructed in the shape of a watch dial which can be fixed on a stopwatch (Fig. A-3). Such stopwatches are now available on the market. In this case the outer scale is read when the exact time is taken for 10 pulse beats. The inner scale is used when the time is taken for 30 beats. Since it takes 15 sec for the watch hand to make a complete revolution, the pulse rate is 120 beats/min (if it takes 30 beats for the hand to go around the dial once). Experience has shown that a person trained in taking pulse rates, whether it be by palpation or by auscultation, obtains values which are in close agreement with those obtained by ECG recordings, providing the pulse rate is not irregular.

Choice of Load

For trained, active individuals, the risk of overstraining in connection with a work test is very slight. For female subjects a suitable load is 450 or 600 kpm/min, and for male test subjects, 600 or 900 kpm/min. If the heart rate exceeds about 130 beats per min, the load can be considered adequate and the test can be discontinued after 6 min. If the pulse rate is slower than about 130 beats per min, the load should be increased by 300 kpm/min after 6 min. If time permits testing with several loads, increase by 300 kpm/min in 6-min periods for as long as the pulse rate remains below about 150 beats per min (time for 30 pulse beats = 12.0 sec). The next working period should be continued for 6 min even if the pulse rate should then exceed 150 beats per min.

For persons who may be expected to have a lower physical work capacity, persons who are completely untrained, or older individuals, lower loads should be chosen and an initial load of 300 kpm/min might be suitable.

If a physician is not present, work tests on persons over 40 years of age should be discontinued if the pulse rate exceeds 150 beats per min (time for 30 pulse beats = 12.0 sec). The load for such persons should not be raised above 600 kpm/min for female subjects or 900 kpm/min for male subjects. In the event

TABLE A-5

Conversion of the time for 30 pulse beats to pulse rate per minute

sec	beats/min	sec	beats/min	sec	beats/min
22.0	82	17.3	104	12.6	143
21.9	82	17.2	105	12.5	144
21.8	83	17.1	105	12.4	145
21.7	83	17.0	106	12.3	146
21.6	83	16.9	107	12.2	148
21.5	84	16.8	107	12.1	149
21.4	84	16.7	108	12.0	150
21.3	85	16.6	108	11.9	151
21.2	85	16.5	109	11.8	153
21.1	85	16.4	110	11.7	154
21.0	86	16.3	110	11.6	155
20.9	86	16.2	111	11.5	157
20.8	87	16.1	112	11.4	158
20.7	87	16.0	113	11.3	159
20.6	87	15.9	113	11.2	161
20.5	88	15.8	114	11.1	162
20.4	88	15.7	115	11.0	164
20.3	89	15.6	115	10.9	165
20.2	89	15.5	116	10.8	167
20.1	90	15.4	117	10.7	168
20.0	90	15.3	118	10.6	170
19.9	90	15.2	118	10.5	171
19.8	91	15.1	119	10.4	173
19.7	91	15.0	120	10.3	175
19.6	92	14.9	121	10.2	176
19.5	92	14.8	122	10.1	178
19.4	93	14.7	122	10.0	180
19.3	93	14.6	123	9.9	182
19.2	94	14.5	124	9.8	184
19.1	94	14.4	125	9.7	186
19.0	95	14.3	126	9.6	188
18.9	95	14.2	127	9.5	189
18.8	96	14.1	128	9.4	191
18.7	96	14.0	129	9.3	194
18.6	97	13.9	129	9.2	196
18.5	97	13.8	130	9.1	198
18.4	98	13.7	131	9.0	200
18.3	98	13.6	132	8.9	202
18.2	99	13.5	133	8.8	205
18.1	99	13.4	134	8.7	207
18.0	100	13.3	135	8.6	209
17.9	101	13.2	136	8.5	212
17.8	101	13.1	137	8.4	214
17.7	102	13.0	138	8.3	217
17.6	102	12.9	140	8.2	220
17.5	103	12.8	141	8.1	222
17.4	103	12.7	142	8.0	225

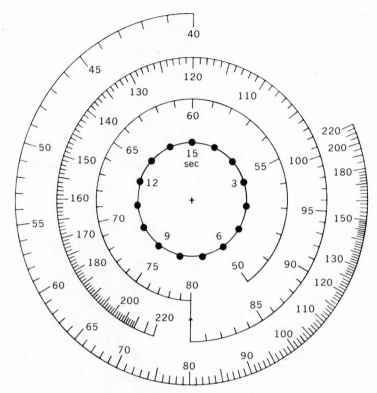

Fig. A-3   The arrangement of the dial of a stopwatch showing the heart rate in beats per minute when the time is taken for 10 (outer dial) or 30 (inner dials) heartbeats.

of pressure or pain in the chest, or marked shortness of breath or distress, the test must be discontinued immediately.

Evaluation of the Work Test

The work test, in the simple form described above, actually gives very limited possibilities of judging the test subject's physical capacity for running, skiing, swimming, etc. In the performance of various kinds of sports and athletics, and in physical work in general, the "motor power" obviously plays an important part, but other critical factors include technique and motivation. However, the test does give some idea of the capacity of the "combustion engine," or the oxygen-transporting capacity, but even here there are sources of error. The maximal pulse rate varies with age, but it can also vary within the same age group. A pulse rate of 150 at work implies an almost maximal effort for a person with a maximal pulse rate of 160, but represents a light load for a person with a pulse ceiling at 200 beats/min.

TABLE A-6
Calculation of maximal oxygen uptake, ml/kg × min

Body weight lb	kg	Maximal oxygen uptake, liters/min																								
		1.5	1.6	1.7	1.8	1.9	2.0	2.1	2.2	2.3	2.4	2.5	2.6	2.7	2.8	2.9	3.0	3.1	3.2	3.3	3.4	3.5	3.6	3.7	3.8	3.9
110	50	30	32	34	36	38	40	42	44	46	48	50	52	54	56	58	60	62	64	66	68	70	72	74	76	78
112	51	29	31	33	35	37	39	41	43	45	47	49	51	53	55	57	59	61	63	65	67	69	71	73	75	76
115	52	29	31	33	35	37	38	40	42	44	46	48	50	52	54	56	58	60	62	63	65	67	69	71	73	75
117	53	28	30	32	34	36	38	40	42	43	45	47	49	51	53	55	57	58	60	62	64	66	68	70	72	74
119	54	28	30	31	33	35	37	39	41	43	44	46	48	50	52	54	56	57	59	61	63	65	67	69	70	72
121	55	27	29	31	33	35	36	38	40	42	44	45	47	49	51	53	55	56	58	60	62	64	65	67	69	71
123	56	27	29	30	32	34	36	38	39	41	43	45	46	48	50	52	54	55	57	59	61	63	64	66	68	70
126	57	26	28	30	32	33	35	37	39	40	42	44	46	47	49	51	53	54	56	58	60	61	63	65	67	68
128	58	26	28	29	31	33	34	36	38	40	41	43	45	47	48	50	52	53	55	57	59	60	62	64	66	67
130	59	25	27	29	31	32	34	36	37	39	41	42	44	46	47	49	51	53	54	56	58	59	61	63	64	66
132	60	25	27	28	30	32	33	35	37	38	40	42	43	45	47	48	50	52	53	55	57	58	60	62	63	65
134	61	25	26	28	30	31	33	34	36	38	39	41	43	44	46	48	49	51	52	54	56	57	59	61	62	64
137	62	24	26	27	29	31	32	34	35	37	39	40	42	44	45	47	48	50	52	53	55	56	58	60	61	63
139	63	24	25	27	29	30	32	33	35	37	38	40	41	43	44	46	48	49	51	52	54	56	57	59	60	62
141	64	23	25	27	28	30	31	33	34	36	38	39	41	42	44	45	47	48	50	52	53	55	56	58	59	61
143	65	23	25	26	28	29	31	32	34	35	37	38	40	42	43	45	46	48	49	51	52	54	55	57	58	60
146	66	23	24	26	27	29	30	32	33	35	36	38	39	41	42	44	45	47	48	50	52	53	55	56	58	59
148	67	22	24	25	27	28	30	31	33	34	36	37	39	40	42	43	45	46	48	49	51	52	54	55	57	58
150	68	22	24	25	26	28	29	31	32	34	35	37	38	40	41	43	44	46	47	49	50	51	53	54	56	57
152	69	22	23	25	26	28	29	30	32	33	35	36	38	39	41	42	43	45	46	48	49	51	52	54	55	57
154	70	21	23	24	26	27	29	30	31	33	34	36	37	39	40	41	43	44	46	47	49	51	52	54	55	56
157	71	21	23	24	25	27	28	30	31	32	34	35	37	38	39	41	42	44	45	46	48	49	51	52	54	55
159	72	21	22	24	25	26	28	29	31	32	33	35	36	38	39	40	42	43	44	46	47	49	50	51	53	54
161	73	21	22	23	25	26	27	29	30	32	33	34	36	37	38	40	41	42	44	45	47	48	49	51	52	53
163	74	20	22	23	24	26	27	28	30	31	32	34	35	36	38	39	41	42	43	45	46	47	49	50	51	53
165	75	20	21	23	24	25	27	28	29	31	32	33	35	36	37	39	40	41	43	44	45	47	48	49	51	52
168	76	20	21	22	24	25	26	28	29	30	32	33	34	36	37	38	39	41	42	43	45	46	47	49	50	51
170	77	19	21	22	23	25	26	27	29	30	31	32	34	35	36	38	39	40	42	43	44	45	47	48	49	51
172	78	19	21	22	23	24	26	27	28	29	31	32	33	35	36	37	38	40	41	42	44	45	46	47	49	50
174	79	19	20	22	23	24	25	27	28	29	30	32	33	34	35	37	38	39	41	42	43	44	46	47	49	49
176	80	19	20	21	23	24	25	26	28	29	30	31	33	34	35	36	38	39	40	41	43	44	45	46	48	49
179	81	19	20	21	22	23	25	26	27	28	30	31	32	33	35	36	37	38	40	41	42	43	44	46	47	48
181	82	18	20	21	22	23	24	26	27	28	29	30	32	33	34	35	37	38	39	40	41	43	45	46	48	48
183	83	18	19	20	22	23	24	25	27	28	29	30	31	33	34	35	36	37	39	40	41	42	43	45	46	47
185	84	18	19	20	21	23	24	25	26	27	29	30	31	32	33	35	36	37	38	39	40	42	43	44	45	46
187	85	18	19	20	21	22	24	25	26	27	28	29	31	32	33	34	35	36	38	39	40	41	42	44	45	46
190	86	17	19	20	21	22	23	24	26	27	28	29	30	31	32	33	35	36	37	38	39	40	42	43	44	45
192	87	17	18	20	21	22	23	24	25	26	28	29	30	31	32	33	34	36	37	38	39	40	41	43	44	45
194	88	17	18	19	20	22	23	24	25	26	27	28	30	31	32	33	34	35	36	38	39	40	41	42	43	44
196	89	17	18	19	20	21	22	24	25	26	27	28	29	30	31	33	34	35	36	37	38	39	40	42	43	44
198	90	17	18	19	20	21	22	23	24	26	27	28	29	30	31	32	33	34	36	37	38	39	40	41	42	43
201	91	16	18	19	20	21	22	23	24	25	26	27	29	30	31	32	33	34	35	36	37	38	40	41	42	43
203	92	16	17	18	20	21	22	23	24	25	26	27	28	29	30	32	33	34	35	36	37	38	39	40	41	43
205	93	16	17	18	19	20	22	23	24	25	26	27	28	29	30	31	32	33	34	35	37	38	39	40	41	42
207	94	16	17	18	19	20	21	22	23	24	26	27	28	29	30	31	32	33	34	35	36	37	38	39	40	41
209	95	16	17	18	19	20	21	22	23	24	25	26	27	28	29	31	32	33	34	35	36	37	38	39	40	41
212	96	16	17	18	19	20	21	22	23	24	25	26	27	28	29	30	31	32	33	34	35	36	38	39	40	41
214	97	15	16	18	19	20	21	22	23	24	25	26	27	28	29	30	31	32	33	34	35	36	37	38	39	40
216	98	15	16	17	18	19	20	21	22	23	24	26	27	28	29	30	31	32	33	34	35	36	37	38	39	40
218	99	15	16	17	18	19	20	21	22	23	24	25	26	27	28	29	30	31	32	33	34	35	36	37	38	39
220	100	15	16	17	18	19	20	21	22	23	24	25	26	27	28	29	30	31	32	33	34	35	36	37	38	39

Body weight, lb	kg	Maximal oxygen uptake, liters/min																				
		4.0	4.1	4.2	4.3	4.4	4.5	4.6	4.7	4.8	4.9	5.0	5.1	5.2	5.3	5.4	5.5	5.6	5.7	5.8	5.9	6.0
110	50	80	82	84	86	88	90	92	94	96	98	100	102	104	106	108	110	112	114	116	118	120
112	51	78	80	82	84	86	88	90	92	94	96	98	100	102	104	106	108	110	112	114	116	118
115	52	77	79	81	83	85	87	88	90	92	94	96	98	100	102	104	106	108	110	112	113	115
117	53	75	77	79	81	83	85	87	89	91	92	94	96	98	100	102	104	106	108	109	111	113
119	54	74	76	78	80	81	83	85	87	89	91	93	94	96	98	100	102	104	106	107	109	111
121	55	73	75	76	78	80	82	84	85	87	89	91	93	95	96	98	100	102	104	105	107	109
123	56	71	78	75	77	79	80	82	84	86	88	89	91	93	95	96	98	100	102	104	106	107
126	57	70	72	74	75	77	79	81	82	84	86	88	89	91	93	95	96	98	100	102	104	105
128	58	69	71	72	74	76	78	79	81	83	84	86	88	90	91	93	95	97	98	100	102	103
130	59	68	69	71	73	75	76	78	80	81	83	85	86	88	90	92	93	95	97	98	100	102
132	60	67	68	70	72	73	75	77	78	80	82	83	85	87	88	90	92	93	95	97	98	100
134	61	66	67	69	70	72	74	75	77	79	80	82	84	85	87	89	90	92	93	95	97	98
137	62	65	66	68	69	71	73	74	76	77	79	81	82	84	85	87	89	90	92	94	95	97
139	63	63	65	67	68	70	71	73	75	76	78	79	81	83	84	86	87	89	90	92	94	95
141	64	63	64	66	67	69	70	72	73	75	77	78	80	81	83	84	86	88	89	91	92	94
143	65	62	63	65	66	68	69	71	72	74	75	77	78	80	82	83	85	86	88	89	91	92
146	66	61	62	64	65	67	68	70	71	73	74	76	77	79	80	82	83	85	86	88	89	91
148	67	60	61	63	64	66	67	69	70	72	73	75	76	78	79	81	82	84	85	87	88	90
150	68	59	60	62	63	65	66	68	69	71	72	74	75	76	78	79	81	82	84	85	87	88
152	69	58	59	61	62	64	65	67	68	70	71	72	74	75	77	78	80	81	83	84	86	87
154	70	57	59	60	61	63	64	66	67	69	70	71	73	74	76	77	79	80	81	83	84	86
157	71	56	58	59	61	62	63	65	66	68	69	70	72	73	75	76	77	79	80	82	83	85
159	72	56	57	58	60	61	63	64	65	67	68	69	71	72	74	75	76	78	79	81	82	83
161	73	55	56	58	59	60	62	63	64	66	67	68	70	71	73	74	75	77	78	79	81	82
163	74	54	55	57	58	59	61	62	64	65	66	68	69	70	72	73	74	76	77	78	80	81
165	75	53	55	56	57	59	60	61	63	64	65	67	68	69	71	72	73	75	76	77	79	80
168	76	53	54	55	57	58	59	61	62	63	64	66	67	68	70	71	72	74	75	76	78	79
170	77	52	53	55	56	57	58	60	61	62	64	65	66	68	69	70	71	73	74	75	77	78
172	78	51	53	54	55	56	58	59	60	62	63	64	65	67	68	69	71	72	73	74	76	77
174	79	51	52	53	54	56	57	58	59	61	62	63	65	66	67	68	70	71	72	73	75	76
176	80	50	51	53	54	55	56	58	59	60	61	63	64	65	66	68	69	70	71	72	74	75
179	81	49	51	52	53	54	56	57	58	59	60	62	63	64	65	67	68	69	70	72	73	74
181	82	49	50	51	52	54	55	56	57	59	60	61	62	63	65	66	67	68	70	71	72	73
183	83	48	49	51	52	53	54	55	57	58	59	60	61	63	64	65	66	67	69	70	71	72
185	84	48	49	50	51	52	54	55	56	57	58	60	61	62	63	64	65	67	68	69	70	71
187	85	47	48	49	51	52	53	54	55	56	58	59	60	61	62	64	65	66	67	68	69	71
190	86	47	48	49	50	51	52	53	55	56	57	58	59	60	62	63	64	65	66	67	69	70
192	87	46	47	48	49	51	52	53	54	55	56	57	59	60	61	62	63	64	66	67	68	69
194	88	45	47	48	49	50	51	52	53	55	56	57	58	59	60	61	63	64	65	66	67	68
196	89	45	46	47	48	49	51	52	53	54	55	56	57	58	60	61	62	63	64	65	66	67
198	90	44	46	47	48	49	50	51	52	53	54	56	57	58	59	60	61	62	63	64	66	67
201	91	44	45	46	47	48	49	51	52	53	54	55	56	57	58	59	60	62	63	64	65	66
203	92	43	45	46	47	48	49	50	51	52	53	54	55	57	58	59	60	61	62	63	64	65
205	93	43	44	45	46	47	48	49	51	52	53	54	55	56	57	58	59	60	61	62	63	65
207	94	43	44	45	46	47	48	49	50	51	52	53	54	55	56	57	59	60	61	62	63	64
209	95	42	43	44	45	46	47	48	49	51	52	53	54	55	56	57	58	59	60	61	62	63
212	96	42	43	44	45	46	47	48	49	50	51	52	53	54	55	56	57	58	59	60	61	63
214	97	41	42	43	44	45	46	47	48	49	51	52	53	54	55	56	57	58	59	60	61	62
216	98	41	42	43	44	45	46	47	48	49	50	51	52	53	54	55	56	57	58	59	60	61
218	99	40	41	42	43	44	45	46	47	48	49	51	52	53	54	55	56	57	58	59	60	61
220	100	40	41	42	43	44	45	46	47	48	49	50	51	52	53	54	55	56	57	58	59	60

TABLE A-7
Suggested form for a record of work tests

Name of test subject: .... Date: .... / ....

Date of birth ..../ .... Place for test ....

Height: .... cm   Weight: .... kg   net/gross ....
                       in.   ft   lb

.... kpm/min	.... kpm/min	.... kpm/min	.... kpm/min
Date /	Date /	Date /	Date /
1'	1'	1'	1'
2'	2'	2'	2'
3'	3'	3'	3'
4'	4'	4'	4'
5'	5'	5'	5'
6'	6'	6'	6'
Pulse Rate:	Pulse Rate:	Pulse Rate:	Pulse Rate:
Corr. max. oxygen uptake Liters/min .... ml/kg × min	Corr. max. oxygen uptake Liters/min .... ml/kg × min	Corr. max. oxygen uptake Liters/min .... ml/kg × min	Corr. max. oxygen uptake Liters/min .... ml/kg × min ......
Training: Remarks:			

The most valuable application of the work test as described above is to test the individual on several different occasions during a period of physical training. In this way it is possible to determine objectively whether the training program has been effective (i.e., a given work load is achieved with a lower pulse rate).

Table A-3 enables the maximal oxygen uptake to be estimated from the heart rate at a certain load. [Example: A male subject has a pulse rate of 147 when working at a rate of 900 kpm/min. According to Table A-3, his maximal oxygen uptake will be about 3.3 liters/min. The maximal oxygen uptake per kilogram of body weight is obtained from Table A-6. In this case, when the test subject weighs 74 kg it will be 45 ml/(kg)(min). If different loads have been used, the mean value of the oxygen uptake rate calculated for each load is applied.]

Obviously, the table values are only approximations. Experience has shown that older persons are generally overestimated in regard to predicted maximal oxygen uptake. The value obtained from Table A-3 must therefore be corrected by multiplication with the age factor given in Table A-4. (Example: A male subject, weight 79 kg, has a pulse rate of 139 beats per min at 900 kpm/min. If he is 50 years of age, the value will be 3.6 × 0.75 = 2.7 liters/min. Then according to Table A-6, the oxygen uptake per kilogram body weight will be 34 ml/kg × min.) A suggested form for the recording of the results of a work test is given in Table A-7.

DEFINITIONS

Kilopond meter (kpm): 1 kp is the force acting on the mass of 1 kg at normal acceleration of gravity.

$$1 \text{ kp} = 9.80665 \text{ newtons (N)}$$
$$1 \text{ kpm} = 9.80665 \text{ joules (J)}$$

Calorie: The amount of heat required at a pressure of one atmosphere to raise the temperature of one gram of water one degree centigrade (from 15° to 16°). Also called *small calorie.*
One kilocalorie, also called *large calorie* = 1000 small calories = 1 kcal.

CONVERSION TABLES

Length and Weight
1 centimeter = 0.39370 in.
1 meter = 39.37 in.
1 kilometer = 0.62137 mile
1 inch = 2.54 cm
1 foot = 30.480 cm
1 milliliter = 0.03381 fl oz
1 liter = 1.0567 U.S. qt
1 kilogram = 2.2046 lb

Power
1 watt = 0.001 kilowatt
1 watt = 0.73756 ft-lb/sec
1 watt = 1×10^7 ergs/sec
1 watt = 0.056884 BTU 1 min = 3.41304 BTU-hr
1 watt = 0.01433 kilocalories/min
1 watt = 1.341×10^{-3} hp (horsepower)
1 watt = 1 joule/sec
1 watt = 6.12 kpm/min

1 kilocalorie per minute = 69.767 watts
1 kilocalorie per minute = 51.457 ft-lb/sec
1 kilocalorie per minute = 6.9770×10^8 ergs/sec
1 kilocalorie per minute = 3.9685 BTU/min
1 kilocalorie per minute = 0.093557 hp

1 horsepower = 745.7 watts
1 horsepower = 550 ft-lb/sec
1 horsepower = 7.457×10^9 ergs/sec
1 horsepower = 42.4176 BTU/min
1 horsepower = 10.688 kcal/min
1 horsepower = 745.7 joules/sec
1 horsepower = 75 kpm/sec

Work and Energy
1 kilocalorie = 4.186×10^{10} ergs
1 kilocalorie = 4,186 joules
1 kilocalorie = 3.9680 BTU
1 kilocalorie = 3087.4 ft-lb
1 kilocalorie = 426.85 kpm
1 kilocalorie = 1.5593×10^{-3} hp-hr

1 erg = 2.3889×10^{-11} kcal
1 erg = 1×10^{-7} joule
1 erg = 9.4805×10^{-14} BTU
1 erg = 7.3756×10^{-8} ft-lb
1 erg = 1.0197×10^{-8} kpm
1 erg = 3.7251×10^{-14} hp-hr

1 joule = 2.3889×10^{-4} kcal
1 joule = 1×10^7 ergs
1 joule = 9.4805×10^{-4} BTU
1 joule = 0.73756 ft-lb
1 joule = 0.10197 kpm
1 joule = 3.7251×10^{-7} hp-hr

1 BTU = 0.25198 kcal
1 BTU = 1.0548 × 10^{10} ergs
1 BTU = 1054.8 joules
1 BTU = 777.98 ft-lb
1 BTU = 107.56 kpm
1 BTU = 3.9292 × 10^{-4} hp-hr

1 foot-pound = 3.2389 × 10^{-4} kcal
1 foot-pound = 1.35582 × 10^{7} ergs
1 foot-pound = 1.3558 joules
1 foot-pound = 1.2854 × 10^{-3} BTU
1 foot-pound = 0.13825 kpm
1 foot-pound = 5.0505 × 10^{-7} hp-hr
1 kilogram-meter = 2.3427 × 10^{-3} kcal
1 kilogram-meter = 9.8066 × 10^{7} ergs
1 kilogram-meter = 9.8066 joules
1 kilogram-meter = 9.2967 × 10^{3} BTU
1 kilogram-meter = 7.2330 ft-lb
1 kilogram-meter = 3.6529 × 10^{-6} hp-hr

49 watts = 300 kpm/min (approx. = 50 watts)
1 watt = 6.12 kpm/min (approx. = 6 kpm/min)
1 kpm/min = 0.1635 watt
1 kp = 9.80665 newtons

Speed

km/hr	mph	m/sec	km/hr	mph	m/sec
10	6.22	2.78	200	124	55.6
20	12.4	5.56	220	137	61.2
30	18.7	8.34	240	149	66.7
40	24.9	11.1	260	162	72.3
50	31.1	13.9	280	174	77.8
60	37.4	16.7	300	187	83.4
70	43.6	19.4	320	199	88.9
80	49.8	22.2	340	211	94.5
90	56.0	25.0	360	224	100
100	62.2	27.8	380	236	106
120	74.7	33.3	400	249	111
140	87.1	38.9	420	261	117
160	99.5	44.5	440	274	122
180	112	50.0	460	286	128

Temperature: Conversion of degrees centigrade into degrees Fahrenheit and vice versa

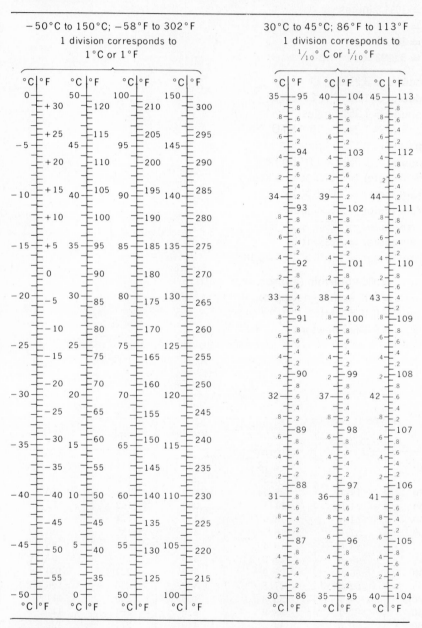

SOURCE: *Documenta Geigy,* "Scientific Tables," 5th ed., Geigy Pharmaceuticals, Ardsley, New York, 1956.

LIST OF SYMBOLS

$\bar{x}$	dash over any symbol indicates a mean value
$\dot{x}$	dot above any symbol indicates time derivate

Respiratory and Hemodynamic Notations

V	gas volume
$\dot{V}$	gas volume/unit time (usually liters/min)
R or RQ	respiratory exchange ratio (volume CO_2/volume O_2)
I	inspired gas
E	expired gas
A	alveolar gas
F	fractional concentration in dry gas phase
f	respiratory frequency (breath/unit time)
TLC	total lung capacity
VC	vital capacity
FRC	functional residual capacity
RV	residual volume
T	tidal gas
D	dead space
FEV	forced expiratory volume
$FEV_{1.0}$	forced expiratory volume in 1 sec
MVV	maximal voluntary ventilation
MVV_{40}	maximal voluntary ventilation at $f = 40$
D_L	diffusing capacity of the lungs (ml/min $\times$ mm Hg)
P	gas pressure
B or Bar	barometric
STPD	0°C, 760 mm Hg, dry
BTPS	body temperature and pressure, saturated with water vapor
ATPD	ambient temperature and pressure, dry
ATPS	ambient temperature and pressure, saturated with water vapor
Q	blood flow or volume
$\dot{Q}$	blood flow/unit time (without other notation, cardiac output; usually liters/min)
SV	stroke volume
HR	heart rate (usually beats/min)
BV	blood volume
THb or Hb_T	total amount of hemoglobin in body
Hb	hemoglobin concentration (g/100 ml)
Hct	hematocrit
BP	blood pressure
R	resistance
C	concentration in blood phase

S	percent saturation of Hb
a	arterial
c	capillary
v	venous

Temperature Notations

T or t	temperature
r or re	rectal
s	skin
e or oe	esophageal (oesophageal)
m	muscle
ty	tympanic
M	metabolic energy yield
C	convective heat exchange
R	radiation heat exchange
E	evaporative heat loss
S	storage of body heat
°C	temperature in degrees centigrade
°F	temperature in degrees Fahrenheit

Dimensions

W	weight
H	height
L	length
LBM	lean body mass
BSA	body surface area

Statistical Notations

M	arithmetic mean
SD or S.D.	standard deviation
SE or S.E.	standard error of the mean
n	number of observations
r	correlation coefficient
range	smallest and largest observed value
Σ	summation
D or d	difference
P	probability
*	denotes a (probably) significant difference; $0.05 \geq P > 0.01$
**	denotes a significant difference; $0.01 \geq P > 0.001$
***	denotes a (highly) significant difference; $P \leq 0.001$

Examples

| V_A | volume of alveolar gas |
| $\dot{V}_E$ | expiratory gas volume/minute |

$\dot{V}_{O_2}$ volume of oxygen/minute (oxygen uptake/min)

V_T tidal volume

P_A alveolar gas pressure

P_B barometric pressure

$F_{I_{O_2}}$ fractional concentration of O_2 in inspired gas

$P_{A_{O_2}}$ alveolar oxygen pressure

pH_a arterial pH

Ca_{O_2} oxygen content in arterial blood

$Ca_{O_2} - C\bar{v}_{O_2}$ difference in oxygen content between arterial and mixed venous blood (often written $a\text{-}\bar{v}O_2$ diff.)

T_r rectal temperature

Index

Index

Page references in italic indicate figures.